Semiconductor Devices

Physics and Technology

S. M. SZE

UMC Chair Professor
National Chiao Tung University
National Nano Device Laboratories
Hsinchu, Taiwan

JOHN WILEY & SONS, INC.

Acquisitions Editor   *William Zobrist*
Marketing Manager   *Katherine Hepburn*
Production Services Manager   *Jeanine Furino*
Production Editor   *Sandra Russell*
Designer   *Harold Nolan*
Production Management Services   *Argosy Publishing Services*

Cover Photography: A transmission-electron micrograph of a floating-gate nonvolatile semiconductor memory with a magnification of 100,000 times. (Photography courtesy of George T. T. Sheng.) For a discussion of the device, see Chapters 1, 6, and 14.

This book was typeset in *New Caledonia* by *Argosy Publishing* and printed and bound by *R. R. Donnelley and Sons, Inc. (Willard).* The cover was printed by *The Lehigh Press.*

The paper in this book was manufactured by a mill whose forest management programs include sustained yield harvesting of its timberlands. Sustained yield harvesting principles ensure that the number of trees cut each year does not exceed the amount of new growth.

The book is printed on acid-free paper.

Library of Congress Cataloging in Publication Data:
Sze, S. M., 1936-
    Semiconductor devices, physics and technology/S.M. Sze.—2nd ed.
        p. cm.
    Includes bibliographical references and index.
    ISBN 0-471-33372-7 (cloth: alk. paper)
    1. Semiconductors. I. Title.
TK7871.85 .S9883 2001

621.3815'2—dc21

2001026003

ISBN 0-471-33372-7

Printed in the United States of America

10 9 8 7 6 5 4 3 2 1

In Memory of My Mentors

Dr. L. J. Chu   *Academia Sinica*

Dr. R. M. Ryder   *Bell Laboratories*

Preface

The book is an introduction to the physical principles of modern semiconductor devices and their advanced fabrication technology. It is intended as a textbook for undergraduate students in applied physics, electrical and electronics engineering, and materials science. It can also serve as a reference for practicing engineers and scientists who need an update on device and technology developments.

▶ WHAT'S NEW IN THE SECOND EDITION

- 50% of the material has been revised or updated. We have added many sections that are of contemporary interest such as flash memory, Pentium chips, copper metallization, and eximer-laser lithography. On the other hand, we have omitted or reduced sections of less important topics to maintain the overall book length.

- We have also made substantial changes in updating the pedagogy. We have adopted a two-color format for all illustrations to enhance their presentation; and all important equations are boxed.

- All device and material parameters have been updated or corrected . For example, the intrinsic carrier concentration in silicon at 300K is 9.65×10^9 cm^{-3}, replacing the old value of 1.45×10^{10} cm^{-3}. This single change has an impact on at least 30% of the problem solutions.

- To improve the development of each subject, sections that contain graduate-level mathematics or physical concepts have been omitted or moved to the Appendixes, at the back of the book.

▶ TOPICAL COVERAGE

- Chapter 1 gives a brief historical review of major semiconductor devices and key technology developments. The text is then organized into three parts.

- Part I, Chapters 2–3, describes the basic properties of semiconductors and their conduction processes, with special emphasis on the two most important semiconductors: silicon (Si) and gallium arsenide (GaAs). The concepts in Part I will be used throughout this book. These concepts requires a background knowledge of modern physics and college calculus.

- Part II, Chapters 4–9, discusses the physics and characteristics of all major semiconductor devices. We begin with the $p–n$ junction which is the key building block of most semiconductor devices. We proceed to bipolar and field-effect devices and then cover microwave, quantum-effect, hot-electron, and photonic devices.

- Part III, Chapters 10–14, deals with processing technology from crystal growth to impurity doping. We present the theoretical and practical aspects of the major steps in device fabrication with an emphasis on integrated devices.

▶ KEY FEATURES

Each chapter includes the following features:
- The chapter starts with an overview of the topical contents. A list of learning goals is also provided.
- The second edition has tripled the worked-out examples that apply basic concepts to specific problems.
- A chapter summary appears at the end of each chapter to summarize the important concepts and to help the student review the content before tackling the homework problems that follow.
- The book includes about 250 homework problems, over 50% of them new to the second edition. Answers to odd-numbered problems, which have numerical solutions are provided in Appendix L at the back of the book.

▶ COURSE DESIGN OPTIONS

The second edition can provide greater flexibility in course design. The book contains enough material for a full-year sequence in device physics and processing technology. Assuming three lectures per week, a two-semester sequence can cover Chapters 1–7 in the first semester, leaving Chapters 8–14 for the second semester. For a three-quarter sequence, the logical break points are Chapters 1–5, Chapters 6-9, and Chapters 10–14.

A two-quarter sequence can cover Chapters 1–5 in the first quarter. The instructor has several options for the second quarter. For example, covering Chapters 6, 11, 12, 13, and 14 produces a strong emphasis on the MOSFET and its related process technologies, while covering Chapters 6–9 emphasizes all major devices. For a one-quarter course on semiconductor device processing, the instructor can cover Section 1.2 and Chapters 10–14.

A one-semester course on basic semiconductor physics and devices can cover Chapters 1–7. A one-semester course on microwave and photonic devices can cover Chapters 1–4, 7–9. If the students already have some familiarity with semiconductor fundamentals, a one-semester course on Submicron MOSFET: Physics and Technology can cover Chapters 1, 6, 10–14. Of course, there are many other course design options depending on the teaching schedule and the instructor's choice of topics.

▶ TEXTBOOK SUPPLEMENTS

- Instructor's Manual. A complete set of detailed solutions to all the end-of-chapter problems has been prepared. These solutions are available free to all adopting faculty.
- The figures used in the text are available, in electronic format, to instructors from the publisher. Instructors can find out more information at the publisher's website at: http://www.wiley.com/college/sze

Acknowledgments

Many people have assisted me in revising this book. I would first like to express my deep appreciation to my colleagues at the National Nano Device Laboratories for their contributions in improving the content of the text, in suggesting state-of-the-art illustrations, and in providing homework problems and solutions: Dr. S. F. Hu on Chapter 2, Dr. W. F. Wu on Chapter 3, Dr. S. H. Chan on Chapter 4, Dr. T. B. Chiou on Chapter 5, Dr. H. C. Lin on Chapter 6, Dr. J. S. Tsang on Chapter 7, Dr. G. W. Huang on Chapter 8, Dr. J. D. Guo on Chapter 9, Dr. S. C. Wu on Chapter 10, Dr. T. C. Chang on Chapter 11, Drs. M. C. Liaw and M. C. Chiang on Chapter 12, Dr. F. H. Ko on Chapter 13, and Dr. T. S. Chao on Chapter 14.

I have benefited significantly from suggestions made by the reviewers: Profs. C. Y. Chang, C. H. Chen, T. Y. Huang, B. Y. Tsui, and T. J. Yang of the National Chiao Tung University, Profs. Y. C. Chen and M. K. Lee of the National Sun Yet-sen University, Prof. C. S. Lai of the Chang Gung University, Prof. W. Y. Liang of the University of Cambridge, Dr. K. K. Ng of Bell Laboratories, Lucent Technologies, Prof. W. J. Tseng of the National Cheng Kung University, Prof. T. C. Wei of the Chung Yuan University, Prof. Y. S. G. Wu of the National Tsing Hwa University, Dr. C. C. Yang of the National Nano Device Laboratories, Prof. W. L. Yang of Feng Chia University, and Dr. A. Yen of the Taiwan Semiconductor Manufacturing Company.

I am further indebted to Mr. N. Erdos for technical editing of the manuscript, Ms. Iris Lin for typing the many revisions of the draft and the final manuscript, and Ms. Y. G. Yang of the Semiconductor Laboratory, the National Chiao Tung University who furnished the hundreds of technical illustrations used in the book. In each case where an illustration was used from another published source, I have received permission from the copyright holder. Even through all illustrations were then adopted and redrawn, I appreciate being granted these permissions. I wish to thank Mr. George T. T. Sheng of the Macronix International Company for providing the flash-memory transmission electron micrography, which is shown in the cover design. I wish also to thank Mr. A. Mutlu, Mr. S. Short, and Ms. R. Steward of Intel Corporation for providing the photographs of the first microprocessor (Intel 4004) and its latest version (Pentium 4).

At John Wiley and Sons, I wish to thank Mr. G. Telecki and Mr. W. Zobrist who encouraged me to undertake the project. I wish also to acknowledge the Spring Foundation of the National Chiao Tung University for the financial support. I would especially like to thank the United Microelectronics Corporation (UMC), Taiwan, ROC, for the UMC Chair Professorship grant that provided the environment to work on this book.

Finally, I am grateful to my wife Therese for her continued support and assistance in this and many previous book projects. I also like to thank my son Raymond (Doctor of Medicine) and my daughter-in-law Karen (Doctor of Medicine), and my daughter Julia (Certified Financial Analyst) and my son-in-law Bob (President, Cameron Global Investment, LLC), who have helped me in their capacities as my medical advisors and financial advisors, respectively.

S. M. Sze
Hsinchu, Taiwan
March 2001

Contents

1

Introduction

- ▶ 1.1 SEMICONDUCTOR DEVICES
- ▶ 1.2 SEMICONDUCTOR TECHNOLOGY
- ▶ SUMMARY

As an undergraduate in applied physics, electrical engineering, electronics engineering, or materials science, you might ask why you need to study semiconductor devices. The reason is that semiconductor devices are the foundation of the electronic industry, which is the largest industry in the world with global sales over one trillion dollars since 1998. A basic knowledge of semiconductor devices is essential to the understanding of advanced courses in electronics. This knowledge will also enable you to contribute to the Information Age, which is based on electronic technology.

Specifically, we cover the following topics:

- Four building blocks of semiconductor devices.
- Eighteen important semiconductor devices and their roles in electronic applications.
- Twenty important semiconductor technologies and their roles in device processing.
- Technology trends toward high-density, high-speed, low-power consumption, and nonvolatility.

▶ 1.1 SEMICONDUCTOR DEVICES

Figure 1 shows the sales volume of the semiconductor-device–based electronic industry in the past 20 years and projects sales to the year 2010. Also shown are the gross world product (GWP) and the sales volumes of automobile, steel, and semiconductor industries.[1,2] We note that the electronic industry has surpassed the automobile industry in 1998. If the current trends continue, in year 2010 the sales volume of the electronic industry will reach three trillion dollars and will constitute about 10% of GWP. The semiconductor industry, which is a subset of the electronic industry, will grow at an even higher rate to surpass the steel industry in the early twenty-first century and to constitute 25% of the electronic industry in 2010.

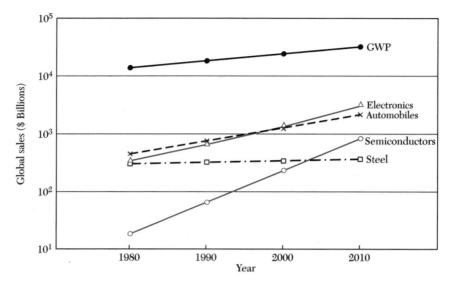

Fig. 1 Gross world product (GWP) and sales volumes of the electronics, automobile, semiconductor, and steel industries from 1980 to 2000 and projected to 2010.[1,2]

1.1.1 Device Building Blocks

Semiconductor devices have been studied for over 125 years.[3] To date, we have about 60 major devices, with over 100 device variations related to them.[4] However, all these devices can be constructed from a small number of device building blocks.

Figure 2*a* is the metal-semiconductor interface, which is an intimate contact between a metal and a semiconductor. This building block was the first semiconductor device ever studied (in the year 1874). This interface can be used as a rectifying contact, that is, the device allows electrical current to flow easily only in one direction, or as an ohmic contact, which can pass current in either direction with a negligibly small voltage drop. We can use this interface to form many useful devices. For example, by using a rectifying contact as the *gate*° and two ohmic contacts as the *source* and *drain*, we can form a *MESFET* (metal-semiconductor field-effect transistor), an important microwave device.

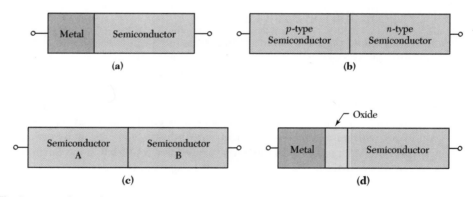

Fig. 2 Basic device building blocks. (*a*) Metal-semiconductor interface; (*b*) *p–n* junction; (*c*) heterojunction interface; and (*d*) metal-oxide-semiconductor structure.

° The italicized terms in this paragraph and in subsequent paragraphs are defined and explained in Part II of the book.

The second building block is the p–n junction (Fig. 2b), which is formed between a p- *type* (with positively charged carriers) and an *n-type* (with negatively charged carriers) semiconductors. The p–n junction is a key building block for most semiconductor devices, and p–n junction theory serves as the foundation of the physics of semiconductor devices. By combining two p–n junctions, that is, by adding another p-type semiconductor, we form the p–n–p *bipolar transistor*, which was invented in 1947 and had an unprecedented impact on the electronic industry. If we combine three p–n junctions to form a p–n–p–n structure, it is a switching device called a *thyristor*.

The third building block (Fig. 2c) is the heterojunction interface, that is, an interface formed between two dissimilar semiconductors. For example, we can use *gallium arsenide* (GaAs) and *aluminum arsenide* (AlAs) to form a heterojunction. Heterojunctions are the key components for high-speed and photonic devices.

Figure 2d shows the metal-oxide-semiconductor (MOS) structure. The structure can be considered a combination of a metal-oxide interface and an oxide-semiconductor interface. By using the MOS structure as the gate and two p–n junctions as the source and drain, we can form a *MOSFET* (MOS field-effect transistor). The MOSFET is the most important device for advanced *integrated circuits*, which contains tens of thousands of devices per integrated circuit chip.

1.1.2 Major Semiconductor Devices

Some major semiconductor devices are listed in Table 1 in chronological order; those with a superscript b are two-terminal devices, otherwise they are three-terminal or four-terminal devices.[3] The earliest systematic study of semiconductor devices (metal-semiconductor contacts) is generally attributed to Braun,[5] who in 1874 discovered that the resistance of contacts between metals and metal sulfides (e.g., copper pyrite) depended on the magnitude and polarity of the applied voltage. The electroluminescence phenomenon (for the *light-emitting diode*) was discovered by Round[6] in 1907. He observed the generation of yellowish light from a crystal of carborundum when he applied a potential of 10 V between two points on the crystals.

TABLE 1 Major Semiconductor Devices

Year	Semiconductor Device[a]	Author(s)/Inventor(s)	Ref.
1874	Metal-semiconductor contact[b]	Braun	5
1907	Light emitting diode[b]	Round	6
1947	Bipolar transistor	Bardeen, Brattain, and Shockley	7
1949	p–n junction[b]	Shockley	8
1952	Thyristor	Ebers	9
1954	Solar cell[b]	Chapin, Fuller, and Pearson	10
1957	Heterojunction bipolar transistor	Kroemer	11
1958	Tunnel diode[b]	Esaki	12
1960	MOSFET	Kahng and Atalla	13
1962	Laser[b]	Hall et al	15
1963	Heterostructure laser[b]	Kroemer, Alferov and Kazarinov	16,17
1963	Transferred-electron diode[b]	Gunn	18
1965	IMPATT diode[b]	Johnston, DeLoach, and Cohen	19

(continued)

TABLE 1 (*continued*)

Year	Semiconductor Device[a]	Author(s)/Inventor(s)	Ref.
1966	MESFET	Mead	20
1967	Nonvolatile semiconductor memory	Kahng and Sze	21
1970	Charge-coupled device	Boyle and Smith	23
1974	Resonant tunneling diode[b]	Chang, Esaki, and Tsu	24
1980	MODFET	Mimura et al.	25
1994	Room-temperature single-electron memory cell	Yano et al.	22
2001	20 nm MOSFET	Chau	14

[a]MOSFET, metal-oxide-semiconductor field-effect transistor; MESFET, metal-semiconductor field-effect transistor; MODFET, modulation-doped field-effect transistor.

[b]Denotes a two-terminal device, otherwise it is a three- or four-terminal device.

In 1947, the point-contact transistor was invented by Bardeen and Brattain.[7] This was followed by Shockley's[8] classic paper on p–n junction and bipolar transistor in 1949. Figure 3 shows the first transistor. The two point contacts at the bottom of the triangular quartz crystal were made from two stripes of gold foil separated by about 50 µm ($1 \mu m = 10^{-4}$ cm) and pressed onto a semiconductor surface. The semiconductor used was germanium. With one gold contact forward biased, that is, positive voltage with respect to the third terminal, and the other reverse biased, the *transistor action* was observed, that is, the input signal was amplified. The bipolar transistor is a key semiconductor device and has ushered in the modern electronic era.

In 1952, Ebers[9] developed the basic model for the thyristor, which is an extremely versatile switching device. The *solar cell* was developed by Chapin, et al.[10] in 1954 using a silicon p–n junction. The solar cell is a major candidate for obtaining energy from the sun because it can convert sunlight directly to electricity and is environmentally benign. In 1957, Kroemer[11] proposed the heterojunction bipolar transistor to improve the transistor performance; this device is potentially one of the fastest semiconductor devices. In 1958, Esaki[12] observed negative resistance characteristics in a heavily doped p–n junction, which led to the discovery of the *tunnel diode*. The tunnel diode and its associated tunneling phenomenon are important for ohmic contacts and carrier transport through thin layers.

The most important device for advanced integrated circuits is the MOSFET, which was reported by Kahng and Atalla[13] in 1960. Figure 4 shows the first device using a thermally oxidized silicon substrate. The device has a gate length of 20 µm and a gate oxide thickness of 100 nm (1 nm $= 10^{-7}$ cm). The two keyholes are the source and drain contacts, and the top elongated area is the aluminum gate evaporated through a metal mask. Although present-day MOSFETs have been scaled down to the deep-submicron regime, the choice of silicon and thermally grown silicon dioxide used in the first MOSFET remains the most important combination of materials. The MOSFET and its related integrated circuits now constitute about 90% of the semiconductor device market. An ultrasmall MOSFET with a channel length of 30 nm has been demonstrated recently.[14] This device can serve as the basis for the most advanced integrated circuit chips containing over one trillion ($>10^{12}$) devices.

In 1962, Hall et al.[15] first achieved lasing in semiconductors. In 1963, Kroemer[16] and Alferov and Kazarinov[17] proposed the *heterostructure laser*. These proposals laid the foundation for modern laser diodes, which can be operated continuously at room temperature. Laser diodes are the key components for a wide range of applications, including

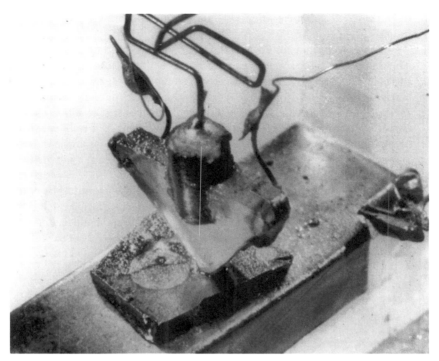

Fig. 3   The first transistor.[7] (Photograph courtesy of Bell Laboratories.)

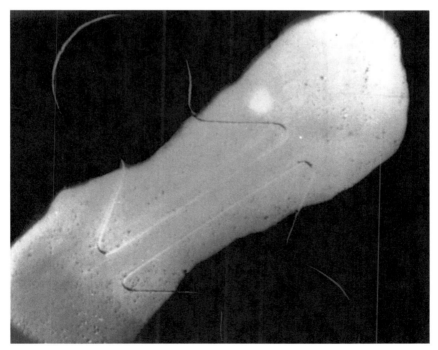

Fig. 4   The first metal-oxide-semiconductor field-effect transistor.[13] (Photograph courtesy of Bell Laboratories.)

digital video disk, optical-fiber communication, laser printing, and atmospheric–pollution monitoring.

Three important microwave devices were invented or realized in the next 3 years. The first device is the *transferred-electron diode* (TED; also called Gunn diode) by Gunn[18] in 1963. The TED is used extensively in such millimeter-wave applications as detection systems, remote controls, and microwave test instruments. The second device is the *IMPATT diode*; its operation was first observed by Johnston et al.[19] in 1965. IMPATT diodes can generate the highest continuous wave (CW) power at millimeter-wave frequencies of all semiconductor devices. They are used in radar systems and alarm systems. The third device is the MESFET, invented by Mead[20] in 1966. It is a key device for monolithic microwave integrated circuits (MMIC).

An important semiconductor memory device was invented by Kahng and Sze[21] in 1967. This is the *nonvolatile semiconductor memory* (NVSM), which can retain its stored information when the power supply is switched off. A schematic diagram of the first NVSM is shown in Fig.5a. Although it is similar to a conventional MOSFET, the major difference is the addition of the *floating gate*, in which semipermanent charge storage is possible. Because of its attributes of nonvolatility, high device density, low-power consumption, and electrical rewritability (e.g., the stored charge can be removed by applying voltage to the control gate), NVSM has become the dominant memory for portable electronic systems such as the cellular phone, notebook computer, digital camera, and smart card.

A limiting case of the floating-gate nonvolatile memory is the *single-electron memory cell* (SEMC) shown in Fig.5b. By reducing the length of the floating gate to ultra-

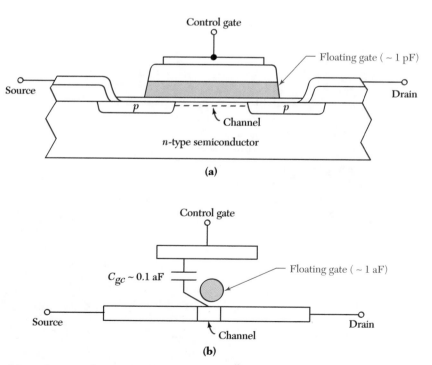

Fig. 5   (*a*) A schematic diagram of the first nonvolatile semiconductor memory (NVSM) with a floating gate.[21] (*b*) A limiting case of the floating-gate NVSM—the single-electron memory cell.[22]

small dimensions (e.g., 10 nm), we obtain the SEMC. At this dimension, when an electron moves into the floating gate, the potential of the gate will be altered so that it will prevent the entrance of another electron. The SEMC is an ultimate floating-gate memory cell, since we need only one electron for information storage. The operation of a SEMC at room temperature was first demonstrated by Yano et al.[22] in 1994. The SEMC can serve as the basis for the most advanced semiconductor memories that can contain over one trillion bits.

The *charge-coupled device* (CCD) was invented by Boyle and Smith[23] in 1970. CCD is used extensively in video cameras and in optical sensing applications. The resonant tunneling diode (RTD) was first studied by Chang et al.[24] in 1974. RTD is the basis for most quantum-effect devices, which offer extremely high density, ultrahigh speed, and enhanced functionality because it permits a greatly reduced number of devices to perform a given circuit function. In 1980, Minura et al.[25] developed the *MODFET* (modulation-doped field-effect transistor). With the proper selection of heterojunction materials, the MODFET is expected to be the fastest field-effect transistor.

Since the invention of the bipolar transistor in 1947, the number and variety of semiconductor devices have increased tremendously as advanced technology, new materials, and broadened comprehension have been applied to the creation of new devices. In Part II of the book, we consider all the devices listed in Table 1. It is hoped that this book can serve as a basis for understanding other devices not included here and perhaps not even conceived of at the present time.

1.2 SEMICONDUCTOR TECHNOLOGY

1.2.1 Key Semiconductor Technologies

Many important semiconductor technologies have been derived from processes invented centuries ago. For example, the lithography process was invented in 1798; in this first process, the pattern, or image, was transferred from a stone plate (litho).[26] In this section, we consider the milestones of technologies that were applied for the first time to semiconductor processing or developed specifically for semiconductor-device fabrication.

Some key semiconductor technologies are listed in Table 2 in chronological order. In 1918, Czochralski[27] developed a liquid-solid monocomponent growth technique. The Czochralski growth is the process used to grow most of the crystals from which silicon wafers are produced. Another growth technique was developed by Bridgman[28] in 1925. The Bridgman technique has been used extensively for the growth of gallium arsenide and related compound semiconductor crystals. Although the semiconductor properties of silicon have been widely studied since early 1940, the study of semiconductor compounds was neglected for a long time. In 1952, Welker[29] noted that gallium arsenide and its related III–V compounds were semiconductors. He was able to predict their characteristics and to prove them experimentally. The technology and devices of these compounds have since been actively studied.

The diffusion of impurity atoms in semiconductors is important for device processing. The basic diffusion theory was considered by Fick[30] in 1855. The idea of using diffusion techniques to alter the type of conductivity in silicon was disclosed in a patent in 1952 by Pfann.[31] In 1957, the ancient lithography process was applied to semiconductor-device fabrication by Andrus.[32] He used photosensitive etch-resistant polymers (photoresist) for pattern transfer. Lithography is a key technology for the semiconductor industry. The continued growth of the industry has been the direct result of improved lithographic technology. Lithography is also a significant economic factor, currently representing over 35% of the integrated-circuit manufacturing cost.

TABLE 2 Key Semiconductor Technologies

Year	Technology[a]	Author(s)/Inventor(s)	Ref.
1918	Czochralski crystal growth	Czochralski	27
1925	Bridgman crystal growth	Bridgman	28
1952	III-V compounds	Welker	29
1952	Diffusion	Pfann	31
1957	Lithographic photoresist	Andrus	32
1957	Oxide masking	Frosch and Derrick	33
1957	Epitaxial CVD growth	Sheftal, Kokorish, and Krasilov	34
1958	Ion implantation	Shockley	35
1959	Hybrid integrated circuit	Kilby	36
1959	Monolithic integrated circuit	Noyce	37
1960	Planar process	Hoerni	38
1963	CMOS	Wanlass and Sah	39
1967	DRAM	Dennard	40
1969	Polysilicon self-aligned gate	Kerwin, Klein, and Sarace	41
1969	MOCVD	Manasevit and Simpson	42
1971	Dry etching	Irving, Lemons, and Bobos	43
1971	Molecular beam epitaxy	Cho	44
1971	Microprocessor (4004)	Hoff et al.	45
1982	Trench isolation	Rung, Momose, and Nagakubo	46
1989	Chemical mechanical polishing	Davari et al.	47
1993	Copper interconnect	Paraszczak et al.	48

[a] CVD, chemical vapor deposition; CMOS, complementary metal-oxide-semiconductor field-effect transistor; DRAM, dynamic random access memory; MOCVD, metalorganic CVD.

The oxide masking method was developed by Frosch and Derrick [33] in 1957. They found that an oxide layer can prevent most impurity atoms from diffusing through it. In the same year, the epitaxial growth process based on chemical vapor deposition technique was developed by Sheftal et al.[34] Epitaxy, derived from the Greek word epi, meaning on, and taxis, meaning arrangement, describes a technique of crystal growth to form a thin layer of semiconductor materials on the surface of a crystal that has a lattice structure identical to that of the crystal. This method is important for the improvement of device performance and the creation of novel device structures.

In 1958, Shockley[35] proposed the method of using ion implantation to dope the semiconductors. Ion implantation has the capability of precisely controlling the number of implanted dopant atoms. Diffusion and ion implantation can complement each other for impurity doping. For example, diffusion can be used for high-temperature, deep-junction processes, whereas ion implantation can be used for lower-temperature, shallow-junction processes.

In 1959, a rudimentary integrated circuit (IC) was made by Kilby.[36] It contained one bipolar transistor, three resistors, and one capacitor, all made in germanium and connected by wire bonding—a hybrid circuit. Also in 1959, Noyce[37] proposed the monolithic IC by fabricating all devices in a single semiconductor substrate (monolith means single stone) and connecting the devices by aluminum metallization. Figure 6 shows the first monolithic IC of a flip-flop circuit containing six devices. The aluminum interconnection lines were obtained by etching evaporated aluminum layer over the entire oxide surface using the lithographic technique. These inventions laid the foundation for the rapid growth of the microelectronics industry.

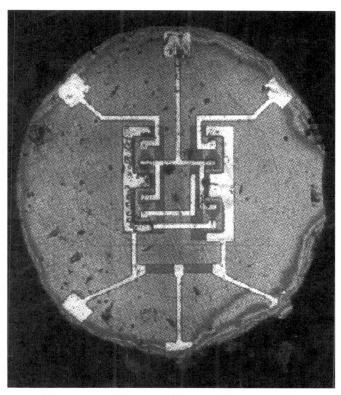

Fig. 6 The first monolithic integrated circuit.[37] (Photograph courtesy of Dr. G. Moore.)

The "planar" process was developed by Hoerni[38] in 1960. In this process, an oxide layer is formed on a semiconductor surface. With the help of a lithography process, portions of the oxide can be removed and windows cut in the oxide. Impurity atoms will diffuse only through the exposed semiconductor surface, and p–n junctions will form in the oxide window areas.

As the complexity of the IC increased, we have moved from *NMOS* (n-channel MOSFET) to *CMOS* (complementary MOSFET) technology, which employs both NMOS and *PMOS* (p-channel MOSFET) to form the logic elements. The CMOS concept was proposed by Wanlass and Sah [39] in 1963. The advantage of CMOS technology is that logic elements draw significant current only during the transition from one state to another (e.g., from 0 to 1) and draw very little current between transitions, allowing power consumption to be minimized. CMOS technology is the dominant technology for advanced ICs.

In 1967, an important two-element circuit, the dynamic random access memory (DRAM), was invented by Dennard.[40] The memory cell contains one MOSFET and one charge-storage capacitor. The MOSFET serves as a switch to charge or discharge the capacitor. Although DRAM is volatile and consumes relatively high power, we expect that DRAM will continue to be the first choice among various semiconductor memories for nonportable electronic systems in the foreseeable future.

To improve the device performance, the polysilicon self-aligned gate process was proposed by Kerwin et al.[41] in 1969. This process not only improved device reliability, it also reduced parasitic capacitances. Also in 1969, the metalorganic chemical vapor deposition

(MOCVD) method was developed by Manasevit and Simpson.[42] This is a very important epitaxial growth technique for compound semiconductors such as GaAs.

As the device dimensions were reduced, a dry etching technique was developed to replace wet chemical etching for high-fidelity pattern transfer. This technique was initiated by Irving et al.[43] in 1971 using a $CF_4 - O_2$ gas mixture to etch silicon wafers. Another important technique developed in the same year is molecular beam epitaxy by Cho.[44] This technique has the advantage of near-perfect vertical control of composition and doping down to atomic dimensions. It is responsible for the creation of numerous photonic devices and quantum-effect devices.

In 1971, the first microprocessor was made by Hoff et al.[45] They put the entire central processing unit (CPU) of a simple computer on one chip. It was a four-bit microprocessor (Intel 4004), shown in Fig. 7, with a chip size of 3 mm × 4 mm, and it contained 2300 MOSFETs. It was fabricated by a p-channel, polysilicon gate process using an 8 μm design rule. This microprocessor performed as well as those in $300,000 IBM computers of the early 1960s—each of which needed a CPU the size of a large desk. This was a major breakthrough for the semiconductor industry. Currently, microprocessors constitute the largest segment of the industry.

Fig. 7 The first microprocessor.[45] (Photograph courtesy of Intel Corp.)

Since early 1980, many new technologies have been developed to meet the requirements of ever-shrinking minimum feature lengths. We consider three key technologies: trench isolation, chemical-mechanical polishing, and the copper interconnect. The trench isolation technology was introduced by Rung et al.[46] in 1982 to isolate CMOS devices. This approach eventually replaced all other isolation methods. In 1989, the chemical-mechanical polishing method was developed by Davari et al.[47] for global planarization of the interlayer dielectrics. This is a key process for multilevel metallization. At submicron dimensions, a widely known failure mechanism is electromigration, which is the transport of metal ions through a conductor due to the passage of an electrical current. Although aluminum has been used since the early 1960s as the interconnect material, it suffers from electromigration at high electrical current. The copper interconnect was introduced in 1993 by Paraszczak et al.[48] to replace aluminum for minimum feature lengths approaching 100 nm. In Part III of this book, we consider all the technologies listed in Table 2.

1.2.2 Technology Trends

Since the beginning of the microelectronics era, the smallest line width or the minimum feature length of an integrated circuit has been reduced at a rate of about 13% per year.[49] At that rate, the minimum feature length will shrink to about 50 nm in the year 2010. Device miniaturization results in reduced unit cost per circuit function. For example, the cost per bit of memory chips has halved every 2 years for successive generations of DRAMs. As device dimension decreases, the intrinsic switching time also decreases. The device speed has improved by four orders of magnitude since 1959. Higher speeds lead to expanded IC functional throughput rates. In the future, digital ICs will be able to perform date processing and numerical computation at terabit-per-second rates. As the device becomes smaller, it consumes less power. Therefore, device miniaturization also reduces the energy used for each switching operation. The energy dissipated per logic gate has decreased by over one million times since 1959.

Figure 8 shows the exponential increase of the actual DRAM density versus the year of first production from 1978 to 2000. The density increases by a factor of 2 every 18 months. If the trends continue, we expect that DRAM density will increase to 8 Gb in the year 2005 and to 64 Gb around the year 2012. Figure 9 shows the exponential increase of the microprocessor computational power. The computational power also increases by a factor of 2 every 18 months. Currently, a Pentium-based personal computer has the same computational power as that of a supercomputer, CRAY 1, of the late 1960s; yet, it is three orders of magnitude smaller. If the trends continue, we will reach 100 GIP (billion instructions per second) in the year 2010.

Figure 10 illustrates the growth curves for different technology drivers.[50] At the beginning of the modern electronic era (1950–1970), the bipolar transistor was the technology driver. From 1970 to 1990, the DRAM and the microprocessor based on MOS devices were the technology drivers because of the rapid growth of personal computers and advanced electronic systems. Since 1990, nonvolatile semiconductor memory has been the technology driver, mainly because of the rapid growth of portable electronic systems.

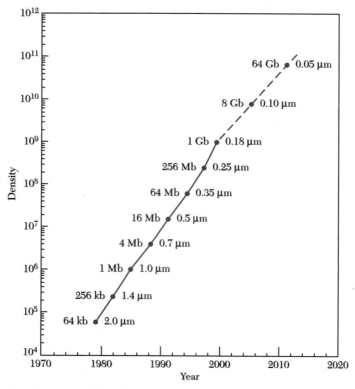

Fig. 8 Exponential increase of dynamic random access memory density versus year based on the Semiconductor Industry Association (SIA) roadmap.[49]

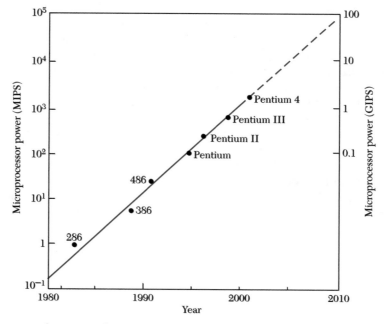

Fig. 9 Exponential increase of microprocessor computational power versus year.

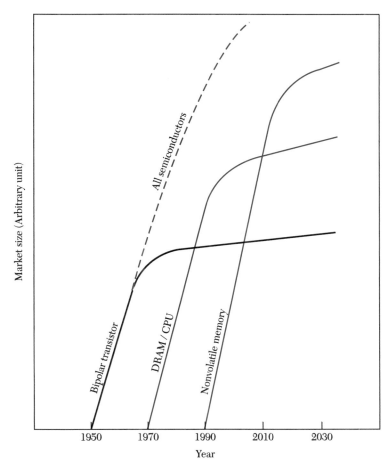

Fig. 10 Growth curves for different technology drivers.[50]

► SUMMARY

Although the *semiconductor-device* field is a relatively new area of study,* it has enormous impact on our society and the global economy. This is because semiconductor devices serve as the foundation of the largest industry in the world—the electronic industry.

In this introductory chapter, we have presented a historical review of major semiconductor devices from the first study of metal-semiconductor contact in 1874 to the fabrication of an ultrasmall 20-nm MOSFET in 2001. Of particular importance are the invention of the bipolar transistor in 1947, which ushered in the modern electronic era; the development of the MOSFET in 1960, which is the most important device for integrated circuits; and the invention of the nonvolalite semiconductor memory in 1967, which has been the technology driver of the electronic industry since 1990.

---

* Semiconductor devices and materials have been studied since the early nineteenth century. However, many traditional devices and materials have been studied for a much longer time. For example, steel was first studied in 1200 BC (over 3000 years ago).

We have also described key semiconductor technologies. The origins of many technologies can be traced back to the late eighteenth and early nineteenth centuries. Of particular importance are the development of the lithographic photoresist in 1957, which established the basic pattern-transfer process for semiconductor devices; the invention of the integrated circuits in 1959, which was seminal to the rapid growth of the microelectronic industry; and the developments of the DRAM in 1967 and the microprocessor in 1971, which constitute the two largest segments of the semiconductor industry.

We have a vast literature on semiconductor-device physics and technology.[51] To date, more than 300,000 papers have been published in this field, and the grand total may reach one million papers in the year 2012. In this book, each chapter deals with a major device or a key technology. Each is presented in a clear and coherent fashion without heavy reliance on the original literature. However, we have selected a few important papers at the end of each chapter for reference and for further reading.

▶ REFERENCES

1. *2000 Electronic Market Data Book,* Electron. Ind. Assoc., Washington, D.C., 2000.

2. *2000 Semiconductor Industry Report*, Ind. Technol. Res. Inst., Hsinchu, Taiwan, 2000.

3. Most of the classic device papers are collected in S. M. Sze, Ed., *Semiconductor Devices: Pioneering Papers*, World Sci., Singapore, 1991.

4. K. K. Ng, *Complete Guide to Semiconductor Devices,* McGraw-Hill, New York, 1995.

5. F. Braun, "Uber die Stromleitung durch Schwefelmetalle," *Ann. Phys. Chem.*, **153**, 556 (1874).

6. H. J. Round, "A Note On Carborundum," *Electron. World*, **19**, 309 (1907).

7. J. Bardeen and W. H. Brattain, "The Transistor, a Semiconductor Triode," *Phys. Rev.*, **71**, 230 (1948).

8. W. Shockley, "The Theory of $p–n$ Junction in Semiconductors and $p–n$ Junction Transistors," *Bell Syst. Tech. J.*, **28**, 435 (1949).

9. J. J. Ebers, "Four Terminal $p–n–p–n$ Transistors," *Proc. IRE*, **40**, 1361 (1952).

10. D. M. Chapin, C. S. Fuller, and G. L. Pearson, "A New Silicon $p–n$ Junction Photocell for Converting Solar Radiation into Electrical Power," *J. Appl. Phys.*, **25**, 676 (1954).

11. H. Kroemer, "Theory of a Wide-Gap Emitter for Transistors," *Proc. IRE*, **45**, 1535 (1957).

12. L. Esaki, "New Phenomenon in Narrow Germanium $p–n$ Junctions," *Phys. Rev.*, **109**, 603 (1958).

13. D. Kahng and M. M. Atalla, "Silicon-Silicon Dioxide Surface Device," in *IRE Device Research Conference*, Pittsburgh, 1960. (The paper can be found in Ref. 3.)

14. R. Chau, "30 nm and 20 nm Physical Gate Length CMOS Transistors. 2001 Silicon Nanoelectronics Workshopt, Kyok, p. 2 (2001).

15. R. N. Hall, et al., "Coherent Light Emission from GaAs Junctions," *Phys. Rev. Lett.*, **9**, 366 (1962).

16. H. Kroemer, "A Proposed Class of Heterojunction Injection Lasers," *Proc. IEEE*, **51**, 1782 (1963).

17. I. Alferov and R. F. Kazarinov, "Semiconductor Laser with Electrical Pumping," U.S.S.R. Patent 181, 737 (1963).

18. J. B. Gunn, "Microwave Oscillations of Current in III–V Semiconductors," *Solid State Commun.*, **1**, 88 (1963).

19. R. L. Johnston, B. C. DeLoach, Jr., and B. G. Cohen, "A Silicon Diode Microwave Oscillator," *Bell Syst. Tech. J.*, **44**, 369 (1965).

20. C. A. Mead, "Schottky Barrier Gate Field Effect Transistor," *Proc. IEEE*, **54**, 307 (1966).

21. D. Kahng and S. M. Sze, "A Floating Gate and Its Application to Memory Devices," *Bell Syst. Tech. J.*, **46,** 1283 (1967).

22. K. Yano, et al. "Room Temperature Single-Electron Memory," *IEEE Trans. Electron Devices*, **41**, 1628 (1994).

23. W. S. Boyle and G. E. Smith, "Charge Coupled Semiconductor Devices," *Bell Syst. Tech. J.*, **49**, 587 (1970).

24. L. L. Chang, L. Esaki, and R. Tsu, "Resonant Tunneling in Semiconductor Double Barriers," *Appl. Phys. Lett*, **24**, 593 (1974).

25. T. Mimura, et al., "A New Field-Effect Transistor with Selectively Doped GaAs/n–Al$_x$Ga$_{1-x}$ as Heterojunction," *Jpn. J. Appl. Phys.*, **19**, L225 (1980).

26. M. Hepher, "The Photoresist Story," *J. Photo. Sci.*, **12**, 181 (1964).

27. J. Czochralski, "Ein neues Verfahren zur Messung der Kristallisationsgeschwindigkeit der Metalle," *Z. Phys. Chem.*, **92**, 219 (1918).

28. P. W. Bridgman, "Certain Physical Properties of Single Crystals of Tungsten, Antimony, Bismuth, Tellurium, Cadmium, Zinc, and Tin," *Proc. Am. Acad. Arts Sci.*, **60**, 303 (1925).

29. H. Welker, "Über Neue Halbleitende Verbindungen," *Z. Naturforsch.*, **7a**, 744 (1952).

30. A. Fick, "Ueber Diffusion," *Ann. Phys. Lpz.*, **170**, 59 (1855).

31. W. G. Pfann, "Semiconductor Signal Translating Device," U.S. Patent 2, 597,028 (1952).

32. J. Andrus, "Fabrication of Semiconductor Devices," U.S. Patent 3,122,817 (filed 1957; granted 1964).

33. C. J. Frosch and L. Derrick, "Surface Protection and Selective Masking During Diffusion in Silicon," *J. Electrochem. Soc.*, **104**, 547 (1957).

34. N. N. Sheftal, N. P. Kokorish, and A. V. Krasilov, "Growth of Single-Crystal Layers of Silicon and Germanium from the Vapor Phase," *Bull. Acad. Sci U.S.S.R., Phys. Ser.*, **21**, 140 (1957).

35. W. Shockley, "Forming Semiconductor Device by Ionic Bombardment," U.S. Patent 2,787,564 (1958).

36. J. S. Kilby, "Invention of the Integrated Circuit," *IEEE Trans. Electron Devices*, **ED-23**, 648 (1976), U.S. Patent 3,138,743 (filed 1959, granted 1964).

37. R. N. Noyce, "Semiconductor Device-and-Lead Structure," U.S. Patent 2,981,877 (filed 1959, granted 1961).

38. J. A. Hoerni, "Planar Silicon Transistors and Diodes," *IRE Int. Electron Devices Meet.*, Washington D.C. (1960).

39. F. M. Wanlass and C. T. Sah, "Nanowatt Logics Using Field-Effect Metal-Oxide Semiconductor Triodes," *Tech. Dig. IEEE Int. Solid-State Circuit Conf.*, p.32, (1963).

40. R. M. Dennard, "Field Effect Transistor Memory," U.S. Patent 3,387,286 (filed 1967, granted 1968).

41. R. E. Kerwin, D. L. Klein, and J. C. Sarace, "Method for Making MIS Structure," U.S. Patent 3,475,234 (1969).

42. H. M. Manasevit and W. I. Simpson, "The Use of Metal–Organic in the Preparation of Semiconductor Materials. I. Epitaxial Gallium-V Compounds," *J. Electrochem. Soc.*, **116**, 1725 (1969).

43. S. M. Irving, K. E. Lemons, and G. E. Bobos, "Gas Plasma Vapor Etching Process," U.S. Patent 3,615,956 (1971).

44. A. Y. Cho, "Film Deposition by Molecular Beam Technique," *J. Vac. Sci. Technol.*, **8**, S 31 (1971).

45. The inventors of the microprocessor are M. E. Hoff, F. Faggin, S. Mazor, and M. Shima. For a profile of M. E. Hoff, see *Portraits in Silicon* by R. Slater, p. 175, MIT Press, Cambridge, 1987.

46. R. Rung, H. Momose, and Y. Nagakubo, "Deep Trench Isolated CMOS Devices," *Tech. Dig. IEEE Int. Electron Devices Meet.*, p.237 (1982).

47. B. Davari, et al., "A New Planarization Technique, Using a Combination of RIE and Chemical Mechanical Polish (CMP)," *Tech. Dig. IEEE Int. Electron Devices Meet.*, p. 61 (1989).

48. J. Paraszczak, et al., "High Performance Dielectrics and Processes for ULSI Interconnection Technologies," *Tech. Dig. IEEE Int. Electron Devices Meet.*, p.261 (1993).

49. *The International Technology Roadmap for Semiconductor*, Semiconductor Ind. Assoc., San Jose, 1999.

50. F. Masuoka, "Flash Memory Technology," *Proc. Int. Electron Devices Mater. Symp.*, 83, Hsinchu, Taiwan (1996).

51. From INSPEC database, National Chaio Tung University, Hsinchu, Taiwan, 2000.

CHAPTER 2

Energy Bands and Carrier Concentration in Thermal Equilibrium

In this chapter, we consider some basic properties of semiconductors. We begin with a discussion of crystal structure, which is the arrangement of atoms in a semiconductor. This is followed by a brief description of the crystal growth technique. We then present the concepts of valence bonds and energy bands, which relate to conduction in semiconductors. Finally, we discuss the concept of carrier concentration in thermal equilibrium. These concepts are used throughout this book.

Specifically, we cover the following topics:

• Element and compound semiconductors and their basic properties.

• The diamond structure and its related crystal planes.

• The bandgap and its impact on electrical conductivity.

• The intrinsic carrier concentration and its dependence on temperature.

• The Fermi level and its dependence on carrier concentration.

▶ 2.1 SEMICONDUCTOR MATERIALS

Solid-state materials can be grouped into three classes—insulators, semiconductors, and conductors. Figure 1 shows the range of electrical conductivities σ (and the corresponding resistivities $\rho = 1/\sigma$)° associated with some important materials in each of the three classes. Insulators such as fused quartz and glass have very low conductivities, on the order of

---

° A list of symbols is given in Appendix A.

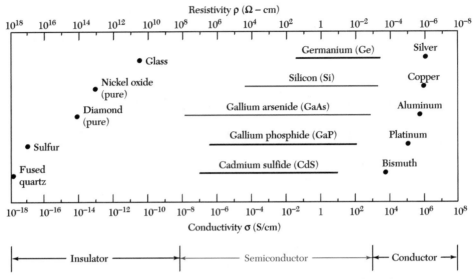

Fig. 1 Typical range of conductivities for insulators, semiconductors, and conductors.

$10^{-18} - 10^{-8}$ S/cm; and conductors such as aluminum and silver have high conductivities, typically from 10^4 to 10^6 S/cm.* Semiconductors have conductivities between those of insulators and those of conductors. The conductivity of a semiconductor is generally sensitive to temperature, illumination, magnetic field, and minute amounts of impurity atoms (typically, about 1 μg to 1 g of impurity atoms in 1 kg of semiconductor materials). This sensitivity in conductivity makes the semiconductor one of the most important materials for electronic applications.

2.1.1 Element Semiconductors

The study of semiconductor materials began in the early nineteenth century.[1] Over the years many semiconductors have been investigated. Table 1 shows a portion of the periodic table related to semiconductors. The element semiconductors, those composed of single

TABLE 1 Portion of the Periodic Table Related to Semiconductors

Period	Column II	III	IV	V	VI
2		B Boron	C Carbon	N Nitrogen	O Oxygen
3	Mg Magnesium	Al Aluminum	Si Silicon	P Phosphorus	S Sulfur
4	Zn Zinc	Ga Gallium	Ge Germanium	As Arsenic	Se Selenium
5	Cd Cadmium	In Indium	Sn Tin	Sb Antimony	Te Tellurium
6	Hg Mercury		Pb Lead		

* The international system of units is presented in Appendix B.

species of atoms, such as silicon (Si) and germanium (Ge), can be found in Column IV. In the early 1950s, germanium was the major semiconductor material. Since the early 1960s silicon has become a practical substitute and has now virtually supplanted germanium as a material for semiconductor fabrication. The main reasons we now use silicon are that silicon devices exhibit better properties at room temperature, and high-quality silicon dioxide can be grown thermally. There is also an economic consideration. Device-grade silicon costs much less than any other semiconductor material. Silicon in the form of silica and silicates comprises 25% of the Earth's crust, and silicon is second only to oxygen in abundance. Currently, silicon is one of the most studied elements in the periodic table; and silicon technology is by far the most advanced among all semiconductor technologies.

2.1.2 Compound Semiconductors

In recent years a number of compound semiconductors have found applications for various devices. The important compound semiconductors as well as the two element semiconductors are listed[2] in Table 2. A binary compound semiconductor is a combination of two elements from the periodic table. For example, gallium arsenide (GaAs) is a III-V compound that is a combination of gallium (Ga) from Column III and arsenic (As) from Column V.

In addition to binary compounds, ternary compounds and quaternary compounds are made for special applications. The alloy semiconductor $Al_xGa_{1-x}As$, which has Al and Ga from Column III and As from Column V is an example of a tenary compound, whereas quaternary compounds of the form $A_xB_{1-x}C_yD_{1-y}$ can be obtained from combination of many binary and ternary compound semiconductors. For example, GaP, InP, InAs, plus GaAs can be combined to yield the alloy semiconductor $Ga_xIn_{1-x}As_yP_{1-y}$. Compared with the element semiconductors, the preparation of compound semiconductors in single-crystal form usually involves much more complex processes.

Many of the compound semiconductors have electrical and optical properties that are different from silicon. These semiconductors, especially GaAs, are used mainly for high-speed electronic and photonic applications. Although we do not know as much about the technology of compound semiconductors as we do about that of silicon, advances in silicon technology have also helped progress in compound semiconductor technology. In this book we are concerned mainly with device physics and processing technology of silicon and gallium arsenide.

► 2.2 BASIC CRYSTAL STRUCTURE

The semiconductor materials we will be studying are single crystals, that is, the atoms are arranged in a three-dimensional periodic fashion. The periodic arrangement of atoms in a crystal is called a *lattice*. In a crystal, an atom never strays far from a single, fixed position. The thermal vibrations associated with the atom are centered about this position. For a given semiconductor, there is a *unit cell* that is representative of the entire lattice; by repeating the unit cell throughout the crystal, one can generate the entire lattice.

2.2.1 Unit Cell

A generalized primitive, three-dimensional unit cell is shown in Fig. 2. The relationship between this cell and the lattice is characterized by three vectors a, b, and c, which need

TABLE 2   Semiconductor Materials[2]

General Classification	Semiconductor	
	Symbol	Name
Element	Si	Silicon
	Ge	Germanium
Binary compound		
IV-IV -	SiC	Silicon carbide
III-V -	AlP	Aluminum phosphide
	AlAs	Aluminum arsenide
	AlSb	Aluminum antimonide
	GaN	Gallium nitride
	GaP	Gallium phosphide
	GaAs	Gallium areside
	GaSb	Gallium antimonide
	InP	Indium phosphide
	InAs	Indium arsenide
	InSb	Indium antimonide
II-VI -	ZnO	Zinc oxide
	ZnS	Zinc sulfide
	ZnSe	Zinc selenide
	ZnTe	Zinc telluride
	CdS	Cadmium sulfide
	CdSe	Cadmium selenide
	CdTe	Cadmium telluride
	HgS	Mercury sulfide
IV-VI -	PbS	Lead sulfide
	PbSe	Lead selenide
	PbTe	Lead telluride
Ternary compound	$Al_xGa_{1-x}As$	Aluminum gallium arsenide
	$Al_xIn_{1-x}As$	Aluminum indium arsenide
	$GaAs_{1-x}P_x$	Gallium arsenic phosphide
	$Ga_xIn_{1-x}As$	Gallium indium arsenide
	$Ga_xIn_{1-x}P$	Gallium indium phosphide
Quaternary compound	$Al_xGa_{1-x}As_ySb_{1-y}$	Aluminum gallium arsenic antimonide
	$Ga_xIn_{1-x}As_{1-y}P_y$	Gallium indium arsenic phosphide

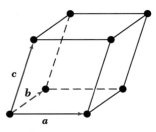

Fig. 2   A generalized primitive unit cell.

not be perpendicular to each other and which may or may not be equal in length. Every equivalent lattice point in the three-dimensional crystal can be found using the set

$$\boldsymbol{R} = m\boldsymbol{a} + n\boldsymbol{b} + p\boldsymbol{c} \tag{1}$$

where m, n, and p are integers.

Figure 3 shows some basic cubic-crystal unit cells. Figure 3a shows a simple cubic (sc) crystal; it has an atom at each corner of the cubic lattice, and each atom has six equidistant nearest-neighbor atoms. The dimension a is called the lattice constant. In the periodic table, only polonium is crystallized in the simple cubic lattice. Figure 3b is a body-centered cubic (bcc) crystal where, in addition to the eight corner atoms, an atom is located at the center of the cube. In a bcc lattice, each atom has eight nearest-neighbor atoms. Crystals exhibiting bcc lattices include those of sodium and tungsten. Figure 3c shows the face-centered cubic (fcc) crystal that has one atom at each of the six cubic faces in addition to the eight corner atoms. In this case, each atom has 12 nearest-neighbor atoms. A large number of elements exhibit the fcc lattice form, including aluminum, copper, gold, and platinum.

▶ **EXAMPLE 1**

If we pack hard spheres in a bcc lattice such that the atom in the center just touches the atoms at the corners of the cube, find the fraction of the bcc unit cell volume filled with hard spheres.

SOLUTION   Each corner sphere in a bcc unit cell is shared with eight neighboring cells; thus, each unit cell contains one-eighth of a sphere at each of the eight corners for a total of one sphere. In addition, each unit cell contains one central sphere. We have the following:

Spheres (atoms) per unit cell = $(1/8) \times 8$ (corner) + 1 (center) = 2;

Nearest-neighbor distance (along the diagonal AE in Fig. 3b) = $a\sqrt{3}/2$;

Radius of each sphere = $a\sqrt{3}/4$;

Volume of each sphere = $4\pi/3 \times (a\sqrt{3}/4)^3 = \pi a^3\sqrt{3}/16$; and

Maximum fraction of unit cell filled = Number of spheres × volume of each sphere/total volume of each unit cell = $2\pi a^3\sqrt{3}/16\ a^3 = \pi\sqrt{3}/8 \approx 0.68$.

Therefore, about 68% of the bcc unit cell volume is filled with hard spheres, and about 32% of the volume is empty.   ◀

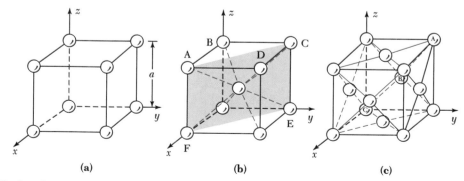

Fig. 3   Three cubic-crystal unit cells. (a) Simple cubic. (b) Body-centered cubic. (c) Face-centered cubic.

2.2.2 The Diamond Structure

The element semiconductors, silicon and germanium, have a diamond lattice structure as shown in Fig. 4a. This structure also belongs to the fcc crystal family and can be seen as two interpenetrating fcc sublattices with one sublattice displaced from the other by one-quarter of the distance along the body diagonal of the cube (i.e., a displacement of $a\sqrt{3}/4$). Although chemically identical, the two sets of atoms belonging to the two sublattices are different from the point view of the crystal structure. It can be seen in Fig. 4a that if a corner atom has one nearest neighbor in the body diagonal direction, then it has no nearest neighbor in the reverse direction. Consequently, it requires two such atoms in the unit cell. Alternatively, a unit cell of a diamond lattice consists of a tetrahedron in which each atom is surrounded by four equidistant nearest neighbors that lie at the corners (refer to the spheres connected by darkened bars in Fig. 4a).

Most of the III-V compound semiconductors (e.g., GaAs) have a *zincblende lattice*, shown in Fig. 4b, which is identical to a diamond lattice except that one fcc sublattice has Column III atoms (Ga) and the other has Column V atoms (As). Appendix F at the end of the book gives a summary of the lattice constants and other properties of important element and binary compound semiconductors.

▶ **EXAMPLE 2**

At 300 K the lattice constant for silicon is 5.43 Å. Calculate the number of silicon atoms per cubic centimeter and the density of silicon at room temperature.

SOLUTION   There are eight atoms per unit cell. Therefore,

$8/a^3 = 8/(5.43 \times 10^{-8})^3 = 5 \times 10^{22}$ atoms/cm^3; and

Density = no. of atoms/cm^3 × atomic weight/Avogadro constant = 5×10^{22} (atoms/cm^3) × 28.09 (g/mol)/6.02 × 10^{23} (atoms/mol) = 2.33 g/cm^3.   ◀

2.2.3 Crystal Planes and Miller Indices

In Fig. 3b we note that there are four atoms in the *ABCD* plane and five atoms in the *ACEF* plane (four atoms from the corners and one from the center) and that the atomic spacing is different for the two planes. Therefore, the crystal properties along different planes are different, and the electrical and other device characteristics can be dependent

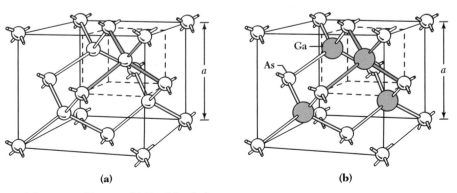

(a)   (b)

Fig. 4   (a) Diamond lattice. (b) Zincblende lattice.

on the crystal orientation. A convenient method of defining the various planes in a crystal is to use *Miller indices*.[3] These indices are obtained using the following steps:

1. Find the intercepts of the plane on the three Cartesian coordinates in terms of the lattice constant.

2. Take the reciprocals of these numbers and reduce them to the smallest three integers having the same ratio.

3. Enclose the result in parentheses (hkl) as the Miller indices for a single plane.

► **EXAMPLE 3**

As shown in Fig. 5, the plane has intercepts at a, $3a$, and $2a$ along the three coordinates. Taking the reciprocals of these intercepts, we get 1, ⅓, and ½. The smallest three integers having the same ratio are 6, 2, and 3 (obtained by multiplying each fraction by 6). Thus, the plane is referred to as a (623)-plane. ◄

Figure 6 shows the Miller indices of important planes in a cubic crystal.° Some other conventions are the following:

1. $(\bar{h}kl)$: For a plane that intercepts the x-axis on the negative side of the origin, such as $(\bar{1}00)$.

2. $\{hkl\}$: For planes of equivalent symmetry, such as $\{100\}$ for (100), (010), (001), $(\bar{1}00)$, $(0\bar{1}0)$, and $(00\bar{1})$ in cubic symmetry.

3. $[hkl]$: For a crystal direction, such as [100] for the x-axis. By definition, the [100]-direction is perpendicular to (100)-plane, and the [111]-direction is perpendicular to the (111)-plane.

4. $\langle hkl \rangle$: For a full set of equivalent directions, such as $\langle 100 \rangle$ for [100], [010], [001], $[\bar{1}00]$, $[0\bar{1}0]$, and $[00\bar{1}]$.

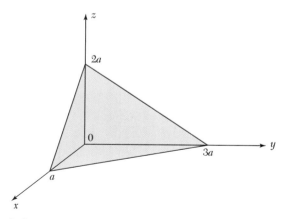

Fig. 5 A (623)-crystal plane.

° In Chapter 6, we show that the $\langle 100 \rangle$ orientation is preferred for silicon metal-oxide-semiconductor field-effect transistors (MOSFETs).

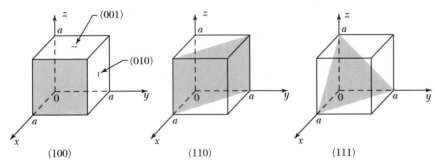

Fig. 6   Miller indices of some important planes in a cubic crystal.

2.3 BASIC CRYSTAL GROWTH TECHNIQUE

In this section, we consider briefly how we grow semiconductor crystals, specifically the silicon crystal because 95% of the semiconductor materials used by the electronic industry is silicon.

The starting material for silicon is a relatively pure form of sand (SiO_2) called quartzite. It is mixed with various forms of carbon and allowed to react to form silicon (98% pure):

$$SiC + SiO_2 \rightarrow Si \text{ (solid)} + SiO \text{ (gas)} + CO \text{ (gas)}. \tag{2}$$

The silicon product is reacted with hydrogen chloride to form trichlorosilane,

$$Si \text{ (solid)} + 3 \text{ HCl (gas)} \rightarrow SiHCl_3 \text{(gas)} + H_2 \text{(gas)}. \tag{3}$$

The trichlorosilane is decomposed using an electric current in a chamber with a controlled ambient, producing rods of ultrapure polycrystallice silicon (i.e., silicon material that contains many single-crystal regions with different size and orientation with respect to one another),

$$SiHCl_3 \text{ (gas)} + H_2 \text{(gas)} \rightarrow Si \text{ (solid)} + 3 \text{ HCl (gas)}. \tag{4}$$

The polycrystalline silicon is now ready for the crystal-growing process. Figure 7 shows many pieces of polycrystalline silicon in a silica (SiO_2) crucible.

Fig. 7   Photograph of polycrystalline silicon in a silica crucible.

The most common crystal growth method is the Czochralski technique. Figure 8 shows a schematic drawing of Czochralski crystal puller. The crucible containing the polycrystalline silicon is heated either by radio-frequency induction or by a thermal resistance method to the melting point of silicon (1412°C). The crucible rotates during the growth to prevent the formation of local hot or cold regions.

The atmosphere around the crystal-growing apparatus or crystal puller is controlled to prevent contamination of the molten silicon. Argon is often used as the ambient gas. When the temperature of the silicon has stabilized, a piece of silicon with a suitable orientation (e.g., ⟨111⟩), which is called the seed crystal, is lowered to the melt and is the starting point for the subsequent growth of a much larger crystal. As the bottom of the seed crystal begins to melt in the molten silicon, the downward motion of the rod holding the seed is reversed. As the seed crystal is slowly withdrawn from the melt (Fig. 9), the molten silicon adhering to the crystal freezes or solidifies, using the crystal structure of the seed crystal as a template. The seed crystal is, therefore, used to initiate the growth of the ingot with the correct crystal orientation. The rod continues its upward movement, forming an even larger crystal. The crystal growth is terminated when the silicon in the crucible is depleted. By carefully controlling the temperature of the crucible and the rotation speeds of the crucible and the rod, precise control of the diameter of the crystal can be maintained. Figure 10 shows a 200 mm diameter silicon crystal ingot. The desired impurity concentration is obtained by adding impurities to the melt in the form of heavily doped silicon prior to crystal growth. A more detailed discussion on the crystal growth of silicon as well as other semiconductors can be founded in Chapter 10.

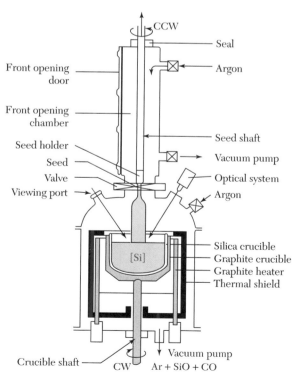

Fig. 8 Simplified schematic drawing of the Czochralski puller. Clockwise (CW), counterclockwise (CCW).

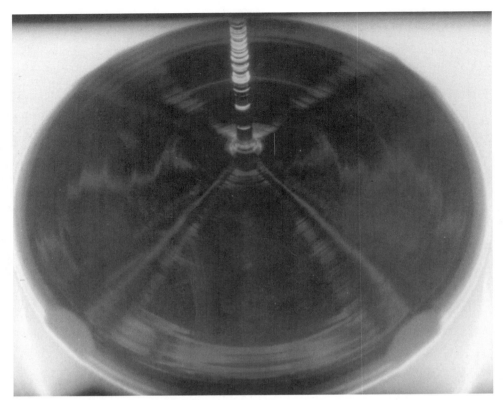

Fig. 9 Photograph of an a 200 mm diameter, (100)-oriented Si crystal being pulled from the melt. (Photograph courtesy of Taisil Electronic Materials Corp., Taiwan.)

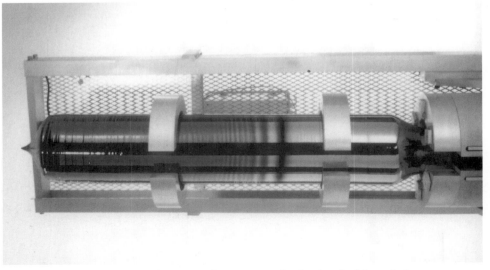

Fig. 10 A 200 mm diameter silicon crystal ingot grown by the Czochralski technique.

▶ 2.4 VALENCE BONDS

As discussed in Section 2.2, each atom in a diamond lattice is surrounded by four nearest neighbors. Figure 11*a* shows the tetrahedron bonds of a diamond lattice. A simplified two-dimensional bonding diagram for the tetrahedron is shown in Fig. 11*b*. Each atom has four electrons in the outer orbit, and each atom shares theses valence electrons with its four neighbors. This sharing of electrons is known as *covalent bonding*; each *electron pair* constitutes a covalent bond. Covalent bonding occurs between atoms of the same element or between atoms of different elements that have similar outer-shell electron configurations. Each electron spends an equal amount of time with each nucleus. However, both electrons spend most of their time between the two nuclei. The force of attraction for the electrons by both nuclei holds the two atoms together.

Gallium arsenide crystallizes in a zincblende lattice, which also has tetrahedron bonds. The major bonding force in GaAs is also due to the covalent bond. However, gallium arsenide has a small ionic contribution that is an electrostatic attractive force between each Ga^+ ion and its four neighboring As^- ions, or between each As^- ion and its four neighboring Ga^+ ions. Electronically, this means that the paired bonding electrons spend slightly more time in the As atom than in the Ga atom.

At low temperatures, the electrons are bound in their respective tetrahedron lattice; consequently, they are not available for conduction. At higher temperatures, thermal vibrations may break the covalent bonds. When a bond is broken or partially broken, a free electron results that can participate in current conduction. Figure 12*a* shows the situation when a valence electron in silicon becomes a free electron. An electron deficiency is left in the covalent bond. This deficiency may be filled by one of the neighboring electrons, which results in a shift of the deficiency location, as from location *A* to location *B* in Fig. 12*b*. We may therefore consider this deficiency as a particle similar to an electron. This fictitious particle is called a *hole*. It carries a positive charge and moves, under the influence of an applied electric field, in the direction opposite to that of an electron. Therefore, both the electron and the hole contribute to the total electric current. The concept of a hole is analogous to that of a bubble in a liquid. Although it is actually the liquid that moves, it is much easier to talk about the motion of the bubble in the opposite direction.

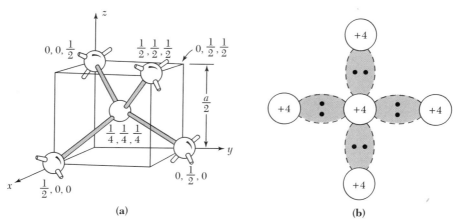

(a)   (b)

Fig. 11 (*a*) A tetrahedron bond. (*b*) Schematic two-dimensional representation of a tetrahedron bond.

▶ 2.5 ENERGY BANDS

2.5.1 Energy Levels of Isolated Atoms

For an isolated atom, the electrons can have discrete energy levels. For example, the energy levels for an isolated hydrogen atom are given by the Bohr model[4]:

$$E_H = -m_0 q^4 / 8\varepsilon_0^2 h^2 n^2 = -13.6/n^2 \text{ eV}, \tag{5}$$

where m_0 is the free-electron mass, q is the electronic charge, ε_0 is the free-space permittivity, h is the Planck constant, and n is a positive integer called the principal quantum number. The quantity eV (electron volt) is an energy unit corresponding to the energy gained by an electron when its potential is increased by one volt. It is equal to the product of q (1.6×10^{-19} coulomb) and one volt, or 1.6×10^{-19} J. The discrete energies are -13.6 eV for the ground state energy level ($n = 1$), -3.4 eV for the first excited-state energy level ($n = 2$), etc. Detailed studies reveal that for higher principle quantum numbers ($n \geq 2$), energy levels are split according to their angular momentum quantum number ($\ell = 0, 1, 2, ..., n - 1$).

We now consider two identical atoms. When they are far apart, the allowed energy levels for a given principal quantum number (e.g., $n = 1$) consist of one doubly degenerate level, that is, both atoms have exactly the same energy. When they are brought closer, the doubly degenerate energy levels will spilt into two levels by the interaction between the atoms. As N isolated atoms are brought together to form a solid, the orbits of the outer electrons of different atoms overlap and interact with each other. This interaction, including those forces of attraction and repulsion between atoms, causes a shift in the energy levels, as in the case of two interacting atoms. However, instead of two levels, N separate but closely spaced levels are formed. When N is large, the result is an essentially continuous band of energy. This band of N levels can extend over a few eV depending on the inter-atomic spacing for the crystal. Figure 13 shows the effect, where the parameter a represents the equilibrium interatomic distance of the crystal.

The actual band splitting in a semiconductor is much more complicated. Figure 14 shows an isolated silicon atom that has 14 electrons. Of the 14 electrons, 10 occupy deep-lying energy levels whose orbital radius is much smaller than the interatomic separation in the crystal. The four remaining valence electrons are relatively weakly bound and can be involved in chemical reactions. Therefore, we only need to consider the outer shell

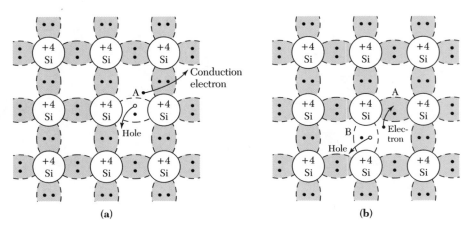

(a)   (b)

Fig. 12 The basic bond representation of intrinsic silicon. (*a*) A broken bond at position A, resulting in a conduction electron and a *hole*. (*b*) A broken bond at position B.

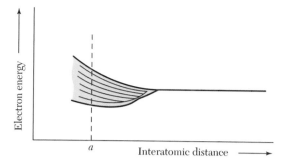

Fig. 13   The splitting of a degenerate state into a band of allowed energies.

(the $n = 3$ level) for the valence electrons, since the two inner shells are completely full and tightly bound to the nucleus. The 3s subshell (i.e., for $n = 3$ and $\ell = 0$) has two allowed quantum states per atom. This subshell will contain two valence electrons at $T = 0$ K. The 3p subshell (i.e., $n = 3$, and $\ell = 1$) has six allowed quantum states per atom. This subshell will contain the remaining two valence electrons of an individual silicon atom.

  Figure 15 is a schematic diagram of the formation of a silicon crystal from N isolated silicon atoms. As the interatomic distance decreases, the 3s and 3p subshell of the N silicon atoms will interact and overlap. At the equilibrium interatomic distance, the bands will again split, with four quantum states per atom in the lower band and four quantum states per atom in the upper band. At a temperature of absolute zero, electrons occupy the lowest energy states, so that all states in the lower band (*the valence band*) will be full and all states in the upper band (*the conduction band*) will be empty. The bottom of the conduction band is called E_C, and the top of the valence band is called E_V. The *bandgap energy* E_g between the bottom of the conduction band and the top of the valence band $(E_C - E_V)$ is the width of the forbidden energy gap, as shown on the far left of Fig. 15. Physically, E_g is the energy required to break a bond in the semiconductor to free an electron to the conduction band and leave a hole in the valence band.

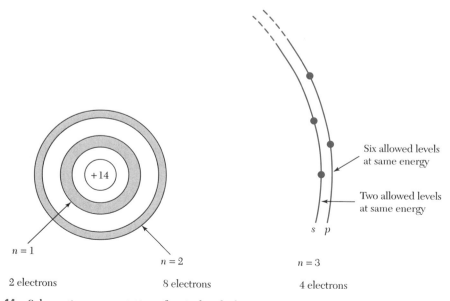

Fig. 14   Schematic representation of an isolated silicon atom.

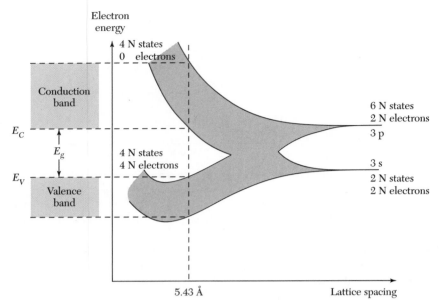

Fig. 15 Formation of energy bands as a diamond lattice crystal is formed by bringing isolated silicon atoms together.

2.5.2 The Energy-Momentum Diagram

The energy E of a free electron is given by

$$E = \frac{p^2}{2m_0},$$ (6)

where p is the momentum and m_0 is the free-electron mass. If we plot E vs. p, we obtain a parabola as shown in Fig. 16. In a semiconductor crystal, an electron in the conduction band is similar to a free electron in that it is relatively free to move about in the crystal. However, because of the periodic potential of the nuclei, Eq. 6 can no longer be valid. However, it turns out that we can still use Eq. 6 if we replace the free-electron mass in Eq. 6 by an effective mass $\dot{m}_n$ (the subscript n refers to the negative charge on an electron), that is,

$$E = \frac{p^2}{2m_n}.$$ (7)

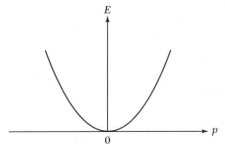

Fig. 16 The parabolic energy (E) vs. momentum (p) curve for a free electron.

The electron effective mass depends on the properties of the semiconductor. If we have an energy-momentum relationship described by Eq.7, we can obtain the effective mass from the second derivative of E with respect to p:

$$m_n \equiv \left(\frac{d^2 E}{dp^2} \right)^{-1}. \qquad (8)$$

Therefore, the narrower the parabola, corresponding to a larger second derivative, the smaller the effective mass. A similar expression can be written for holes (with effective mass m_p where the subscript p refers to the positive charge on a hole). The effective-mass concept is very useful because it enables us to treat electrons and holes essentially as classical charged particles.

Figure 17 shows a simplified energy-momentum relationship of a special semiconductor with an electron effective mass of $m_n = 0.25\,m_0$ in the conduction band (the upper parabola) and a hole effective mass of $m_p = m_0$ in the valence band (the lower parabola). Note that the electron energy is measured upward and the hole energy is measured downward. The spacing at $p = 0$ between these two parabolas is the bandgap E_g, shown previously in Fig. 15.

The actual energy-momentum relationships (also called energy-band diagram) for silicon and gallium arsenide are much more complex. They are shown in Fig. 18 for two crystal directions. We note that the general features in Fig. 18 are similar to those in Fig. 17. First of all, there is a bandgap E_g between the bottom of the conduction band and the top of the valence band. Second, near the minimum of the conduction band or the maximum of the valence band, the E-p curves are essentially parabolic. For silicon, Fig 18a, the maximum in the valence band occurs at $p = 0$, but the minimum in the conduction band occurs along the [100] direction at $p = p_c$. Therefore, in silicon, when an electron makes a transition from the maximum point in the valence band to the minimum point in the conduction band, it requires not only an energy change ($\geq E_g$) but also some momentum change ($\geq p_c$).

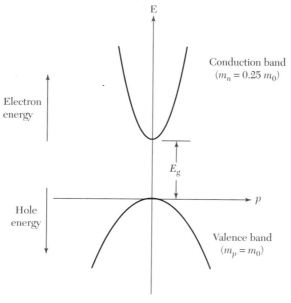

Fig. 17 A schematic energy-momentum diagram for a special semiconductor with $m_n = 0.25\,m_0$ and $m_p = m_0$.

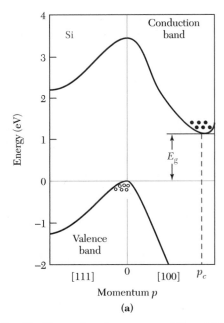 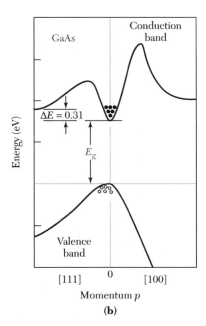

Fig. 18 Energy band structures of Si and GaAs. Circles (○) indicate holes in the valence bands and dots (●) indicate electrons in the conduction bands.

For gallium arsenide, Fig. 18*b*, the maximum in the valence band and the minimum in the conduction band occur at the same momentum ($p = 0$). Thus, an electron making a transition from the valence band to the conduction band can do so without a change in momentum.

Gallium arsenide is called a *direct semiconductor*, because it does not require a change in momentum for an electron transition from the valence band to the conduction band. Silicon is called an *indirect semiconductor,* because a change of momentum is required in a transition. This difference between direct and indirect band structures is very important for light-emitting diodes and semiconductor lasers. These devices require direct semiconductors for efficient generation of photons (refer to Chapter 9).

We can obtain the effective mass from Fig. 18 using Eq. 8. For example, for gallium arsenide with a very narrow conduction–band parabola, the electron effective mass is $0.063\ m_0$, where for silicon, with a wider conduction–band parabola, the electron effective mass is $0.19\ m_0$.

2.5.3 Conduction in Metals, Semiconductors, and Insulators

The enormous variation in electrical conductivity of metals, semiconductors, and insulators as shown in Fig. 1 may be explained qualitatively in terms of their energy bands. We shall see that the electron occupation of the highest band or of the highest two bands determines the conductivity of a solid. Figure 19 shows the energy band diagrams of three classes of solids—metals, semiconductors, and insulators.

Metals
The characteristics of a metal (also called a conductor) include a very low value of resistivity, and the conduction band either is partially filled (such as Cu) or overlaps the valence band (such as Zn or Pb) so that there is no bandgap, as shown in Fig. 19*a*. As a conse-

quence, the uppermost electrons in the partially filled band or electrons at the top of the valence band can move to the next-higher available energy level when they gain kinetic energy (e.g., from an applied electric field). Electrons are free to move with only a small applied field in a metal because there are many unoccupied states close to the occupied energy states. Therefore, current conduction can readily occur in conductors.

Insulators

In an insulator such as silicon dioxide (SiO_2), the valence electrons form strong bonds between neighboring atoms. These bonds are difficult to break, and consequently there are no free electrons to participate in current conduction at or near room temperature. As shown in the energy band diagram Fig. 19c, insulators are characterized by a large bandgap. Note that electrons occupy all energy levels in the valence band and all energy levels in the conduction band are empty. Thermal energy° or the energy of an applied electric field is insufficient to raise the uppermost electron in the valence band to the conduction band. Thus, although an insulator has many vacant states in the conduction band that can accept electrons, so few electrons actually occupy conduction band states that the overall contribution to electrical conductivity is very small, resulting in a very high resistivity. Therefore, silicon dioxide is an insulator; it can not conduct current.

Semiconductors

Now, consider a material that has a much smaller energy gap, on the order of 1 eV (Fig. 19b). Such materials are called semiconductors. At T = 0 K, all electrons are in the valence band, and there are no electrons in the conduction band. Thus, semiconductors are poor conductors at low temperatures. At room temperature and under normal atmosphere, values of E_g are 1.12 eV for Si and 1.42 eV for GaAs. The thermal energy kT at room temperature is a good fraction of E_g, and an appreciable numbers of electrons are thermally excited from the valence band to the conduction band. Since there are many empty states in the conduction band, a small applied potential can easily move these electrons, resulting in a moderate current.

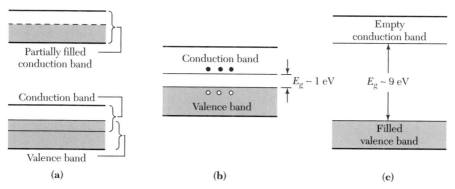

Fig. 19 Schematic energy band representations of (a) a conductor with two possibilities (either the partially filled conduction band shown at the upper portion or the overlapping bands shown at the lower portion), (b) a semiconductor, and (c) an insulator.

° The thermal energy is of the order of kT. At room temperature, kT is 0.026 eV, which is much smaller than the bandgap of an insulator.

▶ ## 2.6 INTRINSIC CARRIER CONCENTRATION

We now derive the carrier concentration in thermal equilibrium condition, that is, the steady-state condition at a given temperature without any external excitations such as light, pressure, or electric field. At a given temperature, continuous thermal agitation results in the excitation of electrons from the valence band to the conduction band and leaves an equal number of holes in the valence band. An *intrinsic semiconductor* is one that contains relatively small amounts of impurities compared with the thermally generated electrons and holes.

To obtain the electron density (i.e., the number of electrons per unit volume) in an intrinsic semiconductor, we first evaluate the electron density in an incremental energy range dE. This density $n(E)$ is given by the product of the density of states $N(E)$, that is, the density of allowed energy states (including electron spin) per energy range per unit volume,° and by the probability of occupying that energy range $F(E)$. Thus, the electron density in the conduction band is given by integrating $N(E)\,F(E)\,dE$ from the bottom of the conduction band (E_C initially taken to be $E = 0$ for simplicity) to the top of the conduction band E_{top}:

$$n = \int_0^{E_{top}} n(E)dE = \int_0^{E_{top}} N(E)F(E)dE, \qquad (9)$$

where n is in cm^{-3}, and $N(E)$ is in (cm^3–eV)$^{-1}$.

The probability that an electron occupies an electronic state with energy E is given by the Fermi–Dirac distribution function, which is also called the Fermi distribution function

$$F(E) = \frac{1}{1 + e^{(E-E_F)/kT}}, \qquad (10)$$

where k is the Boltzmann constant, T is the absolute temperature in degrees Kelvin, and E_F is the energy of the Fermi level. The Fermi energy is the energy at which the probability of occupation by an electron is exactly one-half. The Fermi distribution is illustrated in Fig. 20 for different temperatures. Note that $F(E)$ is symmetrical around the Fermi energy E_F. For energies that are $3kT$ above or below the Fermi energy, the exponential term in Eq. 10 becomes larger than 20 or smaller than 0.05, respectively. The Fermi distribution function can thus be approximated by simpler expressions:

$$F(E) \cong e^{-(E - E_F)/kT} \quad \text{for} \quad (E - E_F) > 3kT, \qquad (11a)$$

and

$$F(E) \cong 1 - e^{-(E - E_F)/kT} \quad \text{for} \quad (E - E_F) < 3kT. \qquad (11b)$$

Equation 11b can be regarded as the probability that a hole occupies a state located at energy E.

Figure 21 shows schematically from left to right the band diagram, the density of states $N(E)$, which varies as $\sqrt{E}$ for a given electron effective mass, the Fermi distribution function, and the carrier concentrations for an intrinsic semiconductor. The carrier concentration can be obtained graphically from Fig. 21 using Eq. 9; that is, the product of $N(E)$ in Fig. 21b and $F(E)$ in Fig. 21c gives the $n(E)$-versus-E curve (upper curve) in Fig. 21d. The upper shaded area in Fig. 21d corresponds to the electron density.

° The density of states $N(E)$ is derived in Appendix H.

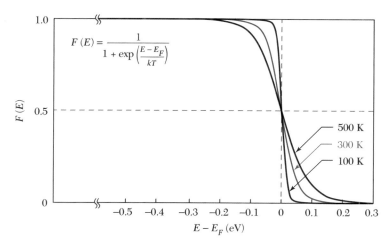

Fig. 20 Fermi distribution function $F(E)$ versus $(E - E_F)$ for various temperatures.

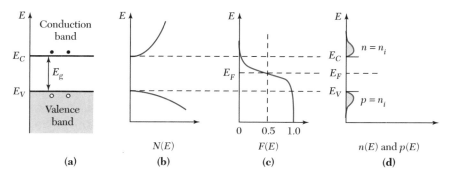

Fig. 21 Instrinsic semiconductor. (*a*) Schematic band diagram. (*b*) Density of states. (*c*) Fermi distribution function. (*d*) Carrier concentration.

There are a large number of allowed states in the conduction band. However, for an intrinsic semiconductor there will not be many electrons in the conduction band. Therefore, the probability of an electron occupying one of these states is small. There also are a large number of allowed states in the valence band. By contrast, most of these are occupied by electrons. Thus, the probability of an electron occupying one of these states in the valence band is nearly unity. There will be only a few unoccupied electron states, that is, holes, in the valence band. As can be seen, the Fermi level is located near the middle of the bandgap (i.e., E_F is many kT below E_C). Substituting the last equation in Appendix H and Eq.11*a* into Eq. 9 yields[*]

$$n = \frac{2}{\sqrt{\pi}} N_C (kT)^{-3/2} \int_0^\infty E^{1/2} \exp\left[-(E - E_F)/kT\right]dE,\tag{12}$$

where
$$N_C \equiv 12(2\pi m_n kT/h^2)^{3/2} \text{ for Si}\tag{13a}$$

$$\equiv 2(2\pi m_n kT/h^2)^{3/2} \text{ for GaAs.}\tag{13b}$$

---

[*] We have taken E_{top} to be ∞, because $F(E)$ becomes very small when $(E - E_C) >> kT.$

If we let $x \equiv E/kT$, Eq. 12 becomes

$$n = \frac{2}{\sqrt{\pi}} N_C \exp\left(E_F / kT\right) \int_0^\infty x^{1/2} e^{-x} dx. \tag{14}$$

The integral in Eq. 14 is of the standard form and equals $\sqrt{\pi}/2$. Therefore, Eq. 14 becomes

$$n = N_C \exp\left(E_F/kT\right). \tag{15}$$

If we refer to the bottom of the conduction band as E_C instead of $E = 0$, we obtain for the electron density in the conduction band

$$\boxed{n = N_C \exp\left[-(E_C - E_F)/kT\right],} \tag{16}$$

where N_C defined in Eq.13 is the *effective density of states* in the conduction band. At room temperature (300 K), N_C is 2.86×10^{19} cm^{-3} for silicon and 4.7×10^{17} cm^{-3} for gallium arsenide.

Similarly, we can obtain the hole density p in the valence band:

$$\boxed{p = N_V \exp\left[-(E_F - E_V)/kT\right],} \tag{17}$$

and

$$N_V \equiv 2\left(2\pi m_p kT/h^2\right)^{3/2}, \tag{18}$$

where N_V is the *effective density of states in the valence band* for both Si and GaAs. At room temperature, N_V is 2.66×10^{19} cm^{-3} for silicon and 7.0×10^{18} cm^{-3} for gallium arsenide.

For an intrinsic semiconductor, the number of electrons per unit volume in the conduction band is equal to the number of holes per unit volume in the valence band, that is, $n = p = n_i$ where n_i is the *intrinsic carrier density*. This relationship of electrons and holes is depicted in Fig. 21d. Note that the shaded area in the conduction band is the same as that in the valence band.

The Fermi level for an intrinsic semiconductor is obtained by equating Eq.16 and Eq. 17:

$$E_F = E_i = (E_C + E_V)/2 + (kT/2) \ln\left(N_V/N_C\right). \tag{19}$$

At room temperature, the second term is much smaller than the bandgap. Hence, the intrinsic Fermi level E_i of an intrinsic semiconductor generally lies very close to the middle of the bandgap.

The intrinsic carrier density is obtained from Eqs. 16, 17, and 19:

$$\boxed{np = n_i^2,} \tag{20}$$

$$\boxed{n_i^2 = N_C N_V \exp\left(-E_g/kT\right),} \tag{21}$$

and

$$\boxed{n_i = \sqrt{N_C N_V} \exp\left(-E_g/2kT\right),} \tag{22}$$

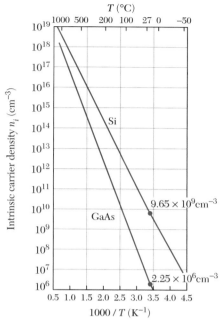

Fig. 22 Intrinsic carrier densities in Si and GaAs as a function of the reciprocal of temperature.[5-7]

where $E_g \equiv E_C - E_V$. Figure 22 shows the temperature dependence of n_i for silicon and gallium arsenide.[5] At room temperature (300 K), n_i is $9.65 \times 10^9 \, \text{cm}^{-3}$ for silicon[6] and $2.25 \times 10^6 \, \text{cm}^{-3}$ for gallium arsenide.[7] As expected, the larger the bandgap, the smaller the intrinsic carrier density.

2.7 DONORS AND ACCEPTORS

When a semiconductor is doped with impurities, the semiconductor becomes *extrinsic* and impurity energy levels are introduced. Figure 23a shows schematically that a silicon atom is replaced (or substituted) by an arsenic atom with five valence electrons. The arsenic atom forms covalent bonds with its four neighboring silicon atoms. The fifth electron has a relatively small binding energy to its host arsenic atom and can be "ionized" to become a conduction electron at a moderate temperature. We say that this electron has been "donated" to the conduction band. The arsenic atom is called a *donor* and the silicon becomes n-type because of the addition of the negative charge carrier. Similarly, Fig. 23b shows that when a boron atom with three valence electrons substitutes for a silicon atom, an additional electron is "accepted" to form four covalent bonds around the boron, and a positively charged "hole" is created in the valence band. This is a p-type semiconductor, and the boron is an *acceptor*.

We can estimate the *ionization energy* for the donor E_D by replacing m_0 with the electron effective mass m_n and taking into account the semiconductor permittivity ε_s in the hydrogen atom model, Eq. 5:

$$E_D = \left(\frac{\varepsilon_0}{\varepsilon_S} \right)^2 \left(\frac{m_n}{m_0} \right) E_H. \tag{23}$$

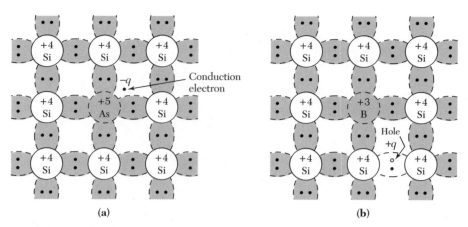

Fig. 23 Schematic bond pictures for (*a*) *n*-type Si with donor (arsenic) and (*b*) *p*-type Si with acceptor (boron).

The ionization energy for donors, measured from the conduction band edge, as calculated from Eq. 23 is 0.025 eV for silicon and 0.007 eV for gallium arsenide. The hydrogen atom calculation for the ionization level of acceptors is similar to that for donors. We consider the unfilled valence band as a filled band plus a hole in the central force field of a negative charged acceptor. The calculated ionization energy, measured from the valence band edge, is 0.05 eV for both silicon and gallium arsenide.

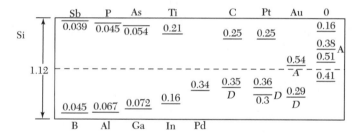

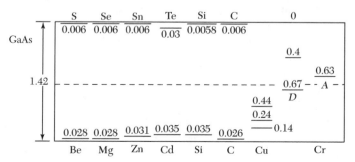

Fig. 24 Measured ionization energies (in eV) for various impurities in Si and GaAs. The levels below the gap center are measured from the top of the valence band and are acceptor levels unless indicated by *D* for donor level. The levels above the gap center are measured from the bottom of the conduction band and are donor levels unless indicated by *A* for acceptor level.[8]

This simple hydrogen atom model cannot account for the details of the ionization energy, particularly for the deep impurity levels in semiconductors (i.e., with ionization energies $\geq 3\,kT$). However, the calculated values do predict the correct order of magnitude of the true ionization energies for shallow impurity levels. Figure 24 shows the measured ionization energies for various impurities in silicon and gallium arsenide.[8] Note that it is possible for a single atom to have many levels; for example, oxygen in silicon has two donor levels and two acceptor levels in the forbidden energy gap.

2.7.1 Nondegenerate Semiconductor

In our previous discussion, we have assumed that the electron or hole concentration is much lower than the effective density of states in the conduction band or the valence band, respectively. In other words, the Fermi level E_F is at least $3kT$ above E_V or $3kT$ below E_C. For such cases, the semiconductor is referred to as a *nondegenerate* semiconductor.

For shallow donors in silicon and gallium arsenide, there usually is enough thermal energy to supply the energy E_D to ionize all donor impurities at room temperature and thus provide the same number of electrons in the conduction band. This condition is called complete ionization. Under a complete ionization condition, we can write the electron density as

$$n = N_D, \tag{24}$$

where N_D is the donor concentration. Figure 25a illustrates complete ionization where the donor level E_D is measured with respect to the bottom of the conduction band and equal concentrations of electrons (which are mobile) and donor ions (which are immobile) are shown. From Eqs. 16 and 24, we obtain the Fermi level in terms of the effective density of states N_C and the donor concentration N_D:

$$E_C - E_F = kT \ln (N_C / N_D). \tag{25}$$

Similarly, for shallow acceptors as shown in Fig. 25b, if there is complete ionization, the concentration of holes is

$$p = N_A, \tag{26}$$

Where N_A is the acceptor concentration. We can obtain the corresponding Fermi level from Eqs. 17 and 26:

$$E_F - E_V = kT \ln (N_V/N_A). \tag{27}$$

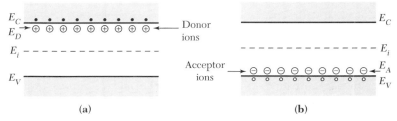

(a)   (b)

Fig. 25 Schematic energy band representation of extrinsic semiconductors with (*a*) donor ions and (*b*) acceptor ions.

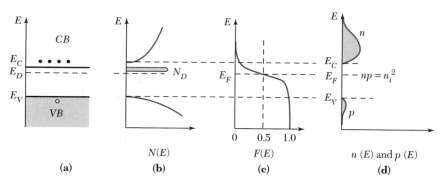

Fig. 26 *n*-Type semiconductor. (*a*) Schematic band diagram. (*b*) Density of states. (*c*) Fermi distribution function. (*d*) Carrier concentration. Note that $np = n_i^2$.

From Eq. 25 we can see that the higher the donor concentration, the smaller the energy difference $(E_C - E_F)$, that is, the Fermi level will move closer to the bottom of the conduction band. Similarly, for higher acceptor concentration, the Fermi level will move closer to the top of the valence band. Figure 26 illustrates the procedure for obtaining the carrier concentration. This figure is similar to that shown in Fig. 21. However, the Fermi level is closer to the bottom of the conduction band, and the electron concentration (upper shaded area) is much larger than the hole concentration (lower shaded area).

It is useful to express electron and hole densities in terms of the intrinsic carrier concentration n_i and the intrinsic Fermi level E_i since E_i is frequently used as a reference level when discussing extrinsic semiconductors. From Eq.16 we obtain

$$n = N_C \exp\left[-(E_C - E_F)/kT\right],$$

$$= N_C \exp\left[-(E_C - E_i)/kT\right] \exp\left[(E_F - E_i)/kT\right],$$

or

$$\boxed{n = n_i \exp\left[(E_F - E_i)/kT\right],} \tag{28}$$

and similarly,

$$\boxed{p = n_i \exp\left[(E_i - E_F)/kT\right].} \tag{29}$$

Note that the product of n and p from Eqs. 28 and 29 equals n_i^2. This result is identical to that for the intrinsic case, Eq. 20. Equation 20 is called the *mass action law*, which is valid for both intrinsic and extrinsic semiconductors under thermal equilibrium conduction. In an extrinsic semiconductor, the Fermi level moves toward either the bottom of the conduction band (*n*-type) or the top of the valence band (*p*-type). Either *n*- or *p*-type carriers will then dominate, but the product of the two types of carriers will remain constant at a given temperature.

► **EXAMPLE 4**

A silicon ingot is doped with 10^{16} arsenic atoms/cm³. Find the carrier concentrations and the Fermi level at room temperature (300 K).

SOLUTION At 300 K, we can assume complete ionization of impurity atoms. We have

$$n \approx N_D = 10^{16}\,\mathrm{cm}^{-3}.$$

From Eq. 20, $\qquad p \approx n_i^2/N_D = (9.65 \times 10^9)^2/10^{16} = 9.3 \times 10^3\,\mathrm{cm}^{-3}.$

The Fermi level measured from the bottom of the conduction band is given by Eq. 25:

$$E_C - E_F = kT \ln(N_C/N_D)$$
$$= 0.0259 \ln(2.86 \times 10^{19}/10^{10}) = 0.205\,\mathrm{eV}.$$

The Fermi level measured from the intrinsic Fermi level is given by Eq. 28:

$$E_F - E_i = kT \ln(N_D/n_i) \approx kT \ln(N_D/n_i)$$
$$= 0.0259 \ln (10^{16}/9.65 \times 10^9) = 0.358\,\mathrm{eV}.$$

These results are shown graphically in Fig. 27. ◄

If both donor and acceptor impurities are present simultaneously, the impurity that is present in a greater concentration determines the type of conductivity in the semiconductor. The Fermi level must adjust itself to preserve charge neutrality, that is, the total negative charges (electrons and ionized acceptors) must equal the total positive charges (holes and ionized donors). Under complete ionization condition, we have

$$n + N_A = p + N_D. \tag{30}$$

Solving Eqs. 20 and 30 yields the equilibrium electron and hole concentrations in an n-type semiconductor:

$$n = \frac{1}{2}\left[N_D - N_A + \sqrt{(N_D - N_A)^2 + 4n_i^2}\, \right], \tag{31}$$

$$p_n = n_i^2/n_n. \tag{32}$$

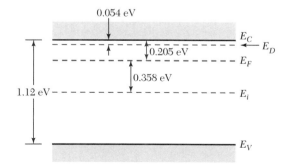

Fig. 27 Band diagram showing Fermi level E_F and intrinsic Fermi level E_i.

The subscript n refers to the n-type semiconductor. Because the electron is the dominant carrier, it is called the *majority carrier*. The hole in the n-type semiconductor is called the *minority carrier*. Similarly, we obtain the concentration of holes (majority carrier) and electrons (minority carrier) in a p-type semiconductor:

$$p_p = \frac{1}{2}\left[N_A - N_D + \sqrt{(N_D - N_A)^2 + 4n_i^2} \right], \tag{33}$$

$$n_p = n_i^2/p_p \tag{34}$$

The subscript p refers to the p-type semiconductor.

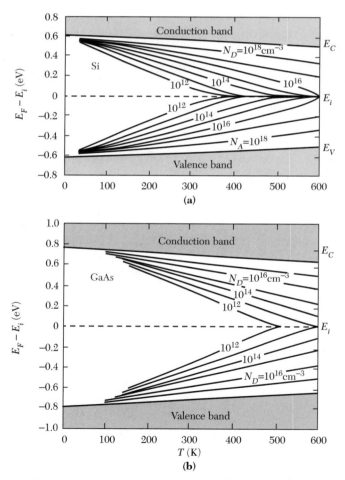

Fig. 28 Fermi level for Si and GaAs as a function of temperature and impurity concentration. The dependence of the bandgap on temperature is shown.[9]

Generally, the magnitude of the net impurity concentration $|N_D - N_A|$ is larger than the intrinsic carrier concentration n_i; therefore, the above relationships can be simplified to

$$n_n \approx N_D - N_A \quad \text{if} \quad N_D > N_A, \tag{35}$$

$$p_p \approx N_A - N_D \quad \text{if} \quad N_A > N_D. \tag{36}$$

From Eqs. 31 to 34 together with Eqs. 16 and 17, we can calculate the position of the Fermi level as a function of temperature for a given acceptor or donor concentration. Figure 28 shows a plot of these calculations for silicon[9] and gallium arsenide. We have incorporated in the figure the variation of the bandgap with temperature (see Problem 8). Note that as the temperature increases, the Fermi level approaches the intrinsic level, that is, the semiconductor becomes intrinsic.

Figure 29 shows electron density in Si as a function of temperature for a donor concentration of $N_D = 10^{15}$ cm^{-3}. At low temperatures, the thermal energy in the crystal is not sufficient to ionize all the donor impurities present. Some electrons are "frozen" at the donor level and the electron density is less than the donor concentration. As the temperature is increased, the condition of complete ionization is reached, (i.e., $n_n = N_D$). As the temperature is further increased, the electron concentration remains essentially the same over a wide temperature range. This is the extrinsic region. However, as the temperature is increased even further, we reach a point where the intrinsic carrier concentration becomes comparable to the donor concentration. Beyond this point, the semiconductor becomes intrinsic. The temperature at which the semiconductor becomes intrinsic depends on the impurity concentrations and the bandgap value and can be obtained from Fig. 22 by setting the impurity concentration equal to n_i.

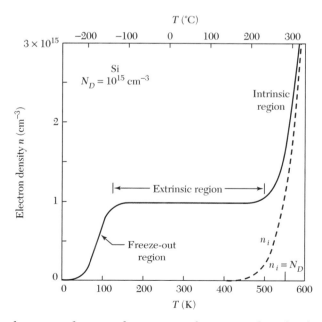

Fig. 29 Electron density as a function of temperature for a Si sample with a donor concentration of 10^{15} cm^{-3}.

2.7.2 Degenerate Semiconductor

When the doping concentration becomes equal or larger than the corresponding effective density of states, we can no longer use the approximation of Eq. 11 and the electron density, Eq. 9, has to be integrated numerically. For very heavily doped n-type or p-type semiconductor, E_F will be above E_C or below E_V. The semiconductor is referred to as a *degenerate* semiconductor.

An important aspect of high doping is the bandgap narrowing effect, that is, high impurity concentration causes a reduction of the bandgap. The bandgap reduction ΔE_g for silicon at room temperature is given by

$$\Delta E_g = 22 \left(\frac{N}{10^{18}} \right)^{1/2} \text{ meV,} \tag{37}$$

where the doping is in cm^{-3}. For example, for $N_D \leq 10^{18}$ cm^{-3}, $\Delta E_g \leq 0.022$ eV, which is less than 2% of the original bandgap. However, for $N_D \geq N_C = 2.86 \times 10^{19} \text{cm}^{-3}$, $\Delta E_g \geq 0.12$ eV, which is a significant fraction of E_g.

▶ SUMMARY

At the beginning of the chapter we listed a few important semiconductor materials. The properties of semiconductors are determined to a large extent by the crystal structure. We have defined the Miller indices to describe the crystal surfaces and crystal orientations and briefly described how semiconductor crystals are grown. More detailed discussions can be found in Chapter 10.

The bonding of atoms and the electron energy-momentum relationship in a semiconductor were considered in connection with the electrical properties. The energy band diagram can be used to understand why some materials are good conductors of electric current whereas others are poor conductors. We have also shown that changing the temperature or the amount of impurities can drastically vary the conductivity of a semiconductor.

▶ REFERENCES

1. R. A. Smith, *Semiconductors*, 2nd ed., Cambridge Univ. Press, London, 1979.

2. R. F. Pierret, *Semiconductor Device Fundamentals*, Addison Wesley, Boston, MA, 1996.

3. C. Kittel, *Introduction to Solid State Physics*, 6th ed., Wiley, New York, 1986.

4. D. Halliday and R. Resnick, *Fundamentals of Physics*, 2nd ed., Wiley, New York, 1981.

5. C. D. Thurmond, "The Standard Thermodynamic Function of the Formation of Electrons and Holes in Ge, Si, GaAs, and GaP," *J. Electrochem. Soc.*, **122**, 1133 (1975).

6. P. P. Altermatt, et al., "The Influence of a New Bandgap Narrowing Model on Measurement of the Intrinsic Carrier Density in Crystalline Silicon," *Tech. Dig., 11th Int. Photovolatic Sci. Eng. Conf.*, Sapporo, p. 719 (1999).

7. J. S. Blackmore, "Semiconducting and Other Major Properties of Gallium Arsenide," *J. Appl. Phys.*, **53**, 123–181 (1982).

8. S. M. Sze, *Physics of Semiconductors Devices*, 2nd ed., Wiley, New York, 1981.

9. A. S. Grove, *Physics and Technology of Semiconductor Devices*, Wiley, New York, 1967.

▶ PROBLEMS (* DENOTES DIFFICULT PROBLEMS)

FOR SECTION 2.2   BASIC CRYSTAL STRUCTURE

1. (a) What is the distance between nearest neighbors in silicon?

    (b) Find the number of atoms per square centimeter in silicon in the (100), (110), and (111) planes.

2. If we project the atoms in a diamond lattice onto the bottom surface with the heights of the atoms in unit of the lattice constant shown in the figure below. Find the heights of the three atoms (X, Y, Z) on the figure.

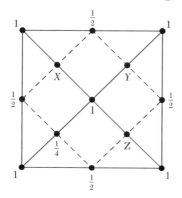

3. Find the maximum fraction of the unit cell volume, which can be filled by identical hard spheres in the simple cubic, face-centered cubic, and diamond lattices.

*4. Calculate the tetrahedral bond angle, the angle between any pair of the four bonds in a diamond lattice. (Hint: represent the four bonds as vectors of equal length. What must be the sum of the four vectors equal? Take components of this vector equation along the direction of one of these vectors.)

5. If a plane has intercepts at $2a$, $3a$, and $4a$ along the three Cartesian coordinates, where a is the lattice constant, find the Miller indices of the planes.

6. (a) Calculate the density of GaAs (the lattice constant of GaAs is 5.65 Å, and the atomic weights of Ga and As are 69.72 and 74.92 g/mol, respectively).

    (b) A gallium arsenide sample is doped with tin. If the tin displaces gallium atoms in the crystal lattice, are donors or acceptors formed? Why? Is the semiconductor n- or p-type?

FOR SECTION 2.3   BASIC CRYSTAL GROWTH TECHNIQUE

7. (a) Does silicon or silicon dioxide have a higher melting point? Why?

    (b) Why is a seed crystal used for crystal growth?

    (c) Why is the crystal orientation of a wafer important?

    (d) What two variables are used to control the diameter of the silicon rod?

FOR SECTION 2.5   ENERGY BANDS

8. The variation of silicon and GaAs bandgaps with temperature can be expressed as $E_g(T) = E_g(0) - \alpha T^2/(T + \beta)$, where $E_g(0) = 1.17$ eV, $\alpha = 4.73 \times 10^{-4}$ eV/K, and $\beta = 636$ K for

silicon; and E_g (0) = 1.519 eV, α = 5.405 × 10^{-4} eV/ K, and β = 204 K for GaAs. Find the bandgaps of Si and GaAs at 100 K and 600 K.

FOR SECTION 2.6   INTRINSIC CARRIER CONCENTRATION

°9. Derive Eq. 17. (Hint: In the valence band, the probability of occupancy of a state by a hole is [1 − F(E)].)

10. At room temperature (300 K) the effective density of states in the valence band is 2.66 × 10^{19} cm^{-3} for silicon and 7 × 10^{18} cm^{-3} for gallium arsenide. Find the corresponding effective masses of holes. Compare these masses with the free-electron mass.

11. Calculate the location of E_i in silicon at liquid nitrogen temperature (77 K), at room temperature (300 K), and at 100°C (let m_p =1.0 m_0 and m_n = 0.19 m_0). Is it reasonable to assume that E_i is in the center of the forbidden gap?

12. Find the kinetic energy of electrons in the conduction band of a nondegenerate n-type semiconductor at 300 K.

13. (a) For a free electron with a velocity of 10^7 cm/s, what is its de Broglie wavelength.

   (b) In GaAs, the effective mass of electrons in the conduction band is 0.063 m_0. If they have the same velocity, find the corresponding de Broglie wavelength.

14. The intrinsic temperature of a semiconductor is the temperatures at which the intrinsic carrier concentration equals the impurity concentration. Find the intrinsic temperature for a silicon sample doped with 10^{15} phosphorus atoms/cm^3.

FOR SECTION 2.7   DONORS AND ACCEPTORS

15. A silicon sample at T = 300 K contains an acceptor impurity concentration of N_A = 10^{16} cm^{-3}. Determine the concentration of donor impurity atoms that must be added so that the silicon is n-type and the Fermi energy is 0.20 eV below the conduction band edge.

16. Draw a simple flat energy band diagram for silicon doped with 10^{16} arsenic atoms/cm^3 at 77 K, 300 K, and 600 K. Show the Fermi level and use the intrinsic Fermi level as the energy reference.

17. Find the electron and hole concentrations and Fermi level in silicon at 300 K (a) for 1 × 10^{15} boron atoms/cm^3 and (b) for 3 × 10^{16} boron atoms/cm^3 and 2.9 × 10^{16} arsenic atoms/cm^3.

18. A Si sample is doped with 10^{17} As atoms/cm^3. What is the equilibrium hole concentration p_0 at 300 K? Where is E_F relative to E_i?

19. Calculate the Fermi level of silicon doped with 10^{15}, 10^{17}, and 10^{19} phosphorus atoms/cm^3 at room temperature, assuming complete ionization. From the calculated Fermi level, check if the assumption of complete ionization is justified for each doping. Assume that

   the ionized donors is given by $n = N_D \, [1 - F(E_D)] = \dfrac{N_D}{1 + e^{(E_F - E_D)/kT}}$.

20. For an n-type silicon sample with 10^{16} cm^{-3} phosphorous donor impurities and a donor level at E_D= 0.045 eV, find the ratio of the neutral donor density to the ionized donor density at 77 K where the Fermi level is 0.0459 below the bottom of the conduction band. The expression for ionized donors is given in Prob. 19.

Carrier Transport Phenomena

In this chapter, we investigate various transport phenomena in semiconductor devices. The transport processes include drift, diffusion, recombination, generation, thermionic emission, tunneling, and impact ionization. We consider the motion of charge carriers (electrons and holes) in semiconductors under the influence of an electric field and a carrier concentration gradient. We also discuss the concept of the nonequilibrium condition where the carrier concentration product pn is different from its equilibrium value n_i^2. Returning to an equilibrium condition through the generation-recombination processes is considered next. We then derive the basic governing equations for semiconductor device operation, which includes the current density equation and the continuity equation. This is followed by a discussion of thermionic emission and tunneling process. The chapter closes with a brief discussion of high-field effects, which include velocity saturation and impact ionization.

Specifically, we cover the following topics:

- The current density equation and its drift and diffusion components.

- The continuity equation and its generation and recombination components.

- Other transport phenomena, including the thermionic emission, tunneling, transferred-electron effect, and impact ionization.

- Methods to measure key semiconductor parameters such as resistivity, mobility, majority-carrier concentration, and minority-carrier lifetime.

▶ ## 3.1 CARRIER DRIFT

3.1.1 Mobility

Consider an *n*-type semiconductor sample with uniform donor concentration in thermal equilibrium. As discussed in Chapter 2, the conduction electrons in the semiconductor conduction band are essentially free particles, since they are not associated with any particular lattice or donor site. The influence of crystal lattices is incorporated in the effective mass of conduction electrons, which differs somewhat from the mass of free electrons. Under thermal equilibrium, the average thermal energy of a conduction electron can be obtained from the theorem for equipartition of energy, $1/2\,kT$ units of energy per degree of freedom, where k is Boltzmann's constant and T is the absolute temperature. The electrons in a semiconductor have three degrees of freedom; they can move about in a three-dimensional space. Therefore, the kinetic energy of the electrons is given by

$$\frac{1}{2}m_n v_{th}^2 = \frac{3}{2}kT, \tag{1}$$

where m_n is the effective mass of electrons and v_{th} is the average thermal velocity. At room temperature (300 K) the thermal velocity of electrons in Eq. 1 is about 10^7 cm/s for silicon and gallium arsenide.

The electrons in the semiconductor are therefore moving rapidly in all directions. The thermal motion of an individual electron may be visualized as a succession of random scattering from collisions with lattice atoms, impurity atoms, and other scattering centers, as illustrated in Fig. 1*a*. The random motion of electrons leads to a zero net displacement of an electron over a sufficiently long period of time. The average distance between collisions is called the *mean free path*, and the average time between collisions is called the *mean free time* τ_c. For a typical value of 10^{-5} cm for the mean free path, τ_c is about 1 ps (ie., $10^{-5}/v_{th} \cong 10^{-12}$ s).

When a small electric field $\mathscr{E}$ is applied to the semiconductor sample, each electron will experience a force $-q\mathscr{E}$ from the field and will be accelerated along the field (in the opposite direction to the field) during the time between collisions. Therefore, an additional velocity component will be superimposed upon the thermal motion of electrons. This additional component is called the *drift velocity*. The combined displacement of an electron due to the random thermal motion and the drift component is illustrated in Fig. 1*b*. Note that there is a net displacement of the electron in the direction opposite to the applied field.

We can obtain the drift velocity v_n by equating the momentum (force × time) applied to an electron during the free flight between collisions to the momentum gained by the electron in the same period. The equality is valid because in a steady state all momen-

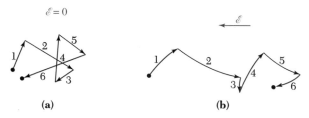

(a)   **(b)**

Fig. 1 Schematic path of an electron in a semiconductor. (*a*) Random thermal motion. (*b*) Combined motion due to random thermal motion and an applied electric field.

tum gained between collisions is lost to the lattice in the collision. The momentum applied to an electron is given by $-q\mathscr{E}\tau_c$, and the momentum gained is $m_n v_n$. We have

$$-q\mathscr{E}\tau_c = m_n v_n \qquad (2)$$

or

$$v_n = -\left(q\tau_c\middle/m_n\right)\mathscr{E} \qquad (2a)$$

Equation *2a* states that the electron drift velocity is proportional to the applied electric field. The proportionality factor depends on the mean free time and the effective mass. The proportionality factor is called the *electron mobility* μ_n in units of cm²/V-s, or

$$\mu_n \equiv \frac{q\tau_c}{m_n}. \qquad (3)$$

Thus,

$$\boxed{v_n = -\mu_n\mathscr{E}.} \qquad (4)$$

Mobility is an important parameter for carrier transport because it describes how strongly the motion of an electron is influenced by an applied electric field. A similar expression can be written for holes in the valence band:

$$\boxed{v_p = \mu_p\mathscr{E},} \qquad (5)$$

where v_p is the hole drift velocity and μ_p is the hole mobility. The negative sign is removed in Eq. 5 because holes drift in the same direction as the electric field.

In Eq. 3 the mobility is related directly to the mean free time between collisions, which in turn is determined by the various scattering mechanisms. The two most important mechanisms are lattice scattering and impurity scattering. Lattice scattering results from thermal vibrations of the lattice atoms at any temperature above absolute zero. These vibrations disturb the lattice periodic potential and allow energy to be transferred between the carriers and the lattice. Since lattice vibration increases with increasing temperature, lattice scattering becomes dominant at high temperatures; hence the mobility decreases with increasing temperature. Theoretical analysis[1] shows that the mobility due to lattice scattering μ_L will decrease in proportion to $T^{-3/2}$.

Impurity scattering results when a charge carrier travels past an ionized dopant impurity (donor or acceptor). The charge carrier path will be deflected owing to Coulomb force interaction. The probability of impurity scattering depends on the total concentration of ionized impurities, that is, the sum of the concentration of negatively and positively charged ions. However, unlike lattice scattering, impurity scattering becomes less significant at higher temperatures. At higher temperatures, the carriers move faster; they remain near the impurity atom for a shorter time and are therefore less effectively scattered. The mobility due to impurity scattering μ_I can theoretically be shown to vary as $T^{3/2}/N_T$, where N_T is the total impurity concentration.[2]

The probability of a collision taking place in unit time, $1/\tau_c$, is the sum of the probabilities of collisions due to the various scattering mechanisms:

$$\frac{1}{\tau_c} = \frac{1}{\tau_{c,\,\text{lattice}}} + \frac{1}{\tau_{c,\,\text{impurity}}}, \qquad (6)$$

or

$$\frac{1}{\mu} = \frac{1}{\mu_L} + \frac{1}{\mu_I}. \tag{6a}$$

Figure 2 shows the measured electron mobility as a function of temperature for silicon with five different donor concentrations.[3] The inset shows the theoretical temperature dependence of mobility due to both lattice and impurity scatterings. For lightly doped samples (e.g., the sample with doping of 10^{14} cm^{-3}), the lattice scattering dominates, and the mobility decreases as the temperature increases. For heavily doped samples, the effect of impurity scattering is most pronounced at low temperatures. The mobility increases as the temperature increases, as can be seen for the sample with doping of 10^{19} cm^{-3}. For a given temperature, the mobility decreases with increasing impurity concentration because of enhanced impurity scatterings.

Figure 3 shows the measured mobilities in silicon and gallium arsenide as a function of impurity concentration at room temperature.[3] Mobility reaches a maximum value at low impurity concentrations; this corresponds to the lattice-scattering limitation. Both electron and hole mobilities decrease with increasing impurity concentration and eventually approach a minimum value at high concentrations. Note also that the mobility of electrons is greater than that of holes. Greater electron mobility is due mainly to the smaller effective mass of electrons.

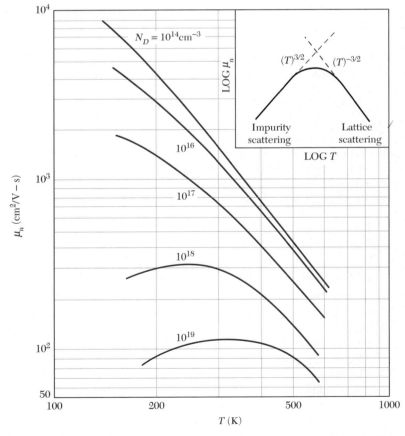

Fig. 2 Electron mobility in silicon versus temperature for various donor concentrations. Insert shows the theoretical temperature dependence of electron mobility.[3]

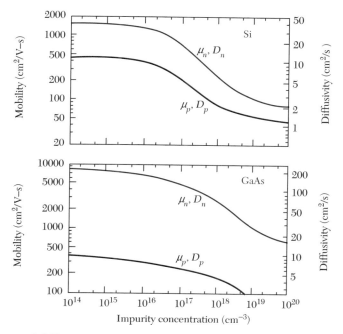

Fig. 3 Mobilities and diffusivities in Si and GaAs at 300 K as a function of impurity concentration.[3]

▶ **EXAMPLE 1**

Calculate the mean free time of an electron having a mobility of 1000 cm²/V-s at 300 K; also calculate the mean free path. Assume $m_n = 0.26\ m_0$ in these calculations.

SOLUTION   From Eq. 3, the mean free time is given by

$$\tau_c = \frac{m_n \mu_n}{q} = \frac{(0.26 \times 0.91 \times 10^{-30}\ \text{kg}) \times (1000 \times 10^{-4}\ \text{m}^2/\text{V-s})}{1.6 \times 10^{-19}\ \text{C}}$$
$$= 1.48 \times 10^{-13}\ \text{s} = 0.148\ \text{ps}.$$

The mean free path is given by

$$l = v_{th}\tau_c = (10^7\ \text{cm/s})(1.48 \times 10^{-13}\text{s}) = 1.48 \times 10^{-6}\ \text{cm}$$
$$= 14.8\ \text{nm}.$$ ◀

3.1.2   Resistivity

We now consider conduction in a homogeneous semiconductor material. Figure 4*a* shows an *n*-type semiconductor and its band diagram at thermal equilibrium. Figure 4*b* shows the corresponding band diagram when a biasing voltage is applied to the right-hand terminal. We assume that the contacts at the left-hand and right-hand terminals are ohmic, that is, there is negligible voltage drop at each of the contacts. The behavior of ohmic contacts are considered in Chapter 7. As mentioned previously, when an electric field $\mathscr{E}$

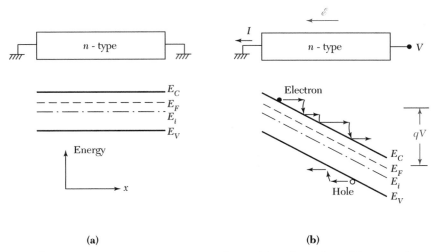

Fig. 4 Conduction process in an n-type semiconductor (a) at thermal equilibrium and (b) under a biasing condition.

is applied to a semiconductor, each electron will experience a force $-q\mathscr{E}$ from the field. The force is equal to the negative gradient of potential energy; that is,

$$-q\mathscr{E} = -(\text{gradient of electron potential energy}) = -\frac{dE_C}{dx}. \qquad (7)$$

Recall that in Chapter 2 the bottom of the conduction band E_C corresponds to the potential energy of an electron. Since we are interested in the gradient of the potential energy, we can use any part of the band diagram that is parallel to E_C (e.g., E_F, E_i, or E_V, as shown in Fig. 4b). It is convenient to use the intrinsic Fermi level E_i because we shall use E_i when we consider p–n junctions in Chapter 4. Therefore, from Eq. 7 we have

$$\mathscr{E} = \frac{1}{q}\frac{dE_C}{dx} = \frac{1}{q}\frac{dE_i}{dx}. \qquad (8)$$

We can define a related quantity ψ as the *electrostatic potential* whose negative gradient equals the electric field:

$$\mathscr{E} \equiv -\frac{d\psi}{dx}. \qquad (9)$$

Comparison of Eqs. 8 and 9 gives

$$\psi = -\frac{E_i}{q}, \qquad (10)$$

which provides a relationship between the electrostatic potential and the potential energy of an electron. For a homogeneous semiconductor shown in Fig. 4b, the potential energy and E_i decrease linearly with distance; thus, the electric field is a constant in the negative x-direction. Its magnitude equals the applied voltage divided by the sample length.

The electrons in the conduction band move to the right side as shown in Fig. 4b. The kinetic energy corresponds to the distance from the band edge (i.e., E_C for electrons). When an electron undergoes a collision, it loses some or all of its kinetic energy to the lattice and drops toward its thermal equilibrium position. After the electron has lost some or all its kinetic energy, it will again begin to move toward the right, and the same process will be repeated many times. Conduction by holes can be visualized in a similar manner but in the opposite direction.

The transport of carriers under the influence of an applied electric field produces a current called the *drift current*. Consider a semiconductor sample shown in Fig. 5, which has a cross-sectional area A, a length L, and a carrier concentration of n electrons/cm³. Suppose we now apply an electric field $\mathscr{E}$ to the sample. The electron current density J_n flowing in the sample can be found by summing the product of the charge $(-q)$ on each electron times the electron's velocity over all electrons per unit volume n:

$$J_n = \frac{I_n}{A} = \sum_{i=1}^{n}(-qv_i) = -qnv_n = qn\mu_n\mathscr{E}, \tag{11}$$

where I_n is the electron current. We have employed Eq. 4 for the relationship between v_n and $\mathscr{E}$.

A similar argument applies to holes. By taking the charge on the hole to be positive, we have

$$J_p = qpv_p = qp\mu_p\mathscr{E}. \tag{12}$$

The total current flowing in the semiconductor sample due to the applied field $\mathscr{E}$ can be written as the sum of the electron and hole current components:

$$J = J_n + J_p = (qn\mu_n + qp\mu_p)\mathscr{E}. \tag{13}$$

The quantity in parentheses is known as *conductivity*:

$$\sigma = q(n\mu_n + p\mu_p). \tag{14}$$

The electron and hole contributions to conductivity are simply additive.

The corresponding resistivity of the semiconductor, which is the reciprocal of σ, is given by

$$\boxed{\rho \equiv \frac{1}{\sigma} = \frac{1}{q(n\mu_n + p\mu_p)}.} \tag{15}$$

Generally, in extrinsic semiconductors, only one of the components in Eq. 13 or 14 is significant because of the many orders-of-magnitude difference between the two carrier densities. Therefore, Eq. 15 reduces to

$$\rho = \frac{1}{qn\mu_n} \tag{15a}$$

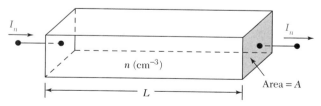

Fig. 5 Current conduction in a uniformly doped semiconductor bar with length L and cross-sectional area A.

for an n-type semiconductor (since $n >> p$), and to

$$\rho = \frac{1}{qp\mu_p} \qquad (15b)$$

for a p-type semiconductor (since $p >> n$).

The most common method for measuring resistivity is the four-point probe method shown in Fig. 6. The probes are equally spaced. A small current I from a constant-current source is passed through the outer two probes and a voltage V is measured between the inner two probes. For a thin semiconductor sample with thickness W that is much smaller than the sample diameter d, the resistivity is given by

$$\rho = \frac{V}{I} \cdot W \cdot CF \qquad \Omega\text{-cm}, \qquad (16)$$

where CF is a well-documented "correction factor." The correction factor depends on the ratio of d/s, where s is the probe spacing. When $d/s > 20$, the correction factor approaches 4.54.

Figure 7 shows the measured room-temperature resistivity as a function of the impurity concentration for silicon and gallium arsenide.[3] At this temperature and for low impurity concentrations, all donor (e.g., P and As in Si) or acceptor (e.g., B in Si) impurities that have shallow energy levels will be ionized. Under these conditions, the carrier concentration is equal to the impurity concentration. From these curves we can obtain the impurity concentration of a semiconductor if the resistivity is known, or vice versa.

▶ **EXAMPLE 2**

Find the room-temperature resistivity of an n-type silicon doped with 10^{16} phosphorus atoms/cm^3.

SOLUTION At room temperature we assume that all donors are ionized; thus,

$$n \approx N_D = 10^{16} \text{ cm}^{-3}.$$

From Fig. 7 we find $\rho \cong 0.5 \ \Omega$–cm. We can also calculate the resistivity from Eq. 15a:

$$\rho = \frac{1}{qn\mu_n} = \frac{1}{1.6 \times 10^{-19} \times 10^{16} \times 1300} = 0.48 \qquad \Omega\text{–cm}.$$

The mobility μ_n is obtained from Fig. 3. ◀

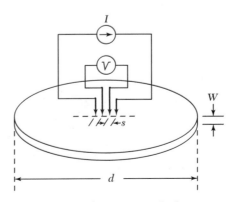

Fig. 6 Measurement of resistivity using a four-point probe.[3]

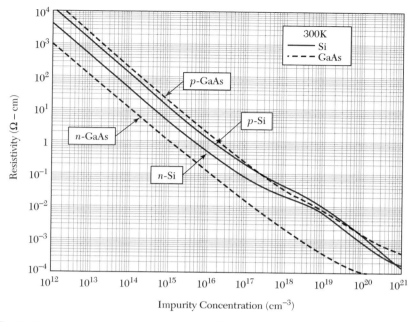

Fig. 7 Resistivity versus impurity concentration[3] for Si and GaAs.

3.1.3 The Hall Effect

The carrier concentration in a semiconductor may be different from the impurity concentration, because the ionized impurity density depends on the temperature and the impurity energy level. To measure the carrier concentration directly, the most commonly used method is the Hall effect. Hall measurement is also one of the most convincing methods to show the existence of holes as charge carriers, because the measurement can give directly the carrier type. Figure 8 shows an electric field applied along the x-axis and a magnetic field applied along the z-axis. Consider a p-type semiconductor sample. The Lorentz force $q\boldsymbol{v} \times \mathbf{B}$ ($= qv_x B_z$) due to the magnetic field will exert an average upward force on the holes flowing in the x-direction. The upward directed current causes an

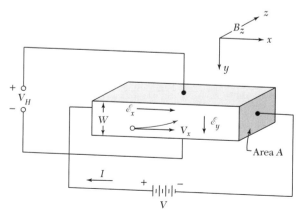

Fig. 8 Basic setup to measure carrier concentration using the Hall effect.

accumulation of holes at the top of the sample that gives rise to a downward-directed electric field $\mathscr{E}_y$. Since there is no net current flow along the y-direction in the steady state, the electric field along the y-axis exactly balances the Lorentz force; that is,

$$q\mathscr{E}_y = qv_x B_z , \tag{17}$$

or

$$\mathscr{E}_y = v_x B_z . \tag{18}$$

Once the electric field $\mathscr{E}_y$ becomes equal to $v_x B_z$, no net force along the y-direction is experienced by the holes as they drift in the x-direction.

The establishment of the electric field is known as the *Hall effect*. The electric field in Eq. 18 is called the *Hall field*, and the terminal voltage $V_H = \mathscr{E}_y W$ (Fig. 8) is called the *Hall voltage*. Using Eq. 12 for the hole drift velocity, the Hall field $\mathscr{E}_y$ in Eq. 18 becomes

$$\mathscr{E}_y = \left(\frac{J_p}{qp}\right) B_z = R_H J_p B_z , \tag{19}$$

where

$$R_H \equiv \frac{1}{qp} . \tag{20}$$

The Hall field $\mathscr{E}_y$ is proportional to the product of the current density and the magnetic field. The proportionality constant R_H is the *Hall coefficient*. A similar result can be obtained for an n-type semiconductor, except that the Hall coefficient is negative:

$$R_H = -\frac{1}{qn} . \tag{21}$$

A measurement of the Hall voltage for a known current and magnetic field yields

$$p = \frac{1}{qR_H} = \frac{J_p B_z}{q\mathscr{E}_y} = \frac{(I/A)B_z}{q(V_H/W)} = \frac{IB_z W}{qV_H A} , \tag{22}$$

where all the quantities in the right-hand side of the equation can be measured. Thus, the carrier concentration and carrier type can be obtained directly from the Hall measurement.

▶ **EXAMPLE 3**

A sample of Si is doped with 10^{16} phosphorus atoms/cm³. Find the Hall voltage in a sample with $W = 500$ μm, $A = 2.5 \times 10^{-3}$ cm², $I = 1$ mA, and $B_z = 10^{-4}$ Wb/cm².

SOLUTION The Hall coefficient is

$$R_H = -\frac{1}{qn} = -\frac{1}{1.6 \times 10^{-19} \times 10^{16}} = -625 \text{ cm}^3/\text{C}.$$

The Hall voltage is

$$V_H = \mathscr{E}_y W = \left(R_H \frac{I}{A} B_z\right) W$$

$$= \left(-625 \cdot \frac{10^{-3}}{2.5 \times 10^{-3}} \cdot 10^{-4}\right) 500 \times 10^{-4}$$

$$= -1.25 \text{ mV}.$$

3.2 CARRIER DIFFUSION

3.2.1 Diffusion Process

In the preceding section, we considered the drift current, that is, the transport of carriers when an electric field is applied. Another important current component can exist if there is a spatial variation of carrier concentration in the semiconductor material. The carriers tend to move from a region of high concentration to a region of low concentration. This current component is called *diffusion current*.

To understand the diffusion process, let us assume an electron density that varies in the x-direction, as shown in Fig. 9. The semiconductor is at uniform temperature, so that the average thermal energy of electrons does not vary with x, only the density $n(x)$ varies. Consider the number of electrons crossing the plane at $x = 0$ per unit time and per unit area. Because of finite temperature, the electrons have random thermal motions with a thermal velocity v_{th} and a mean free path l (note that $l = v_{th}\tau_c$, where τ_c is the mean free time.) The electrons at $x = -l$, one mean free path away on the left side, have equal chances of moving left or right; and in a mean free time τ_c, one half of them will move across the plane $x = 0$. The average rate of electron flow per unit area F_1 of electrons crossing plane $x = 0$ from the left is then

$$F_1 = \frac{\frac{1}{2} n(-l) \cdot l}{\tau_c} = \frac{1}{2} n(-l) \cdot v_{th}. \tag{23}$$

Similarly, the average rate of electron flow per unit area F_2 of electrons at $x = l$ crossing plane $x = 0$ from the right is

$$F_2 = \frac{1}{2} n(l) \cdot v_{th}. \tag{24}$$

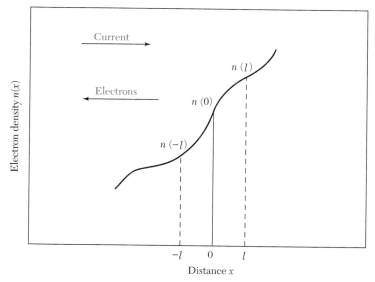

Fig. 9 Electron concentration versus distance; l is the mean free path. The directions of electron and current flows are indicated by arrows.

The net rate of carrier flow from left to right is

$$F = F_1 - F_2 = \frac{1}{2} v_{th}[n(-l) - n(l)]. \tag{25}$$

Approximating the densities at $x = \pm l$ by the first two terms of a Taylor series expansion, we obtain

$$F = \frac{1}{2} v_{th} \left\{ n \left[(0) - l\frac{dn}{dx} \right] - \left[n(0) + l\frac{dn}{dx} \right] \right\}$$

$$= -v_{th} l\frac{dn}{dx} \equiv -D_n \frac{dn}{dx}, \tag{26}$$

where $D_n \equiv v_{th}l$ is called the *diffusion coefficient* also called the *diffusivity*. Because each electron carriers a charge $-q$, the carrier flow gives rise to a current

$$J_n = -qF = qD_n \frac{dn}{dx}. \tag{27}$$

The diffusion current is proportional to the spatial derivative of the electron density. Diffusion current results from the random thermal motion of carriers in a concentration gradient. For an electron density that increases with x, the gradient is positive, and the electrons will diffuse toward the negative x-direction. The current is positive and flows in the direction opposite to that of the electrons as indicated in Fig. 9.

▶ **EXAMPLE 4**

Assume that, in an n-type semiconductor at $T = 300$ K, the electron concentration varies linearly from 1×10^{18} to 7×10^{17} cm^{-3} over a distance of 0.1 cm. Calculate the diffusion current density if the electron diffusion coefficient is $D_n = 22.5$ cm^2/s.

SOLUTION The diffusion current density is given by

$$J_{n, \text{diff}} = qD_n \frac{dn}{dx} \approx qD_n \frac{\Delta n}{\Delta x}$$

$$= (1.6 \times 10^{-19})(22.5)\left(\frac{1 \times 10^{18} - 7 \times 10^{17}}{0.1} \right) = 10.8 \quad \text{A/cm}^2. \quad ◀$$

3.2.2 Einstein Relation

Equation 27 can be written in a more useful form using the theorem for the equipartition of energy for this one-dimensional case. We can write

$$\frac{1}{2} m_n v_{th}^2 = \frac{1}{2} kT. \tag{28}$$

From Eqs. 3, 26, and 28 and using the relationship $l = v_{th}\tau_c$, we obtain

$$D_n = v_{th}l = v_{th}\left(v_{th}\tau_c \right) = v_{th}^2 \left(\frac{\mu_n m_n}{q} \right) = \left(\frac{kT}{m_n} \right)\left(\frac{\mu_n m_n}{q} \right), \tag{29}$$

or

$$D_n = \left(\frac{kT}{q}\right)\mu_n.$$

(30)

Equation 30 is known as the *Einstein relation*. It relates the two important constants (diffusivity and mobility) that characterize carrier transport by diffusion and by drift in a semiconductor. The Einstein relation also applies between D_p and μ_p. Values of diffusivities for silicon and gallium arsenide are shown in Fig. 3.

▷ **EXAMPLE 5**

Minority carriers (holes) are injected into a homogeneous *n*-type semiconductor sample at one point. An electric field of 50 V/cm is applied across the sample, and the field moves these minority carriers a distance of 1 cm in 100 µs. Find the drift velocity and the diffusivity of the minority carriers.

SOLUTION $v_p = \dfrac{1\ \text{cm}}{100 \times 10^{-6}\ \text{s}} = 10^4 \quad \text{cm/s};$

$\mu_p = \dfrac{v_p}{\mathscr{E}} = \dfrac{10^4}{50} = 200 \quad \text{cm}^2/\text{V-s};$

$D_p = \dfrac{kT}{q}\mu_p = 0.0259 \times 200 = 5.18 \quad \text{cm}^2/\text{s}.$ ◀

3.2.3 Current Density Equations

When an electric field is present in addition to a concentration gradient, both drift current and diffusion current will flow. The total current density at any point is the sum of the drift and diffusion components:

$$J_n = q\mu_n n\mathscr{E} + qD_n\frac{dn}{dx},$$

(31)

where $\mathscr{E}$ is the electric field in the *x*-direction.

A similar expression can be obtained for the hole current:

$$J_p = q\mu_p p\mathscr{E} - qD_p\frac{dp}{dx}.$$

(32)

We use the negative sign in Eq. 32 because for a positive hole gradient the holes will diffuse in the negative *x*-direction. This diffusion results in a hole current that also flows in the negative *x*-direction.

The total conduction current density is given by the sum of Eqs. 31 and 32:

$$J_{cond} = J_n + J_p.$$

(33)

The three expressions (Eqs. 31–33) constitute the current density equations. These equations are important for analyzing device operations under low electric fields. However, at sufficiently high electric fields the terms $\mu_n \mathscr{E}$ and $\mu_p \mathscr{E}$ should be replaced by the saturation velocity v_s discussed in Section 3.7.

3.3 GENERATION AND RECOMBINATION PROCESSES

In thermal equilibrium the relationship $pn = n_i^2$ is valid. If excess carriers are introduced to a semiconductor so that $pn > n_i^2$, we have a *nonequilibrium situation*. The process of introducing excess carriers is called *carrier injection*. Most semiconductor devices operate by the creation of charge carriers in excess of the thermal equilibrium values. We can introduce excess carriers by optical excitation or forward-biasing a p–n junction (discussed in Chapter 4).

Whenever the thermal-equilibrium condition is disturbed (i.e., $pn \neq n_i^2$), processes exist to restore the system to equilibrium (i.e., $pn = n_i^2$). In the case of the injection of excess carriers, the mechanism that restores equilibrium is recombination of the injected minority carriers with the majority carriers. Depending on the nature of the recombination process, the released energy that results from the recombination process can be emitted as a photon or dissipated as heat to the lattice. When a photon is emitted, the process is called radiative recombination; otherwise, it is called nonradiative recombination.

Recombination phenomena can be classified as direct and indirect processes. Direct recombination, also called band-to-band recombination, usually dominates in direct-bandgap semiconductors, such as gallium arsenide, whereas indirect recombination via bandgap recombination centers dominates in indirect bandgap semiconductors, such as silicon.

3.3.1 Direct Recombination

Consider a direct-bandgap semiconductor in thermal equilibrium. The continuous thermal vibration of lattice atoms causes some bonds between neighboring atoms to be broken. When a bond is broken, an electron-hole pair is generated. In terms of the band diagram, the thermal energy enables a valence electron to make an upward transition to the conduction band, leaving a hole in the valence band. This process is called carrier generation and is represented by the generation rate G_{th} (number of electron-hole pairs generated per cm³ per second) in Fig. 10a. When an electron makes a transition downward from the conduction band to the valence band, an electron-hole pair is annihilated. This reverse process is called recombination; it is represented by the recombination rate R_{th} in Fig. 10a. Under thermal equilibrium conditions, the generation rate G_{th} must equal

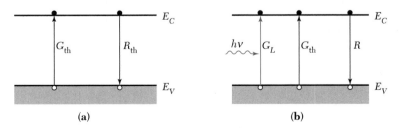

(a)　　　　　　　　　　　　(b)

Fig. 10　Direct generation and recombination of electron-hole pairs: (*a*) at thermal equilibrium and (*b*) under illumination.

the recombination rate R_{th}, so that the carrier concentrations remain constant and the condition $pn = n_i^2$ is maintained.

When excess carriers are introduced to a direct-bandgap semiconductor, the probability is high that electrons and holes will recombine directly, because the bottom of the conduction band and the top of the valence band are lined up and no additional momentum is required for the transition across the bandgap. The rate of the direct recombination R is expected to be proportional to the number of electrons available in the conduction band and the number of holes available in the valence band; that is,

$$R = \beta np, \tag{34}$$

where β is the proportionality constant. As discussed previously, in thermal equilibrium the recombination rate must be balanced by the generation rate. Therefore, for an n-type semiconductor, we have

$$G_{th} = R_{th} = \beta n_{no} p_{no}. \tag{35}$$

In this notation for carrier concentrations the first subscript refers to the type of the semiconductor. The subscript o indicates an equilibrium quantity. The n_{no} and p_{no} represent electron and hole densities, respectively, in an n-type semiconductor at thermal equilibrium. When we shine a light on the semiconductor to produce electron-hole pairs at a rate G_L (Fig. 10b), the carrier concentrations are above their equilibrium values. The recombination and generation rate become

$$R = \beta n_n p_n = \beta(n_{no} + \Delta n)(p_{no} + \Delta p), \tag{36}$$

$$G = G_L + G_{th}, \tag{37}$$

where Δn and Δp are the excess carrier concentrations, given by

$$\Delta n = n_n - n_{no} \tag{38a}$$

$$\Delta p = p_n - p_{no}, \tag{38b}$$

and $\Delta n = \Delta p$ to maintain overall charge neutrality.

The net rate of change of hole concentration is given by

$$\frac{dp_n}{dt} = G - R = G_L + G_{th} - R. \tag{39}$$

In steady state, $dp_n / dt = 0$. From Eq. 39 we have

$$G_L = R - G_{th} \equiv U, \tag{40}$$

where U is the net recombination rate. Substituting Eqs. 35 and 36 into Eq. 40 yields

$$U = \beta(n_{no} + p_{no} + \Delta p)\Delta p. \tag{41}$$

For low-level injection Δp, $p_{no} \ll n_{no}$, Eq. 41 is simplified to

$$U \cong \beta n_{no}\Delta p = \frac{p_n - p_{no}}{\dfrac{1}{\beta n_{no}}}. \tag{42}$$

Therefore, the net recombination rate is proportional to excess minority carrier concentration. Obviously, $U = 0$ in thermal equilibrium. The proportionality constant $1/\beta n_{no}$ is called the *lifetime* τ_p of the excess minority carriers, or

$$U \equiv \frac{p_n - p_{no}}{\tau_p},$$ (43)

where

$$\tau_p \equiv \frac{1}{\beta n_{no}}.$$ (44)

The physical meaning of lifetime can best be illustrated by the transient response of a device after the sudden removal of the light source. Consider an n-type sample, as shown in Fig. 11a, that is illuminated with light and in which the electron-hole pairs are generated uniformly throughout the sample with a generation rate G_L. The time-dependent expression is given by Eq. 39. In steady state, from Eqs. 40 and 43

$$G_L = U = \frac{p_n - p_{no}}{\tau_p}$$ (45)

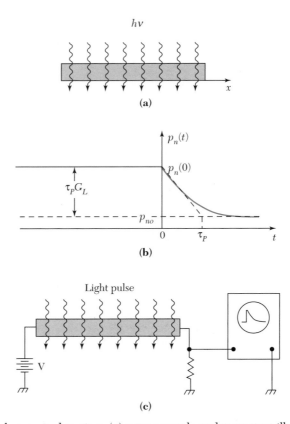

Fig. 11 Decay of photoexcited carriers. (a) n-type sample under constant illumination. (b) Decay of minority carriers (holes) with time. (c) Schematic setup to measure minority carrier lifetime.

or

$$p_n = p_{no} + \tau_p G_L. \tag{45a}$$

If at an arbitrary time, say $t = 0$, the light is suddenly turned off, the boundary conditions are $p_n(t = 0) = p_{no} + \tau_p G_L$, as given by Eq. 45a, and $p_n(t \to \infty) = p_{no}$. The time-dependent expression of Eq. 39 becomes

$$\frac{dp_n}{dt} = G_{th} - R = -U = -\frac{p_n - p_{no}}{\tau_p} \tag{46}$$

and the solution is

$$p_n(t) = p_{no} + \tau_p G_L \exp\left(-t/\tau_p\right). \tag{47}$$

Figure 11b shows the variation of p_n with time. The minority carriers recombine with majority carriers and decay exponentially with a time constant τ_p, which corresponds to the lifetime defined in Eq. 44.

This case illustrates the main idea of measuring the carrier lifetime using the photoconductivity method. Figure 11c shows a schematic setup. The excess carriers, generated uniformly throughout the sample by the light pulse, cause a momentary increase in the conductivity. The increase in conductivity manifests itself by a drop in voltage across the sample when a constant current is passed through it. The decay of the conductivity can be observed on an oscilloscope and is a measure of the lifetime of the excess minority carriers.

▷ **EXAMPLE 6**

A Si sample with $n_{no} = 10^{14}$ cm^{-3} is illuminated with light and 10^{13} electron-hole pairs/cm^3 are created every microsecond. If $\tau_n = \tau_p = 2$ μs, find the change in the minority carrier concentration.

SOLUTION   Before illumination

$$p_{no} = n_i^2 / n_{no} = (9.65 \times 10^9)^2 / 10^{14} \approx 9.31 \times 10^5 \text{ cm}^{-3}.$$

After illumination

$$p_n = p_{no} + \tau_p G_L = 9.31 \times 10^5 + 2 \times 10^{-6} \times \frac{10^{13}}{1 \times 10^{-6}} \approx 2 \times 10^{13} \text{ cm}^{-3}. \qquad ◄$$

3.3.2   Indirect Recombination

For indirect-bandgap semiconductors, such as silicon, a direct recombination process is very unlikely, because the electrons at the bottom of the conduction band have nonzero momentum with respect to the holes at the top of the valence band (see Chapter 2). A direct transition that conserves both energy and momentum is not possible without a simultaneous lattice interaction. Therefore the dominant recombination process in such semiconductors is indirect transition via localized energy states in the forbidden energy gap.[4] These states act as stepping stones between the conduction band and the valence band.

Figure 12 shows various transitions that occur in the recombination process through intermediate-level states (also called recombination centers). We illustrate the charging state of the center before and after each of the four basic transitions takes place. The arrows in the figure designate the transition of the electron in a particular process. The illustration is for the case of a recombination center with a single energy level that is neutral

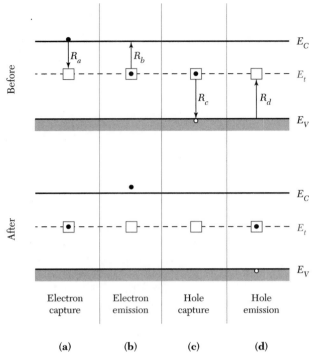

Fig. 12 Indirect generation-recombination processes at thermal equilibrium.

when not occupied by an electron or negative when it is occupied. In indirect recombination, the derivation of the recombination rate is more complicated. The detailed derivation is given in Appendix I, and the recombination rate is given by [4]

$$
U = \frac{v_{th}\sigma_n\sigma_p N_t\left(p_n n_n - n_i^2\right)}{\sigma_p[p_n + n_i e^{(E_i - E_t)/kT}] + \sigma_n[n_n + n_i e^{(E_t - E_i)/kT}]}, \tag{48}
$$

where v_{th} is the thermal velocity of carriers given in Eq. 1, N_t is the concentration of the recombination center in the semiconductor, and σ_n is the capture cross section of electrons. The quantity σ_n describes the effectiveness of the center to capture an electron and is a measure of how close the electron has to come to the center to be captured. σ_p is the capture cross section of holes.

We can simplify the general expression for the dependence of U on E_t by assuming equal electron and hole capture cross sections, that is, $\sigma_n = \sigma_p = \sigma_o$. Equation 48 then becomes

$$
U = v_{th}\sigma_o N_t \frac{\left(p_n n_n - n_i^2\right)}{p_n + n_n + 2n_i \cosh\left(\dfrac{E_t - E_i}{kT}\right)}. \tag{49}
$$

Under a low-injection condition in an n-type semiconductor so that $n_n \gg p_n$, the recombination rate can be written as

$$U \approx v_{th}\sigma_o N_t \frac{p_n - p_{no}}{1 + \left(\dfrac{2n_i}{n_{no}}\right)\cosh\left(\dfrac{E_t - E_i}{kT}\right)} = \frac{p_n - p_{no}}{\tau_p}. \tag{50}$$

The recombination rate for indirect recombination is given by the same expression as Eq. 43; however, τ_p depends on the locations of the recombination centers.

3.3.3 Surface Recombination

Figure 13 shows schematically the bonds at a semiconductor surface.[5] Because of the abrupt discontinuity of the lattice structure at the surface, a large number of localized energy states or generation-recombination centers may be introduced at the surface region. These energy states, called *surface states*, may greatly enhance the recombination rate at the surface region. The kinetics of surface recombination are similar to those considered before for bulk centers. The total number of carriers recombining at the surface *per unit area* and unit time can be expressed in a form analogous to Eq. 48. For a low-injection condition, and for the limiting case where electron concentration at the surface is essentially equal to the bulk majority carrier concentration, the total number of carriers recombining at the surface per unit area and unit time can be simplified to

$$U_s \cong v_{th}\sigma_p N_{st}(p_s - p_{no}), \tag{51}$$

where p_s denotes the hole concentrations at the surface, and N_{st} is the recombination center density per unit area in the surface region. Since the product $v_{th}\sigma_p N_{st}$ has its dimension in centimeters per second, it is called the *low-injection surface recombination velocity* S_{lr}:

$$S_{lr} \equiv v_{th}\sigma_p N_{st}. \tag{52}$$

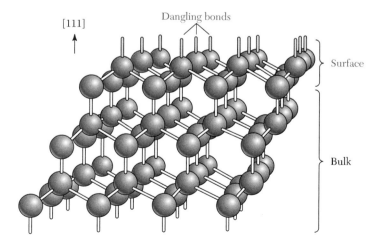

Fig. 13 Schematic diagram of bonds at a clean semiconductor surface. The bonds are anisotropic and differ from those in the bulk.[5]

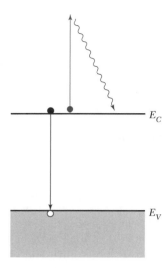

Fig. 14 Auger recombination.

3.3.4 Auger Recombination

The Auger recombination process occurs by the transfer of the energy and momentum released by the recombination of an electron-hole pair to a third particle that can be either an electron or a hole. The example of Auger recombination process is shown in Fig. 14. A second electron in the conduction band absorbs the energy released by the direct recombination. After the Auger process, the second electron becomes an energetic electron. It loses its energy to the lattice by scattering events. Usually, Auger recombination is important when the carrier concentration is very high as a result of either high doping or high injection level. Because the Auger process involves three particles, the rate of Auger recombination can be expressed as

$$R_{Aug} = Bn^2p \text{ or } Bnp^2. \tag{53}$$

The proportionality constant B has a strong temperature dependence.

▶ 3.4 CONTINUITY EQUATION

In the previous sections we considered individual effects such as drift due to an electric field, diffusion due to a concentration gradient, and recombination of carriers through intermediate-level recombination centers. We now consider the overall effect when drift, diffusion, and recombination occur simultaneously in a semiconductor material. The governing equation is called the *continuity equation*.

To derive the one-dimensional continuity equation for electrons, consider an infinitesimal slice with a thickness dx located at x shown in Fig. 15. The number of electrons in the slice may increase due to the *net* current flow into the slice and the *net* carrier generation in the slice. The overall rate of electron increase is the algebraic sum of four components: the number of electrons flowing into the slice at x, minus the number of electrons flowing out at $x + dx$, plus the rate at which electrons are generated, minus the rate at which they are recombined with holes in the slice.

The first two components are found by dividing the currents at each side of the slice by the charge of an electron. The generation and recombination rates are designated by

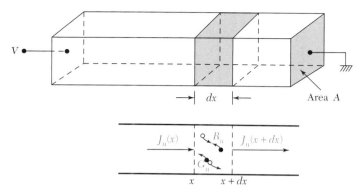

Fig. 15 Current flow and generation-recombination processes in an infinitesimal slice of thickness dx.

G_n and R_n, respectively. The overall rate of change in the number of electrons in the slice is then

$$\frac{\partial n}{\partial t} A dx = \left[\frac{J_n(x)A}{-q} - \frac{J_n(x + dx)A}{-q} \right] + (G_n - R_n)A dx, \tag{54}$$

where A is the cross-sectional area and $A dx$ is the volume of the slice. Expanding the expression for the current at $x + dx$ in Taylor series yields

$$J_n(x + dx) = J_n(x) + \frac{\partial J_n}{\partial x} dx + \ldots \tag{55}$$

We thus obtain the basic *continuity equation* for electrons:

$$\frac{\partial n}{\partial t} = \frac{1}{q} \frac{\partial J_n}{\partial x} + (G_n - R_n). \tag{56}$$

A similar continuity equation can be derived for holes, except that the sign of the first term on the right-hand side of Eq. 56 is changed because of the positive charge associated with a hole:

$$\frac{\partial p}{\partial t} = -\frac{1}{q} \frac{\partial J_p}{\partial x} + (G_p - R_p). \tag{57}$$

We can substitute the current expressions from Eqs. 31 and 32 and the recombination expressions from Eq. 43 into Eqs. 56 and 57. For the one-dimensional case under low-injection condition, the continuity equations for minority carriers (i.e., n_p in a p-type semiconductor or p_n in an n-type semiconductor) are

$$\frac{\partial n_p}{\partial t} = n_p \mu_n \frac{\partial \mathscr{E}}{\partial x} + \mu_n \mathscr{E} \frac{\partial n_p}{\partial x} + D_n \frac{\partial^2 n_p}{\partial x^2} + G_n - \frac{n_p - n_{po}}{\tau_n}, \tag{58}$$

$$\frac{\partial p_n}{\partial t} = -p_n \mu_p \frac{\partial \mathscr{E}}{\partial x} - \mu_p \mathscr{E} \frac{\partial p_n}{\partial x} + D_p \frac{\partial^2 p_n}{\partial x^2} + G_p - \frac{p_n - p_{no}}{\tau_p}. \tag{59}$$

In addition to the continuity equations, Poisson's equation

$$\frac{d\mathscr{E}}{dx} = \frac{\rho_s}{\varepsilon_s} \tag{60}$$

must be satisfied, where ε_s is the semiconductor dielectric permittivity and ρ_s is the space charge density given by the algebraic sum of the charge carrier densities and the ionized impurity concentrations, $q(p - n + N_D^+ - N_A^-)$.

In principle, Eqs. 58 through 60 together with appropriate boundary conditions have a unique solution. Because of the algebraic complexity of this set of equations, in most cases the equations are simplified with physical approximations before a solution is attempted. We solve the continuity equations for three important cases.

3.4.1 Steady-State Injection from One Side

Figure 16a shows an n-type semiconductor where excess carriers are injected from one side as a result of illumination. It is assumed that light penetration is negligibly small (i.e., the assumptions of zero field and zero generation for $x > 0$). At steady state there is a

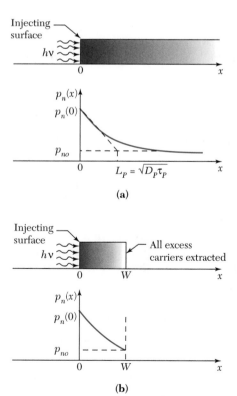

Fig. 16 Steady-state carrier injection from one side. (*a*) Semiinfinite sample. (*b*) Sample with thickness W.

concentration gradient near the surface. From Eq. 59 the differential equation for the minority carriers inside the semiconductor is

$$\frac{\partial p_n}{\partial t} = 0 = D_p \frac{\partial^2 p_n}{\partial x^2} - \frac{p_n - p_{no}}{\tau_p}. \qquad (61)$$

The boundary conditions are $p_n(x = 0) = p_n(0) = $ constant value and $p_n(x \to \infty) = p_{no}$. The solution of $p_n(x)$ is

$$p_n(x) = p_{no} + [p_n(0) - p_{no}]e^{-x/L_p}. \qquad (62)$$

The length L_p is equal to $\sqrt{D_p \tau_p}$ and is called the *diffusion length*. Figure 16a shows the variation of the minority carrier density, which decays with a characteristic length given by L_p.

If we change the second boundary condition as shown in Fig. 16b so that all excess carriers at $x = W$ are extracted, that is, $p_n(W) = p_{no}$, then we obtain a new solution for Eq. 61:

$$p_n(x) = p_{no} + [p_n(0) - p_{no}] \left[\frac{\sinh\left(\dfrac{W - x}{L_p}\right)}{\sinh(W / L_p)} \right]. \qquad (63)$$

The current density at $x = W$ is given by the diffusion current expression, Eq. 32 with $\mathcal{E} = 0$:

$$J_p = -qD_p \frac{\partial p_n}{\partial x}\bigg|_W = q[p_n(0) - p_{no}] \frac{D_p}{L_p} \frac{1}{\sinh(W/L_p)}. \qquad (64)$$

3.4.2 Minority Carriers at the Surface

When surface recombination is introduced at one end of a semiconductor sample under illumination (Fig. 17), the hole current density flowing into the surface from the bulk of

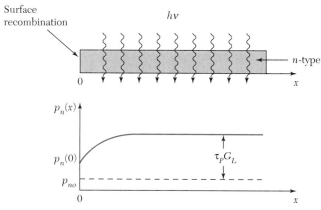

Fig. 17 Surface recombination at $x = 0$. The minority carrier distribution near the surface is affected by the surface recombination velocity.[6]

the semiconductor is given by qU_s. In this example, it is assumed that the sample is uniformly illuminated with uniform generation of carriers. The surface recombination leads to a lower carrier concentration at the surface. This gradient of hole concentration yields a diffusion current density that is equal to the surface recombination current. Therefore, the boundary condition at $x = 0$ is

$$qD_p \frac{dp_n}{dx}\bigg|_{x=0} = qU_s = qS_{lr}[p_n(0) - p_{no}]. \tag{65}$$

The boundary condition at $x = \infty$ is given by Eq. 45a. At steady state the differential equation is

$$\frac{\partial p_n}{\partial t} = 0 = D_p \frac{\partial^2 p_n}{\partial x^2} + G_L - \frac{p_n - p_{no}}{\tau_p}. \tag{66}$$

The solution of the equation, subject to the boundary conditions above, is[6]

$$p_n(x) = p_{no} + \tau_p G_L \left(1 - \frac{\tau_p S_{lr} e^{-x/L_p}}{L_p + \tau_p S_{lr}} \right). \tag{67}$$

A plot of this equation for a finite S_{lr} is shown in Fig. 17. When $S_{lr} \to 0$, then $p_n(x) \to p_{no} + \tau_p G_L$, as obtained previously (Eq. 45a). When $S_{lr} \to \infty$, then

$$p_n(x) = p_{no} + \tau_p G_L (1 - e^{-x/L_p}). \tag{68}$$

From Eq. 68 we can see that at the surface the minority carrier density approaches its thermal equilibrium value p_{no}.

3.4.3 The Haynes–Shockley Experiment

One of the classic experiments in semiconductor physics is the demonstration of drift and diffusion of minority carriers, first made by Haynes and Shockley.[7] The experiment allows independent measurement of the minority carrier mobility μ and diffusion coefficient D. The basic setup of the Haynes–Shockley experiment is shown in Fig. 18a. The voltage source, V_1, establishes an electric field in the $+x$ direction in the n-type semiconductor bar. Excess carriers are produced and effectively injected into the semiconductor bar at contact (1) by a pulse. Contact (2) will collect a fraction of the excess carriers as they drift through the semiconductor. After a pulse, the transport equation is given by Eq. 59 by setting $G_p = 0$ and $\partial\mathscr{E}/\partial x = 0$ (i.e., the applied field is constant across the semiconductor bar):

$$\frac{\partial p_n}{\partial t} = -\mu_p \mathscr{E} \frac{\partial p_n}{\partial x} + D_p \frac{\partial^2 p_n}{\partial x^2} - \frac{p_n - p_{no}}{\tau_p}. \tag{69}$$

If no field is applied along the sample, the solution is given by

$$p_n(x, t) = \frac{N}{\sqrt{4\pi D_p t}} \exp\left(-\frac{x^2}{4D_p t} - \frac{t}{\tau_p} \right) + p_{no}, \tag{70}$$

where N is the number of electrons or holes generated per unit area. Figure 18b shows this solution as the carriers diffuse away from the point of injection and recombine.

If an electric field is applied along the sample, the solution is in the form of Eq. 70, except that x is replaced by $x - \mu_p \mathscr{E} t$ (Fig. 18c). Thus, all the excess carriers move with the drift velocity $\mu_p \mathscr{E}$. At the same time, the carriers diffuse outward and recombine as in the field-free case.

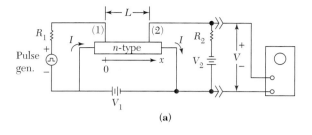

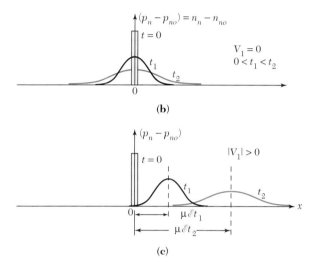

Fig. 18 The Haynes–Shockley experiment. (*a*) Experimental setup. (*b*) Carrier distributions without an applied field. (*c*) Carrier distributions with an applied field.[7]

▶ **EXAMPLE 7**

In a Haynes–Shockley experiment, the maximum amplitudes of the minority carriers at $t_1 = 100$ μs and $t_2 = 200$ μs differ by a factor of 5. Calculate the minority carrier lifetime.

SOLUTION When an electric field is applied, the minority carrier distribution is given by

$$\Delta p \equiv p_n - p_{no} = \frac{N}{\sqrt{4\pi D_p \tau}} \ \exp\left(-\frac{(x - \mu_p \mathscr{E} t)^2}{4 D_p t} - \frac{t}{\tau_p} \right).$$

At the maximum amplitude

$$\Delta p = \frac{N}{\sqrt{4\pi D_p t}} \ \exp\left(-\frac{t}{\tau_p} \right).$$

Therefore

$$\frac{\Delta p(t_1)}{\Delta p(t_2)} = \frac{\sqrt{t_2}}{\sqrt{t_1}} \frac{\exp(-t_1/\tau_p)}{\exp(-t_2/\tau_p)} = \sqrt{\frac{200}{100}} \ \exp\left(\frac{200-100}{\tau_p \ (\mu s)} \right) = 5$$

$$\therefore \tau_p = \frac{200-100}{\ln(5/\sqrt{2})} = 79 \ \mu s.$$

◀

▶ 3.5 THERMIONIC EMISSION PROCESS

In previous sections, we considered carrier transport phenomena inside the bulk semiconductor. At the semiconductor surface, carriers may recombine with the recombination centers due to the dangling bonds of the surface region. In addition, if the carriers have sufficient energy, they may be "thermionically" emitted into the vacuum. This is called the *thermionic emission process*.

Figure 19*a* shows the band diagram of an isolated *n*-type semiconductor. The electron affinity, $q\chi$, is the energy difference between the conduction band edge and the vacuum level in the semiconductor; and the work function, $q\phi_s$, is the energy between the Fermi level and the vacuum level in the semiconductor. From Fig. 19*b*, it is clear that an electron can be thermionically emitted into the vacuum if its energy is above $q\chi$.

The electron density with energies above $q\chi$ can be obtained from an expression similar to that for the electron density in the conduction band (Eqs. 9 and 16 of Chapter 2) except that the lower limit of the integration is $q\chi$ instead of E_C:

$$n_{th} = \int_{q\chi}^{\infty} n(E)dE = N_C \exp\left[-\frac{q(\chi + V_n)}{kT}\right], \tag{71}$$

where N_C is the effective density of states in the conduction band, and V_n is the difference between the bottom of the conduction band and the Fermi level.

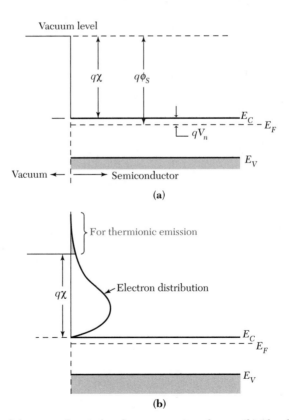

Fig. 19 (*a*) The band diagram of an isolated *n*-type semiconductor. (*b*) The thermionic emission process.

▷ **EXAMPLE 8**

Calculate the thermionically emitted electron density, n_{th}, at room temperature for an n-type silicon sample with an electron affinity of $q\chi = 4.05$ eV and $qV_n = 0.2$ eV. If we reduce the effective $q\chi$ to 0.6 eV, what is n_{th}?

SOLUTION

$$n_{th} \ (4.05 \text{ eV}) = 2.86 \times 10^{19} \ \exp\left(-\frac{4.05 + 0.2}{0.0259}\right) = 2.86 \times 10^{19} \ \exp\left(-164\right)$$

$$\cong 10^{-52} \ \approx 0$$

$$n_{th} \ (0.6 \text{ eV}) = 2.86 \times 10^{19} \ \exp\left(-\frac{0.8}{0.0259}\right) = 2.86 \times 10^{19} \ \exp\left(-30.9\right)$$

$$= 1 \times 10^{6} \text{ cm}^{-3}.$$

◀

From the above example, we see that at 300 K there is no emission of electrons into vacuum for $q\chi = 4.05$. However, if we can lower the effective electron affinity to 0.6 eV, a substantial number of electrons can be thermionically emitted. The thermionic emission process is of particular important for metal-semiconductor contacts to be considered in Chapter 7.

▷ ## 3.6 TUNNELING PROCESS

Figure 20a shows the energy band diagram when two isolated semiconductor samples are brought close together. The distance between them is d, and the potential barrier height qV_0 is equal to the electron affinity $q\chi$. If the distance is sufficient small, the electrons in the left-side semiconductor may transport across the barrier and move to the right-side semiconductor, even if the electron energy is much less than the barrier height. This process is associated with the *quantum tunneling phenomenon*.

Based on Fig. 20a, we have redrawn the one-dimensional potential barrier diagram in Fig. 20b. We first consider the transmission (or tunneling) coefficient of a particle (e.g., electron) through this barrier. In the corresponding classic case, the particle is always reflected if its energy E is less than the potential barrier height qV_0. We see that in the quantum case, the particle has finite probability to transmit or "tunnel" through the potential barrier.

The behavior of a particle (e.g., a conduction electron) in the region where $qV(x) = 0$ can be described by the Schrödinger equation:

$$-\frac{\hbar^2}{2m_n} \frac{d^2\psi}{dx^2} = E\psi \tag{72}$$

or

$$\frac{d^2\psi}{dx^2} = -\frac{2m_n E}{\hbar^2} \psi, \tag{73}$$

where m_n is the effective mass, $\hbar$ is the reduced Planck constant, E is the kinetic energy, and ψ is the wave function of the particle. The solutions are

$$\psi(x) = Ae^{jkx} + Be^{-jkx} \qquad x \le 0, \tag{74}$$

$$\psi(x) = Ce^{jkx} \qquad x \ge d, \tag{75}$$

where $k \equiv \sqrt{2m_n E/\hbar^2}$. For $x \le 0$, we have an incident-particle wave function (with amplitude A) and a reflected wave function (with amplitude B); for $x \ge d$, we have a transmitted wave function (with amplitude C).

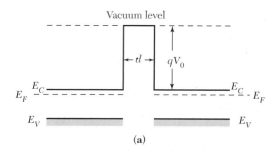

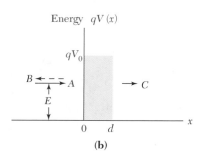

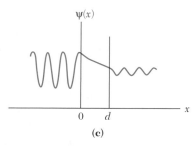

Fig. 20 (*a*) The band diagram of two isolated semiconductors with a distance *d*. (*b*) One-dimensional potential barrier. (*c*) Schematic representation of the wave function across the potential barrier.

Inside the potential barrier, the wave equation is given by

$$-\frac{\hbar^2}{2m_n}\frac{d^2\psi}{dx^2}+qV_0\psi = E\psi \tag{76}$$

or

$$\frac{d^2\psi}{dx^2}=\frac{-2m_n(qV_0-E)}{\hbar^2}\psi. \tag{77}$$

The solution for $E < qV_0$ is

$$\psi(x) = Fe^{\beta x}+Ge^{-\beta x}, \tag{78}$$

where $\beta \equiv \sqrt{2m_n(qV_0-E)/\hbar^2}$. A schematic representation of the wave functions across the barrier is shown in Fig. 20c. The continuity of ψ and $d\psi/dx$ at $x = 0$ and $x = d$, which is required by the boundary conditions, provides four relations between the five coefficients (A, B, C, F, and G). We can solve for $(C/A)^2$, which is the *transmission coefficient*:

$$\left(\frac{C}{A}\right)^2 = \left[1 + \frac{(qV_0 \sinh \beta d)^2}{4E(qV_0 - E)}\right]^{-1}. \tag{79}$$

The transmission coefficient decreases monotonically as E decreases. When $\beta d \gg 1$, the transmission coefficient becomes quite small and varies as

$$\left(\frac{C}{A}\right)^2 \sim \exp(-2\beta d) = \exp\left[-2d\sqrt{2m_n(qV_0 - E)/\hbar^2}\right]. \tag{80}$$

To have a finite transmission coefficient, we require a small tunneling distance d, a low-potential barrier qV_0, and a small effective mass. These results will be used for tunnel diodes in Chapter 8.

3.7 HIGH-FIELD EFFECTS

At low electric-fields, the drift velocity is linearly proportional to the applied field. We assume that the time interval between collision, τ_c, is independent of the applied field. This is a reasonable assumption as long as the drift velocity is small compared with the thermal velocity of carriers, which is about 10^7 cm/s for silicon at room temperature.

As the drift velocity approaches the thermal velocity, its field dependence on the electric field will begin to depart from the linear relationship given in Section 3.1. Figure 21 shows the measured drift velocities of electrons and holes in silicon as a function of the electric field. It is apparent that initially the field dependence of the drift velocity is linear, corresponding to a constant mobility. As the electric field is further increased, the drift velocity increases less rapidly. At sufficiently large fields, the drift velocity approaches a saturation velocity. The experimental results can be approximated by the empirical expression[8]

$$v_n, \ v_p = \frac{v_s}{[1 + (\mathscr{E}_0/\mathscr{E})^\gamma]^{1/\gamma}}, \tag{81}$$

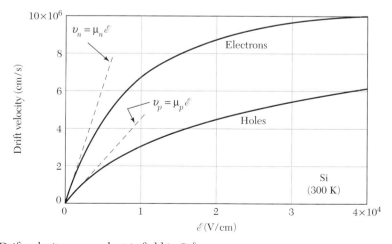

Fig. 21 Drift velocity versus electric field in Si.[8]

where v_s is the saturation velocity (10^7cm/s for Si at 300 K), $\mathscr{E}_0$ is a constant, equal to 7×10^3 V/cm for electrons and 2×10^4 V/cm for holes in high-purity silicon materials, and γ is 2 for electrons and 1 for holes. Velocity saturation at high fields is particularly likely for field-effect transistors (FETs) with very short channels. Even moderate voltage can result in a high field along the channel. This effect is discussed in Chapter 6.

The high-field transport in n-type gallium arsenide is different from that of silicon.[9] Figure 22 shows the measured drift velocity versus field for n-type and p-type gallium arsenide. The results for silicon are also shown in this log-log plot for comparison. Note that for n-type GaAs, the drift velocity reaches a maximum, then decreases as the field further increases. This phenomenon is due to the energy band structure of gallium arsenide that allows the transfer of conduction electrons from a high-mobility energy minimum (called a valley) to low-mobility, higher-energy satellite valleys, that is, electron transfer from the central valley to the satellite valleys along the [111] direction shown in Chapter 2.

To understand this phenomenon, consider the simple two-valley model of n-type gallium arsenide shown in Fig. 23. The energy separation between the two valleys is $\Delta E = 0.31$ eV. The lower valley's electron effective mass is denoted by m_1, the electron mobility by μ_1, and the electron density by n_1. The upper-valley quantities are denoted by m_2, μ_2, and n_2, respectively; and the total electron concentration is given by $n = n_1 + n_2$. The steady-state conductivity of the n-type GaAs can be written as

$$\sigma = q(\mu_1 n_1 + \mu_2 n_2) = qn\overline{\mu}, \tag{82}$$

where the average mobility is

$$\overline{\mu} \equiv (\mu_1 n_1 + \mu_2 n_2)/(n_1 + n_2). \tag{83}$$

The drift velocity is then

$$v_n = \overline{\mu}\mathscr{E}. \tag{84}$$

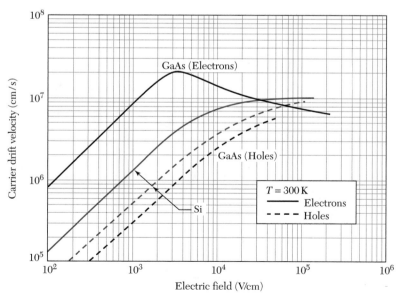

Fig. 22 Drift velocity versus electric field in Si and GaAs. Note that for n-type GaAs, there is a region of negative differential mobility.[8,9]

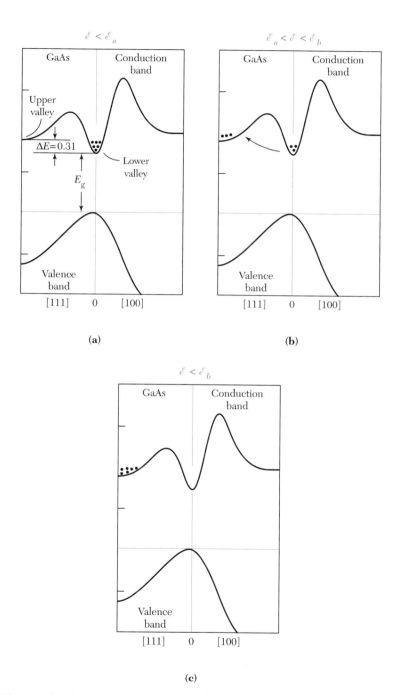

Fig. 23 Electron distributions under various conditions of electric fields for a two-valley semiconductor.

For simplicity we make the following assignments for the electron concentrations in the various ranges of electric-field values illustrated in Fig. 23. In Fig. 23a, the field is low and all electrons remain in the lower valley. In Fig. 23b, the field is higher and some electrons gain sufficient energies from the field to move to the higher valley. In Fig. 23c, the field is high enough to transfer all electrons to the higher valley. Thus, we have

$$n_1 \cong n \qquad \text{and} \qquad n_2 \cong 0 \qquad \text{for } 0 < \mathscr{E} < \mathscr{E}_a,$$
$$n_1 + n_2 \cong n \qquad\qquad\qquad \text{for } \mathscr{E}_a < \mathscr{E} < \mathscr{E}_b, \qquad (85)$$
$$n_1 \cong 0 \qquad \text{and} \qquad n_2 \cong n \qquad \text{for } \mathscr{E} > \mathscr{E}_b.$$

Using these relations, the effective drift velocity takes on the asymptotic values

$$v_n \cong \mu_1 \mathscr{E} \qquad \text{for } 0 < \mathscr{E} < \mathscr{E}_a,$$
$$v_n \cong \mu_2 \mathscr{E} \qquad \text{for } \mathscr{E} > \mathscr{E}_b. \qquad (86)$$

If $\mu_1 \mathscr{E}_a$ is larger than $\mu_2 \mathscr{E}_b$, there is a region in which the drift velocity decreases with an increasing field between $\mathscr{E}_a$ and $\mathscr{E}_b$, as shown in Fig. 24. Because of the characteristics of the drift velocity in n-type gallium arsenide, this material is used in microwave transferred-electron devices discussed in Chapter 8.

When the electric field in a semiconductor is increased above a certain value, the carriers gain enough kinetic energy to generate electron-hole pairs by an *avalanche process* that is shown schematically in Fig. 25. Consider an electron in the conduction band (designated by 1). If the electric field is high enough, this electron can gain kinetic energy before it collides with the lattice. On impact with the lattice, the electron imparts most of its kinetic energy to break a bond, that is, to ionize a valence electron from the valence band to the conduction band and thereby generate an electron-hole pair (designated by 2 and 2′). Similarly, the generated pair now begins to accelerate in the field and collides with the lattice as indicated in the figure. In turn, they will generate other electron-hole pairs (e.g., 3 and 3′, 4 and 4′), and so on. This process is called the avalanche process; it is also referred to as the *impact ionization* process. This process will result in breakdown in p–n junction, which is discussed in Chapter 4.

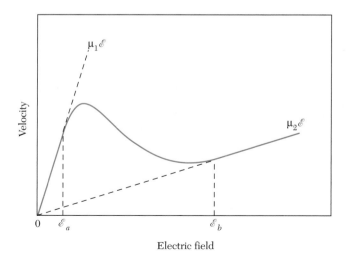

Fig. 24 One possible velocity-field characteristic of a two-valley semiconductor.

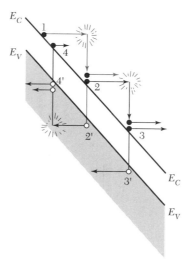

Fig. 25 Energy band diagram for the avalanche process.

To gain some ideas about the ionization energy involved, let us consider the process leading to 2–2′ shown in Fig. 25. Just prior to the collision, the fast-moving electron (no. 1) has a kinetic energy $\frac{1}{2} m_1 v_s^2$ and a momentum $m_1 v_s$, where m_1 is the effective mass and v_s is the saturation velocity. After collision, there are three carriers: the original electron plus an electron–hole pair (no. 2 and no. 2′). If we assume that the three carriers have the same effective mass, the same kinetic energy, and the same momentum, the total kinetic energy is $\frac{3}{2} m_1 v_f^2$, and the total momentum is $3 m_1 v_f$, where v_f is the velocity after collision. To conserve both energy and momentum before and after the collision, we require that

$$\frac{1}{2} m_1 v_s^2 = E_g + \frac{3}{2} m_1 v_f^2 \tag{87}$$

and

$$m_1 v_s = 3 m_1 v_f, \tag{88}$$

where in Eq. 87 the energy E_g is the bandgap corresponding to the minimum energy required to generate an electron-hole pair. Substituting Eq. 88 into Eq. 87 yields the required kinetic energy for the ionization process:

$$E_0 = \frac{1}{2} m_1 v_s^2 = 1.5 E_g. \tag{89}$$

It is obvious that E_0 must be larger than the bandgap for the ionization process to occur. The actual energy required depends on the band structure. For silicon, the value for E_0 is 3.6 eV (3.2 E_g) for electrons and 5.0 eV (4.4 E_g) for holes.

The number of electron-hole pairs generated by an electron per unit distance traveled is called the *ionization rate* for the electron, α_n. Similarly, α_p is the ionization rate for the holes. The measured ionization rates for silicon and gallium arsenide are shown[9] in Fig. 26. We note that both α_n and α_p are strongly dependent on the electric field. For a substantially large ionization rate (say 10^4 cm^{-1}), the corresponding electric field is $\geq 3 \times 10^5$ V/cm for silicon and $\geq 4 \times 10^5$ V/cm for gallium arsenide. The electron-hole pair generation rate G_A from the avalanche process is given by

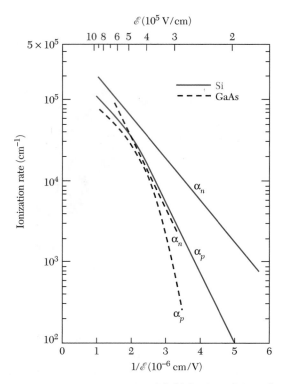

Fig. 26 Measured ionization rates versus reciprocal field for Si and GaAs.[9]

$$G_A = \frac{1}{q}\left(\alpha_n |J_n| + \alpha_p |J_p|\right),\qquad(90)$$

where J_n and J_p are the electron and hole current densities, respectively. This expression can be used in the continuity equation for devices operated under an avalanche condition.

▶ SUMMARY

Various transport processes are at work in semiconductor devices. These include drift, diffusion, generation, recombination, thermionic emission, tunneling, and impact ionization.

One of the key transport processes is the carrier drift under the influence of an electric field. At low fields, the drift velocity is proportional to the electric field. This proportionality constant is called mobility. Another key transport process is the carrier diffusion under the influence of the carrier concentration gradient. The total current is the sum of the drift and diffusion components.

Excess carriers in a semiconductor cause a nonequilibrium condition. Most semiconductor devices operate under nonequilibrium conditions. Carriers can be generated by various means such as forward biasing a p–n junction, incident light, and impact ionization. The mechanism that restores equilibrium is the recombination of the excess minority carriers with the majority carriers by direct band-to-band recombination or via localized energy states in the forbidden energy gap. The governing equation for the rate of change of charge carriers is the continuity equation.

Among other transport processes, thermionic emission occurs when carriers in the surface region gain enough energy to be emitted into the vacuum level. Another, the tunneling process, is based on the quantum tunneling phenomena that results in the transport of electrons across a potential barrier even if the electron energy is less than the barrier height.

As the electric field becomes higher, the drift velocity departs from its linear relationship with the applied field and approaches a saturation velocity. This effect is particularly important in the study of short-channel field-effect transistors discussed in Chapter 6. When the field exceeds a certain value, the carriers gain enough kinetic energy to generate electron-hole pair by colliding with the lattice and breaking a bond. This effect is particularly important in the study of p–n junctions. The high field accelerates these new electron-hole pairs, which collide with the lattice to create more electron-hole pairs. As this process, called impact ionization or the avalanche process, continues, the p–n junction breaks down and conducts a large current. The junction breakdown is discussed in Chapter 4.

► REFERENCES

1. R. A. Smith, *Semiconductors*, 2nd ed., Cambridge Univ. Press, London, 1978.

2. J. L. Moll, *Physics of Semiconductors*, McGraw-Hill, New York, 1964.

3. W. F. Beadle, J. C. C. Tsai, and R. D. Plummer, Eds., *Quick Reference Manual for Semiconductor Engineers*, Wiley, New York, 1985.

4. (a) R. N. Hall, "Electron–Hole Recombination in Germanium," *Phys. Rev.*, **87**, 387 (1952); (b) W. Shockley and W. T. Read, "Statistics of Recombination of Holes and Electrons," *Phys. Rev.*, **87**, 835(1952).

5. M. Prutton, *Surface Physics*, 2nd ed., Clarendon, Oxford, 1983.

6. A. S. Grove, *Physics and Technology of Semiconductor Devices*, Wiley, New York, 1967.

7. J. R. Haynes and W. Shockley, "The Mobility and Life of Injected Holes and Electrons in Germanium," *Phys. Rev.*, **81**, 835 (1951).

8. D. M. Caughey and R. E. Thomas, "Carrier Mobilities in Silicon Empirically Related to Doping and Field," *Proc. IEEE*, **55**, 2192 (1967).

9. S. M. Sze, *Physics of Semiconductor Devices*, 2nd ed., Wiley, New York, 1981.

► PROBLEMS (* INDICATES DIFFICULT PROBLEMS)

FOR SECTION 3.1 CARRIER DRIFT

1. Find the resistivities of intrinsic Si and intrinsic GaAs at 300 K.

2. Assume that the mobility of electrons in silicon at T = 300 K is μ_n = 1300 cm^2/V-s. Also assume that the mobility is mainly limited by lattice scattering. Determine the electron mobility at (a) T = 200 K and (b) T = 400 K.

3. Two scattering mechanisms exist in a semiconductor. If only the first mechanism is present, the mobility will be 250 cm^2/V-s. If only the second mechanism is present, the mobility will be 500 cm^2/V-s. Determine the mobility when both scattering mechanisms exist at the same time.

4. Find the electron and hole concentrations, mobilities, and resistivities of silicon samples at 300 K, for each of the following impurity concentrations: (a) 5×10^{15} boron atoms/cm^3; (b) 2×10^{16} boron atoms/cm^3 and 1.5×10^{16} arsenic atoms/cm^3; and (c) 5×10^{15} boron atoms/cm^3, 10^{17} arsenic atoms/cm^3, and 10^{17} gallium atoms/cm^3.

*5. Consider a compensated n-type silicon at T = 300 K, with a conductivity of σ = 16 (Ω-cm)$^{-1}$ and an acceptor doping concentration of 10^{17} cm^{-3}. Determine the donor concentration and the electron mobility. (A compensated semiconductor is one that contains both donor and acceptor impurity atoms in the same region.)

6. For a semiconductor with a constant mobility ratio $b \equiv \mu_n/\mu_p > 1$ independent of impurity concentration, find the maximum resistivity ρ_m in terms of the intrinsic resistivity ρ_i and the mobility ratio.

7. A four-point probe (with probe spacing of 0.5 mm) is used to measure the resistivity of a p-type silicon sample. Find the resistivity of the sample if its diameter is 200 mm and its thickness is 50 μm. The contact current is 1 mA, and the measured voltage between the inner two probes is 10 mV.

8. Given a silicon sample of unknown doping, Hall measurement provides the following information: $W = 0.05$ cm, $A = 1.6 \times 10^{-3}$ cm^2 (refer to Fig. 8), $I = 2.5$ mA, and the magnetic field is 30 nT (1 T = 10^{-4} Wb/cm^2). If a Hall voltage of +10 mV is measured, find the Hall coefficient, conductivity type, majority carrier concentration, resistivity, and mobility of the semiconductor sample.

9. A semiconductor is doped with N_D ($N_D \gg n_i$) and has a resistance R_1. The same semiconductor is then doped with an unknown amount of acceptors N_A ($N_A \gg N_D$), yielding a resistance of 0.5 R_1. Find N_A in terms of N_D if $D_n/D_p = 50$.

*10. Consider a semiconductor that is nonuniformly doped with donor impurity atoms $N_D(x)$. Show that the induced electric field in the semiconductor in thermal equilibrium is given

by $\mathscr{E}(x) = -\left(\dfrac{kT}{q}\right)\dfrac{1}{N_D(x)}\dfrac{dN_D(x)}{dx}$.

FOR SECTION 3.2 CARRIER DIFFUSION

11. An intrinsic Si sample is doped with donors from one side such that $N_D = N_0 \exp(-ax)$. (a) Find an expression for the built-in field $\mathscr{E}(x)$ at equilibrium over the range for which $N_D \gg n_i$. (b) Evaluate $\mathscr{E}(x)$ when $a = 1\ \mu\text{m}^{-1}$.

12. An n-type Si slice of a thickness L is inhomogeneously doped with phosphorus donor whose concentration profile is given by $N_D(x) = N_0 + (N_L - N_0)(x/L)$. What is the formula for the electric potential difference between the front and the back surfaces when the sample is at thermal and electric equilibria regardless of how the mobility and diffusivity varies with position? What is the formula for the equilibrium electric field at a plane x from the front surface for a constant diffusivity and mobility?

FOR SECTION 3.3 GENERATION AND RECOMBINATION PROCESS

13. Calculate the electron and hole concentration under steady-state illumination in an n-type silicon with $G_L = 10^{16}$ cm^{-3}s^{-1}, $N_D = 10^{15}$ cm^{-3}, and $\tau_n = \tau_p = 10$ μs.

14. An n-type silicon sample has 2×10^{16} arsenic atoms/cm^3, 2×10^{15} bulk recombination centers/cm^3, and 10^{10} surface recombination centers/cm^2. (a) Find the bulk minority carrier lifetime, the diffusion length, and the surface recombination velocity under low-injection conditions. The values of σ_p and σ_s are 5×10^{-15} and 2×10^{-16} cm^2, respectively. (b) If the sample is illuminated with uniformly absorbed light that creates 10^{17} electron-hole pairs/cm^2-s, what is the hole concentration at the surface?

15. Assume that an n-type semiconductor is uniformly illuminated, producing a uniform excess generation rate G. Show that in steady state the change in the semiconductor conductivity is given by $\Delta\sigma = q(\mu_n + \mu_p)\tau_p G$.

FOR SECTION 3.4 CONTINUITY EQUATION

16. The total current in a semiconductor is constant and is composed of electron drift current and hole diffusion current. The electron concentration is constant and equal to 10^{16} cm^{-3}.

The hole concentration is given by

$$p(x) = 10^{15} \exp\left(\frac{-x}{L}\right) \text{ cm}^{-3} \qquad (x \geq 0),$$

where $L = 12$ μm. The hole diffusion coefficient is $D_p = 12$ cm²/s and the electron mobility is $\mu_n = 1000$ cm²/V-s. The total current density is $J = 4.8$ A/cm². Calculate (a) the hole diffusion current density versus x, (b) the electron current density versus x, and (c) the electric field versus x.

*17. Excess carriers are injected on one surface of a thin slice of n-type silicon with thickness W and extracted at the opposite surface where $p_n(W) = p_{no}$. There is no electric field in the region $0 < x < W$. Derive the expression for current densities at the two surfaces.

18. In Prob. 17, if carrier lifetime is 50 μs and $W = 0.1$ mm, calculate the portion of injected current that reaches the opposite surface by diffusion ($D = 50$ cm²/s).

*19. An n-type semiconductor has excess carrier holes 10^{14} cm⁻³, and a bulk minority carrier lifetime 10^{-6} s in the bulk material, and a minority carrier lifetime 10^{-7} s at the surface. Assume zero applied electric field and let $D_p = 10$ cm²/s. Determine the steady-state excess carrier concentration as a function of distance from the surface ($x = 0$) of the semiconductor.

FOR SECTION 3.5 THERMIONIC EMISSION PROCESS

20. A metal, with a work function $\phi_m = 4.2$ V, is deposited on an n-type silicon semiconductor with affinity $\chi = 4.0$ V and $E_g = 1.12$ eV. What is the potential barrier height seen by electrons in the metal moving into the semiconductor?

21. Consider a tungsten filament with metal work function ϕ_m inside a high vacuum chamber. Show that if a current is passed through the filament to heat it up sufficiently, the electrons with enough thermal energy will escape into the vacuum and the resulted thermionic current density is

$$J = A^\circ T^2 \exp\left(\frac{-q\phi_m}{kT}\right)$$

where A° is $4\pi qmk^2 / h^3$ and m is free electron mass. The definite integral

$$\int_{-\infty}^{\infty} e^{-ax^2} dx = \left(\frac{\pi}{a}\right)^{1/2}.$$

FOR SECTION 3.6 TUNNELING PROCESS

22. Consider a electron with an energy of 2 eV impinging on a potential barrier with 20 eV and a width of 3 Å. What is the tunneling probability?

23. Evaluate the transmission coefficient for an electron of energy 2.2 eV impinging on a potential barrier of height 6.0 eV and thickness 10^{-10} meters. Repeat the calculation for a barrier thickness of 10^{-9} meters.

FOR SECTION 3.7 HIGH FIELD EFFECTS

24. Use the velocity-field relations for Si and GaAs shown in Fig. 22 to determine the transit time of electrons through a 1 μm distance in these materials for an electric field of (a) 1 kV/cm and (b) 50 kV/cm.

25. Assume that a conduction electron in Si ($\mu_n = 1350$ cm²/V-s) has a thermal energy kT, related to its mean thermal velocity by $E_{th} = m_0 v_{th}^2/2$. This electron is placed in an electric field of 100 V/cm. Show that the drift velocity of the electron in this case is small compared to its thermal velocity. Repeat for a field of 10^4 V/cm, using the same value of μ_n. Comment on the actual mobility effects at this higher value of field.

p–n Junction

In the preceding chapters we considered the carrier concentrations and transport phenomena in homogeneous semiconductor materials. In this chapter we discuss the behavior of single-crystal semiconductor material containing both *p*- and *n*-type regions that form a *p–n* junction. Most modern *p–n* junctions are made by planar technology to be described in Section 4.1.

A *p–n* junction serves an important role both in modern electronic applications and in understanding other semiconductor devices. It is used extensively in rectification, switching, and other operations in electronic circuits. It is a key building block for the bipolar transistor and thyristor (Chapter 5), as well as for metal-oxide-semiconductor field-effect transisitors (MOSFETs) (Chapter 6). Given proper biasing conditions or when exposed to light, *p–n* junction also functions as either microwave (Chapter 8) or photonic device (Chapter 9).

We also consider a related device—the heterojunction, which is a junction formed between two dissimilar semiconductors. It has many unique features that are not readily available from the conventional *p–n* junctions. The heterojunction is an important building block for heterojunction bipolar transistors (Chapter 5), modulation doped field-effect transistors (Chapter 7), quantum-effect devices (Chapter 8), and photonic devices (Chapter 9).

Specifically, we cover the following topics:

- The formation of a *p–n* junction physically and electrically.
- The behavior of the junction depletion layer under voltage biases.
- The current transport in a *p–n* junction and the influence of the generation and recombination processes.
- The charge storage in a *p–n* junction and its influence on the transient behavior.

- The avalanche multiplication in a *p–n* junction and its impact on the maximum reverse voltage.
- The heterojunction and its basic characteristics.

4.1 BASIC FABRICATION STEPS

Today, planar technology is used extensively for integrated circuit (IC) fabrication. Figures 1 and 2 show the major steps of a planar process. These steps include oxidation, lithography,

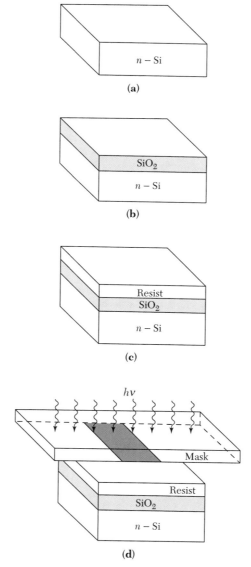

Fig. 1 (*a*) A bare *n*-type Si wafer. (*b*) An oxidized Si wafer by dry or wet oxidation. (*c*) Application of resist. (*d*) Resist exposure through the mask.

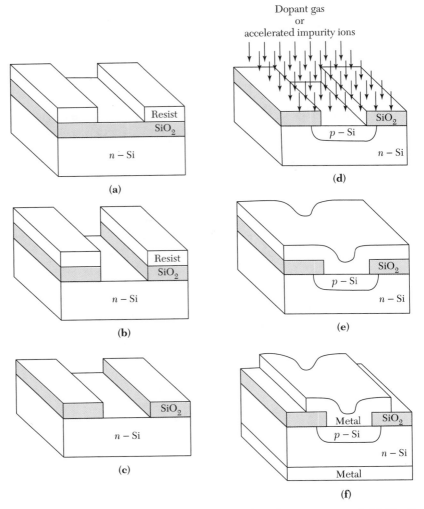

Fig. 2 (*a*) The wafer after the development. (*b*) The wafer after SiO$_2$ removal. (*c*) The final result after a complete lithography process. (*d*) A *p–n* junction is formed in the diffusion or implantation process. (*e*) The wafer after metallization. (*f*) A *p–n* junction after the complete processes.

ion implanation, and metallization. We describe these steps briefly in this section. More detailed discussions can be found in Chapters 10–14.

4.1.1 Oxidation

The development of a high-quality silicon dioxide (SiO$_2$) has helped to establish the dominance of Si in the production of commercial ICs. Generally, SiO$_2$ functions as an insulator in a number of device structures or as a barrier to diffusion or implantation during device fabrication. In the fabrication of a *p–n* junction (Fig. 1), the SiO$_2$ film is used to define the junction area.

There are two SiO$_2$ growth methods, dry and wet oxidation, depending on whether dry oxygen or water vapor is used. Dry oxidation is usually used to form thin oxides in a device structure because of its good Si-SiO$_2$ interface characteristics, whereas wet oxidation is used for thicker layers because of its higher growth rate. Figure 1*a* shows a section of a bare Si wafer ready for oxidation. After the oxidation process, a SiO$_2$ layer is formed all over the wafer surface. For simplicity, Fig. 1*b* shows only the upper surface of an oxidized wafer.

4.1.2 Lithography

Another technology, called photolithography, is used to define the geometry of the *p–n* junction. After the formation of SiO$_2$, the wafer is coated with an ultraviolet (UV)-light–sensitive material called a photoresist, which is spun on the wafer surface by a high-speed spinner. After spinner (Fig. 1*c*), the wafer is baked at about 80°–100°C to drive the solvent out of the resist and to harden the resist for improved adhesion. Figure 1*d* shows the next step, which is to expose the wafer through a patterned mask using an UV-light source. The exposed region of the photoresist-coated wafer undergoes a chemical reaction depending on the type of resist. The area exposed to light become polymerized and difficult to remove in an etchant.[*] The polymerized region remains when the wafer is placed in a developer, whereas the unexposed region (under the opaque area) dissolves and washes away. Figure 2*a* shows the wafer after the development. The wafer is again baked to 120°–180°C for 20 minutes to enhance the adhesion and improve the resistance to the subsequent etching process. Then, an etch using buffered hydrofluoric acid (HF) removes the unprotected SiO$_2$ surface (Fig. 2*b*). Last, the resist is stripped away by a chemical solution or an oxygen plasma system. Figure 2*c* shows the final result of a region without oxide (a window) after the lithography process. The wafer is now ready for forming the *p–n* junction by a diffusion or ion-implantation process.

4.1.3 Diffusion and Ion Implantation

In the diffusion method, the semiconductor surface not protected by the oxide is exposed to a source with a high concentration of opposite-type impurity. The impurity moves into the semiconductor crystal by solid-state diffusion. In the ion-implantation method, the intended impurity is introduced into the semiconductor by accelerating the impurity ions to a high-energy level and then implanting the ions in the semiconductor. The SiO$_2$ layer serves as barrier to impurity diffusion or ion implantation. After the diffusion or implantation process, the *p–n* junction is formed as shown in Fig. 2*d*. Due to lateral diffusion of impurities or lateral straggle of implanted ions, the width of the *p* region is slightly wider than the window opening.

4.1.4 Metallization

After diffusion or ion-implantation process, a metallization process is used to form ohmic contacts and interconnections (Fig. 2*e*). Metal films can be formed by physical vapor deposition and chemical vapor deposition. The lithography process is again used to define the front contact which is shown in Fig. 2*f*. A similar metallization step is done on the back contact without using a lithography process. Normally a low-temperature (≤ 500°C) anneal would also be performed to promote low-resistance contacts between the metal layers

---

[*] This is a negative photoresist. We can also use positive photoresist. Lithography and photoresists are considered in detail in Chapter 12.

and the semiconductor. With the completion of the metallization, the *p–n* junctions become functional.

4.2 THERMAL EQUILIBRIUM CONDITION

The most important characteristic of *p–n* junctions is that they rectify, that is, they allow current to flow easily in only one direction. Figure 3 shows the current-voltage characteristics of a typical silicon *p–n* junction. When we apply "forward bias" to the junction (i.e., positive voltage on the *p*-side), the current increases rapidly as the voltage increases. However, when we apply a "reverse bias," virtually no current flows initially. As the reverse bias is increased the current remains very small until a critical voltage is reached, at which point the current suddenly increases. This sudden increase in current is referred to as the junction breakdown. The applied forward voltage is usually less than 1 V, but the reverse critical voltage, or breakdown voltage, can vary from just a few volts to many thousands of volts depending on the doping concentration and other device parameters.

4.2.1 Band Diagram

In Fig. 4*a*, we see two regions of *p*- and *n*-type semiconductor materials that are uniformly doped and physically separated before the junction is formed. Note that the Fermi level E_F is near the valence band edge in the *p*-type material and near the conduction band edge in the *n*-type material. While *p*-type material contains a large concentration of holes with few electrons, the opposite is true for *n*-type material.

When the *p*- and *n*-type semiconductors are jointed together, the large carrier concentration gradients at the junction cause carrier diffusion. Holes from the *p*-side diffuse into the *n*-side, and electrons from the *n*-side diffuse into the *p*-side. As holes continue to leave the *p*-side, some of the negative acceptor ions (N_A^-) near the junction are left uncompensated, since the acceptors are fixed in the semiconductor lattice, whereas the holes are mobile. Similarly, some of the positive donor ions (N_D^+) near the junction are left uncompensated as the electrons leave the *n*-side. Consequently, a negative space charge

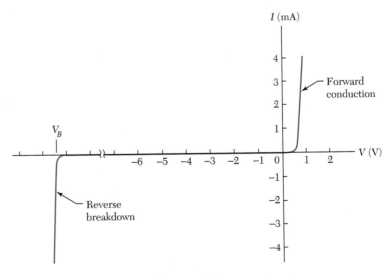

Fig. 3   Current-voltage characteristics of a typical silicon *p–n* junction.

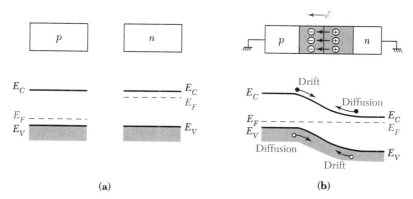

Fig. 4 (*a*) Uniformly doped *p*-type and *n*-type semiconductors before the junction is formed. (*b*) The electric field in the depletion region and the energy band diagram of a *p–n* junction in thermal equilibrium.

forms near the *p*-side of the junction and a positive space charge forms near the *n*-side. This space charge region creates an electric field that is directed from the positive charge toward the negative charge, as indicated in the upper illustration of Fig. 4*b*.

The electric field is in the direction opposite to the diffusion current for each type of charge carrier. The lower illustration of Fig. 4*b* shows that the hole diffusion current flows from left to right, whereas the hole drift current due to the electric field flows from right to left. The electron diffusion current also flows from left to right, whereas the electron drift current flows in the opposite direction. Note that because of their negative charge, electrons diffuse from right to left, opposite the direction of electron current.

4.2.2 Equilibrium Fermi Levels

At thermal equilibrium, that is, the steady-state condition at a given temperature without any external excitations, the individual electron and hole current flowing across the junctions are identically zero. Thus, for each type of carrier the drift current due to the electric field must exactly cancel the diffusion current due to the concentration gradient. From Eq. 32 in Chapter 3,

$$J_p = J_p(\text{drift}) + J_p(\text{diffusion})$$

$$= q\mu_p p \mathscr{E} - qD_p \frac{dp}{dx}$$

$$= q\mu_p p \left(\frac{1}{q} \frac{dE_i}{dx} \right) - kT\mu_p \frac{dp}{dx} = 0, \tag{1}$$

where we have used Eq. 8 of Chapter 3 for the electric field and the Einstein relation $D_p = (kT/q)\mu_p$. Substituting the expression for hole concentration

$$p = n_i e^{(E_i - E_F)/kT} \tag{2}$$

and its derivative

$$\frac{dp}{dx} = \frac{p}{kT} \left(\frac{dE_i}{dx} - \frac{dE_F}{dx} \right) \tag{3}$$

into Eq. 1 yields the net hole current density

$$J_p = \mu_p p \frac{dE_F}{dx} = 0 \tag{4}$$

or

$$\frac{dE_F}{dx} = 0. \tag{5}$$

Similarly, we obtain for the net electron current density

$$J_n = J_n(\text{drift}) + J_n(\text{diffusion})$$

$$= q\mu_n n \mathscr{E} + qD_n \frac{dn}{dx}$$

$$= \mu_n n \frac{dE_F}{dx} = 0. \tag{6}$$

Thus, for the condition of zero net electron and hole currents, the Fermi level must be constant (i.e., independent of *x*) throughout the sample as illustrated in the energy band diagram of Fig. 4*b*.

The constant Fermi level required at thermal equilibrium results in a unique space charge distribution at the junction. We repeat the one-dimensional *p–n* junction and the corresponding equilibrium energy band diagram in Figs. 5*a* and 5*b*, respectively. The unique space charge distribution and the electrostatic potential ψ are given by Poisson's equation:

$$\boxed{\frac{d^2\psi}{dx^2} \equiv -\frac{d\mathscr{E}}{dx} = -\frac{\rho_s}{\varepsilon_s} = -\frac{q}{\varepsilon_s}\left(N_D - N_A + p - n\right).} \tag{7}$$

Here we assume that all donors and acceptors are ionized.

In regions far away from the metallurgical junction, charge neutrality is maintained and the total space charge density is zero. For these neutral regions we can simplify Eq. 7 to

$$\frac{d^2\psi}{dx^2} = 0 \tag{8}$$

and

$$N_D - N_A + p - n = 0. \tag{9}$$

For a *p*-type neutral region, we assume $N_D = 0$ and $p \gg n$. The electrostatic potential of the *p*-type neutral region with respect to the Fermi level, designated as ψ_p in Fig. 5*b*, can be obtained by setting $N_D = n = 0$ in Eq. 9 and by substituting the result ($p = N_A$) into Eq. 2:

$$\psi_p \equiv -\frac{1}{q}\left(E_i - E_F\right)\Big|_{x \leq -x_p} = -\frac{kT}{q}\ln\left(\frac{N_A}{n_i}\right). \tag{10}$$

Similarly, we obtain the electrostatic potential of the *n*-type neutral region with respect to the Fermi level:

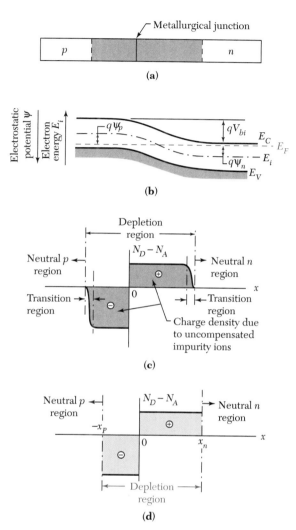

Fig. 5 (*a*) A *p–n* junction with abrupt doping changes at the metallurgical junction. (*b*) Energy band diagram of an abrupt junction at thermal equilibrium. (*c*) Space charge distribution. (*d*) Rectangular approximation of the space charge distribution.

$$\psi_n \equiv -\frac{1}{q}\left(E_i - E_F\right)\Big|_{x \geq x_n} = -\frac{kT}{q}\ln\left(\frac{N_D}{n_i}\right). \tag{11}$$

The total electrostatic potential difference between the *p*-side and the *n*-side neutral regions at thermal equilibrium is called the *built-in potential* V_{bi}:

$$\boxed{V_{bi} = \psi_n - \psi_p = \frac{kT}{q}\ln\left(\frac{N_A N_D}{n_i^2}\right).} \tag{12}$$

4.2.3 Space Charge

Moving from a neutral region toward the junction, we encounter a narrow transition region, shown in Fig. 5c. Here the space charge of impurity ions is partially compensated by the mobile carriers. Beyond the transition region we enter the completely depleted region where the mobile carrier densities are zero. This is called the *depletion region* (also called the space-charge region). For typical *p–n* junctions in silicon and gallium arsenide, the width of each transition region is small compared with the width of the depletion region. Therefore, we can neglect the transition region and represent the depletion region by the rectangular distribution shown in Fig. 5d, where x_p and x_n denote the depletion layer widths of the *p*- and *n*-sides for the completely depleted region with $p = n = 0$. Equation 7 becomes

$$\frac{d^2\psi}{dx^2} = \frac{q}{\varepsilon_s}\left(N_A - N_D\right). \tag{13}$$

The magnitudes of $|\psi_p|$ and ψ_n as calculated from Eqs.10 and 11 are plotted in Fig. 6 as a function of the doping concentration of silicon and gallium arsenide. For a given doping concentration, the electrostatic potential of gallium arsenide is higher because of its smaller intrinsic concentration n_i.

▶ **EXAMPLE 1**

Calculate the built-in potential for a silicon *p–n* junction with $N_A = 10^{18}$ cm^{-3} and $N_D = 10^{15}$ cm^{-3} at 300 K.

SOLUTION From Eq. 12 we obtain

$$V_{bi} = \left(0.0259\right)\left[\frac{10^{18} \times 10^{15}}{\left(9.65 \times 10^9\right)^2}\right] = 0.774 \text{ V}.$$

Also from Fig. 6,

$$V_{bi} = \psi_n + |\psi_p| = 0.30 \text{ V} + 0.47 \text{ V} = 0.77 \text{ V}. \qquad \blacktriangleleft$$

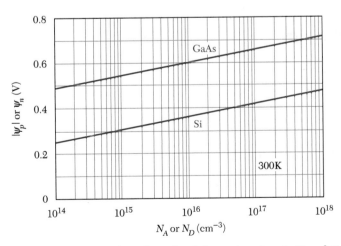

Fig. 6 Built-in potentials on the *p*-side and *n*-side of abrupt junctions in Si and GaAs as a function of impurity concentration.

▶ 4.3 DEPLETION REGION

To solve Possion's equation, Eq.13, we must know the impurity distribution. In this section we consider two important cases—the abrupt junction and the linearly graded junction. Figure 7a shows an *abrupt junction*, that is, a *p–n* junction formed by shallow diffusion or low-energy ion implantation. The impurity distribution of the junction can be approximated by an abrupt transition of doping concentration between the *n*- and *p*-type regions. Figure 7b shows a linearly graded junction. For either deep diffusions or high-energy ion implantations, the impurity profiles may be approximated by linearly graded junctions, that is, the impurity distribution varies linearly across the junction. We consider the depletion regions of both types of junction.

4.3.1 Abrupt Junction

The space charge distribution of an abrupt junction is shown in Fig. 8a. In the depletion region, free carriers are totally depleted so that Possion's equation, Eq.13, simplifies to

$$\frac{d^2\psi}{dx^2} = +\frac{qN_A}{\varepsilon_s} \qquad \text{for} \quad -x_p \leq x < 0, \tag{14a}$$

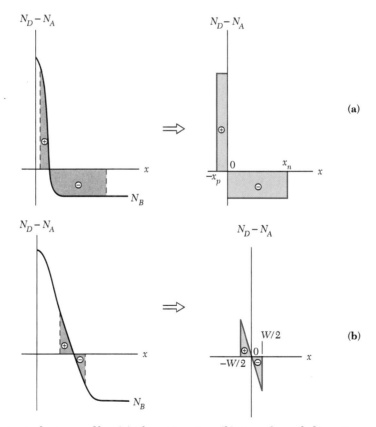

Fig. 7 Approximate doping profiles. (*a*) Abrupt junction. (*b*) Linearly graded junction.

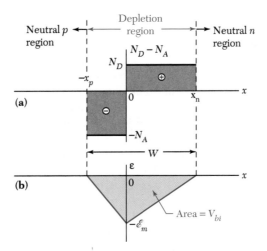

Fig. 8 (a) Space charge distribution in the depletion region at thermal equilibrium. (b) Electric-field distribution. The shaded area corresponds to the built-in potential.

$$\frac{d^2\psi}{dx^2} = -\frac{qN_D}{\varepsilon_s} \qquad \text{for} \qquad 0 < x \le x_n. \tag{14b}$$

The overall space charge neutrality of the semiconductor requires that the total negative space charge per unit area in the p-side must precisely equal the total positive space charge per unit area in the n-side:

$$N_A x_p = N_D x_n. \tag{15}$$

The total depletion layer width W is given by

$$W = x_p + x_n. \tag{16}$$

The electric field shown in Fig. 8b is obtained by integrating Eqs. 14a and 14b, which gives

$$\mathscr{E}(x) = -\frac{d\psi}{dx} = -\frac{qN_A(x + x_p)}{\varepsilon_s} \qquad \text{for} \qquad -x_p \le x < 0 \tag{17a}$$

and

$$\mathscr{E}(x) = -\mathscr{E}_m + \frac{qN_D x}{\varepsilon_x} = \frac{qN_D}{\varepsilon_s}(x - x_n) \qquad \text{for} \quad 0 < x \le x_n, \tag{17b}$$

where $\mathscr{E}_m$ is the maximum field that exists at $x = 0$ and is given by

$$\mathscr{E}_m = \frac{qN_D x_n}{\varepsilon_s} = \frac{qN_A x_p}{\varepsilon_s}. \tag{18}$$

Integating Eqs. 17a and 17b over the depletion region gives the total potential variation, namely, the built-in potential V_{bi}:

$$V_{bi} = -\int_{-x_p}^{x_n} \mathscr{E}(x)dx = -\int_{-x_p}^{0} \mathscr{E}(x)dx \Big|_{p\text{-side}} - \int_{0}^{x_n} \mathscr{E}(x)dx \Big|_{n\text{-side}}$$

$$= \frac{qN_A x_p^2}{2\varepsilon_s} + \frac{qN_D x_n^2}{2\varepsilon_s} = \frac{1}{2}\mathscr{E}_m W. \tag{19}$$

Therefore, the area of the field triangle in Fig. 8*b* corresponds to the built-in potential.

Combining Eqs. 15 to 19 gives the total depletion layer width as a function of the built-in potential,

$$W = \sqrt{\frac{2\varepsilon_s}{q}\left(\frac{N_A + N_D}{N_A N_D}\right)V_{bi}} \, . \tag{20}$$

When the impurity concentration on one side of an abrupt junction is much higher than that of the other side, the junction is called a *one-sided abrupt junction* (Fig. 9*a*). Figure 9*b* shows the space charge distribution of a one-sided abrupt p^+–n junction, where $N_A \gg N_D$. In this case, the depletion layer width of the *p*-side is much smaller than the *n*-side (i.e., $x_p \ll x_n$), and the expression for W can be simplified to

$$W \cong x_n = \sqrt{\frac{2\varepsilon_s V_{bi}}{q N_D}} \, . \tag{21}$$

The expression for the electric-field distribution is the same as Eq. 17b:

$$\mathscr{E}(x) = -\mathscr{E}_m + \frac{q N_B x}{\varepsilon_s}, \tag{22}$$

where N_B is the lightly doped bulk concentration (i.e., N_D for a p^+–n junction). The field decreases to zero at $x = W$. Therefore,

$$\mathscr{E}_m = \frac{q N_B W}{\varepsilon_s} \tag{23}$$

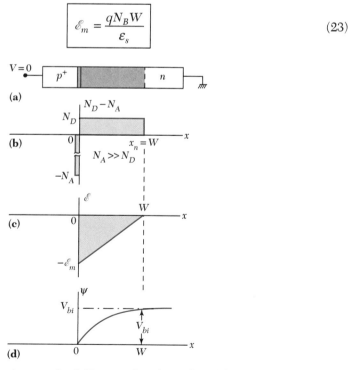

Fig. 9 (*a*) One-sided abrupt junction (with $N_A \gg N_D$) in thermal equilibrium. (*b*) Space charge distribution. (*c*) Electric-field distribution. (*d*) Potential distribution with distance, where V_{bi} is the built-in potential.

and

$$\mathscr{E}\left(x\right) = \frac{qN_B}{\varepsilon_s}\left(-W + x\right) = -\mathscr{E}_m\left(1 - \frac{x}{W}\right), \tag{24}$$

which is shown in Fig. 9c.

Integrating Possion's equation once more gives the potential distribution

$$\psi\left(x\right) = -\int_0^x \mathscr{E}\,dx = \mathscr{E}_m\left(x - \frac{x^2}{2W}\right) + \text{constant.} \tag{25}$$

With zero potential in the neutral *p*-region as a reference, or $\psi(0) = 0$, and employing Eq. 19,

$$\boxed{\psi\left(x\right) = \frac{V_{bi}\,x}{W}\left(2 - \frac{x}{W}\right).} \tag{26}$$

The potential distribution is shown in Fig. 9d.

▶ **EXAMPLE 2**

For a silicon one-sided abrupt junction with $N_A = 10^{19}$ cm^{-3} and $N_D = 10^{16}$ cm^{-3}, calculate the depletion layer width and the maximum field at zero bias ($T = 300$ K).

SOLUTION From Eqs. 12, 21, and 23, we obtain

$$V_{bi} = 0.0259\ln\left[\frac{10^{19} \times 10^{16}}{\left(9.65 \times 10^9\right)^2}\right] = 0.895 \text{ V},$$

$$W \cong \sqrt{\frac{2\varepsilon_s V_{bi}}{qN_D}} = 3.41 \times 10^{-5} = 0.343 \ \mu\text{m},$$

$$\mathscr{E}_m = \frac{qN_B W}{\varepsilon_s} = 0.52 \times 10^4 \text{ V/cm}. \qquad \blacktriangleleft$$

The previous discussions are for a *p–n* junction at thermal equilibrium without external bias. The equilibrium energy band diagram, shown again in Fig. 10a, illustrates that the total electrostatic potential across the junction is V_{bi}. The corresponding potential energy difference from the *p*-side to the *n*-side is qV_{bi}. If we apply a positive voltage V_F to the *p*-side with respect to the *n*-side, the *p–n* junction becomes forward-biased, as shown in Fig. 10b. The total electrostatic potential across the junction decreases by V_F, that is, it is replaced with $V_{bi} - V_F$. Thus, forward bias reduces the depletion layer width.

By contrast, as shown in Fig. 10c, if we apply positive voltage V_R to the *n*-side with respect to the *p*-side, the *p–n* junction now becomes reverse-biased and the total electrostatic potential across the junction increases by V_R, that is, it is replaced by $V_{bi} + V_R$. Here, we find that reverse bias increases the depletion layer width. Substituting these voltage values in Eq. 21 yields the depletion layer widths as a function of the applied voltage for a one-sided abrupt junction:

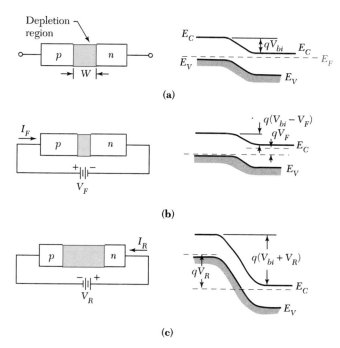

Fig. 10 Schematic representations of depletion layer width and energy band diagrams of a p–n junction under various biasing conditions. (*a*) Thermal-equilibrium condition. (*b*) Forward-bias condition. (*c*) Reverse-bias condition.

$$W = \sqrt{\frac{2\varepsilon_s \left(V_{bi} - V\right)}{qN_B}}, \tag{27}$$

where N_B is the lightly doped bulk concentration, and V is positive for forward bias and negative for reverse bias. Note that the depletion layer width W varies as the square root of the total electrostatic potential difference across the junction.

4.3.2 Linearly Graded Junction

We first consider the case of thermal equilibrium. The impurity distribution for a linearly graded junction is shown in Fig. 11*a*. The Possion equation for the case is

$$\frac{d^2\psi}{dx^2} = \frac{-d\mathscr{E}}{dx} = \frac{-\rho_s}{\varepsilon_s} = \frac{-q}{\varepsilon_s}ax \qquad -\frac{W}{2} \leq x \leq \frac{W}{2}, \tag{28}$$

where a is the impurity gradient (in cm^{-4}) and W is the depletion-layer width.

We have assumed that mobile carriers are negligible in the depletion region. By integrating Eq. 28 once with the boundary conditions that the electric field is zero at $\pm W/2$, we obtain the electric-field distribution shown in Fig. 11*b*

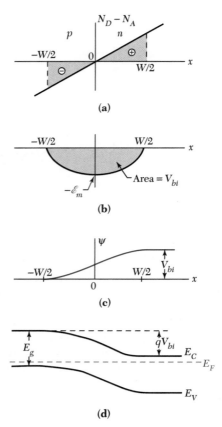

Fig. 11 Linearly graded junction in thermal equilibrium. (*a*) Impurity distribution. (*b*) Electric-field distribution. (*c*) Potential distribution with distance. (*d*) Energy band diagram.

$$\mathscr{E}\left(x\right) = -\frac{qa}{\varepsilon_s}\left[\frac{\left(W/2\right)^2 - x^2}{2}\right]. \tag{29}$$

The maximum field at $x = 0$ is

$$\mathscr{E}_m = \frac{qaW^2}{8\varepsilon_s}. \tag{29a}$$

Integrating Eq. 28 once again yields both the potential distribution and the corresponding energy band diagram shown in Figs. 11*c* and 11*d*, respectively. The built-in potential and the depletion layer width are given by

$$V_{bi} = \frac{qaW^3}{12\varepsilon_s} \tag{30}$$

and

$$W = \left(\frac{12\varepsilon_s V_{bi}}{qa}\right)^{1/3}. \tag{31}$$

Since the values of the impurity concentrations at the edges of the depletion region (–W/2 and W/2) are the same and both are equal to $aW/2$, the built-in potential for a linearly graded junction may be expressed in a form similar to Eq. 12*:

$$V_{bi} = \frac{kT}{q} \ln\left[\frac{(aW/2)(aW/2)}{n_i^2}\right] = \frac{2kT}{q} \ln\left(\frac{aW}{2n_i}\right). \tag{32}$$

Solving the transcendental equation that results when W is eliminated from Eqs. 31 and 32 yields the built-in potential as a function of a. The results for silicon and gallium arsenide linearly graded junctions are shown in Fig. 12.

When either forward or reverse bias is applied to the linearly graded junction, the variations of the depletion layer width and the energy band diagram will be similar to those shown in Fig. 10 for abrupt junctions. However, the depletion layer width will vary as $(V_{bi} – V)^{1/3}$, where V is positive for forward bias and negative for reverse bias.

▷ **EXAMPLE 3**

For a silicon linearly graded junction with a impurity gradient of 10^{20} cm^{-4}, the depletion-layer width is 0.5 μm. Calculate the maximum field and built-in voltage $(T = 300 \text{ K})$:

SOLUTION From Eq. 29a and Eq. 32, we obtain

$$\mathscr{E}_m = \frac{qaW^2}{8\varepsilon_s} = \frac{1.6\times10^{-19}\times10^{20}\times\left(0.5\times10^{-4}\right)^2}{8\times11.9\times8.85\times10^{-14}} = 4.75\times10^3 \text{ V / cm},$$

$$V_{bi} = \frac{2kT}{q}\left(\frac{aW}{2n_i}\right) = 2\times0.0259\left(\frac{10^{20}\times0.5\times10^{-4}}{2\times9.65\times10^9}\right) = 0.645 \text{ V}.$$

◀

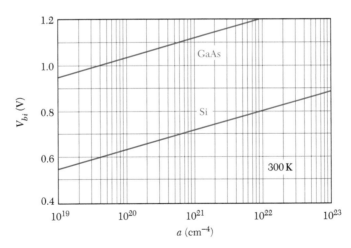

Fig. 12 Built-in potential for a linearly graded junction in Si and GaAs as a function of impurity gradient.

* Based on an accurate numerical technique, the built-in potential is given by $V_{bi} = \frac{2}{3}\frac{kT}{q}\ln\left(\frac{a^2\varepsilon_s kT/q}{8qn_i^3}\right)$.

For a given impurity gradient, the V_{bi} is smaller than that calculated from Eq.32 by about 0.05 – 0.1 V.

▶ 4.4 DEPLETION CAPACITANCE

The junction depletion layer capacitance per unit area is defined as $C_j = dQ/dV$, where dQ is the incremental change in depletion layer charge per unit area for an incremental change in the applied voltage dV.[*]

Figure 13 illustrates the depletion capacitance of a *p–n* junction with an arbitrary impurity distribution. The charge and electric-field distributions indicated by the solid lines correspond to a voltage V applied to the *n*-side. If this voltage is increased by an amount dV, the charge and field distributions will expand to those regions bounded by the dashed lines. In Fig. 13*b*, the incremental charge dQ corresponds to the colored area between the two charge distribution curves on either side of the depletion region. The incremental space charges on the *n*- and *p*-sides of the depletion region are equal but with opposite charge polarity, thus maintaining overall charge neutrality. This incremental charge dQ causes an increase in the electric field by an amount $d\mathscr{E} = dQ/\varepsilon_s$ (from Possion's equation). The corresponding change in the applied voltage dV, represented by the cross-hatched area in Fig. 13*c* is approximately $Wd\mathscr{E}$, which equals WdQ/ε_s. Therefore, the depletion capacitance per unit area is given by

$$C_j \equiv \frac{dQ}{dV} = \frac{dQ}{W\dfrac{dQ}{\varepsilon_s}} = \frac{\varepsilon_s}{W} \tag{33}$$

or

$$\boxed{C_j = \frac{\varepsilon_s}{W} \quad \text{F/cm}^2.} \tag{33a}$$

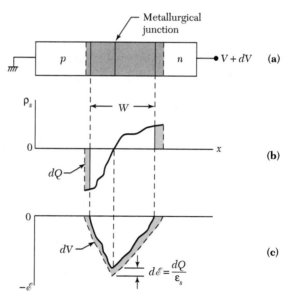

Fig. 13 (*a*) *p–n* junction with an arbitrary impurity profile under reverse bias. (*b*) Change in space charge distribution due to change in applied bias. (*c*) Corresponding change in electric-field distribution.

---

[*]The capacitance is also referred to as the transition region capacitance.

4.4.1 Capacitance-Voltage Characteristics

Equation 33 for the depletion capacitance per unit area is the same as the standard expression for a parallel-plate capacitor where the spacing between the two plates represents the depletion-layer width. The equation is valid for any arbitrary impurity distribution.

In deriving Eq. 33 we have assumed that only the variation of the space charge in the depletion region contributes to the capacitance. This certainly is a good assumption for the reverse-bias condition. For forward biases, however, a large current can flow across the junction corresponding to a large number of mobile carriers present within the neutral region. The incremental change of these mobile carriers with respect to the biasing voltage contributes an additional term, called the diffusion capacitance, which is considered in Section 4.6.

For a one-sided abrupt junction, we obtain, from Eqs. 27 and 33,

$$C_j = \frac{\varepsilon_s}{W} = \sqrt{\frac{q\varepsilon_s N_B}{2(V_{bi} - V)}} \qquad (34)$$

or

$$\boxed{\frac{1}{C_j^2} = \frac{2(V_{bi} - V)}{q\varepsilon_s N_B}.} \qquad (35)$$

It is clear from Eq. 35 that a plot of $1/C_j^2$ versus V produces a straight line for a one-sided abrupt junction. The slope gives the impurity concentration N_B of the substrate, and the intercept (at $1/C_j^2 = 0$) gives V_{bi}.

▶ **EXAMPLE 4**

For a silicon one-sided abrupt junction with $N_A = 2 \times 10^{19}$ cm^{-3} and $N_D = 8 \times 10^{15}$ cm^{-3}, calculate the junction capacitance at zero bias and reversed bias of 4 V ($T = 300$ K).

SOLUTION   From Eqs. 12, 27, and 34, we obtain at zero bias

$$V_{bi} = 0.0259\ln\frac{2 \times 10^{19} \times 8 \times 10^{15}}{\left(9.65 \times 10^9\right)^2} = 0.906 \text{ V},$$

$$W\big|_{V=0} \cong \sqrt{\frac{2\varepsilon_s V_{bi}}{qN_D}} = \sqrt{\frac{2 \times 11.9 \times 8.85 \times 10^{-14} \times 0.906}{1.6 \times 10^{-19} \times 8 \times 10^{15}}} = 3.86 \times 10^{-5} = 0.386 \text{ } \mu\text{m},$$

$$C_j\big|_{V=0} = \frac{\varepsilon_s}{W\big|_{V=0}} = \sqrt{\frac{q\varepsilon_s N_B}{2V_{bi}}} = 2.728 \times 10^{-8} \text{ F / cm}^2.$$

From Eqs. 27 and 34, we can obtain at reverse bias of 4 V:

$$W\big|_{V=-4} \cong \sqrt{\frac{2\varepsilon_s(V_{bi} - V)}{qN_D}} = \sqrt{\frac{2 \times 11.9 \times 8.85 \times 10^{-14} \times (0.906 + 4)}{1.6 \times 10^{-19} \times 8 \times 10^{15}}} = 8.99 \times 10^{-5} = 0.899 \text{ } \mu\text{m},$$

$$C_j\big|_{V=-4} = \frac{\varepsilon_s}{W\big|_{V=-4}} = \sqrt{\frac{q\varepsilon_s N_B}{2(V_{bi} - V)}} = 1.172 \times 10^{-8} \text{ F / cm}^2.$$

◀

4.4.2 Evaluation of Impurity Distribution

The capacitance-voltage characteristics can be used to evaluate an arbitrary impurity distribution. We consider the case of p^+–n junction with a doping profile on the n-side, as shown in Fig. 14b. As before, the incremental change in depletion layer charge per unit area dQ for an incremental change in the applied voltage dV is given by $qN(W)\,dW$ (i.e., the shaded area in Fig. 14b). The corresponding change in applied voltage (shaded area in Fig. 14c) is

$$dV \cong \left(d\mathscr{E}\right) W = \left(\frac{dQ}{\varepsilon_s}\right) W = \frac{qN(W)\,dW^2}{2\varepsilon_s}. \tag{36}$$

By substituting W from Eq. 33, we obtain an expression for the impurity concentration at the edge of the depletion region:

$$N\!\left(W\right) = \frac{2}{q\varepsilon_s}\left[\frac{1}{d\!\left(1/C_j{}^2\right)/dV}\right]. \tag{37}$$

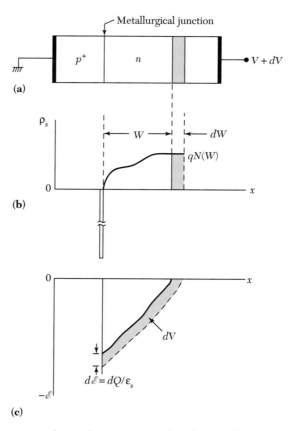

Fig. 14 (*a*) p^+–n junction with an arbitary impurity distribution. (*b*) Change in space charge distribution in the lightly doped side due to a change in applied bias. (*c*) Corresponding change in the electric-field distribution.

Thus, we can measure the capacitance per unit area versus reverse-bias voltage and plot $1/C_j^2$ versus V. The slope of the plot, that is, $d\,(1/C_j^2)/dV$, yields $N(W)$. Simultaneously, W is obtained from Eq. 33. A series of such calculations produces a complete impurity profile. This approach is referred to as the *C–V* method for measuring impurity profiles.

For a linearly graded junction, the depletion layer capacitance is obtained from Eqs. 31 and 33:

$$C_j = \frac{\varepsilon_s}{W} = \left[\frac{qa\varepsilon_s^2}{12(V_{bi} - V)} \right]^{1/3} \text{F/cm}^2. \tag{38}$$

For such a junction we can plot $1/C^3$ versus V and obtain the impurity gradient and V_{bi} from the slope and the intercept, respectively.

4.4.3 Varactor

Many circuit applications employ the voltage-variable properties of reverse-biased *p–n* junctions. A *p–n* junction designed for such a purpose is called a *varactor*, which is a shortened form of variable reactor. As previously derived, the reverse-biased depletion capacitance is given by

$$C_j \propto \left(V_{bi} + V_R \right)^{-n} \tag{39}$$

or

$$C_j \propto \left(V_R \right)^{-n} \quad \text{for} \quad V_R \gg V_{bi}, \tag{39a}$$

where $n = \frac{1}{3}$ for a linearly graded junction and $n = \frac{1}{2}$ for an abrupt junction. Thus, the voltage sensitivity of C (i.e., variation of C with V_R) is greater for an abrupt junction than for a linearly graded junction. We can further increase the voltage sensitivity by using a hyperabrupt junction having an exponent n (Eq. 39) greater than $\frac{1}{2}$.

Figure 15 shows three p^+–n doping profiles with the doner distribution $N_D(x)$ given by $B(x/x_0)^m$, where B and x_0 are constants, $m = 1$ for a linearly graded junction, $m = 0$ for an abrupt junction, and $m = -\frac{3}{2}$ for a hyperabrupt junction. The hyperabrupt profile can

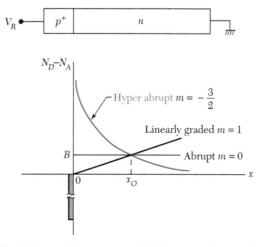

Fig. 15 Impurity profiles for hyperabrupt, one-sided abrupt, and one-sided linearly graded junctions.

be achieved by epitaxial growth techniques discussed in Chapter 10. To obtain the capacitance-voltage relationship, we solve Possion's equation:

$$\frac{d^2\psi}{dx^2} = -B\left(\frac{x}{x_0}\right)^m.$$

(40)

Integrating Eq. 40 twice with appropriate boundary conditions gives the dependence of the depletion layer width on the reverse bias:

$$W \propto \left(V_R\right)^{1/(m+2)}$$

(41)

Therefore,

$$C_j = \frac{\varepsilon_s}{W} \propto \left(V_R\right)^{-1/(m+2)}$$

(42)

Comparing Eq. 42 with Eq. 39a yields $n = 1/(m + 2)$. For hyperabrupt junctions with $n > \frac{1}{2}$, m must be a negative number.

By choosing different values for m, we can obtain a wide variety of C_j-versus-V_R dependencies for specific applications. One interesting example, shown in Fig. 15, is the case for $m = -\frac{3}{2}$. For this case, $n = 2$. When this varactor is connected to an inductor L in a resonant circuit, the resonant frequency varies linearly with the voltage applied to the varactor:

$$\omega_r = \frac{1}{\sqrt{LC_j}} \propto \frac{1}{\sqrt{V_R^{-n}}} = V_R \qquad \text{for } n = 2.$$

(43)

▶ 4.5 CURRENT-VOLTAGE CHARACTERISTICS

A voltage applied to a *p–n* junction will disturb the precise balance between the diffusion current and drift current of electrons and holes. Under forward bias, the applied voltage reduces the electrostatic potential across the depletion region, as shown in the middle of Fig. 16. The drift current is reduced in comparison to the diffusion current. We have an enhanced hole diffusion from the *p*-side to the *n*-side and electron diffusion from the *n*-side to the *p*-side. Therefore, minority carrier injections occur, that is, electrons are injected into the *p*-side, whereas holes are injected into the *n*-side. Under reverse bias, the applied voltage increases the electrostatic potential across the depletion region as shown in the middle of Fig. 16*b*. This greatly reduces the diffusion currents, resulting in a small reverse current. In this section, we first consider the ideal current-voltage characteristics. We then discuss departures from these ideal characteristics due to generation and recombination and other effects.

4.5.1 Ideal Characteristics

We now derive the ideal current-voltage characteristics based on the following assumptions: (a) The depletion region has abrupt boundaries and, outside the boundaries, the semiconductor is assumed to be neutral; (b) the carrier densities at the boundaries are related by the electrostatic potential difference across the junction. (c) the low-injection condition, that is, the injected minority carrier densities, are small compared with the majority carrier densities; in other words, the majority carrier densities are changed negligibly at the boundaries of neutral regions by the applied bias; and (d) neither genera-

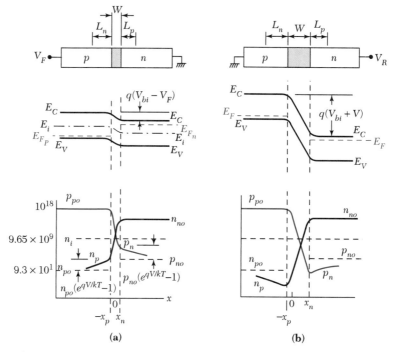

Fig. 16 Depletion region, energy band diagram and carrier distribution. (*a*) Forward bias. (*b*) Reverse bias.

tion nor recombination current exists in the depletion region, and the electron and hole currents are constant throughout the depletion region. Departures from these idealized assumptions are considered in the next section.

At thermal equilibrium, the majority carrier density in the neutral regions is essentially equal to the doping concentration. We use the subscripts n and p to denote the semiconductor type and the subscript o to specify the condition of thermal equilibrium. Hence, n_{no} and n_{po} are the equilibrium electron densities in the n- and p-sides, respectively. The expression for the built-in potential in Eq. 12 can be rewritten as

$$V_{bi} = \frac{kT}{q} \ln \frac{p_{po} n_{no}}{n_i^2} = \frac{kT}{q} \ln \frac{n_{no}}{n_{po}}, \tag{44}$$

where the mass action law $p_{po} n_{po} = n_i^2$ has been used. Rearranging Eq. 44 gives

$$n_{no} = n_{po} e^{qV_{bi}/kT}. \tag{45}$$

Similarly, we have

$$p_{po} = p_{no} e^{qV_{bi}/kT}. \tag{46}$$

We note from Eqs. 45 and 46 that the electron density and the hole density at the two boundaries of the depletion region are related through the electrostatic potential difference V_{bi} at thermal equilibrium. From our second assumption we expect that the same relation holds when the electrostatic potential difference is changed by an applied voltage.

When a forward bias is applied, the electrostatic potential difference is reduced to $V_{bi} - V_F$; but when a reverse bias is applied, the electrostatic potential difference is increased to $V_{bi} + V_R$. Thus, Eq. 45 is modified to

$$n_n = n_p e^{q(V_{bi} - V)/kT},\qquad (47)$$

where n_n and n_p are the nonequilibrium electron densities at the boundaries of the depletion region in the *n*- and *p*-sides, respectively, with V positive for forward bias and negative for reverse bias. For the low-injection condition, the injected minority carrier density is much smaller than the majority carrier density; therefore, $n_n \cong n_{no}$. Substituting this condition and Eq. 45 into Eq. 47 yields the electron density at the boundary of the depletion region on the *p*-side ($x = -x_p$):

$$n_p = n_{po} e^{qV/kT} \qquad (48)$$

or

$$n_p - n_{po} = n_{po}\left(e^{qV/kT} - 1\right). \qquad (48a)$$

Similarly, we have

$$p_n = p_{no} e^{qV/kT} \qquad (49)$$

or

$$p_n - p_{no} = p_{no}\left(e^{qV/kT} - 1\right) \qquad (49a)$$

at $x = x_n$ for the *n*-type boundary. Figures 16a and 16b show band diagrams and carrier concentrations in a *p–n* junction under forward-bias and reverse-bias conditions, respectively. Note that the minority carrier densities at the boundaries ($-x_p$ and x_n) increase substantially above their equilibrium values under forward bias, whereas they decrease below their equilibrium values under reverse bias. Equations 48 and 49 define the minority carrier densities at the boundaries of depletion region. These equations are the most important boundary conditions for the ideal current-voltage characteristics.

Under our idealized assumptions, no current is generated within the depletion region; all currents come from the neutral regions. In the neutral *n*-region, there is no electric field, thus the steady-state continuity equation reduces to

$$\frac{d^2 p_n}{dx^2} - \frac{p_n - p_{no}}{D_p \tau_p} = 0. \qquad (50)$$

The solution of Eq. 50 with the boundary conditions of Eq. 49 and p_n ($x = \infty$) = p_{no} gives

$$p_n - p_{no} = p_{no}\left(e^{qV/kT} - 1\right) e^{-(x-x_n)/L_p}, \qquad (51)$$

where L_p, which is equal to $\sqrt{D_p \tau_p}$, is the diffusion length of holes (minority carriers) in the *n*-region. At $x = x_n$,

$$J_p(x_n) = -qD_p \frac{dp_n}{dx}\bigg|_{x_n} = \frac{qD_p p_{no}}{L_p}\left(e^{qV/kT} - 1\right). \qquad (52)$$

Similarly, we obtain for the neutral p-region

$$n_p - n_{po} = n_{po}\left(e^{qV/kT} - 1\right) e^{(x+x_p)/L_n} \tag{53}$$

and

$$J_n\left(-x_p\right) = qD_n \left.\frac{dn_p}{dx}\right|_{-x_p} = \frac{qD_n n_{po}}{L_n}\left(e^{qV/kT} - 1\right), \tag{54}$$

where L_n, which is equal to $\sqrt{D_n \tau_n}$, is the diffusion length of electrons. The minority carrier densities (Eqs. 51 and 53) are shown in the middle of Fig. 17.

The graphs illustrate that the injected minority carriers recombine with the majority carriers as the minority carriers move away from the boundaries. The electron and hole currents are shown at the bottom of Fig. 17. The hole and electron currents at the boundaries are given by Eqs. 52 and 54, respectively. The hole diffusion current will decay

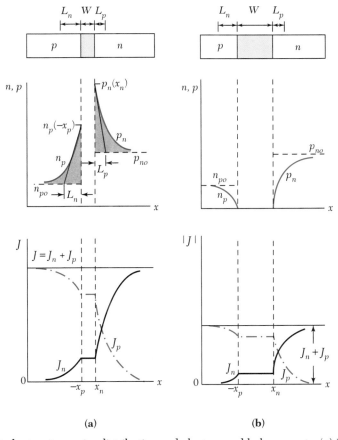

Fig. 17 Injected minority carrier distribution and electron and hole currents. (*a*) Forward bias. (*b*) Reverse bias. The figure illustrates idealized currents. For practical devices, the currents are not constant across the space charge layer.

exponentially in the *n*-region with diffusion length L_p, and the electron diffusion current will decay exponentially in the *p*-region with diffusion length L_n.

The total current is constant throughout the device and is the sum of Eqs. 52 and 54:

$$J = J_p(x_n) + J_n(-x_p) = J_s\left(e^{qV/kT} - 1\right), \tag{55}$$

$$J_s \equiv \frac{qD_p p_{no}}{L_p} + \frac{qD_n n_{po}}{L_n}, \tag{55a}$$

where J_s is the saturation current density. Equation 55 is the *ideal diode equation*.[1] The ideal current-voltage characteristic is shown in Figs. 18*a* and 18*b* in the Cartesian and

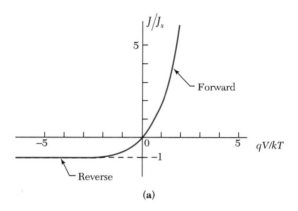

(a)

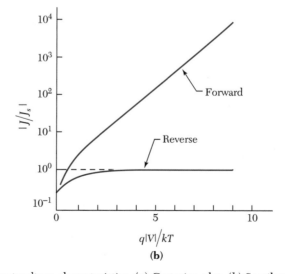

(b)

Fig. 18 Ideal current-voltage characteristics. (*a*) Cartesian plot. (*b*) Semilog plot.

semilog plots, respectively. In the forward direction with positive bias on the p-side, for $V \geq 3kT/q$, the rate of current increase is constant, as shown in Fig. 18b. At 300 K for every decade change of current, the voltage change for an ideal diode is 60 mV (= 2.3 kT/q). In the reverse direction, the current density saturates at $-J_s$.

▶ **EXAMPLE 5**

Calculate the ideal reverse saturation current in a Si p–n junction diode with a cross-sectional area of 2×10^{-4} cm². The parameters of the diode are

$$N_A = 5 \times 10^{16} \text{ cm}^{-3}, \quad N_D = 10^{16} \text{ cm}^{-3}, \quad n_i = 9.65 \times 10^9, \text{ cm}^{-3}$$
$$D_n = 21 \text{ cm}^2/\text{s}, \quad D_p = 10 \text{ cm}^2/\text{s}, \quad \tau_p = \tau_n = 5 \times 10^{-7} \text{ s}.$$

SOLUTION From Eq. 55a and $L_p = \sqrt{D_p \tau_p}$, we can obtain

$$J_s = \frac{qD_p p_{n0}}{L_p} + \frac{qD_n n_{p0}}{L_n} = qn_i^2 \left(\frac{1}{N_D} \sqrt{\frac{D_p}{\tau_p}} + \frac{1}{N_A} \sqrt{\frac{D_n}{\tau_n}} \right),$$

$$= 1.6 \times 10^{-19} \times \left(9.65 \times 10^9\right)^2 \left(\frac{1}{10^{16}} \sqrt{\frac{10}{5 \times 10^{-7}}} + \frac{1}{5 \times 10^{16}} \sqrt{\frac{21}{5 \times 10^{-7}}} \right),$$

$$= 8.58 \times 10^{-12} \text{ A/cm}^2.$$

From the cross-sectional area A = 2×10^{-4} cm², we obtain

$$I_s = A \times J_s = 2 \times 10^{-4} \times 8.58 \times 10^{-12} = 1.72 \times 10^{-15} \text{ A}.$$

◀

4.5.2 Generation-Recombination and High-Injection Effects

The ideal diode equation, Eq. 55, adequately describes the current-voltage characteristics of germanium p–n junctions at low current densities. For silicon and gallium arsenide p–n junctions, however, the ideal equation can only give qualitative agreement because of the generation or recombination of carriers in the depletion region.

Consider the reverse-bias condition first. Under reverse bias, carrier concentrations in the depletion region fall far below their equilibrium concentrations. The dominant generation-recombination processes discussed in Chapter 3 are those of electron and hole emissions through bandgap generation-recombination centers. The capture processes are not important because their rates are proportional to the concentration of free carriers, which is very small in the reverse-biased depletion region.

The two emission processes operate in the steady state by alternately emitting electrons and holes. The rate of electron-hole pair generation can be obtained from Eq. 48 of Chapter 3 with the conditions $p_n < n_i$ and $n_n < n_i$:

$$G = -U = \left[\frac{\sigma_p \sigma_n \upsilon_{th} N_t}{\sigma_n \exp\left(\dfrac{E_t - E_i}{kT}\right) + \sigma_p \exp\left(\dfrac{E_i - E_t}{kT}\right)} \right] n_i$$

$$\equiv \frac{n_i}{\tau_g}, \tag{56}$$

where τ_g, the generation lifetime, is the reciprocal of the expression in the square brackets. We can arrive at an important conclusion about electron-hole generation from this expression. Let us consider a simple case where $\sigma_n = \sigma_p = \sigma_o$. For this case, Eq. 56 reduces to

$$G = \frac{\sigma_o v_{th} N_t n_i}{2 \cosh\left(\dfrac{E_t - E_i}{kT}\right)}. \tag{57}$$

The generation rate reaches a maximum value at $E_t = E_i$ and falls off exponentially as E_t moves in either direction away from the middle of the bandgap. Thus, only those centers with an energy level of E_t near the intrinsic Fermi level can contribute significantly to the generation rate.

The current due to generation in the depletion region is

$$J_{gen} = \int_0^W qG\,dx \cong qGW = \frac{qn_i W}{\tau_g} \tag{58}$$

where W is the depletion layer width. The total reverse current for a p^+–n junction, that is, for $N_A \gg N_D$ and for $V_R > 3kT/q$, can be approximated by the sum of both the diffusion current in the neutral regions and the generation current in the depletion region:

$$\boxed{J_R \cong q\sqrt{\frac{D_p}{\tau_p}}\frac{n_i^2}{N_D} + \frac{qn_i W}{\tau_g}.} \tag{59}$$

For semiconductors with large values of n_i, such as germanium, the diffusion current dominates at room temperature, and the reverse current follows the ideal diode equation. But if n_i is small, such as for silicon and gallium arsenide, the generation current in the depletion region may dominate.

▶ **EXAMPLE 6**

Consider the Si *p–n* junction diode in Example 5 and assume $\tau_g = \tau_p = \tau_n$, calculate the generation current density for a reverse bias of 4 V.

SOLUTION From Eq. 20, we obtain

$$W = \sqrt{\frac{2\varepsilon_s}{q}\left(\frac{N_A + N_D}{N_A N_D}\right)(V_{bi} + V)} = \sqrt{\frac{2\varepsilon_s}{q}\left(\frac{N_A + N_D}{N_A N_D}\right)\left(\frac{kT}{q}\ln\frac{N_A N_D}{n_i^2} + V\right)}$$

$$= \sqrt{\frac{2 \times 11.9 \times 8.85 \times 10^{-14}}{1.6 \times 10^{-19}}\left(\frac{5 \times 10^{16} + 10^{16}}{5 \times 10^{16} \times 10^{16}}\right)\left(0.0259\ \ln\frac{5 \times 10^{16} \times 10^{16}}{(9.65 \times 10^9)^2} + V\right)}$$

$$= 3.97 \times \sqrt{0.758 + V} \times 10^{-5}\ \text{cm}.$$

Hence the generation current density is

$$J_{gen} = \frac{qn_i W}{\tau_g} = \frac{1.6 \times 10^{-19} \times 9.65 \times 10^9}{5 \times 10^{-7}} \times 3.97 \times \sqrt{0.758 + V} \times 10^{-5}\ \text{A/cm}^2$$

$$= 1.22 \times \sqrt{0.758 + V} \times 10^{-7}\ \text{A/cm}^2.$$

If we apply a reversed bias of 4 V, the generation current density is 2.66×10^{-7} A/cm². ◀

Under forward bias, the concentrations of both electrons and holes exceed their equilibrium values. The carriers will attempt to return to their equilibrium values by recombination. Therefore, the dominant generation-recombination processes in the depletion region are the capture processes. From Eq. 49 we obtain

$$p_n n_n \cong p_{no} n_{no} e^{qV/kT} = n_i^2 e^{qV/kT}. \tag{60}$$

Substituting Eq. 60 in Eq. 48 of Chapter 3 and assuming $\sigma_n = \sigma_p = \sigma_o$ yields

$$U = \frac{\sigma_o \upsilon_{th} N_t n_i^2 \left(e^{qV/kT} - 1\right)}{n_n + p_n + 2n_i \cosh \dfrac{E_i - E_t}{kT}}. \tag{61}$$

In either recombination or generation, the most effective centers are those located near E_i. As practical examples, gold and copper yield effective generation-recombination centers in silicon where the values of $E_t - E_i$ are 0.02 V for gold and –0.02 eV for copper. In gallium arsenide, chromium gives an effective center with an $E_t - E_i$ value of 0.08 eV.

Equation 61 can be simplified for the case $E_t = E_i$:

$$U = \sigma_o \upsilon_{th} N_t \frac{n_i^2 \left(e^{qV/kT} - 1\right)}{n_n + p_n + 2n_i} \tag{62}$$

For a given forward bias, U reaches its maximum value at a location in the depletion region either where the denominator $n_n + p_n + 2n_i$ is a minimum or where the sum of the electron and hole concentrations, $n_n + p_n$, is at its minimum value. Since the product of these concentrations is a constant given by Eq. 60, the condition $d(p_n + n_n) = 0$ leads to

$$dp_n = -dn_n = \frac{p_n n_n}{p_n^2} dp_n \tag{63}$$

or

$$p_n = n_n \tag{64}$$

as the condition for the minimum. This condition exists at the location in the depletion region, where E_i is halfway between E_{Fp} and E_{Fn}, as illustrated in the middle of Fig. 16a. Here, the carrier concentrations are

$$p_n = n_n = n_i e^{qV/2kT} \tag{65}$$

and, therefore,

$$U_{\max} = \sigma_o \upsilon_{th} N_t \frac{n_i^2 \left(e^{qV/kT} - 1\right)}{2n_i \left(e^{qV/2kT} + 1\right)}. \tag{66}$$

For $V > 3kT/q$,

$$U_{\max} \cong \frac{1}{2} \sigma_o \upsilon_{th} N_t n_i e^{qV/2kT}. \tag{67}$$

The recombination current is then

$$J_{rec} = \int_0^W qU dx \cong \frac{qW}{2} \sigma_o \upsilon_{th} N_t n_i e^{qV/2kT} = \frac{qW n_i}{2\tau_r} e^{qV/2kT}, \tag{68}$$

where τ_r is the effective recombination lifetime given by $1/(\sigma_o v_{th} N_t.)$ The total forward current can be approximated by the sum of Eqs. 55 and 68. For $p_{no} \gg n_{po}$ and $V > 3kT/q$ we have

$$J_F = q\sqrt{\frac{D_p}{\tau_p}}\,\frac{n_i^2}{N_D}\,e^{qV/kT} + \frac{qWn_i}{2\tau_r}\,e^{qV/2kT}. \tag{69}$$

In general, the experimental results can be represented empirically by

$$J_F \approx \exp\left(\frac{qV}{\eta kT}\right), \tag{70}$$

where the factor η is called the *ideality factor*. When the ideal diffusion current dominates, $\eta = 1$; whereas when the recombination current dominates, $\eta = 2$. When both currents are comparable, η has a value of between 1 and 2.

Figure 19 shows the measured forward characteristics of a silicon and gallium arsenide p–n junction at room temperature.[2] At low current levels, recombination current dominates and $\eta = 2$. At higher current levels, diffusion current dominates and η approaches 1.

At even higher current levels, we notice that the current departs from the ideal $\eta = 1$ situation and increases more gradually with forward voltage. This phenomenon is associated with two effects: series resistance and high injection. We first consider the series resistance effect. At both low- and medium-current levels, the IR drop across the neutral regions is usually small compared with kT/q (26 mV at 300 K), where I is the forward current and R is the series resistance. For example, for a silicon diode with $R =$

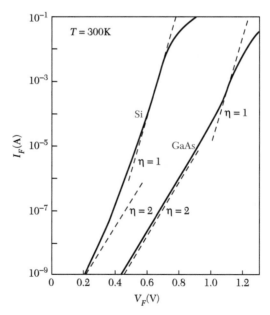

Fig. 19 Comparison of the forward current-voltage characteristics of Si and GaAs diodes[2] at 300 K. Dashed lines indicate slopes of different ideality factors η.

1.5 ohms, the *IR* drop at 1 mA is only 1.5 mV. However, at 100 mA, the *IR* drop becomes 0.15 V, which is six times larger than kT/q. This *IR* drop reduces the bias across the depletion region; therefore, the current becomes

$$I \cong I_s \exp\left[\frac{q(V - IR)}{kT}\right] = \frac{I_s \exp(qV / kT)}{\exp\left[\frac{q(IR)}{kT}\right]} \qquad (71)$$

and the ideal diffusion current is reduced by the factor $\exp[q(IR)/kT]$.

At high-current densities, the injected minority carrier density is compariable to the majority concentration, that is at the *n*-side of the junction $p_n\,(x = x_n) \cong n_n$. This is the high-injection condition. By substituting the high-injection condition in Eq. 60, we obtain $p_n\,(x = x_n) \cong n_i \exp\,(qV/2kT)$. Using this as a boundary condition, the current becomes roughly proportional to $\exp\,(qV/2kT)$. Thus, the current increases at a slower rate under the high-injection condition.

4.5.3 Temperature Effect

Operating temperature has a profound effect on device performance. In both the forward-bias and reverse-bias conditions, the magnitudes of the diffusion and the recombination-generation currents depend strongly on temperature. We consider the forward-bias case first. The ratio of hole diffusion current to the recombination is given by

$$\frac{I_{\text{diffusion}}}{I_{\text{recombination}}} = 2\,\frac{n_i}{N_D}\,\frac{L_p}{W}\,\frac{\tau_r}{\tau_p}\,e^{qV/2kT} \approx \exp\left(-\frac{E_g - qV}{2kT}\right). \qquad (72)$$

This ratio depends on both the temperature and the semiconductor bandgap. Figure 20*a* shows the temperature dependence of the forward characteristics of a silicon diode. At

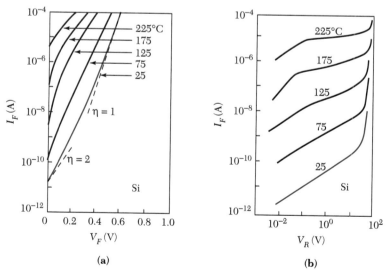

Fig. 20 Temperature dependence of the current-voltage characteristics of a Si diode[2]. (*a*) Forward bias. (*b*) Reverse bias.

room temperature for small forward voltages, the recombination current generally dominates, whereas at higher forward voltages the diffusion current usually dominates. At a given forward bias, as the temperature increases, the diffusion current will increase more rapidly than the recombination current. Therefore, the ideal diode equation will be followed over a wide range of forward biases as the temperature increases.

The temperature dependence of the saturation current density J_s (Eq. 55a) for a one-sided p^+–n junction in which diffusion current dominates is given by

$$J_s \cong \frac{qD_p p_{no}}{L_p} \approx n_i^2 \approx \exp\left(-\frac{E_g}{kT}\right). \tag{73}$$

Thus, the activation energy obtained from the slope of a plot of J_s versus $1/T$ corresponds to the energy bandgap E_g.

In the reverse-bias condition for a p^+–n junction, the ratio of the diffusion current to the generation current is

$$\frac{I_{\text{diffusion}}}{I_{\text{generation}}} = \frac{n_i L_p}{N_D W} \frac{\tau_g}{\tau_p}. \tag{74}$$

This ratio is proportional to the intrinsic carrier density n_i. As the temperature increases, the diffusion current eventually dominates. Figure 20*b* shows the effects of temperature on the reverse characteristics of a silicon diode. At low temperatures, the generation current dominates and the reverse current varies as $\sqrt{V_R}$ in accordance with Eq. 58 for an abrupt junction (i.e., $W \sim \sqrt{V_R}$). As the temperature increases beyond 175°C, the current demonstrates a saturation tendency for $V_R \geq 3kT/q$, at which point the diffusion current becomes dominant.

▶ 4.6 CHARGE STORAGE AND TRANSIENT BEHAVIOR

Under forward bias, electrons are injected from the n-region into the p-region and holes are injected from the p-region into the n-region. Once injected across the junction, the minority carriers recombine with the majority carriers and decay exponentially with distance, as shown in Fig. 17*a*. These minority-carrier distributions lead to current flow and to charge storage in the p-n junction. We consider the stored charge, its effect on junction capacitance, and the transient behavior of the p–n junction due to sudden changes of bias.

4.6.1 Minority-Carrier Storage

The charge of injected minority carriers per unit area stored in the neutral n-region can be found by integrating the excess holes in the neutral region, shown as the shaded area in the middle of Fig. 17*a*, using Eq. 51:

$$Q_p = q\int_{x_n}^{\infty} (p_n - p_{no})\ dx,$$

$$= q\int_{x_n}^{\infty} p_{no}\left(e^{qV/kT} - 1\right) e^{-(x-x_n)/L_p} dx,$$

$$= qL_p p_{no}\left(e^{qV/kT} - 1\right). \tag{75}$$

A similar expression can be obtained for the stored electrons in the neutral p-region. The number of stored minority carriers depends on both the diffusion length and the charge density at the boundary of the depletion region. We can express the stored charge in terms of the injected current. From Eqs. 52 and 75, we have

$$Q_p = \frac{L_p^2}{D_p} J_p(x_n) = \tau_p J_p(x_n) \tag{76}$$

Equation 76 states that the amount of stored charge is the product of the current and lifetime of the minority carriers. This is because the injected holes diffuse farther into the n-region before recombining if their lifetime is longer; thus, more holes are stored.

▶ **EXAMPLE 7**

For an ideal abrupt silicon p^+–n junction with $N_D = 8 \times 10^{15}$ cm^{-3}, calculate the stored minority carriers per unit area in the neutral n-region when a forward bias of 1V is applied. The diffusion length of the holes is 5 μm.

SOLUTION   From Eq. 75, we obtain

$$Q_p = qL_p p_{no}\left(e^{qV/kT} - 1\right) = 1.6 \times 10^{-19} \times 5 \times 10^{-4}\, \text{cm} \times \frac{\left(9.65 \times 10^9\right)^2}{8 \times 10^{15}} \times \left(e^{\frac{1}{0.0259}} - 1\right)$$

$$= 4.69 \times 10^{-2}\ \text{C/cm}^2. \qquad\qquad ◀$$

4.6.2   Diffusion Capacitance

The depletion-layer capacitance considered previously accounts for most of the junction capacitance when the junction is reverse biased. When the junction is forward biased, there is an additional significant contribution to junction capacitance from the rearrangement of the stored charges in the neutral regions. This is called the *diffusion capacitance*, denoted C_d, a term derived from the ideal-diode case in which minority carriers move across the neutral region by diffusion.

The diffusion capacitance of the stored holes in the neutral n-region is obtained by applying the definition $C_d = AdQ_p/dV$ to Eq. 75:

$$C_d = \frac{Aq^2 L_p p_{no}}{kT} e^{qV/kT}, \tag{77}$$

where A is the device cross-section area. We may add the contribution to C_d of the stored electrons in the neutral p-region in cases of significant storage. For a p^+–n junction, however, $n_{po} \ll p_{no}$, and the contribution to C_d of the stored electrons becomes insignificant. Under reverse bias (i.e., V is negative), Eq. 77 shows that C_d is inconsequential because of negligible minority-carrier storage.

In many applications we prefer to represent a p–n junction by an equivalent circuit. In addition to diffusion capacitance C_d and depletion capacitance C_j, we must include conductance to account for the current through the device. In the ideal diode the conductance can be obtained from Eq. 55:

$$G = \frac{AdJ}{dV} = \frac{qA}{kT} J_s e^{qV/kT} = \frac{qA}{kT}\left(J + J_s\right) \cong \frac{qI}{kT}. \tag{78}$$

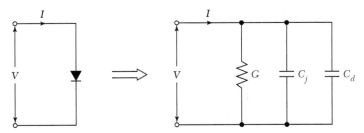

Fig. 21 Small-signal equivalent circuit of a *p–n* junction.

The diode equivalent circuit is shown in Fig. 21, where C_j stands for the total depletion capacitance (i.e., the result in Eq. 33 times the device area A). For low-voltage, sinusoidal excitation of a diode that is biased quiescently (i.e., at *dc*), the circuit shown in Fig. 21 provides adequate accuracy. Therefore, we refer to it as the diode small-signal equivalent circuit.

4.6.3 Transient Behavior

For switching applications, the forward-to-reverse-bias transition must be nearly abrupt and the transient time short. Figure 22*a* shows a simple circuit where a forward current I_F flows through a *p–n* junction. At time $t = 0$, switch S is suddenly thrown to the right and an initial reverse current $I_R \cong V/R$ flows. The transient time t_{off}, plotted in Fig. 22*b*, is the time required for the current to reach 10% of the initial reverse current I_R.

  The transient time may be estimated as follows. Under the forward-bias condition, the stored minority carriers in the *n*-region for a p^+–*n* junction is given by Eq. 76:

$$Q_p = \tau_p J_p = \tau_p \frac{I_F}{A}, \tag{79}$$

where I_F is the total forward current and A is the device area. If the average current flowing during the turn-off period is $I_{R, ave}$, the turn-off time is the length of time required to remove the total stored charge Q_p:

$$t_{off} \cong \frac{Q_p A}{I_{R,ave}} = \tau_p \left(\frac{I_F}{I_{R,ave}} \right). \tag{80}$$

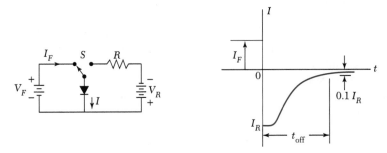

Fig. 22 Transient behavior of a *p–n* junction. (*a*) Basic switching circuit. (*b*) Transient response of the current switched from forward bias to reverse bias.

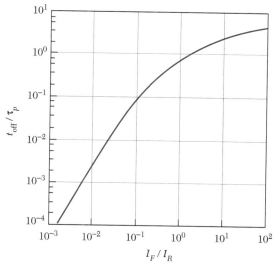

Fig. 23 Normalized transient time versus the ratio of forward current to reverse current.[3]

Thus the turn-off time depends on both the ratio of forward to reverse currents and the lifetime of the minority carriers. The result of a more precise turn-off time calculation[3] accounting for the time-dependent minority-carrier diffusion problem is shown in Fig. 23. For fast-switching devices, we must reduce the lifetime of the minority carriers. Therefore, recombination-generation centers that have energy levels located near mid-bandgap, such as gold in silicon, are usually introduced.

► 4.7 JUNCTION BREAKDOWN

When a sufficiently large reverse voltage is applied to a *p–n* junction, the junction breaks down and conducts a very large current. Although the breakdown process is not inherently destructive, the maximum current must be limited by an external circuit to avoid excessive junction heating. Two important breakdown mechanisms are the tunneling effect and avalanche multiplication. We consider the first mechanism briefly and then discuss avalanche multiplication in detail, because avalanche breakdown imposes an upper limit on the reverse bias for most diodes. Avalanche breakdown also limits the collector voltage of a bipolar transistor (Chapter 5) and the drain voltage of a MOSFET (Chapter 6). In addition, the avalanche multiplication mechanisms can generate microwave power, as in an IMPATT diode (Chapter 8), and detect optical signals, as in an avalanche photodetector (Chapter 9).

4.7.1 Tunneling Effect

When a high electric field is applied to a *p–n* junction in the reverse direction, a valence electron can make a transition from the valence band to the conduction band, as shown in Fig. 24*a*. This process, in which an electron penetrates through the energy bandgap, is called tunneling.

The tunneling process is discussed in Chapter 3. Tunneling occurs only if the electric field is very high. The typical field for silicon and gallium arsenide is about 10^6 V/cm or higher. To achieve such a high field, the doping concentrations for both *p*- and *n*-regions

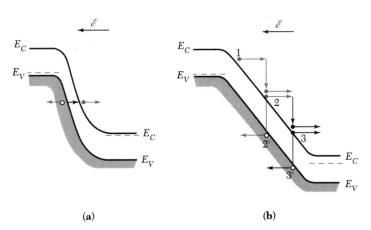

Fig. 24 Energy band diagrams under junction-breakdown conditions. (*a*) Tunneling effect. (*b*) Avalanche multiplication.

must be quite high ($>5 \times 10^{17}$ cm^{-3}). The breakdown mechanisms for silicon and gallium arsenide junctions with breakdown voltages of less than about $4E_g/q$, where E_g is the bandgap, are the result of the tunneling effect. For junctions with breakdown voltages in excess of $6E_g/q$, the breakdown mechanism is the result of avalanche multiplication. At voltages between 4 and $6E_g/q$, the breakdown is due to a mixture of both avalanche multiplication and tunneling.[4]

4.7.2 Avalanche Multiplication

The avalanche multiplication process is illustrated in Fig. 24*b*. The *p–n* junction such as a p^+–n one-sided abrupt junction with a doping concentration of $N_D \cong 10^{17}$ cm^{-3} or less is under reverse bias. This figure is essentially the same as Fig. 25 in Chapter 3. A thermally generated electron in the depletion region (designated by 1) gains kinetic energy from the electric field. If the field is sufficiently high, the electron can gain enough kinetic energy that on collision with an atom, it can break the lattice bonds, creating an electron-hole pair (2 and 2′). These newly created electron and hole both acquire kinetic energy from the field and create additional electron-hole pairs (e.g., 3 and 3′). These in turn continue the process, creating other electron-hole pairs. This process is therefore called *avalanche multiplication*.

To derive the breakdown condition, we assume that a current I_{no} is incident at the left-hand side of the depletion region of width W, as shown in Fig. 25. If the electric field in the depletion region is high enough to initiate the avalanche multiplication process, the electron current I_n will increase with distance through the depletion region to reach a value $M_n I_{no}$ at W, where M_n, the multiplication factor, is defined as

$$M_n \equiv \frac{I_n(W)}{I_{no}}. \tag{81}$$

Similarly, the hole current I_p increases from $x = W$ to $x = 0$. The total current $I = (I_p + I_n)$ is constant at steady state. The incremental electron current at x equals the number of electron-hole pairs generated per second in the distance dx:

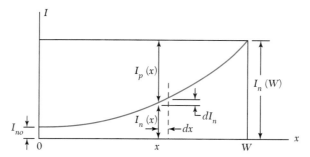

Fig. 25 Depletion region in a *p–n* junction with multiplication of an incident current.

$$d\left(\frac{I_n}{q}\right) = \left(\frac{I_n}{q}\right)\left(\alpha_n dx\right) + \left(\frac{I_p}{q}\right)\left(\alpha_p dx\right) \tag{82}$$

or

$$\frac{dI_n}{dx} + \left(\alpha_p - \alpha_n\right)I_n = \alpha_p I, \tag{82a}$$

where α_n and α_p are the electron and hole ionization rates, respectively. If we use the simplified assumption that $\alpha_n = \alpha_p = \alpha$, the solution of Eq. 82a is

$$\frac{I_n(W) - I_n(0)}{I} = \int_0^W \alpha dx. \tag{83}$$

From Eqs. 81 and 83, we have

$$1 - \frac{1}{M_n} = \int_0^W \alpha dx. \tag{83a}$$

The avalanche breakdown voltage is defined as the voltage where M_n approaches infinity. Hence, the breakdown condition is given by

$$\boxed{\int_0^W \alpha dx = 1.} \tag{84}$$

From both the breakdown condition described above and the field dependence of the ionization rates, we may calculate the critical field (i.e., the maximum electric field at breakdown) at which the avalanche process takes place. Using measured α_n and α_p (Fig. 26 in Chapter 3) the critical field $\mathscr{E}_c$ are calculated for silicon and gallium arsenide one-sided abrupt junctions and shown in Fig. 26 as functions of the impurity concentration of the substrate. Also indicated is the critical field for the tunneling effect. It is evident that tunneling occurs only in semiconductors having high doping concentrations.

With the critical field determined, we may calculate the breakdown voltages. As discussed previously, voltages in the depletion region are determined from the solution of Possion's equation:

$$\boxed{V_B\left(\text{breakdown voltage}\right) = \frac{\mathscr{E}_c W}{2} = \frac{\varepsilon_s \mathscr{E}_c^2}{2q}\left(N_B\right)^{-1}} \tag{85}$$

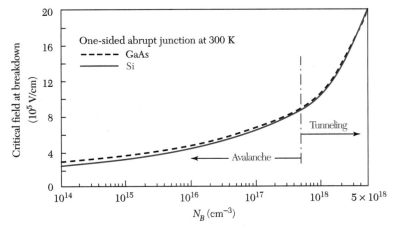

Fig. 26 Critical field at breakdown versus background doping for Si and GaAs one-sided abrupt junctions.[5]

for one-sided abrupt junctions and

$$V_B = \frac{2\mathscr{E}_c W}{3} = \frac{4\mathscr{E}_c^{3/2}}{3}\left(\frac{2\varepsilon_s}{q}\right)^{1/2}(a)^{-1/2} \tag{86}$$

for linearly graded junctions, where N_B is the background doping of the lightly doped side, ε_s is the semiconductor permittivity, and a is the impurity gradient. Since the critical field is a slowly varying function of either N_B or a, the breakdown voltage, as a first-order approximation, varies as N_B^{-1} for abrupt junctions and as $a^{-1/2}$ for linearly graded junctions.

Figure 27 shows the calculated avalanche breakdown voltages for silicon and gallium arsenide junctions.[5] The dash-dot line (to the right) at high dopings or high-impurity gradients indicates the onset of the tunneling effect. Gallium arsenide has higher breakdown voltages than silicon for a given N_B or a, mainly because of its larger bandgap. The larger the bandgap, the larger the critical field must be for sufficient kinetic energy to be gained between collisions. As Eqs. 85 and 86 demonstrate, the larger critical field, in turn, gives rise to higher breakdown voltage.

The inset of Fig. 28 shows the space-charge distribution of a diffused junction with a linear gradient near the surface and a constant doping inside the semiconductor. The breakdown voltage lies between the two limiting cases of abrupt junction and linearly graded junction considered previously.[6] For large a and low N_B, the breakdown voltage of the diffused junctions is given by the abrupt junction results shown on the bottom line in Fig. 28, whereas for small a and high N_B, V_B is given by the linearly graded junction results indicated by the parallel lines in Fig. 28.

▶ **EXAMPLE 8**

Calculate the breakdown voltage for a Si one-sided p^+–n abrupt junctions with $N_D = 5 \times 10^{16}$ cm^{-3}.

SOLUTION From Fig. 26, we see that the critical field at breakdown for a Si one-sided abrupt junction is about 5.7×10^5 V/cm. Then from Eq. 85, we obtain

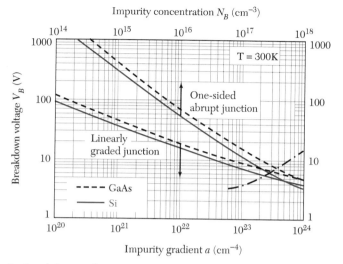

Fig. 27 Avalanche breakdown voltage versus impurity concentration for one-sided abrupt junction and avalanche breakdown voltage versus impurity gradient for linearly graded junction in Si and GaAs. Dash-dot line indicates the onset of the tunneling mechanism.[5]

$$V_B \left(\text{breakdown voltage}\right) = \frac{\mathscr{E}_c W}{2} = \frac{\varepsilon_s \mathscr{E}_c^{\,2}}{2q} \left(N_B\right)^{-1},$$

$$= \frac{11.9 \times 8.85 \times 10^{-14} \times \left(5.7 \times 10^5\right)^2}{2 \times 1.6 \times 10^{-19}} \left(5 \times 10^{16}\right)^{-1}$$

$$= 21.4 \ \ \text{V.} \qquad \qquad \blacktriangleleft$$

In Figs. 27 and 28 we assume that the semiconductor layer is thick enough to support the reverse-biased depletion layer width W_m at breakdown. If the semiconductor

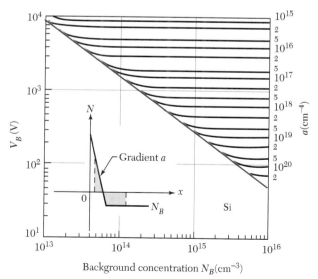

Fig. 28 Breakdown voltage for diffused junctions. Inset shows the space charge distribution.[6]

layer W is smaller than W_m, as shown in the inset of Fig. 29, the device will be punched through; that is, the depletion layer will reach the n–n^+ interface prior to breakdown. Increase the reverse bias further and the device will break down. The critical field $\mathscr{E}_c$ is essentially the same as that shown in Fig. 26. Therefore, the breakdown voltage V'_B for the punch-through diode is

$$\frac{V'_B}{V_B} = \frac{\text{shaded area in Fig. 29 inset}}{\left(\mathscr{E}_c W_m\right)/2},$$

$$= \left(\frac{W}{W_m}\right)\left(2 - \frac{W}{W_m}\right). \tag{87}$$

Punch-through occurs when the doping concentration N_B becomes sufficiently low, as in a p^+–π–n^+ or p^+–v–n^+ diode, where π stands for a lightly doped p-type and v stands for a lightly doped n-type semiconductor. The breakdown voltages for such diodes calculated from Eqs. 85 and 87 are shown in Fig. 29. For a given thickness, the breakdown voltage approaches a constant value as the doping decreases.

▶ **EXAMPLE 9**

For a GaAs p^+–n one-sided abrupt junction with $N_D = 8 \times 10^{14}$ cm^{-3}, calculate the depletion width at breakdown. If the n-type region of this structure is reduced to 20 μm, calculate the breakdown voltage.

SOLUTION From Fig. 27, we can find the breakdown voltage (V_B) is about 500 V, which is much larger than the built-in voltage (V_{bi}). And from Eq. 27, we obtain

$$W = \sqrt{\frac{2\varepsilon_s\left(V_{bi} - V\right)}{qN_B}} \cong \sqrt{\frac{2 \times 12.4 \times 8.85 \times 10^{-14} \times 500}{1.6 \times 10^{-19} \times 8 \times 10^{14}}} = 2.93 \times 10^{-3} \quad = 29.3 \text{ μm}.$$

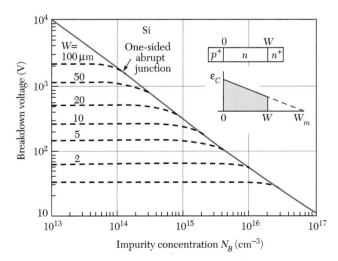

Fig. 29 Breakdown voltage for p^+–π–n^+ and p^+–v–n^+ junctions. W is the thickness of the lightly doped p-type (π) or the lightly doped n-type (v) region.

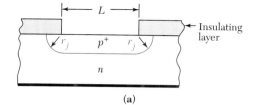

(a)

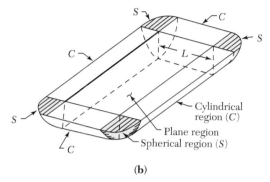

(b)

Fig. 30 (*a*) Planar diffusion process that forms junction curvature near the edge of the diffusion mask, where r_j is the radius of curvature. (*b*) Cylindrical and spherical regions formed by diffusion through a rectangular mask.

When the *n*-type region reduces to 20 μm, the punch-through will occur first. From Eq. 87, we can obtain

$$\frac{V_B'}{V_B} = \frac{\text{shaded area in Fig. 29 inset}}{\left(\mathscr{E}_c W_m\right)/2} = \left(\frac{W}{W_m}\right)\left(2 - \frac{W}{W_m}\right),$$

$$V_B' = V_B\left(\frac{W}{W_m}\right)\left(2 - \frac{W}{W_m}\right) = 500 \times \left(\frac{20}{29.3}\right)\left(2 - \frac{20}{29.3}\right) = 449 \text{ V.} \quad \blacktriangleleft$$

Another important consideration of breakdown voltage is the junction curvature effect.[7] When a *p–n* junction is formed by diffusion through a window in the insulating layer on a semiconductor, the impurities diffuse downward and sideways (see Chapter 13). Hence, the junction has a plane (or flat) region with nearly cylindrical edges, as shown in Fig. 30*a*. If the diffusion mask contains sharp corners, the corner of the junction will acquire the roughly spherical shape shown in Fig. 30*b*. Because the spherical or cylindrical regions of the junction have a higher field intensity, they determine the avalanche breakdown voltage. The calculated results for silicon one-sided abrupt junctions are shown in Fig. 31. The solid line represents the plane junctions considered previously. Note that as the junction radius r_j becomes smaller, the breakdown voltage decreases dramatically, especially for spherical junctions at low impurity concentrations.

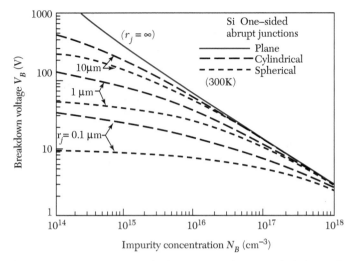

Fig. 31 Breakdown voltage versus impurity concentration for one-sided abrupt doping profile with cylindrical and spherical junction geometries,[7] where r_j is the radius of curvature indicated in Fig. 30.

▶ 4.8 HETEROJUNCTION

A heterojunction is defined as a junction formed between two dissimilar semiconductors. Figure 32*a* shows the energy band diagram of two isolated pieces of semiconductors prior to the formation of a heterojunction. The two semiconductors are assumed to have different energy bandgaps E_g, different dielectric permittivities ε_s, different work function $q\phi_s$, and different electron affinities $q\chi$. The work function is defined as the energy required to remove an electron from Fermi level E_F to a position just outside the material (the vacuum level). The electron affinity is the energy required to remove an electron from the bottom of the conduction band E_C to the vacuum level. The difference in energy of the conduction band edges in the two semiconductors is represented by ΔE_C, and the difference in energy in the valence band edges is represented by ΔE_V. From Fig. 32*a*, ΔE_C and ΔE_V can be expressed by

$$\Delta E_C = q\left(\chi_2 - \chi_1\right) \tag{88a}$$

and

$$\Delta E_V = E_{g1} + q\chi_1 - \left(E_{g2} + q\chi_2\right) = \Delta E_g - \Delta E_C, \tag{88b}$$

where ΔE_g is the energy band difference and $\Delta E_g = E_{g1} - E_{g2}$.

Figure 32*b* shows the equilibrium band diagram of an ideal abrupt heterojunction formed between these semiconductors.[8] In this diagram it is assumed that there is a negligible number of traps or generation-recombination centers at the interface of the two dissimilar semiconductors. Note that this assumption is valid only when heterojunctions are formed between semiconductors with closely matched lattice constants. Therefore, we must choose lattice-matched materials to satisfy the assumption.[§] For example, the $Al_xGa_{1-x}As$ materials, with x from 0 to 1, is the most important material for heterojunc-

---

[§]The lattice-mismatched epitaxy, also called the strained-layer epitaxy, is considered in Section 10.6.

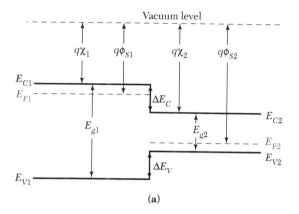

(a)

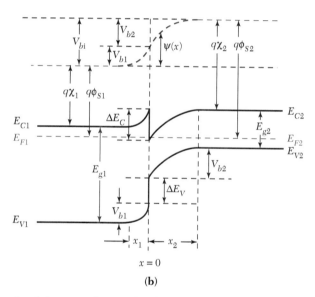

(b)

Fig. 32 (a) Energy band diagram of two isolated semiconductors. (b) Energy band diagram of an ideal n–p heterojunction at thermal equilibrium.

tions. When $x = 0$, we have GaAs, with a bandgap of 1.42 eV and a lattice constant of 5.6533 Å at 300 K. When $x = 1$, we have AlAs, with a bandgap of 2.17 eV and a lattice constant of 5.6605 Å. The bandgap for the ternary $Al_xGa_{1-x}As$ increases with x; however, the lattice constant remains essentially a constant. Even for the extreme cases where $x = 0$ and $x = 1$, the lattice constant mismatch is only 0.1%.

There are two basic requirements in the construction of the energy band diagram: (a) the Fermi level must be the same on both sides of the interface in thermal equilibrium, and (b) the vacuum level must be continuous and parallel to the band edges. Because of these requirements, the discontinuity in conduction band edges ΔE_C and valence band edges ΔE_V will be unaffected by doping as long as the bandgap E_g and electron affinity $q\chi$ are not functions of doping (i.e., as in nondegenerate semiconductors). The total built-in potential V_{bi} can be expressed by

$$V_{bi} = V_{b1} + V_{b2}, \tag{89}$$

where V_{b1} and V_{b2} are the electrostatic potentials at equilibrium in semiconductors 1 and 2, respectively.

Under the conditions that the potential and the *free-carrier flux density* (defined as the rate of free-carrier flow through a unit area) are continuous at the heterointerface, we can derive the depletion widths and capacitance from the Poisson equation using the conventional depletion approximation. One boundary condition is the continuity of electric displacement, that is, $\varepsilon_1 \mathscr{E}_1 = \varepsilon_2 \mathscr{E}_2$, where $\mathscr{E}_1$ and $\mathscr{E}_2$ are the electric fields at the interface $(x = 0)$ in semiconductors 1 and 2, respectively. V_{b1} and V_{b2} are given by

$$V_{b1} = \frac{\varepsilon_2 N_2 \left(V_{bi} - V\right)}{\varepsilon_1 N_1 + \varepsilon_2 N_2}, \tag{90a}$$

$$V_{b2} = \frac{\varepsilon_1 N_1 \left(V_{bi} - V\right)}{\varepsilon_1 N_1 + \varepsilon_2 N_2}, \tag{90b}$$

where N_1 and N_2 are the doping concentrations in semiconductors 1 and 2, respectively. The depletion widths x_1 and x_2 can be obtained by

$$x_1 = \sqrt{\frac{2\varepsilon_1 \varepsilon_2 N_2 \left(V_{bi} - V\right)}{qN_1 \left(\varepsilon_1 N_1 + \varepsilon_2 N_2\right)}} \tag{91a}$$

and

$$x_2 = \sqrt{\frac{2\varepsilon_1 \varepsilon_2 N_1 \left(V_{bi} - V\right)}{qN_2 \left(\varepsilon_1 N_1 + \varepsilon_2 N_2\right)}} \tag{91b}$$

▶ **EXAMPLE 10**

Consider an ideal abrupt heterojunction with a built-in potential of 1.6 V. The impurity concentrations in semiconductor 1 and 2 are 1×10^{16} donors/cm^3 and 3×10^{19} acceptors/cm^3, and the dielectric constants are 12 and 13, respectively. Find the electrostatic potential and depletion width in each material at thermal equilibrium.

SOLUTION From Eq. 90, the electrostatic potentials of a heterojunction at thermal equilibrium or $V = 0$ are

$$V_{b1} = \frac{13 \times \left(3 \times 10^{19}\right) \times 1.6}{12 \times \left(1 \times 10^{16}\right) + 13 \times \left(3 \times 10^{19}\right)} = 1.6 \text{ V}$$

and

$$V_{b2} = \frac{12 \times \left(1 \times 10^{16}\right) \times 1.6}{12 \times \left(1 \times 10^{16}\right) + 13 \times \left(3 \times 10^{19}\right)} = 4.9 \times 10^{-4} \text{ V.}$$

The depletion widths can be calculated by Eq. 91:

$$x_1 = \sqrt{\frac{2 \times 12 \times 13 \times \left(8.85 \times 10^{-14}\right) \times \left(3 \times 10^{19}\right) \times 1.6}{\left(1.6 \times 10^{-19}\right) \times \left(1 \times 10^{16}\right) \times \left(12 \cdot 1 \times 10^{16} + 13 \cdot 3 \times 10^{19}\right)}} = 4.608 \times 10^{-5} \text{ cm,}$$

$$x_2 = \sqrt{\frac{2 \times 12 \times 13 \times \left(8.85 \times 10^{-14}\right) \times \left(1 \times 10^{16}\right) \times 1.6}{\left(1.6 \times 10^{-19}\right) \times \left(3 \times 10^{19}\right) \times \left(12 \cdot 1 \times 10^{16} + 13 \cdot 3 \times 10^{19}\right)}} = 1.536 \times 10^{-8} \text{ cm.}$$

We see that most of the built-in potential is in the semiconductor with a lower doping concentration. The depletion width there is also much wider. ◀

SUMMARY

A p–n junction is formed when a p-type and an n-type semiconductor are brought into intimate contact. The p–n junction, in addition to being a device that is used in many application, is the basic building block for other semiconductor devices. Therefore, an understanding of junction theory serves as the foundation to understanding other semiconductor devices.

Most modern p–n junctions are fabricated using "planar technology." This technology includes the thermal oxidation process to grow an oxide layer on the semiconductor surface, the lithography process to open an window in the oxide, the diffusion or ion implantaion process to form a p–n junction in the window area, and the metallization process to provide contacts to connect the junction to other circuit elements. Planar technology is covered in detail in Chapters 10–14.

When a p–n junction is formed, there are uncompensated negative ions (N_A^-) on the p-side and uncompensated positive ions (N_D^+) on the n-side. Therefore, a depletion region (i.e., depletion of mobile carriers) is formed at the junction. This region, in turn, creates an electric field. At thermal equilibrium, the drift current due to the electric field is exactly balanced by the diffusion current due to concentration gradients of the mobile carriers on the two sides of the junction. When a positive voltage is applied to the p-side with respect to the n-side, a large current will flow through the junction. However, when a negative voltage is applied, virtually no current flows. This "rectifying" behavior is the most important characteristic of p–n junctions.

The basic equations presented in Chapters 2 and 3 have been used to develop the ideal static and dynamic behaviors of p–n junctions. We derived expressions for the depletion region, the depletion capacitance, and the ideal current-voltage characteristics of p–n junctions. However, practical devices depart from these ideal characteristics because of carrier generation and recombination in the depletion layers, high injection under forward bias, and series-resistance effects. The theory and methods of calculating the effects of these departures from the ideal are discussed in detail. We also considered other factors that influence p–n junctions, such as minority-carrier storage, diffusion capacitance, and transient behavior in high-frequency and switching applications.

A limiting factor in the operation of p–n junctions is junction breakdown—especially that due to avalanche multiplication. When a sufficiently large reverse voltage is applied to a p–n junction, the junction breaks down and conducts a very large current. Therefore, the breakdown voltage imposes an upper limit on the reverse bias for p–n junctions. We derived equations for the breakdown condition of the p–n junction and have shown the effect of device geometry and doping on the breakdown voltage.

A related device is the heterojunction formed between two dissimilar semiconductors. We obtained expressions for its electrostatic potentials and depletion widths. These expressions are simplified to that for a conventional p–n junction when these two semiconductors become identical.

▶ REFERENCES

1. W. Schockly, *Electrons and Holes in Semiconductors,* Van Nostrand, Princeton, NJ, 1950.

2 A. S. Grove, *Physics and Technology of Semiconductor Devices,* Wiley, New York, 1967.

3. R. H. Kingston, "Switching Time in Junction Diodes and Junction Transistors," *Proc. IRE,* **42**, 829 (1954).

4. J. L. Moll, *Physics of Semiconductors,* McGraw-Hill, New York, 1964.

5. S. M. Sze and G. Gibbons, "Avalanche Breakdown Voltages of Abrupt and Linearly Graded *p–n* Junctions in Ge, Si, GaAs and GaP," *Appl. Phys. Lett.,* **8**, 111 (1966).

6. S. K. Ghandhi, *Semiconductor Power Devices,* Wiley, New York, 1977.

7. S. M. Sze and G. Gibbons, "Effect of Junction Curvature on Breakdown Voltages in Semiconductors," *Solid State Electron.,* **9**, 831 (1966).

8. H. Kroemer, " Critique of Two Recent Theories of Heterojunction Lineups," *IEEE Electron Device Lett.,* **EDL-4**, 259 (1983).

▶ PROBLEMS (* DENOTES DIFFICULT PROBLEMS)

FOR SECTION 4.3 DEPLETION REGION

*1. A diffused silicon *p–n* junction has a linearly graded junction on the *p*-side with $a = 10^{19}$ cm^{-4}, and a uniform doping of 3×10^{14} cm^{-3} on the *n*-side. If the depletion layer width of the *p*-side is 0.8μm at zero bias, find the total depletion layer width, built-in potential, and maximum field at zero bias.

*2. Sketch the potential distribution in the Si *p–n* junction in Prob. 1.

3. For an ideal silicon *p–n* abrupt junction with $N_A = 10^{17}$ cm^{-3} and $N_D = 10^{15}$ cm^{-3}, (a) calculate V_{bi} at 250, 300, 350, 400, 450, and 500 K and plot V_{bi} versus T; (b) comment on your result in terms of energy band diagram; and (c) find the depletion layer width and the maximum field at zero bias for $T = 300$ K.

4. Determine the *n*-type doping concentration to meet the following specifications for a Si *p–n* junction:

$N_A = 10^{18}$ cm^{-3}, $\mathscr{E}_{max} = 4 \times 10^5$ V/cm at $V_R = 30$ V, $T = 300$ K.

FOR SECTION 4.4 DEPLETION CAPACITANCE

*5. An abrupt *p–n* junction has a doping concentration of 10^{15}, 10^{16}, or 10^{17} cm^{-3} on the lightly doped *n*-side and of 10^{19} cm^{-3} on the heavily doped *p*-side. Obtain series of curves of $1/C^2$ versus V, where V ranges from –4 V to 0 V in steps of 0.5 V. Comment on the slopes and the interceptions at the voltage axis of these curves.

6. For a silicon linearly graded junction with a impurity gradient of 10^{20} cm^{-4}, calculate the built-in potential and the junction capacitance at reverse bias of 4 V ($T = 300$ K).

7. A one-sided p^+–*n* Si junction at 300 K is doped with $N_A = 10^{19}$cm^{-3}. Design the junction so that $C_j = 0.85$ pF at $V_R = 4.0$ V.

FOR SECTION 4.5 CURRENT-VOLTAGE CHARACTERISTICS

8. Assume that the *p–n* junction considered in Prob. 3 contains 10^{15} cm^{-3} generation-recombination centers located 0.02 eV above the intrinsic Fermi level of silicon with $\sigma_n = \sigma_p = 10^{-15}$ cm^2. If $v_{th} \cong 10^7$ cm/s, calculate the generation and recombination current at –0.5 V.

9. Consider a Si *p–n* junction with *n*-type doping concentration of 10^{16} cm^{-3} and is forward biased with $V = 0.8$ V at 300 K. Calculate the minority-carrier hole concentration at the edge of the space charge region.

10. Calculate the applied reverse-bias voltage at which the ideal reverse current in a *p–n* junction diode at T = 300 K reaches 95% of its reverse saturation current value.

11. Design the Si *p–n* diode such that J_n = 25 A/cm² and J_p = 7 A/cm² at V_a = 0.7 V. The remaining parameters are given in Ex. 5.

12. An ideal silicon *p–n* junction has $N_D = 10^{18}$ cm⁻³, $N_A = 10^{16}$ cm⁻³, $\tau_p = \tau_n = 10^{-6}$ s, and a device area of 1.2×10^{-5} cm². (*a*) Calculate the theoretical saturation current at 300 K. (*b*) Calculate the forward and reverse currents at ±0.7 V.

13. In Prob. 12, assume the widths of the two sides of the junction are much greater than the respective minority-carrier diffusion length. Calculate the applied voltage at a forward current of 1 mA at 300 K.

14. A silicon p^+–n junction has the following parameters at 300 K: $\tau_p = \tau_g = 10^{-6}$ s, $N_D = 10^{15}$ cm⁻³, $N_A = 10^{19}$ cm⁻³. (*a*) Plot diffusion current density, J_{gen}, and total current density versus applied reverse voltage. (*b*) Repeat the above results for $N_D = 10^{17}$ cm⁻³.

FOR SECTION 4.6   CHARGE STORAGE AND TRANSIENT BEHAVIOR

15. For an ideal abrupt silicon p^+–n junction with $N_D = 10^{16}$ cm⁻³, find the stored minority carriers per unit area in the neutral *n*-region when a forward bias of 1 V is applied. The length of neutral region is 1 μm and the diffusion length of the holes is 5 μm.

FOR SECTION 4.7   JUNCTION BREAKDOWN

16. For a silicon p^+–n one-sided abrupt junction with $N_D = 10^{15}$ cm⁻³, find the depletion layer width at breakdown. If the *n*-region is reduced to 5 μm, calculate the breakdown voltage and compare your result with Fig. 29.

17. Design an abrupt Si p^+–n junction diode that has a reverse breakdown voltage of 130 V and has a forward-bias current of 2.2 mA at V_a = 0.7 volt. Assume $\tau_{po} = 10^{-7}$s.

18. In Fig. 20*b*, the avalanche breakdown voltage increases with increasing temperature. Give a qualitative argument for the result.

19. If $\alpha_n = \alpha_p = 10^4 (\mathscr{E}/4 \times 10^5)^6$ cm⁻¹ in gallium arsenide, where $\mathscr{E}$ is in V/cm, find the breakdown voltage of (*a*) a *p-i-n* diode with an intrinsic-layer width of 10μm and (*b*) p^+–n junction with a doping of 2×10^{16} cm⁻³ for the lightly doped side.

20. Consider that a Si *p–n* junction at 300 K with a linearly doping profile varies from $N_A = 10^{18}$ cm⁻³ to $N_D = 10^{18}$cm⁻³ over a distance of 2 μm. Calculate the breakdown voltage.

FOR SECTION 4.8   HETEROJUNCTION

21. For the ideal heterojunction in Ex. 10, find the electrostatic potential and depletion width in each material for applied voltage of 0.5 V and –5 V.

22. For an *n*-type GaAs/ *p*-type $Al_{0.3}Ga_{0.7}As$ heterojunction at room temperature, ΔE_C = 0.21 eV. Find the total depletion width at thermal equilibrium when both sides have impurity concentration of 5×10^{15} cm⁻³. (Hint: the bandgap of $Al_xGa_{1-x}As$ is given by $E_g(x)$ = 1.424 + 1.247x eV, and the dielectric constant is 12.4 – 3.12x. Assume N_C and N_V are the same for $Al_xGa_{1-x}As$ with 0 < x < 0.4.

Bipolar Transistor and Related Devices

The transistor (contraction for *transfer resistor*) is a multijunction semiconductor device. Normally, the transistor is integrated with other circuit elements for voltage gain, current gain, or signal-power gain. The bipolar transistor, also called the bipolar junction transistor (BJT), is one of the most important semiconductor devices. It has been used extensively in high-speed circuits, analog circuits, and power applications. Bipolar devices are semiconductor devices in which both electrons and holes participate in the conduction process. This is in contrast to the field-effect devices, discussed in Chapter 6 and 7, in which predominantly only one kind of carrier participates.

The bipolar transistor was invented by a research team at Bell Laboratories[1] in 1947. The device had two metal wires with sharp points making contact with a germanium substrate (see Fig. 3 in Chapter 1). The first transistor was primitive by today's standards, yet it revolutionized the electronics industry and changed our way of life.

For modern bipolar transistors, we have replaced the germanium with silicon and replaced the point contacts with two closely coupled *p–n* junctions in the form of *p-n-p* or *n-p-n* structures. We consider the transistor action of the coupled junctions and derive the static characteristics from the minority carrier distributions in the regions. We also discuss the frequency response and switching behavior of the transistor. In addition, we briefly consider the heterojunction bipolar transistor in which one or both *p–n* junctions are formed between dissimilar semiconductors.

In the final section, a related bipolar device called a thyristor is introduced. The basic thyristor has three closely coupled *p–n* junctions in the form of a *p-n-p-n* structure.[2] The device exhibits bistable characteristics and can be switched between a high-impedance "off" state and a low-impedance "on" state. The name thyristor is derived from *gas thyratron*, which is a gas-filled tube with similar bistable characteristics. Because of the two stable state (on and off) and the low power dissipation in these states, thyristors are useful in many applications. We consider the physical operation of the thyristor and a few related switching devices. Furthermore, the various thyristor types and their applications are briefly introduced.

Specifically, we cover the following topics:

- The current gain and modes of operation of bipolar transistors.
- The cutoff frequency and switching time of a bipolar transistor.
- The advantages of heterojunction bipolar transistor.
- The power handling capability of thyristor and related bipolar devices.

5.1 THE TRANSISTOR ACTION

A perspective view of a discrete *p-n-p* bipolar transistor is shown Fig. 1. The transistor is formed by starting with a *p*-type substrate. An *n*-type region is thermally diffused through an oxide window into the *p*-type substrate. A very heavily doped *p*⁺ region is then diffused into the *n*-type region. Metallic contacts are made to the *p*⁺- and *n*-regions through the windows opened in the oxide layer and to the *p*-region at the bottom. The details of transistor fabrication processes are considered in later chapters.

An idealized, one-dimensional structure of a *p-n-p* bipolar transistor is shown in Fig. 2*a*. Normally, the bipolar transistor has three separately doped regions and two *p–n* junctions. The heavily doped *p*⁺-region is called the *emitter* (defined as symbol *E* in the figure). The narrow central *n*-region, with moderately doped concentration, is called the *base* (symbol *B*). The width of the base is small compared with the minority-carrier diffusion length. The lightly doped *p*-region is called the *collector* (symbol *C*). The doping concentration in each region is assumed to be uniform. It should be noted that the concepts developed for the *p–n* junction can be applied directly to the transistor.

Figure 2*b* illustrates the circuit symbol for a *p-n-p* transistor. The current components and voltage polarities are shown in the figure. The arrows of the various currents indicate the direction of current flow under normal operating conditions (also called the *active mode*). The + and – signs are used to define the voltage polarities. We can also denote the voltage polarity by a double subscript on the voltage symbol. In the active mode, the emitter-base junction is forward biased ($V_{EB} > 0$) and the base-collector junction

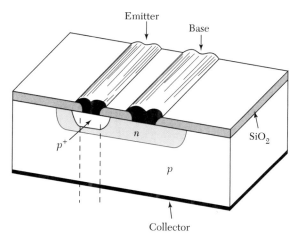

Fig. 1   Perspective view of a silicon *p-n-p* bipolar transistor.

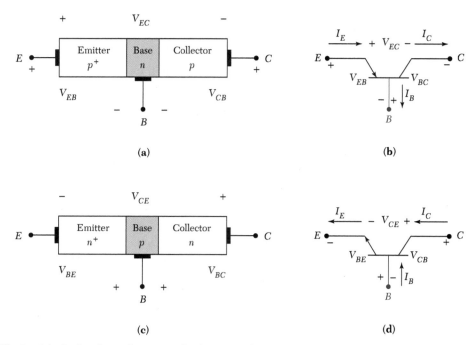

Fig. 2 (*a*) Idealized one-dimensional schematic of a *p-n-p* bipolar transistor and (*b*) its circuit symbol. (*c*) Idealized one-dimensional schematic of an *n-p-n* bipolar transistor and (*d*) its circuit symbol.

is reverse biased ($V_{CB} > 0$). According to Kirchhoff's circuit laws, there are only two independent currents for this three-terminal device. If two currents are known, the third current can be obtained.

The *n-p-n* bipolar transistor is the complementary structure of the *p-n-p* bipolar transistor. The structure and circuit symbol of an ideal *n-p-n* transistor are shown in Figs. 2*c* and 2*d*, respectively. The *n-p-n* structure can be obtained by interchanging *p* for *n* and *n* for *p* from the *p-n-p* transistor. As a result the current flow and voltage polarity are all reversed. In subsequent sections, we concentrate on the *p-n-p* type because the direction of minority-carrier (hole) flow is the same as current flow. It provides a more intuitive base for understanding the mechanisms of charge transport. Once we understand the *p-n-p* transistor, we need only to reverse the polarities and conduction types to describe the *n-p-n* transistor.

5.1.1 Operation in the Active Mode

Figure 3*a* show an idealized *p-n-p* transistor in thermal equilibrium, that is, where all three leads are connected together or all are grounded. The depletion regions near the two junctions are illustrated by colored areas. Figure 3*b* shows the impurity densities in the three doped regions, where the emitter is more heavily doped than the collector. However, the base doping is less than the emitter doping, but greater than the collector doping. Figure 3*c* shows the corresponding electric-field profiles in the two depletion regions.

Figure 3*d* illustrates the energy band diagram, which is a simple extension of the thermal-equilibrium situation for the *p–n* junction as applied to a pair of closely coupled

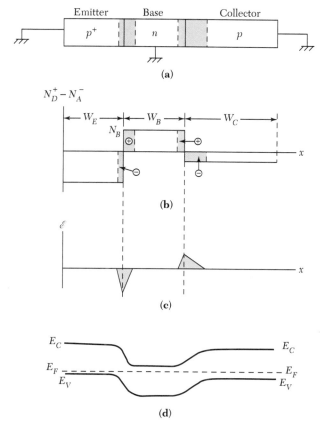

Fig. 3 (*a*) A *p-n-p* transistor with all leads grounded (at thermal equilibrium). (*b*) Doping profile of a transistor with abrupt impurity distributions. (*c*) Electric-field profile. (*d*) Energy band diagram at thermal equilibrium.

p^+-n and n-p junctions. The results obtained for the $p–n$ junction in Chapter 4 are equally applicable to the emitter-base and base-collector junctions. At thermal equilibrium there is no net current flow, hence the Fermi level is a constant in the regions.

Figure 4 shows the corresponding cases when the transistor in Fig. 3 is biased in the active mode. Figure 4*a* is a schematic of the transistor connected as an amplifier with the *common-base configuration*, that is, the base lead is common to the input and output circuits.[3] Figures 4*b* and 4*c* show the charge densities and the electric fields, respectively, under biasing conditions. Note that the depletion layer width of the emitter-base junction is narrower and the collector-base junction is wider, compared with the equilibrium case shown in Fig. 3.

Figure 4*d* shows the corresponding energy band diagram under the active mode. Since the emitter-base junction is forward biased, holes are injected (or emitted) from the p^+ emitter into the base and electrons are injected from the n base into the emitter. Under the ideal-diode condition, there is no generation-recombination current in the depletion region; these two current components constitute the total emitter current. The collector-base junction is reverse biased, and a small reverse saturation current will flow across the junction. However, if the base width is sufficiently narrow, the holes injected from the emitter can diffuse through the base to reach the base-collector depletion edge

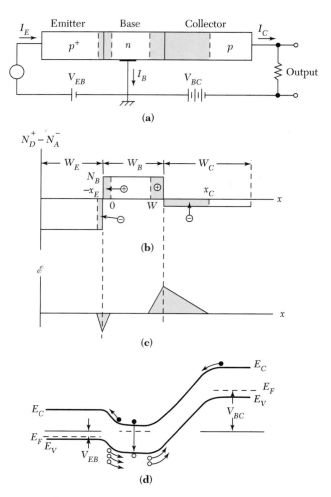

Fig. 4 (*a*) The transistor shown in Fig. 3 under the active mode of operation.[3] (*b*) Doping profiles and the depletion regions under biasing conditions. (*c*) Electric-field profile. (*d*) Energy band diagram.

and then "float up" into the collector (recall the "bubble analogy"). This transport mechanism gives rise to the terminology of *emitter*, which emits or injects carriers, and of *collector*, which collects these carriers injected from a nearby junction. If most of the injected holes can reach the collector without recombining with electrons in the base region, then the collector hole current will be very close to the emitter hole current.

Therefore, carriers injected from a nearby emitter junction can result in a large current flow in a reverse-biased collector junction. This is the *transistor action*, and it can be realized only when the two junctions are physically close enough to interact in the manner described. Therefore, the two junctions are called the *interacting p–n junctions*. If, on the other hand, the two junctions are so far apart that all the injected holes are recombined in the base before reaching the base-collector junction, then the transistor action is lost and the *p-n-p* structure becomes merely two diodes connected back to back.

5.1.2 Current Gain

Figure 5 shows the various current components in an ideal p-n-p transistor biased in the active mode. Note that we assume that there are no generation-recombination currents in the depletion regions. The holes injected from the emitter constitute the current I_{Ep}, which is the largest current component in a well-designed transistor. Most of the injected holes will reach the collector junction and give rise to the current I_{Cp}. There are three base current components, labeled I_{BB}, I_{En}, and I_{Cn}. I_{BB} corresponds to electrons that must be supplied by the base to replace electrons recombined with the injected holes (i.e., I_{BB} = $I_{Ep} - I_{Cp}$). I_{En} corresponds to the current arising from electrons being injected from the base to the emitter. However, I_{En} is not desirable, as is shown later. It can be minimized by using heavier emitter doping (Section 5.2) or a heterojunction (Section 5.4). I_{Cn} corresponds to thermally generated electrons that are near the base-collector junction edge and drift from the collector to the base. As indicated in the figure, the direction of the electron current is opposite the direction of the electron flow.

We can now express the terminal currents in terms of the various current components described above:

$$I_E = I_{Ep} + I_{En}, \tag{1}$$

$$I_C = I_{Cp} + I_{Cn}, \tag{2}$$

$$I_B = I_E - I_C = I_{En} + (I_{Ep} - I_{Cp}) - I_{Cn}. \tag{3}$$

An important parameter in the characterization of bipolar transistors is the *common-base current gain* α_0. This quantity is defined by

$$\alpha_0 \equiv \frac{I_{Cp}}{I_E}. \tag{4}$$

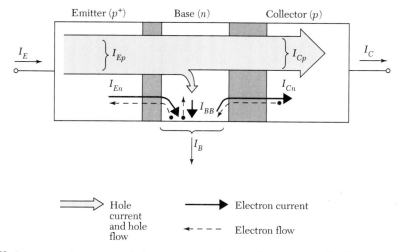

Fig. 5 Various current components in a p-n-p transistor under active mode of operation. The electron flow is in the opposite direction to the electron current.

Substituting Eq. 1 into Eq. 4 yields

$$\alpha_0 = \frac{I_{Cp}}{I_{Ep} + I_{En}} = \left(\frac{I_{Ep}}{I_{Ep} + I_{En}}\right)\left(\frac{I_{Cp}}{I_{Ep}}\right). \tag{5}$$

The first term on the right-hand side is called the *emitter efficiency γ*, which is a measure of the injected hole current compared with the total emitter current:

$$\gamma \equiv \frac{I_{Ep}}{I_E} = \frac{I_{Ep}}{I_{Ep} + I_{En}}. \tag{6}$$

The second term is called the *base transport factor α_T*, which is the ratio of the hole current reaching the collector to the hole current injected from the emitter:

$$\alpha_T \equiv \frac{I_{Cp}}{I_{Ep}}. \tag{7}$$

Therefore, Eq. 5 becomes

$$\alpha_0 = \gamma\alpha_T \tag{8}$$

For a well-designed transistor, because I_{En} is small compared with I_{Ep} and I_{Cp} is close to I_{Ep}, both γ and α_T approach unity. Therefore, α_0 is close to 1.

We can express the collector current in terms of α_0. The collector current can be described by substituting Eqs. 6 and 7 into Eq. 2:

$$I_C = I_{Cp} + I_{Cn} = \alpha_T I_{Ep} + I_{Cn} = \gamma\alpha_T\left(\frac{I_{Ep}}{\gamma}\right) + I_{Cn} = \alpha_0 I_E + I_{Cn}, \tag{9}$$

where I_{Cn} corresponds to the collector-base current flowing with the emitter open-circuited ($I_E = 0$). We designate I_{Cn} as I_{CBO}, where the first two subscripts (CB) refer to the two terminals between which the current (or voltage) is measured and the third subscript (O) refers to the state of the third terminal with respect to the second. In the present case, I_{CBO} designates the leakage current between the collector and the base with the emitter-base junction open. The collector current for the common-base configuration is then given by

$$I_C = \alpha_0 I_E + I_{CBO}. \tag{10}$$

▶ **EXAMPLE 1**

For an ideal *p-n-p* transistor, the current components are given by $I_{Ep} = 3$ mA , $I_{En} = 0.01$ mA, $I_{Cp} = 2.99$ mA, and $I_{Cn} = 0.001$ mA. Determine (a) the emitter efficiency γ, (b) the base transport factor α_T, (c) the common-base current gain α_0, and (d) I_{CBO}.

SOLUTION

(a) Using Eq. 6, the emitter efficiency is

$$\gamma = \frac{I_{Ep}}{I_{Ep} + I_{En}} = \frac{3}{3 + 0.01} = 0.9967.$$

(b) The base transport factor can be obtained from Eq. 7:

$$\alpha_T = \frac{I_{Cp}}{I_{Ep}} = \frac{2.99}{3} = 0.9967.$$

(c) The common-base current gain is given by Eq. 8:

$$\alpha_0 = \gamma\alpha_T = 0.9967 \times 0.9967 = 0.9934.$$

(d) $I_E = I_{Ep} + I_{En} = 3 + 0.01 = 3.01$ mA

$I_C = I_{Cp} + I_{Cn} = 2.99 + 0.001 = 2.991$

Using Eq. 10, we find

$$I_{CBO} = I_C - \alpha_0 I_E = 2.991 - 0.9934 \times 3.01 = 0.87 \ \mu\text{A}. \qquad ◄$$

► 5.2 STATIC CHARACTERISTICS OF BIPOLAR TRANSISTOR

In this section, we study the static current-voltage characteristics for an ideal transistor and derive equations for the terminal currents. The current equations are based on the minority-carrier concentration in each region and therefore are described by semiconductor parameters such as doping and minority-carrier lifetime.

5.2.1 Carrier Distribution in Each Region

To derive the current-voltage expression for an ideal transistor, we assume the following:

1. The device has uniform doping in each region.
2. The hole drift current in the base region as well as the collector saturation current is negligible.
3. There is low-level injection.
4. There are no generation-combination currents in the depletion regions.
5. There are no series resistances in the device.

Basically, we assume that holes are injected from the emitter into the base under forward-biased condition. These holes then diffuse across the base region and reach the collector junction. Once we determine the minority-carrier distribution (i.e., holes in the n-type base region), we can obtain the current from the minority-carrier gradient.

Base Region
Figure 4c shows the electric-field distributions across the junction depletion regions. The minority-carrier distribution in the neutral base region can be described by the field-free, steady-state continuity equation

$$D_p\left(\frac{d^2 p_n}{dx^2}\right) - \frac{p_n - p_{no}}{\tau_p} = 0, \qquad (11)$$

where D_p and τ_p are the diffusion constant and the lifetime of minority carriers, respectively. The general solution of Eq. 11 is

$$p_n(x) = p_n + C_1 e^{x/L_p} + C_2 e^{-x/L_p}, \tag{12}$$

where $L_p = \sqrt{D_p \tau_p}$ is the diffusion length of holes. The constants C_1 and C_2 can be determined by the boundary conditions for the active mode:

$$p_n(0) = p_{no} e^{qV_{EB}/kT} \tag{13a}$$

and

$$p_n(W) = 0, \tag{13b}$$

where p_{no} is the equilibrium minority-carrier concentration in the base, given by $p_{no} = n_i^2/N_B$, and N_B denotes the uniform donor concentration in the base. The first boundary condition (Eq. 13a) states that under forward bias, the minority-carrier concentration at the edge of the emitter-base depletion region ($x = 0$) is increased above the equilibrium value by the exponential factor $e^{qV_{EB}/kT}$. The second boundary condition (Eq. 13b) states that under reverse bias, the minority carrier concentration at the edge of the base-collector depletion region ($x = W$) is zero.

Substituting Eq. 13 into the general solution expressed in Eq. 12 yields

$$p_n(x) = p_{no} (e^{qV_{EB}/kT} - 1) \left[\frac{\sinh\left(\dfrac{W-x}{L_p}\right)}{\sinh\left(\dfrac{W}{L_p}\right)} \right] + p_{no} \left[1 - \frac{\sinh\left(\dfrac{x}{L_p}\right)}{\sinh\left(\dfrac{W}{L_p}\right)} \right]. \tag{14}$$

The sinh function, $\sinh(\Lambda)$, can be approximately expressed by Λ when $\Lambda \ll 1$. For example, when $\Lambda < 0.3$, the difference between $\sinh(\Lambda)$ and Λ is less than 1.5 percent. Therefore, when $W/L_p \ll 1$, the distribution equation can be simplified as

$$p_n(x) = p_{no} e^{qV_{EB}/kT} \left(1 - \frac{x}{W} \right) = p_n(0) \left(1 - \frac{x}{W} \right). \tag{15}$$

The distribution approaches a straight line. The approximation is reasonable because the width of the base region is designed to be much smaller than the diffusion length of the minority carrier. Figure 6 shows a linear minority-carrier distribution in a typical transistor operated under active mode. Note that assuming linear minority-carrier distribution can simplify the derivation of current-voltage characteristics. Therefore, we use the assumption hereafter to derive equations for the current-voltage characteristics.

Emitter and Collector Regions

The minority-carrier distributions in the emitter and collector can be obtained in a manner similar to the one used to obtain the distributions for the base region. In Fig. 6, the boundary conditions in the neutral emitter and collector regions are

$$n_E(x = -x_E) = n_{EO} e^{qV_{EB}/kT} \tag{16}$$

and

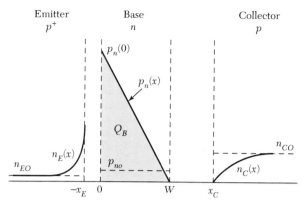

Fig. 6 Minority carrier distribution in various regions of a *p-n-p* transistor under the active mode of operation.

$$n_C(x = x_C) = n_{CO}\, e^{-q|V_{CB}|} = 0, \qquad (17)$$

where n_{EO} and n_{CO} are the equilibrium electron concentrations in the emitter and collector, respectively. We assume that the emitter depth and the collector depth are much larger than their corresponding diffusion lengths L_E and L_C, respectively. Substituting these boundary conditions into expressions similar to Eq. 12 yields

$$n_E(x) = n_{EO} + n_{EO}\left(e^{qV_{EB}/kT} - 1\right) e^{\frac{x+x_E}{L_E}} \qquad x \le -x_E, \qquad (18)$$

$$n_C(x) = n_{CO} - n_{CO}\, e^{-\frac{x-x_C}{L_C}} \qquad x \ge x_C . \qquad (19)$$

5.2.2 Ideal Transistor Currents for Active Mode Operation

Once the minority-carrier distributions are known, the various current components shown in Fig. 6 can be calculated. The hole current I_{Ep}, injected from the emitter at $x = 0$, is proportional to the gradient of the minority carrier concentration. For $W/L_p \ll 1$, the hole current I_{Ep} can be expressed by using Eq. 15:

$$I_{Ep} = A\left(-qD_p \frac{dp_n}{dx}\bigg|_{x=0}\right) \cong \frac{qAD_p\, p_{no}}{W}\, e^{qV_{EB}/kT} . \qquad (20)$$

Similarly, the hole current collected by the collector at $x = W$ is

$$I_{Cp} = A\left(-qD_p \frac{dp_n}{dx}\bigg|_{x=W}\right)$$

$$\cong \frac{qAD_p\, p_{no}}{W}\, e^{qV_{EB}/kT} . \qquad (21)$$

Note that I_{Ep} is equal to I_{Cp} when $W/L_p \ll 1$. The electron current I_{En}, which is due to electron flow from the base to the emitter, and I_{Cn}, which is due to electron flow from the collector to the base, are

$$I_{En} = A\left(-qD_E\frac{dn_E}{dx}\Big|_{x=-x_E}\right) = \frac{qAD_E n_{EO}}{L_E}\left(e^{qV_{EB}/kT} - 1\right), \tag{22}$$

$$I_{Cn} = A\left(-qD_C\frac{dn_C}{dx}\Big|_{x=x_C}\right) = \frac{qAD_C n_{CO}}{L_C}, \tag{23}$$

where D_E and D_C are the diffusion constants in the emitter and collector, respectively.

The terminal currents can now be obtained from these equations. The emitter current is the sum of Eqs. 20 and 22:

$$I_E = a_{11}\left(e^{qV_{EB}/kT} - 1\right) + a_{12} \tag{24}$$

where

$$a_{11} \equiv qA\left(\frac{D_p p_{no}}{W} + \frac{D_E n_{EO}}{L_E}\right), \tag{25}$$

$$a_{12} \equiv \frac{qAD_p p_{no}}{W} . \tag{26}$$

The collector current is the sum of Eqs. 21 and 23:

$$I_C = a_{21}\left(e^{qV_{EB}/kT} - 1\right) + a_{22}, \tag{27}$$

where

$$a_{21} \equiv \frac{qAD_p p_{no}}{W}, \tag{28}$$

$$a_{22} \equiv qA\left(\frac{D_p p_{no}}{W} + \frac{D_C n_{CO}}{L_C}\right). \tag{29}$$

Note that $a_{12} = a_{21}$. The base current for the ideal transistor is the difference between the emitter current (I_E) and the collector current (I_C). Therefore, the base current can be obtained by subtracting Eq. 27 from Eq. 24:

$$I_B = \left(a_{11} - a_{21}\right)\left(e^{qV_{EB}/kT} - 1\right) + \left(a_{12} - a_{22}\right). \tag{30}$$

From these discussions, we see that the currents in the three terminals of a transistor are mainly determined by the minority carrier distribution in the base region. Once we derive the current components, the common-base current gain α_0 can be obtained by using Eqs. 6 through 8.

▶ **EXAMPLE 2**

An ideal p^+-n-p transistor has impurity concentrations of 10^{19}, 10^{17}, and 5×10^{15} cm^{-3} in the emitter, base, and collector regions, respectively; the corresponding lifetimes are 10^{-8}, 10^{-7}, and 10^{-6} s. Assume that an effective cross section area A is 0.05 mm^2 and the emitter-base junction is forward-biased to 0.6 V. Find the common-base current gain of the transistor. Note that the other device parameters are $D_E = 1$ cm^2/s, $D_p = 10$ cm^2/s, $D_C = 2$ cm^2/s, and $W = 0.5$ μm.

SOLUTION In the base region,

$$L_p = \sqrt{D_p \tau_p} = \sqrt{10 \cdot 10^{-7}} = 10^{-3} \text{ cm},$$

$$p_{no} = n_i^2 / N_B = \left(9.65 \times 10^9\right)^2 / 10^{17} = 9.31 \times 10^2 \text{ cm}^{-3}.$$

Similarly, in the emitter region, $L_E = \sqrt{D_E \tau_E} = 10^{-4}$ cm and $n_{EO} = n_i^2 / N_E = 9.31$ cm^{-3}. Since $W/L_p = 0.05 \ll 1$, the current components are given by

$$I_{Ep} = \frac{1.6 \times 10^{-19} \times 5 \times 10^{-4} \times 10 \cdot 9.31 \times 10^2}{0.5 \times 10^{-4}} \times e^{0.6/0.0259} \times 10^{-4} \text{A} = 1.7137 \times 10^{-4} \text{A},$$

$$I_{Cp} = 1.7137 \times 10^{-4} \text{ A},$$

$$I_{En} = \frac{1.6 \times 10^{-19} \times 5 \times 10^{-4} \times 1 \times 9.31}{10^{-4}} \left(e^{0.6/0.0259} - 1\right) = 8.5687 \times 10^{-8} \text{A}.$$

Therefore, the common-base current gain α_0 is

$$\alpha_0 = \frac{I_{Cp}}{I_{Ep} + I_{En}} = \frac{1.7137 \times 10^{-4}}{1.7137 \times 10^{-4} + 8.5687 \times 10^{-8}} = 0.9995.$$

◀

For the case of $W/L_p \ll 1$, we can simplify the emitter efficiency from Eqs. 20 and 22:

$$\gamma \equiv \frac{I_{Ep}}{I_{Ep} + I_{En}} \cong \frac{\dfrac{D_p p_{no}}{W}}{\dfrac{D_p p_{no}}{W} + \dfrac{D_E n_{EO}}{L_E}} = \frac{1}{1 + \dfrac{D_E}{D_P} \dfrac{n_{EO}}{p_{no}} \dfrac{W}{L_E}} \qquad (31)$$

or

$$\gamma = \frac{1}{1 + \dfrac{D_E}{D_p} \cdot \dfrac{N_B}{N_E} \cdot \dfrac{W}{L_E}}, \qquad (31a)$$

where $N_B(=n_i^2/p_{no})$ is the impurity doping in the base and $N_E(=n_i^2/n_{EO})$ is the impurity doping in the emitter. This equation shows that to improve γ, we should decrease the ratio N_B/N_E, that is, there should be much heavier doping in the emitter than in the base. This is the reason why we use p^+-doping in the emitter.

5.2.3 Modes of Operation

A bipolar transistor has four modes of operation, depending on the voltage polarities on the emitter-base junction and the collector-base junction. Figure 7 shows the V_{EB} and V_{CB} voltages for the four modes of operations of a p-n-p transistor. The corresponding minority carrier distributions are also shown. So far in this chapter we have considered the *active mode* of transistor operation. In the active mode, the emitter-base junction is forward biased and the base-collector junction is reverse biased.

In the *saturation mode*, both junctions are forward biased, leading to the nonzero minority-carrier distribution at the edge of each depletion region. Therefore, the boundary condition at $x = W$ becomes $p_n(W) = p_{no}e^{qV_{CB}/kT}$ instead of the one given by Eq. 13b. The saturation mode corresponds to small biasing voltage and large output current, that is, the transistor is in a conducting state and acts as a closed (or on) switch.

In the *cutoff mode*, both junctions are reverse biased. The boundary conditions of Eq. 13 become $p_n(0) = p_n(W) = 0$. The cutoff mode corresponds to the open (or off) state of the transistor as a switch.

The fourth mode of operation is the *inverted mode*, which is sometimes called the inverted active mode. In this mode, the emitter-base junction is reverse biased and the collector-base junction is forward biased. The inverted mode corresponds to the case where the collector acts like the emitter and the emitter acts like a collector. In such condition, the device is used backward. However, the current gain for the inverted mode is generally lower than that for the active mode. It is because of poor "emitter efficiency" resulting from low collector doping with respect to the base doping (Eq. 31).

The current-voltage relationships for the various modes of operation can be obtained by following the same procedures used for the active mode, with an appropriate change

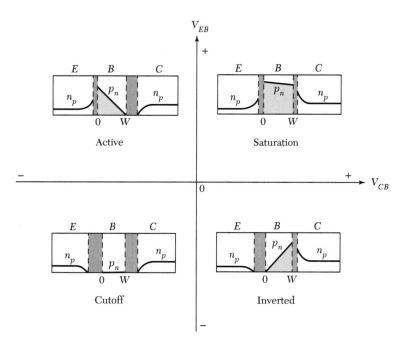

Fig. 7 Junction polarities and minority carrier distributions of a p-n-p transistor under four modes of operation.

in the boundary conditions as in Eq. 13. The general expressions applicable to all modes of operations are

$$I_E = a_{11} \left(e^{qV_{EB}/kT} - 1 \right) - a_{12} \left(e^{qV_{CB}/kT} - 1 \right) \qquad (32a)$$

and

$$I_C = a_{21} \left(e^{qV_{EB}/kT} - 1 \right) - a_{22} \left(e^{qV_{CB}/kT} - 1 \right) \qquad (32b)$$

where the coefficients a_{11}, a_{12}, a_{21}, and a_{22} are given by Eqs. 25, 26, 28, and 29, respectively. Note that in Eqs. 32a and 32b the biasing voltages for the junctions can be positive or negative depending on the mode of operation.

5.2.4 Current-Voltage Characteristics of Common-Base and Common-Emitter Configurations

Using Eq. 32, we can obtain the current-voltage characteristics for a transitor in common-base configuration. We note that in this configuration, V_{EB} and V_{BC} are the input and output voltages, and I_E and I_C are the input and output currents, respectively.

However, in circuit applications the common-emitter configuration is most often used, where the emitter lead is common to the input and output circuits. The general expressions of the currents, shown in Eq. 32, are also applicable to the common-emitter configuration. In this case, to generate the current-voltage characteristics, V_{EB} and I_B are the input parameters and V_{EC} and I_C are the output parameters.

Common-Base Configuration

Figure 8a shows the common-base configuration of a p-n-p transistor. Figure 8b shows the measured results of output current-voltage characteristics for the common-base configuration. The various modes of operation are indicated on the figure. Note that the collector current is practically equal to emitter current (i.e., $\alpha_0 \cong 1$) and virtually independent of V_{BC}. This is in close agreement with the ideal transistor behavior given by Eqs. 10 and 27. The collector current remains practically constant, even down to zero volts for V_{BC}, where the holes are still extracted by the collector. This is indicated by the hole distributions

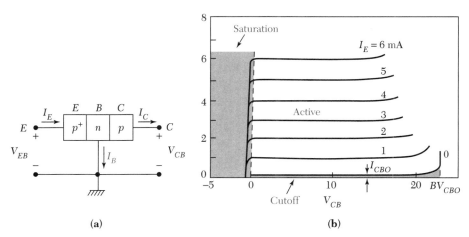

(a)   (b)

Fig. 8 (a) Common-base configuration of a p-n-p transistor. (b) Its output current-voltage characteristics.

shown in Fig. 9a. Since the hole gradient at $x = W$ changes only slightly from $V_{BC} > 0$ to $V_{BC} = 0$, the collector current remains essentially the same over the entire active mode of operation. To reduce the collector current to zero, we have to apply a small forward bias, about 1 V for silicon, to the base-collector junction (in the saturation mode), as shown in Fig. 9b. The forward bias will sufficiently increase the hole density at $x = W$ to make it equal to that of the emitter at $x = 0$ (see the horizontal line in Fig. 9b). Therefore, the hole gradient at $x = W$ as well as the collector current will be reduced to zero.

Common-Emitter Configuration

Figure 10a shows the common-emitter configuration for a p-n-p transistor. The collector current for the common-emitter configuration can be obtained by substituting Eq. 3 into Eq. 10:

$$I_C = \alpha_0 \left(I_B + I_C\right) + I_{CBO} \ . \tag{33}$$

Solving for I_C, we obtain

$$I_C = \frac{\alpha_0}{1 - \alpha_0} I_B + \frac{I_{CBO}}{1 - \alpha_0} \ . \tag{34}$$

We now designate β_0 as the *common-emitter current gain*, which is the incremental change of I_C with respect to an incremental change of I_B. From Eq. 34, we obtain

$$\beta_0 \equiv \frac{\Delta I_C}{\Delta I_B} = \frac{\alpha_0}{1 - \alpha_0} \ . \tag{35}$$

We can also designate I_{CEO} as

$$I_{CEO} \equiv \frac{I_{CBO}}{1 - \alpha_0} \ . \tag{36}$$

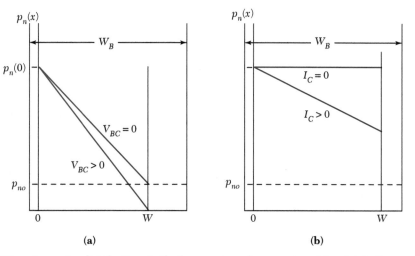

(a)   **(b)**

Fig. 9 Minority carrier distributions in the base region of a p-n-p transistor. (a) Active mode for $V_{BC} = 0$ and $V_{BC} > 0$. (b) Saturation mode with both junctions forward biased.

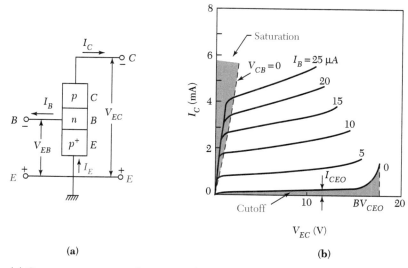

(a)   (b)

Fig. 10   (*a*) Common-emitter configuration of a *p-n-p* transistor. (*b*) Its output current-voltage characteristics.

This current corresponds to the collector-emitter leakage current for $I_B = 0$. Equation 34 becomes

$$I_C = \beta_0 I_B + I_{CEO} \, .$$

(37)

Because the value of α_0 is generally close to unity, β_0 is much larger than 1. For example, if $\alpha_0 = 0.99$, β_0 is 99; and if α_0 is 0.998, β_0 is 499. Therefore, a small change in the base current can give rise to a much larger change in the collector current. Figure 10*b* shows the measured results of output current-voltage characteristics with various input base currents. Note that the figure shows nonzero collector-emitter leakage current I_{CEO} when $I_B = 0$.

▷ **EXAMPLE 3**

Refering to Ex. 1, find the common-emitter current gain β_0. Express I_{CEO} in terms of β_0 and I_{CBO} and find the value of I_{CEO}.

SOLUTION   The common-base current gain α_0 in Example 1 is 0.9934. Hence, we can obtain β_0 by

$$\beta_0 = \frac{0.9934}{1 - 0.9934} = 150.5.$$

Equation 36 can be expressed by

$$I_{CEO} = \left(\frac{\alpha_0}{1 - \alpha_0} + 1 \right) I_{CBO}$$

$$= \left(\beta_0 + 1 \right) I_{CBO} \, .$$

Therefore, $I_{CEO} = \left(150.5 + 1 \right) \times 0.87 \times 10^{-6} = 1.32 \times 10^{-4}$ A.   ◀

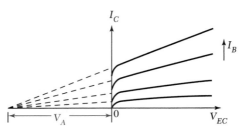

Fig. 11 Schematic diagram of the Early effect and Early voltage V_A. The collector currents for different base currents meet at $-V_A$.

In an ideal transistor with the common-emitter configuration, the collector current for a given I_B is expected to be independent of V_{EC} for $V_{EC} > 0$. This is true when we assume that the neutral base width (W) is constant. However, since the width of the space charge region extending into the base region varies with the base-collector voltage, the base width is a function of the base-collector voltage. The collector current, therefore, is dependent on V_{EC}. As the base-collector reverse-bias voltage increases, the base width will be reduced. The reduced base width causes the gradient in the minority-carrier concentration to increase, which causes an increase in the diffusion current. As a result, β_0 will be increased. Figure 11 shows pronounced slopes, and I_C increasing with increasing V_{EC}. This deviation is known as the *Early effect*[4] or the *base width modulation*. By extrapolating the collector currents and intersecting the V_{EC} axis, we can obtain the voltage V_A, which is called the *Early voltage*.

▷ 5.3 FREQUENCY RESPONSE AND SWITCHING OF BIPOLAR TRANSISTOR

In Section 5.2 we discussed four possible modes of operation that depend on the biasing conditions of the emitter-base and collector-base junctions. Generally, in analog or linear circuits the transistors are operated in the active mode only. However, in digital circuits all four modes of operation may be involved. In this section we consider the frequency response and switching characteristics of bipolar transistors.

5.3.1 Frequency Response

High-Frequency Equivalent Circuit
In previous discussions, we were concerned with the static [or direct current (dc)] characteristics of the bipolar transistor. We now study its alternating current (ac) characteristics when a small-signal voltage or current is superimposed upon the dc values. The term small-signal means that the peak values of the ac signal current and voltage are smaller than the dc values. Consider an amplifying circuit shown in Fig. 12a, where the transistor is connected in a common-emitter configuration. For a given dc input voltage V_{EB}, a dc base current I_B and dc collector current I_C flow in the transistor. These currents correspond to the operating point shown in Fig. 12b. The load line, determined by the applied voltage V_{CC} and the load resistance R_L, intercepts the V_{EC} axis at V_{CC} and has a slope of $(-1/R_L)$. When a small ac signal is superimposed on the input voltage, the base current i_B will vary as a function of time, as illustrated in Fig. 12b. This variation, in turn, brings about a corresponding variation in the output current i_C, which is β_0 times larger than the input current variation. Thus, the transistor amplifies the input signal.

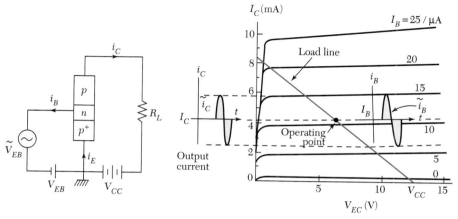

Fig. 12 (*a*) Bipolar transistor connected in the common-emitter configuration. (*b*) Small-signal operation of the transistor circuit.

The equivalent circuit for this low-frequency amplifier is shown in Fig. 13*a*. At higher frequencies, we extend the equivalent circuit by adding the appropriate capacitances. Since the emitter-base junction is forward biased, we expect to have a depletion capacitance C_{EB} and a diffusion capacitance C_d similar to that of a forward-biased *p–n* junction. For the reverse-biased collector-base junction, we expect to have only a depletion capacitance

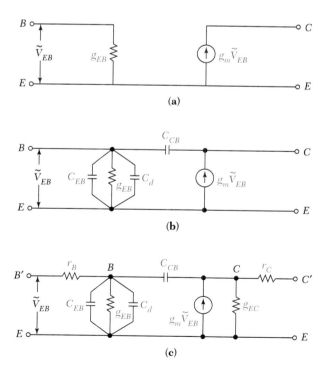

Fig. 13 (*a*) Basic transistor equivalent circuit. (*b*) Basic circuit with the addition of depletion and diffusion capacitances. (*c*) Basic circuit with the addition of resistance and conductance.

C_{CB}. The high-frequency equivalent circuit with the three added capacitances is shown in Fig. 13b. Note that g_m ($\equiv \tilde{i}_C/\tilde{v}_{EB}$) is called the *transconductance* and g_{EB} ($\equiv \tilde{i}_B/\tilde{v}_{EB}$) is called the *input conductance*. To account for the base width modulation effect, there is a finite output conductance $g_{EC} \equiv \tilde{i}_C/\tilde{v}$. In addition, we have a base resistance r_B and a collector resistance r_C. Figure 13c represents the high-frequency equivalent circuit incorporating all of the above elements.

Cutoff Frequency

In Fig. 13c, the transconductance g_m and the input conductance g_{EC} are dependent on the common-base current gain. At low frequencies, the current gain is a constant, independent of the operating frequency. However, the current gain will decrease after a certain critical frequency is reached. A typical plot of the current gain versus operating frequency is shown in Fig. 14. The common-base current gain α can be described as

$$\alpha = \frac{\alpha_0}{1 + j\left(f/f_\alpha\right)} \tag{38}$$

where α_0 is the low-frequency (or dc) common-base current gain and f_α is the *common-base cutoff frequency*. At $f = f_\alpha$ the magnitude of α is 0.707 α_0 (3 dB down).

Figure 14 also shows the common-emitter current gain β. From Eq. 38 we have

$$\beta \equiv \frac{\alpha}{1 - \alpha} = \frac{\beta_0}{1 + j\left(f/f_\beta\right)}, \tag{39}$$

where the f_β is the *common-emitter cutoff frequency* and is given by

$$f_\beta = \left(1 - \alpha_0\right)f_\alpha \tag{40}$$

Since $\alpha_0 \approx 1$, f_β is much smaller than f_α. Another cutoff frequency is f_T when $|\beta|$ becomes unity. By setting the magnitude of the right-hand side of Eq. 39 equal to 1, we obtain

$$f_T = \sqrt{\beta_0^2 - 1}f_\beta \cong \beta_0\left(1 - \alpha_0\right)f_\alpha \cong \alpha_0 f_\alpha \tag{41}$$

Hence, f_T is very close to but is smaller than f_α.

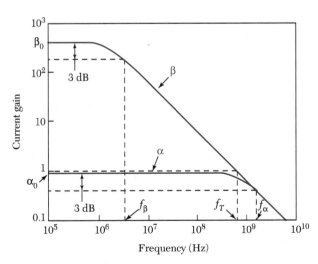

Fig. 14 Current gain as a function of operating frequency.

The cutoff frequency f_T can also be expressed as $(2\pi\tau_T)^{-1}$ where τ_T is the total time of the carrier transit from the emitter to the collector. τ_T includes the emitter delay time τ_E, the base transit time τ_B, and the collector transit time τ_C. The most important delay time is τ_B. The distance traveled by the minority carriers in the base in a time interval dt is $dx = v(x)\,dt$, where $v(x)$ is the effective minority-carrier velocity in the base. This velocity is related to the current as

$$I_p = qv(x)p(x)A, \tag{42}$$

where A is the device area and $p(x)$ is the distribution of the minority carriers. The transit time τ_B required for a hole to traverse the base is given by

$$\tau_B = \int_0^W \frac{dx}{v(x)} = \int_0^W \frac{q\,p(x)\,A}{I_p}\,dx \tag{43}$$

For a straight-line hole distribution, as given by Eq. 15, the integration of Eq. 43 using Eq. 21 for I_p leads to

$$\boxed{\tau_B = \frac{W^2}{2D_p}.} \tag{44}$$

To improve the frequency response, the transit time of minority carriers across the base must be short. Therefore, high-frequency transistors are designed with a small base width. Because the electron diffusion constant in silicon is about three times larger than that of holes, all high-frequency silicon transistors are of the *n-p-n* type (i,e., the minority carrier in the base is electron). Another way to reduce the base transit time is to use a graded base with a built-in field. For a large doping variation (i.e., high base doping near the emitter and low base doping near the collector), the built-in field in the base will aid the motion of carriers toward the collector and reduce the base transit time.

5.3.2 Switching Transients

For digital applications, a transistor is designed to function as a switch. In these applications, we use a small base current to change the collector current from an *off* condition to an *on* condition (or vice versa), in a very short time. The off condition corresponds to a high-voltage and low-current state, and the on condition corresponds to a low-voltage and high-current state. A basic setup of a switching circuit is shown in Fig. 15a, where the emitter-base voltage V_{EB} is suddenly changed from a negative value to a positive value. The output current of the transistor is shown in Fig. 15b. The collector current is initially very low because both the emitter-base junction and the collector-base junction are reverse biased. The current will follow the load line through the active region and will finally reach a high current level, where both junctions become forward biased. Thus, the transistor is virtually open-circuited between the emitter and collector terminals in the *off* condition, which corresponds to the cutoff mode, and short-circuited in the *on* condition , which corresponds to the saturation mode. Therefore, a transistor operated in this mode can nearly duplicate the function of an ideal switch.

The switching time is the time required for a transistor to switch from the off condition to the on condition, or vice versa. Figure 16a shows that when an input current pulse is applied to the emitter-base terminal at time $t = 0$, the transistor is being turned on. At $t = t_2$, the base current is suddenly switched to zero and the transistor is being turned off. The transient behavior of the collector current I_C can be determined by the

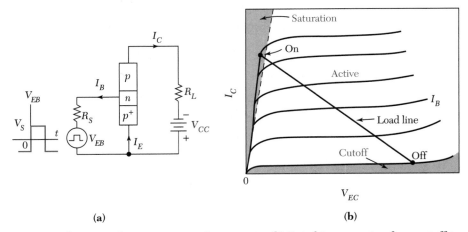

(a)   (b)

Fig. 15 (*a*) Schematic of a transistor switching circuit. (*b*) Switching operation from cutoff to saturation.

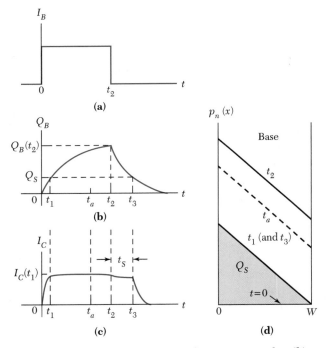

Fig. 16 Transistor switching characteristics. (*a*) Input base current pulse. (*b*) Variations of the base-stored charge with time. (*c*) Variation of the collector current with time. (*d*) Minority-carrier distributions in the base at different times.

variation of the total excess minority carrier charge stored in the base, $Q_B(t)$. A plot of $Q_B(t)$ as a function of time is shown in Fig. 16*b*. During the turn-on transient, the base-stored charge will increase from zero to $Q_B(t_2)$. During the turn-off transient, the base-stored charge will decrease from $Q_B(t_2)$ to zero. For $Q_B(t) < Q_S$, where Q_S is the base charge when $V_{CB} = 0$ (i.e., at the edge of saturation, as shown in Fig. 16*d*), the transistor is in the active mode.

The variation of I_C with time is plotted in Fig. 16c. In the turn-on transient, the stored base charge reaches Q_S, the charge at the edge of saturation at $t = t_1$. For $Q_B > Q_S$ the device is operated in saturation mode, and both the emitter and collector currents remain essentially constant. Figure 16d shows that for any $t > t_1$ (say $t = t_a$), the hole distribution $p_n(x)$ will be parallel to that for $t = t_1$. Therefore, the gradients at $x = 0$ and $x = W$, as well as the currents, remain the same. In the turn-off transient, since the device is initially in the saturation mode, the collector current remains relatively unchanged until Q_B is reduced to Q_S (Fig. 16d). The time from t_2 to t_3 when $Q_B = Q_S$ is called the *storage time delay* t_S. When $Q_B = Q_S$, the device enters the active mode at $t = t_3$. After that time, the collector current will decay exponentially toward zero.

The turn-on time depends on how fast we can add holes (minority carries in the p-n-p transistor) to the base region. The turn-off time depends on how fast we can remove the holes by recombination. One of the most important parameters for switching transistors is the minority carrier lifetime τ_p. One effective method to reduce τ_p for faster switching is to introduce efficient generation-recombination centers near the midgap.

5.4 THE HETEROJUNCTION BIPOLAR TRANSISTOR

We have considered the heterojunction in Sec. 4.8. A heterojunction bipolar transistor (HBT) is a transistor in which one or both p–n junctions are formed between dissimilar semiconductors. The primary advantage of an HBT is its high emitter efficiency (γ). The circuit applications of the HBT are essentially the same as those of bipolar transistors. However, the HBT has higher-speed and higher-frequency capability in circuit operation. Because of these features, the HBT has gained popularity in photonic, microwave, and digital applications. For example, in microwave applications, HBT is used in solid-state microwave and millimeter-wave power amplifiers, oscillators, and mixers.

5.4.1 Current Gain in HBT

Let semiconductor 1 be the emitter and semiconductor 2 be the base of a HBT. We now consider the impact of the bandgap difference between these two semiconductors on the current gain of a HBT.

When the base-transport factor α_T is very close to unity, the common-emitter current gain can be expressed from Eqs. 8 and 35 as

$$\beta_0 \equiv \frac{\alpha 0}{1 - \alpha_0} \equiv \frac{\gamma \alpha_T}{1 - \gamma \alpha_T} = \frac{\gamma}{1 - \gamma} \quad \text{(for } \alpha_T = 1\text{)}. \tag{45}$$

Substituting γ from Eq. 31 in Eq. 45 yields (for n-p-n transistors)

$$\beta_0 = \frac{1}{\dfrac{D_E}{D_n} \dfrac{p_{EO}}{n_{po}} \dfrac{W}{L_E}} \approx \frac{n_{po}}{p_{EO}} \tag{46}$$

The minority carrier concentrations in the emitter and the base are given by

$$p_{EO} = \frac{n_i^2 (\text{emitter})}{N_E (\text{emitter})} = \frac{N_C N_V \exp(-E_{gE}/kT)}{N_E}, \tag{47}$$

$$n_{po} = \frac{n_i^2 (\text{base})}{N_B (\text{base})} = \frac{N_C' N_V' \exp(-E_{gB}/kT)}{N_B}, \tag{48}$$

where N_C and N_V are the densities of states in the conduction band and the valence band, respectively, and E_{gE} is the bandgap of the emitter semiconductor. N'_C, N'_V, and E_{gB} are the corresponding parameters for the base semiconductor. Therefore

$$\beta_0 \sim \frac{N_E}{N_B} \exp\left(\frac{E_{gE} - E_{gB}}{kT}\right) = \frac{N_E}{N_B} \exp\left(\frac{\Delta E_g}{kT}\right). \tag{49}$$

▶ **EXAMPLE 4**

A HBT has a bandgap of 1.62 eV for the emitter, and a bandgap of 1.42 eV for the base. A BJT has a bandgap of 1.42 eV for both the emitter and base materials; it has an emitter doping of 10^{18} cm^{-3} and a base doping of 10^{15} cm^{-3}.
(a) If the HBT has the same dopings as the BJT, find the improvement of β_0. (b) If the HBT has the same emitter doping and the same β_0 as the BJT, how much can we increase the base doping of the HBT? Assume that all other device parameters are the same.

SOLUTION

(a) $\dfrac{\beta_0(\text{HBT})}{\beta_0(\text{BJT})} = \dfrac{\exp\left(\dfrac{E_{gE} - E_{gB}}{kT}\right)}{1} = \exp\left(\dfrac{1.62 - 1.42}{0.0259}\right) = \exp\left(\dfrac{0.2}{0.0259}\right) = \exp(7.722) = 2257.$

We have an improvement of 2257 times in β_0.

(b) $\beta_0(\text{HBT}) = \dfrac{N_E}{N'_B} \exp(7.722) = \beta_0(\text{BJT}) = \dfrac{N_E}{N_B}$

$\therefore N'_B = N_B \exp(7.722) = 2257 \times 10^{15} = 2.26 \times 10^{18}$ cm^{-3}.

The base doping of the heterojunction can be increased to 2.26×10^{18} cm^{-3} to maintain the same β_0. ◀

5.4.2 Basic HBT Structures

Most developments of HBT technology are for the $Al_xGa_{1-x}As$/GaAs material system. Figure 17*a* shows a schematic structure of a basic *n-p-n* HBT. In this device, the *n*-type emitter is formed in the wide bandgap $Al_xGa_{1-x}As$ whereas the *p*-type base is formed in the lower bandgap GaAs. The *n*-type collector and *n*-type subcollector are formed in GaAs with light doping and heavy doping, respectively. To facilitate the formation of ohmic contacts, a heavily doped *n*-type GaAs layer is formed between the emitter contact and the AlGaAs layer. Due to the large bandgap difference between the emitter and the base materials, the common-emitter current gain can be extremely large. However, in homojunction bipolar transistors, there is essentially no bandgap difference; instead the ratio of the doping concentration in the emitter and base must be very high. This is the fundamental difference between the homojunction and the heterojunction bipolar transistors (see Ex. 4).

Figure 17*b* shows the energy band diagram of the HBT under the active mode of operation. The bandgap difference between the emitter and the base will provide band offsets at the heterointerface. In fact, the superior performance of the HBT results directly from the valence-band discontinuity ΔE_V at the heterointerface. ΔE_V increases the valence-band barrier height in the emitter-base heterojunction and thus reduces the injection of holes from the base to the emitter. This effect in the HBT allows the use of a heavily

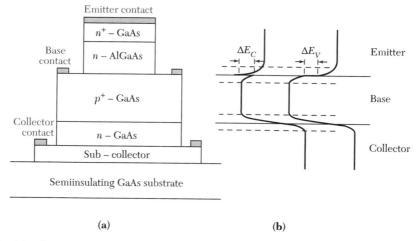

(a)　　　　　　　　　　　　(b)

Fig. 17　(*a*) Schematic cross section of an *n-p-n* heterojunction bipolar transistor (HBT) structure. (*b*) Energy band diagram of a HBT operated under active mode.

doped base while maintaining a high emitter efficiency and current gain. The heavily doped base can reduce the base sheet resistance.[5] In addition, the base can be made very thin without concern about the *punch-through* effect in the narrow base region. The punch-through effect arises when the base-collector depletion region penetrates completely through the base and reaches the emitter-base depletion region. A thin base region is desirable because it reduces the base transit time and increases the cutoff frequency.[6]

5.4.3 Advanced HBTs

In recent years the InP-based (InP/InGaAs or AlInAs/InGaAs) material systems have been extensively studied. The InP-based heterostructures had several advantages.[7] The InP/InGaAs structure has very low surface recombination, and because of a higher electron mobility in InGaAs than in GaAs, superior high-frequency performance is expected. A typical performance curve for an InP-based HBT is shown [8] in Fig. 18. A very high cutoff frequency of 254 GHz is obtained. In addition, the InP collector region has higher

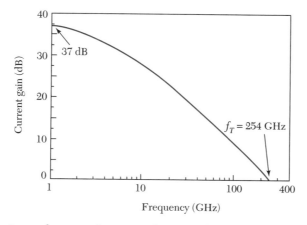

Fig. 18　Current gain as a function of operating frequency for an InP-based HBT.[8]

drift velocity at high fields than that in the GaAs collector. The InP collector breakdown voltage is also higher than that in the GaAs.

Another heterojunction is in the Si/SiGe material system. This system has several properties that are attractive for HBT applications. Like AlGaAs/GaAs HBTs, Si/SiGe HBTs have high-speed capability since the base can be heavily doped because of the bandgap difference. The small trap density at the silicon surface minimizes the surface recombination current and ensures a high current gain even at low collector current. Compatability with the standard silicon technology is another attractive feature. Figure 19a shows a typical Si/SiGe HBT structure. A comparison of the base and collector currents measured from a Si/SiGe HBT and a Si homojunction bipolar transistor is given by Fig. 19b. The results indicate that the Si/SiGe HBT has a higher current gain than the Si homojunction bipolar transistor. Compared with GaAs- and InP-based HBTs, the Si/SiGe HBT, however, has a lower cutoff frequency because of the lower mobilities in Si.

The conduction band discontinuity ΔE_C shown in Fig. 17b is not desirable, since the discontinuity will make it necessary for the carriers in the heterojunction to transport by means of thermionic emission across a barrier or by tunneling through it. Therefore, the emitter efficiency and the collector current will suffer. The problems can be alleviated by improved structures such as the graded-layer and the graded-base heterojunctions. Figure 20 shows an energy band diagram in which the ΔE_C is eliminated by a graded

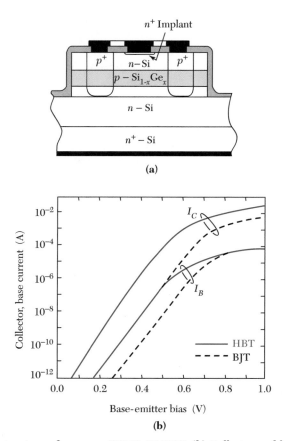

Fig. 19 (a) Device structure of an *n-p-n* Si/SiGe/Si HBT. (b) Collector and base current versus V_{EB} for a HBT and bipolar junction transistor (BJT).

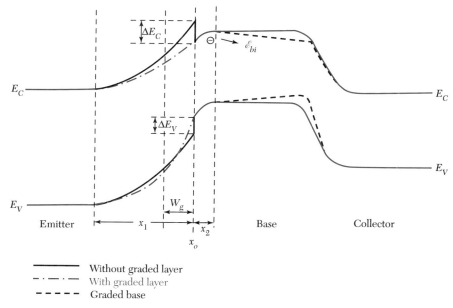

Fig. 20 Energy band diagrams for a heterojunction bipolar transistor with and without graded layer in the junction, and with and without a graded-base layer.

layer placed between the emitter and base heterojunction. The thickness of the graded layer is W_g.

The base region can also have a graded profile, which results in a reduction of the bandgap from the emitter side to the collector side. The energy band diagram of the graded base HBT is illustrated in Fig. 20 (dotted line). Note that there is a built-in electric field $\mathscr{E}_{bi}$ in the quasi-neutral base. It results in a reduction in the minority-carrier transit time and, thus, an increase in the common-emitter current gain and the cutoff frequency of the HBT. $\mathscr{E}_{bi}$ can be obtained, for example, by varying linearly the Al mole fraction x of $Al_xGa_{1-x}As$ in the base from $x = 0.1$ to $x = 0$.

For the design of the collector layer, it is necessary to consider the collector transit time delay and the breakdown voltage requirement. A thicker collector layer will improve the breakdown voltage of the base-collector junction but proportionally increase the transit time. In most devices for high-power applications, the carriers move through the collector at their saturation velocities because very large electric fields are maintained in this layer.

It is possible, however, to increase the velocities by lowering the electric field with certain doping profile in the collector layer. One way is to use p^- collectors with a p^+ pulse-doped layer near the subcollector for an n-p-n HBT. Therefore, electrons entering the collector layer can maintain their higher mobility of the lower valley during most of the collector transit time. Such a device is called a *ballistic collector transistor* (BCT).[9] An energy band diagram of a BCT is shown in Fig. 21. The BCT has been shown to have more favorable frequency response characteristics compared with conventional HBTs over a narrow range of bias voltages. Because of its advantages at relatively low collector voltage and current conditions, the BCT is used for switching applications and microwave-power amplifications.

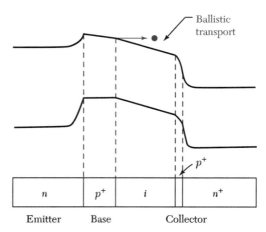

Fig. 21   Energy band diagram for the ballistic collector transistor (BCT). [9]

▶   ## 5.5 THE THYRISTOR AND RELATED POWER DEVICES

The thyristor is an important power device that is designed for handling high voltages and large currents. The thyristor is mainly used for switching applications that require the device to change from an *off* or blocking state to an *on* or conducting state, or vice versa.[10] We have considered the use of bipolar transistors in this application, in which the base current drives the transistor from cutoff to saturation for the on-state, and from saturation to cutoff for the off-state. The operation of a thyristor is intimately related to the bipolar transistor, in which both electrons and holes are involved on the transport processes. However, the switching mechanisms in a thyristor are different from those of a bipolar transistor. Also, because of the device construction, thyristors have a much wider range of current- and voltage-handling capabilities. Thyristors are now available[11] with current ratings from a few milliamperes to over 5000 A and voltage ratings extending above 10,000 V. We first consider the operating principles of basic thyristors and discuss some related high-power and high-frequency thyristors.

5.5.1   Basic Characteristics

Figure 22*a* shows a schematic cross-sectional view of a thyristor structure, which is a four-layer *p-n-p-n* device with three *p–n* junctions in series: *J*1, *J*2, and *J*3. The contact electrode to the outer *p*-layer is called the *anode* and that to the outer *n*-layer is called the *cathode*. This structure without any additional electrode is a two-terminal device and is called the *p-n-p-n* diode. If an additional electrode, called the *gate* electrode, is connected to the inner *p*-layer (*p*2), the resulting three-terminal device is commonly called the *semiconductor-controlled rectifier* (SCR) or *thyristor*.

A typical doping profile of a thyristor is shown in Fig. 22*b*. An *n*-type, high-resistivity silicon wafer is chosen as the starting material (*n*1-layer). A diffusion step is used to form the *p*1- and *p*2-layers simultaneously. Finally, an *n*-type layer is alloyed (or diffused) into one side of the wafer to form the *n*2-layer. Figure 22*c* shows the energy band diagram of a thyristor in thermal equilibrium. Note that at each junction there is a depletion region with a built-in potential that is determined by the impurity doping profile.

The basic current-voltage characteristics of a *p-n-p-n* diode is shown in Fig. 23. It exhibits five distinct regions:

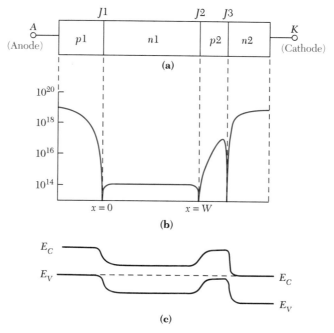

Fig. 22 (*a*) Four-layer *p-n-p-n* diode. (*b*) Typical doping profile of a thyristor. (*c*) Energy band diagram of a thyristor in thermal equilibrium.

0-1: The device is in the forward-blocking or off-state and has very high impedance. Forward breakover (or switching) occurs where *dV/dI* = 0; and at point 1 we define a forward-breakover voltage V_{BF} and a switching current I_S.

1-2: The device is in a negative-resistance region, that is, the current increases as the voltage decreases sharply.

2-3: The device is in the forward-conducting or on-state and has low impedance. At point 2, where *dV/dI* = 0, we define the holding current I_h and holding voltage V_h.

0-4: The device is in the reverse-blocking state.

4-5: The device is in the reverse-breakdown region.

Thus, a *p-n-p-n* diode operated in the forward region is a bistable device that can switch from a high-impedance, low-current off-state to a low-impedance, high-current on-state, or vice versa.

To understand the forward-blocking characteristics, we consider the device as two bipolar transistors, a *p-n-p* transistor and an *n-p-n* transistor, connected in a special way, as shown in Fig. 24. They are connected with the base of one transistor attached to the collector of the other, and vice versa. The relationship between the emitter, collector, and base currents and the dc common-base current gain were given in Eqs. 3 and 10. The base current of the *p-n-p* transistor (transistor 1 with current gain α_1) is

$$
\begin{aligned}
I_{B1} &= I_{E1} - I_{C1} \\
&= \left(1 - \alpha_1\right)I_{E1} - I_1 \\
&= \left(1 - \alpha_1\right)I - I_1,
\end{aligned}
\tag{50}
$$

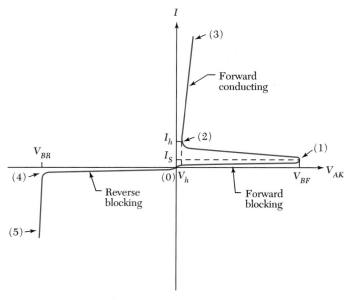

Fig. 23 Current-voltage characteristics of a p-n-p-n diode.

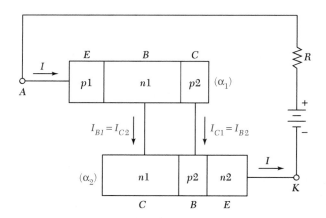

Fig. 24 Two-transistor representation of a thyristor.[2]

where I_1 is the leakage current I_{CBO} for the transistor 1. This base current is supplied by the collector of the n-p-n transistor (transistor 2 with the current gain α_2). The collector current of the n-p-n transistor is

$$I_{C2} = \alpha_2 I_{E2} + I_2 = \alpha_2 I + I_2, \tag{51}$$

where I_2 is the leakage current I_{CBO} for the transistor 2. By equating I_{B1} and I_{C2}, we obtain

$$I = \frac{I_1 + I_2}{1 - \left(\alpha_1 + \alpha_2\right)} . \tag{52}$$

▷ **EXAMPLE 5**

Consider a thyristor in which the leakage currents I_1 and I_2 are 0.4 and 0.6 mA, respectively. Explain the forward-blocking characteristics when $(\alpha_1 + \alpha_2)$ is 0.01 and 0.9999.

SOLUTION The current gains are functions of the current I and generally increase with increasing current. At low currents both α_1 and α_2 are much less than 1, and we have

$$I = \frac{0.4 \times 10^{-3} + 0.6 \times 10^{-3}}{1 - 0.01} = 1.01 \text{ mA}$$

In this case, the current flowing through the device is the sum of the leakage currents I_1 and I_2 ($\cong 1$ mA).

As the applied voltage increases, the current I also increases, as do α_1 and α_2. This in turn causes I to increase further—a regenerative behavior. When $\alpha_1 + \alpha_2 = 0.9999$,

$$I = \frac{0.4 \times 10^{-3} + 0.6 \times 10^{-3}}{1 - 0.9999} = 10 \text{ A} .$$

This value is 10,000 times larger than $I_1 + I_2$. Therefore, as $(\alpha_1 + \alpha_2)$ approaches 1, the current I increases without limit, that is, the device is at forward breakover. ◀

The variations of the depletion layer widths of a *p-n-p-n* diode biased in different regions are shown in Fig. 25. At thermal equilibrium, Fig. 25*a*, there is no current flowing and the depletion layer widths are determined by the impurity doping profiles. In

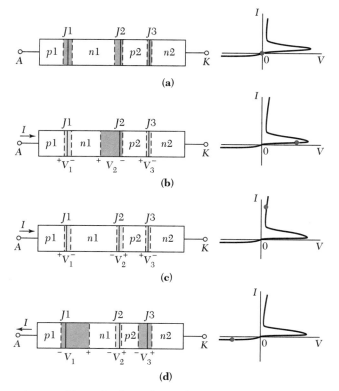

Fig. 25 Depletion layer widths and voltage drops of a thyristor operated under (*a*) equilibrium, (*b*) forward blocking, (*c*) forward conducting, and (*d*) reverse blocking.

the forward-blocking state, Fig. 25b, junction J1 and J3 are forward biased and J2 is reverse biased. Most of the voltage drop occurs across the central junction J2. In the forward-conduction state, Fig. 25c, all three junctions are forward biased. The two transistors (p1-n1-p2 and n1-p2-n2) are in saturation mode of operation. Therefore, the voltage drop across the device is very low, given by $(V_1 - |V_2| + V_3)$, which is approximately equal to the voltage drop across one forward-biased p–n junction. In the reverse-blocking state, Fig. 25d, junction J2 is forward biased but both J1 and J3 are reversed biased. For the doping profile shown in Fig. 22b, the reverse-breakdown voltage will be mainly determined by J1 because of the lower impurity concentration in the n1-region.

Figure 26a shows the device configuration of a thyristor that is fabricated by planar processes with a gate electrode connected to the p2-region. A cross section of the thyristor along the dashed lines is shown in Fig. 26b. The current-voltage characteristic of the thyristor is similar to that of the p-n-p-n diode, except that the gate current I_g causes an increase of $\alpha_1 + \alpha_2$ and results in a breakover at a lower voltage. Figure 27 shows the effect of gate current on the current-voltage characteristics of a thyristor. As the gate current increases, the forward breakover voltage decreases.

A simple application of a thyristor is shown in Fig. 28a, where a variable power is delivered to a load from a constant line source. The load R_L may be a light bulb or a heater, such as furnace. The amount of power delivered to the load during each cycle depends on the timing of the gate-current pulses of the thyristor (Fig. 28b). If the current pulses are delivered to the gate near the beginning of each cycle, more power will be delivered to the load. However, if the current pulses are delayed, the thyristor will not turn on until later in the cycle, and the amount of power delivered to the load will be substantially reduced.

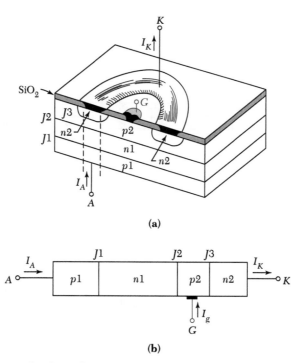

Fig. 26 (a) Schematic of a planar three-terminal thyristor. (b) One-dimensional cross section of the planar thyristor.

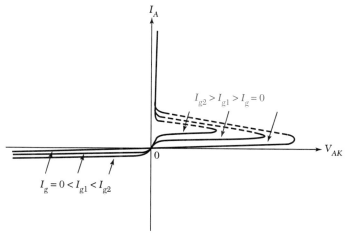

Fig. 27   Affect of gate current on current-voltage characteristics of a thyristor.

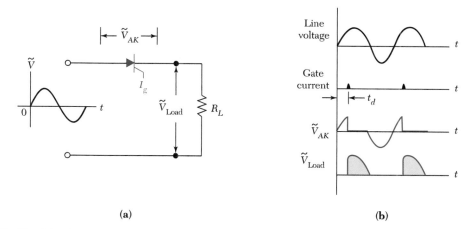

(a)

(b)

Fig. 28   (*a*) Schematic circuit for a thyristor application. (*b*) Wave forms of voltages and gate current.

5.5.2   Bidirectional Thyristor

A bidirectional thyristor is a switching device that has on- and off-states for positive and negative anode voltages and is therefore useful in ac applications. The bidirectional *p-n-p-n* diode switch is called a *diac* (*diode ac* switch). It behaves like two conventional *p-n-p-n* diodes with the anode of the first diode connected to the cathode of the second, and vice versa. Figure 29*a* illustrates such a structure where *M*1 stands for main terminal 1 and *M*2 for main terminal 2. When we integrate this arrangement into a single two-terminal device, we have a diac, as shown in Fig. 29*b*. The symmetry of this structure will result in identical performance for either polarity of applied voltage.

When a positive voltage is applied to *M*1 with respect to *M*2, junction *J*4 is reverse biased so that the *n*2′ region does not contribute to the functioning of the device. Therefore, the *p*1-*n*1-*p*2-*n*2 layers constitute a *p-n-p-n* diode that produces the forward portion of the *I-V* characteristic shown in Fig. 29*c*. If a positive voltage is applied to *M*2, a current

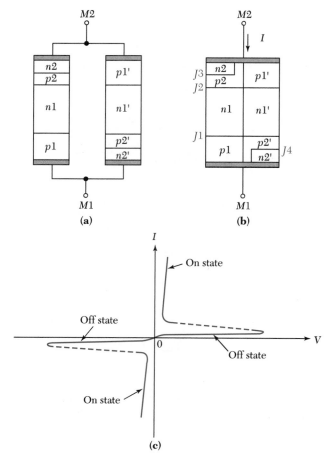

Fig. 29 (*a*) Two reverse-connected *p-n-p-n* diodes. (*b*) Integration of the diodes into a single two-terminal diode ac switch (diac). (*c*) Current-voltage characteristics of a diac.

will conduct in the opposite direction and *J3* will be reverse biased. Therefore, the *p1'-n1'-p2'-n2'* layers from the reverse *p-n-p-n* diode produce the reverse portion of the *I-V* characteristics shown in Fig. 29*c*.

A bidirectional three-terminal thyristor is called a *triac* (*triode ac* switch). The triac can switch the current in either direction by applying a low-voltage, low-current pulse of either polarity between the gate and one of the two main terminals, *M1* and *M2*, as shown in Fig. 30. The operational principles and the *I-V* characteristics of a triac are similar to those of a diac. By adjusting the gate current, the breakover voltage can be varied in either polarity.

5.5.3 Related Thyristor Types and Applications

In addition to the bidirectional thyristors, there are many other types of thyristors for different applications.[12] The majority of these types are based on the four-layer thyristor structure but with special designs.

The *conventional thyristors* considered in Section 5.5.2 are usually designed to have similar forward and reverse blocking capabilities. Two broad classes exist in the conven-

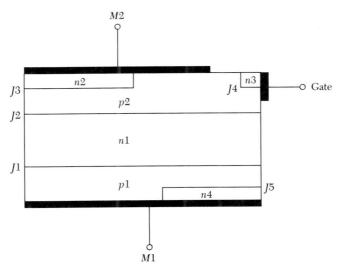

Fig. 30 Cross section of a triode ac switch, a six-layer structure having five *p–n* junctions.

tional thyristor: one is the converter grade thyristor, and the other is the inverter grade thyristor. The former is used for low frequency and is designed to have the lowest possible on-state voltage drop but can only switch slowly. The latter is designed for high frequency and has a fast turn-off but it generally has a larger on-state voltage drop.

The *asymmetric thyristor* is shown in Fig. 31*a*. It does not have a reverse blocking capability since the *n*-base contains an additional n^+ layer adjacent to junction *J*1. Figure 31*b* shows the structure of a conventional thyristor for comparison. Note that the

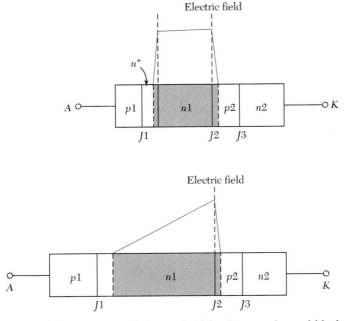

Fig. 31 Comparison of the structure and electric field for the same forward-blocking voltage: (*a*) the asymmetric thyristor and (*b*) the conventional thyristor.

asymmetric thyristor has a higher-resistivity $n1$-base and an adjacent low-resistivity $n^+=$ layer. In forward blocking the depletion layer on the n-side of the blocking junction spreads to the n^+-layer. However, the depletion layer is then effectively constrained from further spreading by the high-doping level of this layer. Consequently, as shown in Fig. 31, the electric-field distribution is approximately constant for the asymmetric thyristor compared with the triangular distribution for the conventional thyristor. The additional n^+-layer allows the use of a much thinner n-base than the conventional thyristor. The reduced base width greatly helps to reduce the on-state voltage drop and shortens the turn-on time of the device through the narrower base. The turn-off performance is also improved since the narrower n-base results in a lower value of stored charge.

The *gate turn-off* (GTO) thyristor is a device that overcomes one of the basic thyristor's limitations, since the GTO can be turned on and off by the gate control.[13] This is achieved by a tight control on the current gain and by distributing the gate over the whole cathode area. Figure 32 shows a GTO with a negative voltage applied to the p-gate. To turn off the current, a reverse-gate voltage and current are applied instead of reversing the polarity of the main terminals. As a result, the holes from the anode begin to be partially removed by the gate. On the other hand, the electrons injected by the cathode emitter are pushed away from the negative gate voltage toward the active center of the device.

The *light activated thyristor* is a thyristor in which the gate terminal is not electrically contacted but is designed to respond to an optical signal. Usually the optical signal is very weak and consequently the device has a high gain.

The power switching applications of the thyristors cover a broad range of both power and frequency.[10] Figure 33 illustrates the major application areas for thryristors categorized by switching frequency and power. At the low-frequency and high-power end of the spectrum, we have the high-voltage direct current (HVDC) devices, which demand the highest power. On the other hand, for high-frequency and low-power applications, we have the lighting control, ultrasonic generators, high frequency power conversion equipment, and switched-mode power supplies (SMPS). However, the major applications are

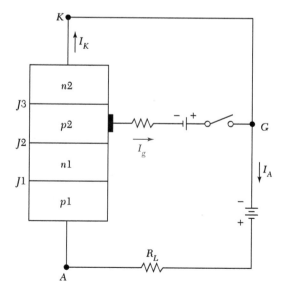

Fig. 32   The gate turn-off thyristor with a negative voltage applied to the gate. The main applications of thyristors.[10]

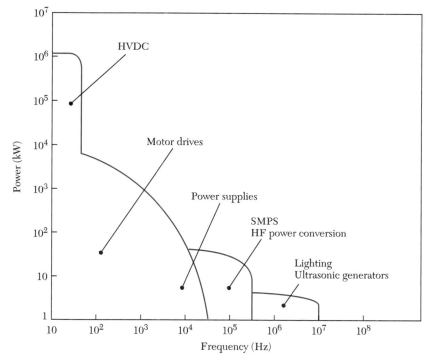

Fig. 33 The main applications of thyristors.[10]

for motor drives and power supplies, which cover a wide range of power and frequency. The motor drive applications, for example, range from small motors used in domestic appliances to the high-power motor drives used in rolling mills.

SUMMARY

The bipolar transistor, developed in 1947, remains one of the most important semiconductor devices. A bipolar transistor is formed when two p–n junctions of the same semiconductor materials are physically close enough to interact. Charge carriers injected from the forward-biased first junction result in a large current flow in the reverse-biased second junction.

We have considered the static characteristics of bipolar transistors, such as the modes of operation and the current-voltage characteristics of the common-emitter configuration. We have also considered the frequency response and switching behavior. A key device parameter of bipolar transistor is the base width, which must be very small compared with the minority-carrier diffusion length to improve the current gain and to increase the cutoff frequency.

Bipolar transistors are used extensively as discrete devices or in integrated circuits for current-gain, voltage-gain, and power-gain applications. They are also used in bipolar-CMOS combination circuits (BiCMOS), covered in Chapters 6 and 14, for high-density, high-speed operations.

The frequency limitations of a conventional bipolar transistor are the result of its low-base doping and relatively wide base. To overcome these limitations, a heterojunction bipolar transistor (HBT) formed between two dissimilar semiconductors can have much higher-base doping and a much narrower base. The HBT has, therefore, gained popularity in millimeter-wave and high-speed digital applications.

Another important bipolar device is the thyristor, which is formed of three or more p–n junctions. The thysistor is used mainly for switching applications. These devices can have current ratings from a few milliamperes to over 5000 A and voltage ratings extending above 10,000 V. We have considered the basic characteristic of thyristor operation. In addition, we discussed the bidirectional thyristor (diac and triac) that has on-off states with either positive or negative terminal voltages, the asymmetric thyristor that has very short turn-on and turn-off times, the gate turn-off thyristor (GTO), and the light-activated thyristor. Thyristors can cover a wide range of applications from low-frequency high-current power supplies to high-frequency, low-power applications, including lighting controls, home appliances, and industrial equipment.

▶ REFERENCES

1. (a) J. Bardeen and W. H. Brattain, "The Transistor, A Semiconductor Triode," *Phys. Rev.*, **74**, 230 (1948). (b) W. Shockley, "The Theory of p–n Junction in Semiconductors and p–n Junction Transistor," *Bell Syst. Tech. J.*, **28**, 435 (1949).

2. J. J. Ebers, "Four-Terminal p-n-p-n Transistor," *Proc. IEEE*, **40**, 1361 (1952).

3. S. M. Sze, *Physics of Semiconductor Devices*, 2nd ed., Wiley, New York, 1981.

4. J. M. Early, "Effects of Space-Charge Layer Widening in Junction Transistors," *Proc. IRE*, **40**, 1401 (1952).

5. J. S. Yuan and J. J. Liou, "Circuit Modeling for Transient Emitter Crowding and Two-Dimensional Current and Charge Distribution Effects," *Solid-State Electron.*, **32**, 623 (1989).

6. J. J. Liou, "Modeling the Cutoff Frequency of Heterojunction Bipolar Transistors Subjected to High Collector-Layer Currents," *J. Appl. Phys.*, **67**, 7125 (1990).

7. B. Jalali and S. J. Pearton, Eds., *InP HBTs: Growth, Processing, and Application*, Artech House, Norwood, MA, 1995.

8. D. Mensa, et al., "Transferred-Substrate HBTs with 250 GHz Current-Gain Cutoff frequency," *Tech. Dig. IEEE Int. Electron Devices Meet.*, p. 657 (1998).

9. T. Ishibashi and Y. Yamauchi, "A Possible Near-Ballistic Collection in an AlGaAa/GaAs HBT with a Modified Collector Structure," *IEEE Trans. Electron Devices*, ED-35, 401 (1988).

10. P. D. Taylor, *Thyristor Design and Realization*, Wiley, New York, 1993.

11. H. P. Lips, "Technology Trends for HVDC Thyristor Valves," *1998 Int. Conf. Power Syst. Tech. Proc.*, **1**, 446 (1998).

12. E. I. Carroll, "Power Electronics for Very High Power Applications," *Seventh Int. Conf. Power Elect. Vari. Speed Drives*, p. 218 (1998).

13. B. K. Bose, "Evaluation of Modern Power Semiconductor Devices and Future Trends of Converters," *IEEE Trans. Ind. Appl.*, **28(2)**, 403 (1992).

▶ PROBLEMS (* INDICATES DIFFICULT PROBLEMS)

FOR SECTION 5.2   STATIC CHARACTERISTICS OF BIPOLAR TRANSISTOR

1. An n-p-n transistor has a base transport factor α_T of 0.998, an emitter efficiency of 0.997, and an I_{Cp} of 10 nA. (a) Calculate α_0 and β_0 for the device. (b) If $I_B = 0$, what is the emitter current?

2. Given that an ideal transistor has an emitter efficiency of 0.999 and the collector-base leakage current is 10 μA, calculate the active region emitter current due to holes if $I_B = 0$.

3. A silicon p-n-p transistor has impurity concentrations of 5×10^{18}, 2×10^{17}, and 10^{16} cm^{-3} in the emitter, base, and collector, respectively. The base width is 1.0 μm, and the device cross-sectional area is 0.2 mm^2. When the emitter-base junction is forward biased to 0.5 V and the base-collector junction is reverse biased to 5 V, calculate (a) the neutral base width and (b) the minority carrier concentration at the emitter-base junction.

4. For the transistor in Prob. 3, the diffusion constants of minority carriers in the emitter, base, and collector are 52, 40, and 115 cm^2/s, respectively; and the corresponding lifetimes are 10^{-8}, 10^{-7}, and 10^{-6} s. Find the current components I_{Ep}, I_{Cp}, I_{En}, I_{Cn}, and I_{BB} illustrated in Fig. 5.

5. Using the results obtained from Prob. 3 and 4, (a) find the terminal currents I_E, I_C, and I_B of the transistor; (b) calculate emitter efficiency, base transport factor, common-base current gain, and common-emitter current gain; and (c) comment on how the emitter efficiency and base transport factor can be improved.

6. Referring to the minority carrier concentration shown in Eq. 14, sketch $p_n(x)/p_n(0)$ curves as a function of x with different W/L_p. Show that the distribution will approach a straight line when W/L_p is small enough (say $W/L_p < 0.1$).

*7. For a transistor under the active mode of operation, use Eq. 14 to find the exact solutions of I_{Ep} and I_{Cp}.

8. Derive the expression for total excess minority-carrier charge Q_B, if the transistor is operated under the active mode and $p_n(0) >> p_{no}$. Explain how the charge can be approximated by the triangle area in the base shown in Fig. 6. In addition, using the parameters in Prob. 3, find Q_B.

9. Using Q_B derived from Prob. 8, show that the collector current expressed in Eq. 27 can be approximated by $I_C \cong (2D_p/W^2)Q_B$.

10. Show that the base transport factor α_T can be simplified to $1 - (W^2/2L_p^2)$.

11. If the emitter efficiency is very close to unity, show that the common-emitter current gain β_0 can be given by $2L_p^2/W^2$. (Hint: Use α_T in Prob. 10.)

12. For a p^+-n-p transistor with high emitter efficiency, find the common-emitter current gain β_0. If the base width is 2 μm and the diffusion constant of minority carrier in the base region is 100 cm^2/s, assume that the lifetime of the carrier in the base region is 3×10^{-7} s. (Hint: Refer to β_0 derived in Prob. 11.)

13. A silicon n-p-n bipolar transistor has impurity concentrations of 3×10^{18}, 2×10^{16}, and 5×10^{15} cm^{-3} in the emitter, base, and collector, respectively. Determine the diffusion constants of minority carrier in the three regions by using Einstein's relationship, $D = (kT/q)\mu$. Assume that he mobilities of electrons and holes, μ_n and μ_p, can be expressed as

$$\mu_n = 88 + \frac{1252}{\left(1 + 0.698 \times 10^{-17}\, N\right)} \quad \text{and} \quad \mu_p = 54.3 + \frac{407}{\left(1 + 0.374 \times 10^{-17}\, N\right)} \quad \text{at } T = 300 \text{ K.}$$

*14. Using the results obtained from Prob. 13, determine the current components in each region with $V_{BE} = 0.6$V (operated under active mode). The device cross-sectional area is 0.01 mm^2 and the neutral-base width is 0.5 μm. Assume the minority-carrier lifetime in each region is the same and equals to 10^{-6} s.

15. Based on the results obtain from Prob. 14, find the emitter efficiency, base transport factor, common-base current gain, and common-emitter current gain.

16. For an ion implanted n-p-n transistor the net impurity doping in the neutral base is given by $N(x) = N_{AO}e^{-x/l}$, where $N_{AO} = 2 \times 10^{18}$ cm^{-3} and $l = 0.3$ μm. (a) Find the total number of impurities in the neutral-base region per unit area (b) Find the average impurity concentration in the neutral-base region for a neutral-base width of 0.8 μm.

17. Referring to Problem 16, if $L_E = 1\mu m$, $N_E = 10^{19}\,cm^{-3}$, $D_E = 1\,cm^2/s$, the average lifetime is 10^{-6} s in the base, and the average diffusion coefficient in the base corresponds to the impurity concentration in Prob. 16, find the common-emitter current gain.

18. Estimate the collector current level for the transistor in Probs. 16 and 17 that has an emitter area of $10^{-4}\,cm^2$. The base resistance of the transistor can be expressed as $10^{-3}\,\overline{\rho}_B/W$, where W is the neutral-base width and $\overline{\rho}_B$ is the average base resistivity.

*19. Plot the common-emitter current gain as a function of the base current I_B from 0 to 25 μA at a fixed V_{EC} of 5 V for the transistor shown in Fig. 10b. Explain why the current gain is not a constant.

20. The general equations of the emitter and collector currents for the basic Ebers-Moll model [J. J. Ebers and J. L. Moll, "Large-Single Behavior of Junction Transistors, " *Proc.IRE.*, **42**,1761(1954)] are

$$I_E = I_{FO}\left(e^{qV_{EB}/kT} - 1\right) - \alpha_R I_{RO}\left(e^{qV_{CB}/kT} - 1\right),$$

$$I_C = \alpha_F I_{FO}\left(e^{qV_{EB}/kT} - 1\right) - I_{RO}\left(e^{qV_{CB}/kT} - 1\right),$$

where α_F and α_R are the *forward common-base current gain* and the *reverse common-base current gain*, respectively. I_{FO} and I_{RO} are the saturation currents of the normally forward- and reverse-biased diodes, respectively. Find α_F and α_R in terms of the constants in Eqs. 25, 26, 28, and 29.

*21. Referring to the transistor in Example 2, find α_F, α_R, I_{FO}, and I_{RO} by using the equations derived in Problem 20.

22. Derive Eq. 32b for the collector current starting with the field-free steady-state continuity equation. (Hint: Consider the minority carrier distribution in the collector region.)

FOR SECTION 5.3   FREQUENCY RESPONSE AND SWITCHING OF BIPOLAR TRANSISTOR

23. A Si transistor has D_p of 10 cm^2/s and W of 0.5 μm. Find the cutoff frequencies for the transistor with a common-base current gain α_0 of 0.998. Neglect the emitter and collector delays.

24. If we want to design a bipolar transistor with 5 GHz cutoff frequency f_T, what the neutral base width W will be? Assume D_p is 10 cm^2/s and neglect the emitter and collector delays.

FOR SECTION 5.4   THE HETEROJUNCTION BIPOLAR TRANSISTOR

25. Consider a $Si_{1-x}Ge_x$/Si HBT with $x = 10\%$ in the base region (and 0% in emitter and collector region). The bandgap of the base region is 9.8% smaller than that of Si. If the base current is due to emitter injection efficiency only, what is the expected change in the common-emitter current gain between 0° and 100°C?

26. For an $Al_xGa_{1-x}As$/GaAs HBT, the bandgap of the $Al_xGa_{1-x}As$ is a function of x and can be expressed as $1.424 + 1.247x\,eV$ (when $x \leq 0.45$) and $1.9 + 0.125x + 0.143x^2\,eV$ (when $0.45 < x \leq 1$). Plot $\beta_0(HBT)/\beta_0(BJT)$ as a function of x.

FOR SECTION 5.5   THE THYRISTOR AND RELATED POWER DEVICES

27. For the doping profile shown in Fig. 22, find the width W ($> 10\mu m$) of the $n1$-region so that the thyristor has a reverse blocking voltage of 120 V. If the current gain α_2 for the $n1$-$p2$-$n2$ transistor is 0.4 independent of current, and α_1 of the $p1$-$n1$-$p2$ transistor can be expressed as $0.5\sqrt{L_p/W}\,\ln(J/J_0)$, where L_p is 25 μm and J_0 is 5×10^{-6} A/cm^2, find the cross-sectional area of the thyristor that will switch at a current I_S of 1 mA.

28. For a GTO thryistor shown in Fig. 32, find the minimum gate current I_g to turn off the thyristor. Assume the current gains of the $p1$-$n1$-$p2$ and $n1$-$p2$-$n2$ transistors are α_1 and α_2, respectively.

MOSFET and Related Devices

The metal-oxide-semiconductor field-effect transistor (MOSFET) is composed of an MOS diode and two p–n junctions placed immediately adjacent to the MOS diode. Since its first demonstration in 1960, the MOSFET has developed quickly and has become the most important device for advanced integrated circuits such as microprocessors and semi-conductor memories. This is because the MOSFET consumes very low power and has a high yield of working devices. Of particular importance is the fact that MOSFET can be readily scaled down and will take up less space than a bipolar transistor using the same design rule.

Specifically, we cover the following topics:

- The inversion condition and threshold voltage of an MOS diode.
- Basic characteristics of a MOSFET.
- MOSFET scaling and its associated short-channel effects.
- Low-power complementary MOS (CMOS) logic structure.
- MOS memory structures.
- Related devices including charge-coupled device, silicon-on-insulator device, and power MOSFET.

▶ 6.1 THE MOS DIODE

The MOS diode is of paramount important in semiconductor device physics because the device is extremely useful in the study of semiconductor surfaces.[§] In practical application, the MOS diode is the heart of the MOSFET—the most important device for advanced integrated circuits. The MOS diode can also be used as a storage capacitor in integrated circuits and it forms the basic building block for charge-coupled devices (CCD). In this section we consider its characteristics in the ideal case, then we extend our consideration to include the effect of metal-semiconductor work-function difference, interface traps, and oxide charges.[1]

6.1.1 The Ideal MOS Diode

A perspective view of an MOS diode is shown in Fig. 1a. The cross section of the device is shown in Fig.1b, where d is the thickness of the oxide and V is the applied voltage on the metal field plate. Throughout this section we use the convention that the voltage V is positive when the metal plate is positively biased with respect to the ohmic contact and V is negative when the metal plate is negatively biased with respect to the ohmic contact.

The energy band diagram of an ideal p-type semiconductor MOS at $V = 0$ is shown in Fig. 2. The work function is the energy difference between the Fermi level and the vacuum level (i.e., $q\phi_m$ for the metal and $q\phi_s$ for the semiconductor). Also shown are the electron affinity $q\chi$, which is the energy difference between the conduction band edge and the vacuum level in the semiconductor, and $q\psi_B$, which is the energy difference between the Fermi level E_F and the intrinsic Fermi level E_i.

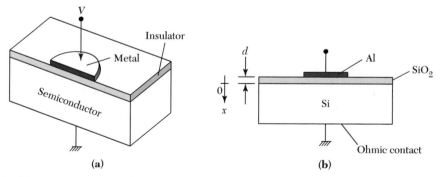

Fig. 1 (*a*) Perspective view of a metal-oxide-semiconductor (MOS) diode. (*b*) Cross-section of an MOS diode.

[§] A more general class of device is the metal-insulator-semiconductor (MIS) diode. However, because in most experimental studies the insulator has been silicon dioxide, in this text the term MOS diode is used interchangeably with MIS diode.

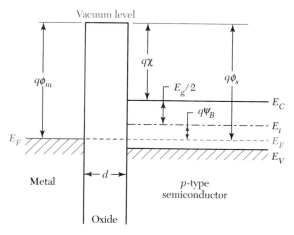

Fig. 2 Energy band diagram of an ideal MOS diode at $V = 0$.

An ideal MOS is defined as follows. (a) At zero applied bias, the energy difference between the metal work function $q\phi_m$ and the semiconductor work function $q\phi_s$ is zero, or the work function difference $q\phi_{ms}$ is zero.[§]

$$q\phi_{ms} \equiv \left(q\phi_m - q\phi_s\right) = q\phi_m - \left(q\chi + \frac{E_g}{2} + q\psi_B\right) = 0, \tag{1}$$

where the sum of the three items in the brackets equals to $q\phi_s$. In other words, the energy band is flat (flat-band condition) when there is no applied voltage. (b) The only charges that exist in the diode under any biasing conditions are those in the semiconductor and those with equal but opposite sign on the metal surface adjacent to the oxide. (c) There is no carrier transport through the oxide under direct current (dc)–biasing conditions, or the resistivity of the oxide is infinite. This ideal MOS diode theory serves as a foundation for understanding practical MOS devices.

When an ideal MOS diode is biased with positive or negative voltages, three cases may exist at the semiconductor surface. For the case of p-type semiconductor, when a negative voltage ($V < 0$) is applied to the metal plate, excess positive carriers (holes) will be induced at the SiO_2-Si interface. In this case, the bands near the semiconductor surface are bent upward, as shown in Fig. 3a. For an ideal MOS diode, no current flows in the device regardless of the value of the applied voltage; thus, the Fermi level in the semiconductor will remain constant. Previously, we determined that the carrier density in the semiconductor depends exponentially on the energy difference $E_i - E_F$, that is,

$$p_p = n_i e^{\left(E_i - E_F\right)/kT}. \tag{2}$$

The upward bending of the energy band at the semiconductor surface causes an increase in the energy $E_i - E_F$ there, which in turn gives rise to an enhanced concentration, an accumulation of holes near the oxide-semiconductor interface. This is called the *accumulation* case. The corresponding charge distribution is shown on the right side of Fig. 3a,

---

[§] This is for a p-type semiconductor. For an n-type semiconductor, the term $q\psi_B$ will be replaced by $-q\psi_B$.

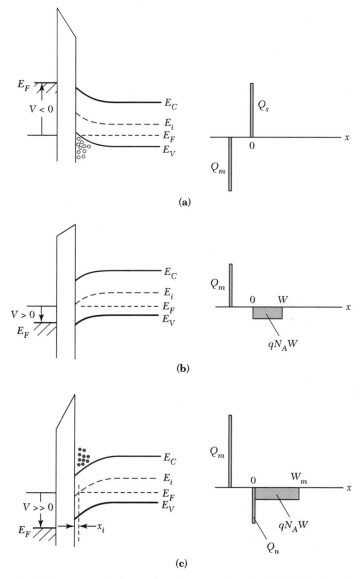

Fig. 3 Energy band diagrams and charge distributions of an ideal MOS diode in (a) accumulation, (b) depletion, and (c) inversion cases.

where Q_s is the positive charge per unit area in the semiconductor and Q_m is the negative charge per unit area ($|Q_m| = Q_s$) in the metal.

When a small positive voltage ($V > 0$) is applied to an ideal MOS diode, the energy bands near the semiconductor surface are bent downward, and the majority carriers (holes) are depleted (Fig. 3b). This is called the *depletion* case. The space charge per unit area, Q_{sc}, in the semiconductor is equal to $-qN_AW$, where W is the width of the surface depletion region.

When a larger positive voltage is applied, the energy bands bend downward even more so that the intrinsic level E_i at the surface crosses over the Fermi level, as shown in Fig. 3c. That means, the positive gate voltage starts to induce excess negative carriers (electrons) at the SiO_2-Si interface. The electron concentration in the semiconductor depends exponentially on the energy difference $E_F - E_i$, and is given by

$$n_p = n_i e^{(E_F - E_i)/kT}. \tag{3}$$

In the case shown in Fig. 3c, $(E_F - E_i) > 0$. Therefore, the electron concentration n_p at the interface is larger than n_i, and the hole concentration given by Eq. 2 is less than n_i. The number of electrons (minority carriers) at the surface is greater than holes (majority carriers); the surface is thus inverted. This is called the *inversion* case.

Initially, the surface is in a *weak inversion* condition since the electron concentration is small. As the bands are bent further, eventually the conduction band edge comes close to the Fermi level. The onset of *strong inversion* occurs when the electron concentration near the SiO_2-Si interface is equal to the substrate doping level. After this point most of the additional negative charges in the semiconductor consist of the charge Q_n (Fig. 3c) in a very narrow n-type inversion layer $0 \leq x \leq x_i$, where x_i is the width of the inversion region. Typically, the value of x_i ranges from 1 to 10 nm and is always much smaller than the surface depletion-layer width.

Once strong inversion occurs, the surface depletion-layer width reaches a maximum. This is because when the bands are bent downward far enough for strong inversion to occur, even a very small increase in band bending corresponding to a very small increase in depletion-layer width results in a large increase in the charge Q_n in the inversion layer. Thus, under a strong inversion condition the charge per unit area Q_s in the semiconductor is the sum of the charge Q_n in the inversion layer and the charge Q_{sc} in the depletion region:

$$Q_s = Q_n + Q_{sc} = Q_n - qN_A W_m, \tag{4}$$

where W_m is the maximum width of the surface depletion region.

The Surface Depletion Region

Figure 4 shows a more detailed band diagram at the surface of a p-type semiconductor. The electrostatic potential ψ is defined as zero in the bulk of the semiconductor. At the semiconductor surface, $\psi = \psi_s$; ψ_s is called the surface potential. We can express electron and hole concentrations in Eqs. 2 and 3 as a function of ψ:

$$n_p = n_i e^{q(\psi - \psi_B)/kT}, \tag{5a}$$

$$p_p = n_i e^{q(\psi_B - \psi)/kT}, \tag{5b}$$

where ψ is positive when the band is bent downward, as shown in Fig. 4. At the surface the densities are

$$n_s = n_i e^{q(\psi_s - \psi_B)/kT}, \tag{6a}$$

$$p_s = n_i e^{q(\psi_B - \psi_s)/kT}. \tag{6b}$$

From this discussion and with the help of Eq. 6, the following regions of surface potential can be distinguished:

$\psi_s < 0$ Accumulation of holes (bands bend upward).

$\psi_s = 0$ Flat-band condition.

$\psi_B > \psi_s > 0$ Depletion of holes (bands bend downward).

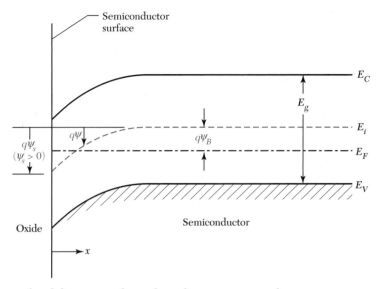

Fig. 4 Energy band diagrams at the surface of a p-type semiconductor.

$\psi_s = \psi_B$ Midgap with $n_s = n_p = n_i$ (intrinsic concentration).

$\psi_s > \psi_B$ Inversion (bands bend downward).

The potential ψ as a function of distance can be obtained by using the one-dimensional Poisson's equation:

$$\frac{d^2\psi}{dx^2} = \frac{-\rho_s(x)}{\varepsilon_s}, \tag{7}$$

where $\rho_s(x)$ is the charge density per unit volume at position x and ε_s is the dielectric permittivity. We use the depletion approximation that we have employed in the study of p–n junctions. When the semiconductor is depleted to a width of W and the charge within the semiconductor is given by $\rho_s = -qN_A$, integration of Poisson's equation gives the electrostatic potential distribution as a function of distance x in the surface depletion region:

$$\psi = \psi_s\left(1 - \frac{x}{W}\right)^2. \tag{8}$$

The surface potential ψ_s is

$$\psi_s = \frac{qN_AW^2}{2\varepsilon_s}. \tag{9}$$

Note that the potential distribution is identical to that for a one-sided n^+-p junction.

The surface is inverted whenever ψ_s is larger than ψ_B. However, we need a criterion for the onset of strong inversion, after which the charges in the inversion layer become significant. A simple criterion is that the electron concentration at the surface is equal to the substrate impurity concentration, i.e., $n_s = N_A$. Since $N_A = n_i e^{q\psi_B/kT}$, from Eq. 6a we obtain

$$\psi_s(inv) \cong 2\psi_B = \frac{2kT}{q}\ln\left(\frac{N_A}{n_i}\right). \tag{10}$$

Equation 10 states that a potential ψ_B is required to bend the energy bands down to the intrinsic condition at the surface ($E_i = E_F$), and bands must then be bent downward by another $q\psi_B$ at the surface to obtain the condition of strong inversion.

As discussed previously, the surface depletion layer reaches a maximum when the surface is strongly inverted. Accordingly, the maximum width of the surface depletion region W_m is given by Eq. 9 in which ψ_s equals $\psi_s(inv)$, or

$$W_m = \sqrt{\frac{2\varepsilon_s\psi_s(inv)}{qN_A}} \cong \sqrt{\frac{2\varepsilon_s(2\psi_B)}{qN_A}} \tag{11}$$

or

$$W_m = 2\sqrt{\frac{\varepsilon_s kT \ln\left(\dfrac{N_A}{n_i}\right)}{q^2 N_A}} \tag{11a}$$

and

$$Q_{sc} = -qN_A W_m \cong -\sqrt{2q\varepsilon_s N_A(2\psi_B)}. \tag{12}$$

▷ EXAMPLE 1

For an ideal metal-SiO$_2$-Si diode having $N_A = 10^{17}$ cm^{-3}, calculate the maximum width of the surface depletion region.

SOLUTION   At room temperature $kT/q = 0.026$ V and $n_i = 9.65 \times 10^9$ cm^{-3}, the dielectric permittivity of Si is $11.9 \times 8.85 \times 10^{-14}$ F/cm. From Eq. 11a

$$W_m = 2\sqrt{\frac{11.9 \times 8.85 \times 10^{-14} \times 0.026 \ln\left(10^{17} \middle/ 9.65 \times 10^9\right)}{1.6 \times 10^{-19} \times 10^{17}}}$$

$$= 10^{-5} \text{ cm} = 0.1 \text{ }\mu\text{m}. \qquad \blacktriangleleft$$

The relationship between W_m and the impurity concentration is shown in Fig. 5 for silicon and gallium arsenide, where N_B is equal to N_A for p-type and N_D for n-type semiconductors.

Ideal MOS Curves

Figure 6a shows the energy band diagram of an ideal MOS diode with band bending identical to that shown in Fig. 4. The charge distribution is shown in Fig. 6b. In the absence of any work function differences, the applied voltage will appear partly across the oxide and partly across the semiconductor. Thus,

$$V = V_o + \psi_s, \tag{13}$$

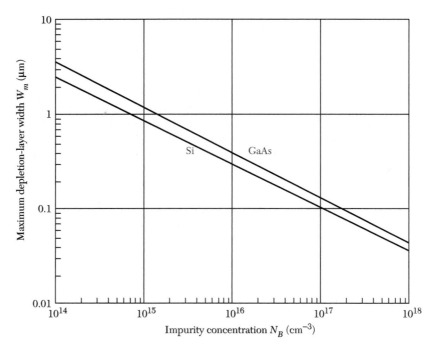

Fig. 5 Maximum depletion-layer width versus impurity concentration of Si and GaAs under strong-inversion condition.

where V_o is the potential across the oxide and is given (Fig. 6c) by

$$V_o = \mathcal{E}_o d = \frac{|Q_s| d}{\varepsilon_{ox}} \equiv \frac{|Q_s|}{C_o}, \tag{14}$$

where $\mathcal{E}_o$ is the field in the oxide, Q_s is the charge per unit area in the semiconductor, and C_o $(= \varepsilon_{ox}/d)$ is the oxide capacitance per unit area. The corresponding electrostatic potential distribution is shown in Fig. 6d.

The total capacitance C of the MOS diode is a series combination (Fig. 7a, inset) of the oxide capacitance C_o and the semiconductor depletion-layer capacitance C_j:

$$C = \frac{C_o C_j}{\left(C_o + C_j \right)} \ \text{F} / \text{cm}^2, \tag{15}$$

where $C_j = \varepsilon_s/W$, the same as for an abrupt p–n junction.

From Eqs. 9, 13, 14, and 15, we can eliminate W and obtain the formula for the capacitance:

$$\frac{C}{C_o} = \frac{1}{\sqrt{1 + \dfrac{2\varepsilon_{ox}^2 V}{qN_A\varepsilon_s d^2}}}, \tag{16}$$

which predicts that the capacitance will decrease with increasing metal-plate voltage while the surface is being depleted. When the applied voltage is negative, there is no depletion region, and we have an accumulation of holes at the semiconductor surface. As a result, the total capacitance is close to the oxide capacitance ε_{ox}/d.

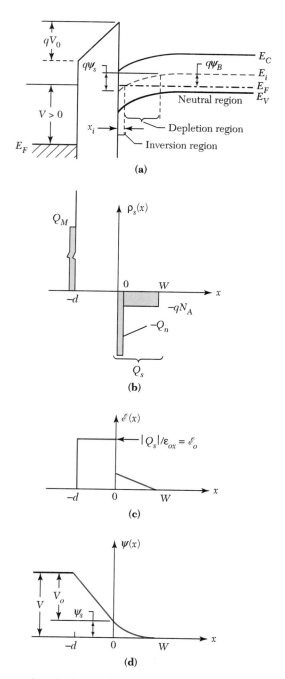

Fig. 6 (*a*) Band diagram of an ideal MOS diode. (*b*) Charge distributions under inversion condition. (*c*) Electric-field distribution. (*d*) Potential distribution.

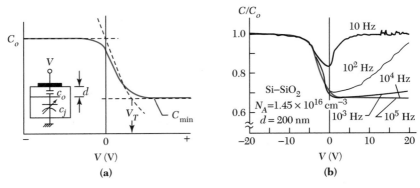

Fig. 7 (*a*) High-frequency MOS *C-V* curve showing its approximated segments (dashed lines). Inset shows the series connection of the capacitors. (*b*) Effect of frequency on the *C-V* curve.[2]

In the other extreme, when strong inversion occurs, the width of the depletion region will not increase with a further increase in applied voltage. This condition takes place at a metal-plate voltage that causes the surface potential ψ_s to reach $\psi_s(inv)$, as given in Eq. 10. Substituting $\psi_s(inv)$ into Eq. 13 and noting that the corresponding charge per unit area is qN_AW_m yields the metal-plate voltage at the onset of strong inversion. This voltage is called the threshold voltage:

$$V_T = \frac{qN_AW_m}{C_o} + \psi_s(inv) \cong \frac{\sqrt{2\varepsilon_s qN_A(2\psi_B)}}{C_o} + 2\psi_B. \qquad (17)$$

Once the strong inversion takes place, the total capacitance will remain at a minimum value given by Eq. 15 with $C_j = \varepsilon_s/W_m$,

$$C_{min} = \frac{\varepsilon_{ox}}{d + \left(\varepsilon_{ox}/\varepsilon_s\right)W_m}. \qquad (18)$$

A typical capacitance-voltage characteristics of an ideal MOS diode is shown in Fig. 7*a* based on both the depletion approximation (Eqs. 16–18) and exact calculations (solid curve). Note the close correlation between the depletion approximation and the exact calculations.

Although we have considered only the *p*-type substrate, all of the considerations are equally valid for an *n*-type substrate with the proper changes in signs and symbols (e.g., Q_p for Q_n). The capacitance-voltage characteristics will have identical shapes but will be mirror images of each other, and the threshold voltage is a negative quantity for an ideal MOS diode on an *n*-type substrate.

In Fig. 7*a* we assumed that when the voltage on the metal plate changes, all the incremental charge appears at the edge of the depletion region. Indeed, this happens when the measurement frequency is high. If, however, the measurement frequency is low enough so that generation-recombination rates in the surface depletion region are equal to or faster than the voltage variation, then the electron concentration (minority carrier) can follow the alternating current (ac) signal and lead to charge exchange with the inversion layer in step with the measurement signal. As a result the capacitance in strong inversion will be that of the oxide layer alone, C_o. Figure 7b shows the measured MOS C-V curves at different frequencies.[2] Note that the onset of the low-frequency curves occurs at $f \leq 100$ Hz.

▷ **EXAMPLE 2**

For an ideal metal-SiO$_2$-Si diode having $N_A = 10^{17}$ cm^{-3} and $d = 5$ nm, calculate the minimum capacitance of the C-V curve in Fig. 7a. The relative dielectric constant of SiO$_2$ is 3.9.

SOLUTION

$$C_o = \frac{\varepsilon_{ox}}{d} = \frac{3.9 \times 8.85 \times 10^{-14}}{5 \times 10^{-7}} = 6.90 \times 10^{-7} \text{ F/cm}^2.$$

$$Q_{sc} = -qN_A W_m = -1.6 \times 10^{-19} \times 10^{17} \times (1 \times 10^{-5}) = -1.6 \times 10^{-7} \text{ C/cm}^2.$$

W_m is obtained in Example 1.

$$\psi_s(inv) \approx 2\psi_B = \frac{2kT}{q}\ln\left(\frac{N_A}{n_i}\right) = 2 \times 0.026 \times \ln\left(\frac{10^{17}}{9.65 \times 10^9}\right) = 0.84 \text{ V}.$$

The minimum capacitance $C_{\min}$ at V_T is

$$C_{\min} = \frac{\varepsilon_{ox}}{d + (\varepsilon_{ox}/\varepsilon_s)W_m} = \frac{3.9 \times 8.85 \times 10^{-14}}{(5 \times 10^{-7}) + (3.9/11.9)(1 \times 10^{-5})}$$

$$= 9.1 \times 10^{-8} \text{ F/cm}^2.$$

Therefore, $C_{\min}$ is about 13% of C_o.   ◀

6.1.2   The SiO$_2$-Si MOS Diode

Of all the MOS diodes, the metal-SiO$_2$-Si is the most extensively studied. The electrical characteristics of the SiO$_2$-Si system approach those of the ideal MOS diode. However, for commonly used metal electrodes, the work function difference $q\phi_{ms}$ is generally not zero, and there are various charges inside the oxide or at the SiO$_2$-Si interface that will, in one way or another, affect the ideal MOS characteristics.

The Work Function Difference
The work function of a semiconductor $q\phi_s$, which is the energy difference between the vacuum level and the Fermi level (Fig. 2), varies with the doping concentration. For a given metal with a fixed work function $q\phi_m$ we expect that the work function difference $q\phi_{ms} \equiv (q\phi_m - q\phi_s)$ will vary depending on the doping of the semiconductor. One of the most common metal electrodes is aluminum, with $q\phi_m = 4.1$ eV. Another material also used extensively is the heavily doped polycrystalline silicon (also called polysilicon). The work function for n^+- and p^+-polysilicon are 4.05 and 5.05 eV, respectively. Figure 8 shows the work function differences for aluminum, n^+-, and p^+-polysilicon on silicon as the doping is varied. It is interesting to note that ϕ_{ms} can vary over a 2 V range depending on the electrode materials and the silicon doping concentration.

To construct the energy band diagram of an MOS diode, we start with an isolated metal and an isolated semiconductor with an oxide layer sandwiched between them (Fig. 9a). In this isolated situation, all bands are flat; this is the flat-band condition. At thermal equilibrium, the Fermi level must be a constant and the vacuum level must be

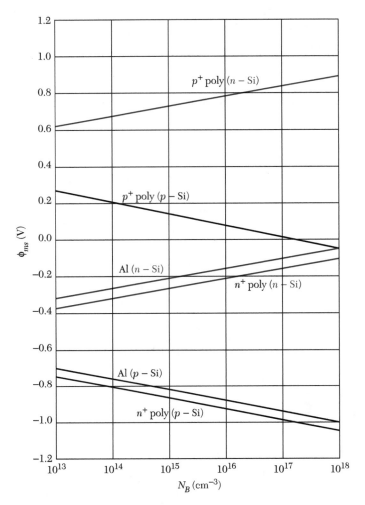

Fig. 8 Work function difference as a function of background impurity concentration for Al, n^+-, and p^+-polysilicon gate materials.

continuous. To accommodate the work function difference, the semiconductor bands bend downward, as shown in Fig. 9b. Thus, the metal is positively charged and the semiconductor surface is negatively charged at thermal equilibrium. To achieve the ideal flat-band condition of Fig. 2, we have to apply a voltage equal to the work function difference $q\phi_{ms}$. This corresponds exactly to the situation shown in Fig. 9a, where we must apply a negative voltage V_{FB} to the metal, and this voltage is called the flat-band voltage $(V_{FB} = \phi_{ms})$.

Interface Traps and Oxide Charges
In addition to the work function difference, the MOS diode is affected by charges in the oxide and traps at the SiO_2-Si interface. The basic classification of these traps and charges are shown in Fig. 10. They are the interface-trapped charge, fixed-oxide charge, oxide-trapped charge, and mobile ionic charge.[3]

Interface-trapped charges Q_{it} are due to the SiO_2-Si interface properties and dependent on the chemical composition of this interface. The traps are located at the SiO_2-Si

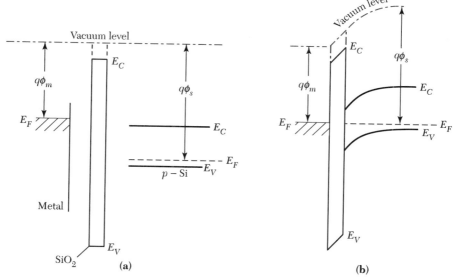

Fig. 9 (*a*) Energy band diagram of an isolated metal and an isolated semiconductor with an oxide layer between them. (*b*) Energy band diagram of an MOS diode in thermal equilibrium.

interface with energy states in the silicon forbidden bandgap. The interface trap density, i.e., number of interface traps per unit area and per eV, is orientation dependent. In $\langle 100 \rangle$ orientation, the interface trap density is about an order of magnitude smaller than that in $\langle 111 \rangle$ orientation. Present-day MOS diodes with thermally grown silicon dioxide on silicon have most of the interface-trapped charges passivated by low-temperature (450°C) hydrogen annealing. The value of Q_{it} for $\langle 100 \rangle$-oriented silicon can be as low as $10^{10} \, cm^{-2}$,

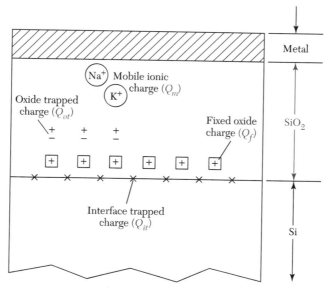

Fig. 10 Terminology for the charges associated with thermally oxidized silicon.[3]

which amounts to about one interface-trapped charge per 10^5 surface atoms. For <111>-oriented silicon, Q_{it} is about 10^{11} cm^{-2}.

The fixed charge Q_f is located within approximately 3 nm of the SiO$_2$-Si interface. This charge is fixed and cannot be charged or discharged over a wide variation of surface potential ψ_s. Generally, Q_f is positive and depends on oxidation and annealing conditions and on silicon orientation. It has been suggested that when the oxidation is stopped, some ionic silicon is left near the interface. These ions, along with uncompleted silicon bonds (e.g., Si-Si or Si-O bonds) at the surface, may result in the positive interface charge $Q_f \cdot Q_f$ can be regarded as a charge sheet located at the SiO$_2$-Si interface. Typical fixed-oxide charge densities for a carefully treated SiO$_2$-Si interface system are about 10^{10} cm^{-2} for a $\langle 100 \rangle$ surface and about 5×10^{10} cm^{-2} for a $\langle 111 \rangle$ surface. Because of the lower values of Q_{it} and Q_f, the $\langle 100 \rangle$ orientation is preferred for silicon MOSFETs.

The oxide-trapped charges Q_{ot} are associated with defects in the silicon dioxide. These charges can be created, for example, by X-ray radiation or high-energy electron bombardment. The traps are distributed inside the oxide layer. Most of process-related Q_{ot} can be removed by low-temperature annealing.

The mobile ionic charges Q_m, such as sodium or other alkali ions, are mobile within the oxide under raised-temperature (e.g., >100°C) and high-electric field operations. Trace contamination by alkali metal ions may cause stability problems in semiconductor devices operated under high-bias and high-temperature conditions. Under these conditions mobile ionic charges can move back and forth through the oxide layer and cause shifts of the C-V curves along the voltage axis. Therefore, special attention must be paid to the elimination of mobile ions in device fabrication.

The charges are the effective net charges per unit area in C/cm^2. We now evaluate the influence of these charges on the flat-band voltage. Consider a positive sheet charge per unit area, Q_o, within the oxide, as shown in Fig. 11. This positive sheet charge will

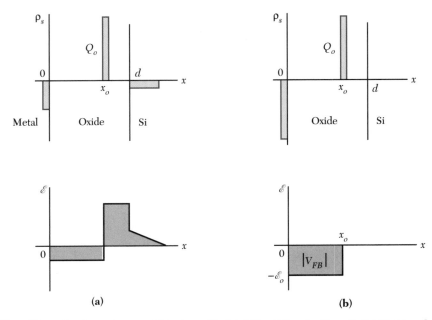

Fig. 11 Effect of a sheet charge within the oxide.[2] (*a*) Condition for $V_G = 0$. (*b*) Flat-band condition.

induce negative charges partly in the metal and partly in the semiconductor as shown in the upper part of Fig. 11a. The resulting field distribution, obtained from integrating Poisson's equation once, is shown in the lower part of Fig. 11a where we have assumed that there is no work function difference, or $q\phi_{ms} = 0$.

To reach the flat-band condition (i.e., no charge induced in the semiconductor), we must apply a negative voltage to the metal, as shown in Fig. 11b. As the negative voltage increases, more negative charges are put on the metal and thereby the electric-field distribution shift downward until the electric field at the semiconductor surface is zero. Under this condition the area contained under the electric-field distribution corresponds to the flat-band voltage V_{FB}:

$$V_{FB} = -\mathscr{E}_o x_o = -\frac{Q_o}{\varepsilon_{ox}} x_o = -\frac{Q_o}{C_o} \frac{x_o}{d}. \tag{19}$$

The flat-band voltage is, thus, dependent on both the density of the sheet charge Q_o and its location x_o within the oxide. When the sheet charge is located very close to the metal— that is , if $x_o = 0$—it will induce no charges in the silicon and therefore have no effect on the flat-band voltage. On the other hand, when Q_o is located very close to the semi-conductor—$x_o = d$, such as the fixed-oxide charge Q_f, it will exert its maximum influence and give rise to a flat-band voltage

$$V_{FB} = -\frac{Q_o}{C_o} \frac{d}{d} = -\frac{Q_o}{C_o}. \tag{20}$$

For the more general case of an arbitrary space charge distribution within the oxide, the flat-band voltage is given by

$$V_{FB} = \frac{-1}{C_o}\left[\frac{1}{d}\int_o^d x\rho(x)dx\right], \tag{21}$$

where $\rho(x)$ is the volume charge density in the oxide. Once we know $\rho_{ot}(x)$, the volume charge density for oxide-trapped charge, and $\rho_m(x)$, the volume charge density for mobile ionic charges, we can obtain Q_{ot} and Q_m and their corresponding contribution to the flat-band voltage:

$$Q_{ot} \equiv \frac{1}{d}\int_o^d \rho_{ot}(x)dx, \tag{22a}$$

$$Q_m \equiv \frac{1}{d}\int_o^d x\rho_m(x)dx. \tag{22b}$$

If the value of the work function difference $q\phi_{ms}$ is not zero and if the value of the interface-trapped charges is negligible, the experimental capacitance-voltage curve will be shifted from the ideal theoretical curve by an amount

$$V_{FB} = \phi_{ms} - \frac{(Q_f + Q_m + Q_{ot})}{C_o}. \tag{23}$$

The curve in Fig. 12a shows the C-V characteristics of an ideal MOS diode. Due to nonzero ϕ_{ms}, Q_f, Q_m, or Q_{ot}, the C-V curve will be shifted by an amount given by Eq. 23. The

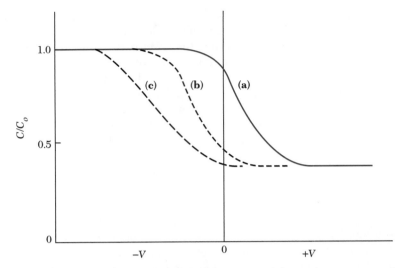

Fig. 12 Effect of a fixed oxide charge and interface traps on the C-V characteristics of an MOS diode.

parallel shift of the C-V curve is illustrated in Fig. 12b. If, in addition, there are large amounts of interface-trapped charges, the charges in the interface traps will vary with the surface potential. The C-V curve will be displaced by an amount that itself changes with the surface potential. Therefore, Fig. 12c is distorted as well as shifted because of interface-trapped charges.

▶ **EXAMPLE 3**

Calculate the flat-band voltage for an n^+-polysilicon-SiO$_2$-Si diode having $N_A = 10^{17}$ cm^{-3} and $d = 5$ nm. Assume the Q_t and Q_m are negligible in the oxide, and Q_f/q is 5×10^{11} cm^{-2}.

SOLUTION From Fig. 8, ϕ_{ms} is -0.98 V for n^+ polysilicon (p-Si) system with $N_A = 10^{17}$ cm^{-3}. C_o is obtained from Ex. 2.

$$V_{FB} = \phi_{ms} - \frac{\left(Q_f + Q_m + Q_{ot}\right)}{C_o}$$

$$= -0.98 - \frac{\left(1.6 \times 10^{-19} \times 5 \times 10^{11}\right)}{6.9 \times 10^{-7}} = -1.10 \text{ V.} \qquad \blacktriangleleft$$

▶ **EXAMPLE 4**

Assume that the volume charge density, $\rho_{ot}(x)$, for oxide-trapped charge Q_{ot} in an oxide layer has a triangular distribution. The distribution is described by the function $(10^{18} - 5 \times 10^{23} \times x)$ cm^{-3}, where x is the distance from the location to the metal-oxide interface. The thickness of the oxide layer is 20 nm. Find the change in the flat-band voltage due to Q_{ot}.

SOLUTION From Eqs. 21 and 22a,

$$\Delta V_{FB} = \frac{Q_{ot}}{C_o} = \frac{d}{\varepsilon_{ox}} \frac{1}{d} \int_0^{2 \times 10^{-6}} x \rho_{ot}\left(x\right) dx,$$

$$= \frac{1.6 \times 10^{-19}}{3.9 \times 8.85 \times 10^{-14}} \left[\frac{1}{2} \times 10^{18} \times \left(2 \times 10^{-6}\right)^2 - \frac{1}{3} \times 5 \times 10^{23} \times \left(2 \times 10^{-6}\right)^3 \right]$$

$$= \frac{1.6 \times 10^{-19} \times \left(2 \times 10^6 - 1.33 \times 10^6\right)}{3.45 \times 10^{-13}}$$

$$= 0.31 \text{ V.}$$ ◀

6.1.3 The Charge-Coupled Device (CCD)

A schematic view of a CCD is shown[4] in Fig. 13. The basic device consists of a closely spaced array of MOS diodes on a continuous insulator (oxide) layer that covers the semiconductor substrate. A CCD can perform a wide range of electronic functions, including

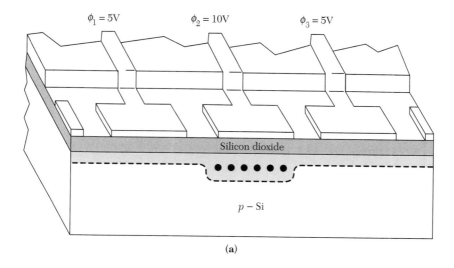

(a)

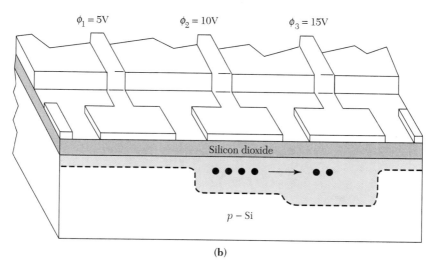

(b)

Fig. 13 Cross section of a three-phase charge-coupled device.[4] (a) High voltage on ϕ_2. (b) ϕ_3 pulsed to a higher voltage for charge transfer.

image sensing and signal processing. The operating principle of the CCD involves the charge storage and transfer actions controlled by the gate electrodes. Figure 13a shows a CCD to which sufficiently large, positive bias pulses have been applied to all the electrodes to produce surface depletion. A slightly higher bias has been applied to the center electrode so that the center MOS structure is under greater depletion and a potential well is formed there, i.e., the potential distribution is shaped like a well because of the larger depletion-layer width under the center electrode. If minority carriers (electrons) are introduced, they will be collected in the potential well. If the potential of the right-hand electrode is increased to exceed that of the central electrode, we obtain the potential distribution shown in Fig. 13b. In this case, the minority carriers will be transferred from the central electrode to the right-hand electrode. Subsequently, the potential on the electrodes can be readjusted so that the quiescent storage site is located at the right-hand electrode. By continuing the process, we can transfer the carriers successively along a linear array.

6.2 MOSFET FUNDAMENTALS

The MOSFET has many acronyms, including IGFET (insulating-gate field-effect transistor), MISFET (metal-insulator-semiconductor field-effect transistor), and MOST (metal-oxide-semiconductor transistor). A perspective view for an n-channel MOSFET is shown in Fig. 14. It is a four-terminal device and consists of a p-type semiconductor substrate in which two n^+ regions, the source and drain, are formed.[§] The metal plate on the oxide is called the gate. Heavily doped polysilicon or a combination of a silicide such as WSi_2 and polysilicon can be used as the gate electrode. The fourth terminal is an ohmic con-

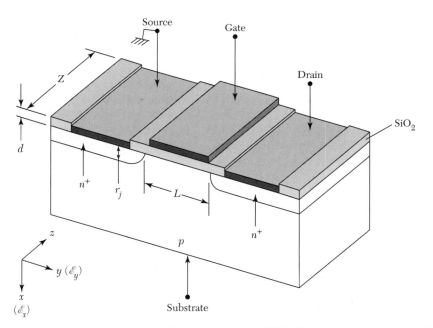

Fig. 14   Perspective view of a metal-oxide-semiconductor field-effect transistor (MOSFET).

[§] For p-channel MOSFETs, doping types in substrate and source/drain regions become n and p^+, respectively.

tact to the substrate. The basic device parameters are the channel length L, which is the distance between the two metallurgical n^+-p junctions, the channel width Z, the oxide thickness d, the junction depth r_j, and the substrate doping N_A. Note that the central section of the device corresponds to the MOS diode discussed in Section 6.1.

The first MOSFET was fabricated in 1960 using a thermally oxidized silicon substrate.[5] The device had a channel length of 20 μm and a gate oxide thickness of 100 nm.[*] Although present-day MOSFETs have been scaled down considerably, the choice of silicon and thermally grown silicon dioxide used in the first MOSFET remains the most important combination. Most of the results in this section are obtained from the Si-SiO$_2$ system.

6.2.1   Basic Characteristics

The source contact is used as the voltage reference throughout this section. When no voltage is applied to the gate, the source-to-drain electrodes correspond to two p–n junctions connected back to back. The only current that can flow from the source to drain is the reverse-leakage current.[†] When we apply a sufficiently large positive bias to the gate, the MOS structure is inverted so that a surface inversion layer (or channel) is formed between the two n^+-region. The source and drain are then connected by a conducting surface n-channel through which a large current can flow. The conductance of this channel can be modulated by varying the gate voltage. The substrate contact can be at the reference voltage or is reverse biased with respect to the source; the substrate bias voltage will also affect the channel conductance.

Linear and Saturation Regions

We now present a qualitative discussion of MOSFET operation. Let us consider that a voltage is applied to the gate, causing an inversion at the semiconductor surface (Fig. 15). If a small drain voltage is applied, electrons will flow from the source to the drain (the corresponding current will flow from drain to source) through the conducting channel. Thus, the channel acts as a resistor, and the drain current I_D is proportional to the drain voltage. This is the *linear region*, as indicated by the constant-resistor line in the right-hand diagram of Fig. 15a.

When the drain voltage increases, eventually it reaches V_{Dsat}, at which the thickness of the inversion layer x_i near $y = L$ is reduced to zero; this is called the pinch-off point, P (Fig. 15b). Beyond the pinch-off point, the drain current remains essentially the same, because for $V_D > V_{Dsat}$, at point P the voltage V_{Dsat} remains the same. Thus, the number of carriers arriving at point P from the source or the current flowing from the drain to the source remains the same. This is the *saturation region*, since I_D is a constant regardless of an increase in the drain voltage. The major change is a decrease of L to the value L' shown in Fig. 15c. Carrier injection from P into the drain depletion region is similar to that of carrier injection from an emitter-base junction to the base-collector depletion region of a bipolar transistor.

We now derive the basic MOSFET characteristics under the following ideal conditions. (a) The gate structure corresponds to an ideal MOS diode, as defined in Section 6.1, that is, there are no interface traps, fixed-oxide charges, or work function differences. (b) Only drift current is considered. (c) Carrier mobility in the inversion layer is constant.

---

[*] A photograph of the first MOSFET is shown in Fig. 4 of Chapter 1.

[†] This is true for the n-channel, normally off MOSFET. Other types of MOSFET are discussed in Section 6.2.2.

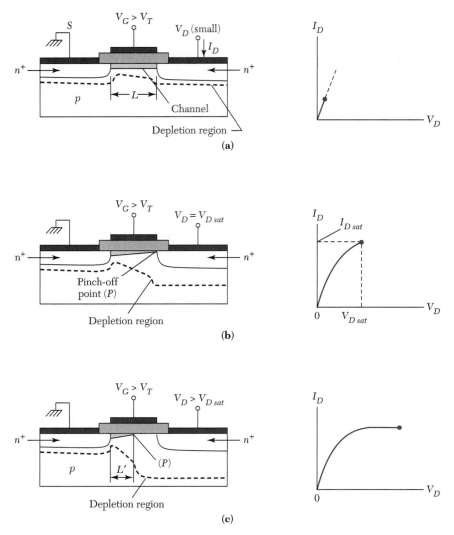

Fig. 15 Operations of the MOSFET and output *I-V* characteristics. (*a*) Low drain voltage. (*b*) Onset of saturation. Point *P* indicates the pinch-off point. (*c*) Beyond saturation.

(d) Doping in the channel is uniform. (e) Reverse-leakage current is negligibly small. (f) The transverse field created by the gate voltage ($\mathcal{E}_x$ in the *x*-direction, shown in Fig. 14, which is perpendicular to the current flow) in the channel is much larger than the longitudinal field created by the drain voltage ($\mathcal{E}_y$ in the *y*-direction, which is in parallel to the current flow). The last condition is called the gradual-channel approximation and generally is valid for long-channel MOSFETs. Under this approximation, the charges contained in the surface depletion region of the substrate are induced solely from the field created by the gate voltage.

Figure 16*a* shows MOSFET operated in the linear region. Under the above ideal conditions, the total charge induced in the semiconductor per unit area, Q_s, at a distance *y* from the source is shown in Fig. 16*b*, which is an enlarged central section of Fig. 16*a*. Q_s is given from Eqs. 13 and 14 by

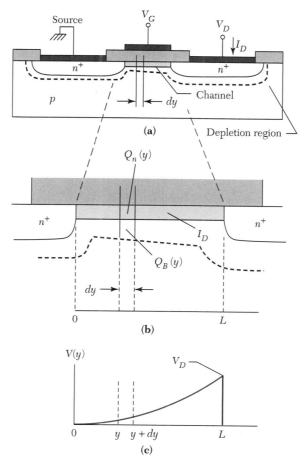

Fig. 16 (*a*) MOSFET operated in the linear region. (*b*) Enlarged view of the channel. (*c*) Drain voltage drop along the channel.

$$Q_s(y) = -\left[V_G - \psi_s(y)\right] C_o, \tag{24}$$

where $\psi_s(y)$ is the surface potential at y and $C_o = \varepsilon_{ox}/d$ is the gate capacitance per unit area. Since Q_s is the sum of the charge in the inversion layer per unit area, Q_n, and the charge in surface depletion region per unit area, Q_{sc}, we can obtain Q_n

$$
\begin{aligned}
Q_n(y) &= Q_s(y) - Q_{sc}(y), \\
&= -\left[V_G - \psi_s(y)\right] C_o - Q_{sc}(y). \tag{25}
\end{aligned}
$$

The surface potential $\psi_s(y)$ at inversion can be approximated by $2\psi_B + V(y)$, where $V(y)$ as shown in Fig. 16c is the reverse bias between the point y and the source electrode (which is assumed to be grounded). The charge within the surface depletion region $Q_{sc}(y)$ was given previously as

$$Q_{sc}(y) = -qN_A W_m \cong -\sqrt{2\varepsilon_s q N_A\left[2\psi_B + V(y)\right]}. \tag{26}$$

Substituting Eq. 26 in Eq. 25 yields

$$Q_n(y) \cong -\left[V_G - V(y) - 2\psi_B \right] C_o + \sqrt{2\varepsilon_s q N_A \left[2\psi_B + V(y) \right]}. \tag{27}$$

The conductivity of the channel at position y can be approximated by

$$\sigma(x) = q n(x) \mu_n(x). \tag{28}$$

For a constant mobility the channel conductance is then given by

$$g = \frac{Z}{L} \int_0^{x_i} \sigma(x) dx = \frac{Z \mu_n}{L} \int_0^{x_i} q n(x) dx. \tag{29}$$

The integral $\int_0^{x_i} q n(x) dx$ corresponds to the total charge per unit area in the inversion layer and is therefore equal to $|Q_n|$, or

$$g = \frac{Z \mu_n}{L} |Q_n|. \tag{30}$$

The channel resistance of an elemental section dy (Fig. 16b) is

$$dR = \frac{dy}{gL} = \frac{dy}{Z \mu_n |Q_n(y)|}, \tag{31}$$

and the voltage drop across the elemental section is

$$dV = I_D dR \frac{I_D dy}{Z \mu_n |Q_n(y)|}, \tag{32}$$

where I_D is the drain current, which is independent of y. Substituting Eq. 27 into Eq. 32 and integrating from the source ($y = 0$, $V = 0$) to the drain ($y = L$, $V = V_D$) yield

$$I_D \approx \frac{Z}{L} \mu_n C_o \left\{ \left(V_G - 2\psi_B - \frac{V_D}{2} \right) V_D - \frac{2}{3} \frac{\sqrt{2\varepsilon_s q N_A}}{C_o} \left[\left(V_D + 2\psi_B \right)^{3/2} - \left(2\psi_B \right)^{3/2} \right] \right\}. \tag{33}$$

Figure 17 shows the current-voltage characteristics of an idealized MOSFET based on Eq. 33. For a given V_G, the drain current first increases linearly with drain voltage (the linear region), then gradually levels off, approaching a saturated value (the saturation region). The dashed line indicates the locus of the drain voltage (V_{Dsat}) at which the current reaches a maximum value.

We now consider the linear and saturation regions. For a small V_D, Eq. 33 reduces to

$$I_D \cong \frac{Z}{L} \mu_n C_o \left(V_G - V_T \right) V_D \quad \text{for} \quad V_D \ll (V_G - V_T), \tag{34}$$

where V_T is the threshold voltage given previously in Eq. 17:

$$V_T = \frac{\sqrt{2\varepsilon_s q N_A \left(2\psi_B \right)}}{C_o} + 2\psi_B. \tag{35}$$

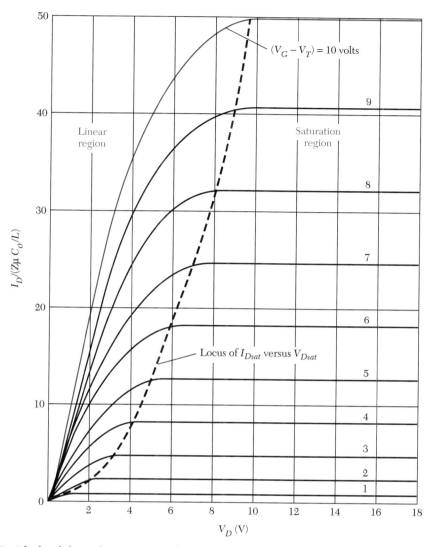

Fig. 17 Idealized drain characteristics of a MOSFET. For $V_D \geq V_{D\text{sat}}$, the drain current remains constant.

By plotting I_D versus V_G (for a given small V_D), the threshold voltage can be deduced from the linearly extrapolated value at the V_G axis. In the linear region, Eq. 34, the channel conductance g_D and the transconductance g_m are given as

$$g_D \equiv \frac{\partial I_D}{\partial V_D} \Big|_{V_G = \text{constant}} \cong \frac{Z}{L} \mu_n C_o \left(V_G - V_T \right), \tag{36}$$

$$g_m \equiv \frac{\partial I_D}{\partial V_G} \Big|_{V_D = \text{constant}} \cong \frac{Z}{L} \mu_n C_o V_D. \tag{37}$$

When the drain voltage is increased to a point that the charge $Q_n(y)$ in the inversion layer at $y = L$ becomes zero, the number of mobile electrons at the drain are reduced drastically. This point is called pinch-off. The drain voltage and the drain current at this point are designated as V_{Dsat} and I_{Dsat}, respectively. For drain voltages larger than V_{Dsat}, we have the saturation region. We can obtain the value of V_{Dsat} from Eq. 27 under the condition $Q_n(L) = 0$:

$$V_{Dsat} \cong V_G - 2\psi_B + K^2 \left(1 - \sqrt{1 + \frac{2V_G}{K^2}}\right), \tag{38}$$

where $K \equiv \dfrac{\sqrt{\varepsilon_s q N_A}}{C_o}$. The saturation current can be obtained by substituting Eq. 38 into Eq. 33:

$$I_{Dsat} \cong \left(\frac{Z\mu_n C_o}{2L}\right)\left(V_G - V_T\right)^2. \tag{39}$$

The threshold voltage V_T in the saturation region for low substrate doping and thin oxide layers is the same as that from Eq. 35. At higher doping levels, V_T becomes V_G dependent.

For an idealized MOSFET in the saturation region, the channel conductance is zero, and the transconductance can be obtained from Eq. 39:

$$g_m \equiv \left.\frac{\partial I_D}{\partial V_G}\right|_{V_D = \text{constant}} = \frac{Z\mu_n \varepsilon_{ox}}{dL}\left(V_G - V_T\right). \tag{40}$$

▶ **EXAMPLE 5**

For an n-channel n^+-polysilicon-SiO$_2$-Si MOSFET with gate oxide = 8 nm, $N_A = 10^{17}$ cm^{-3} and $V_G = 3$V, calculate V_{Dsat}.

SOLUTION

$$C_o = \frac{\varepsilon_{ox}}{d} = \frac{3.9 \times 8.85 \times 10^{-14}}{8 \times 10^{-7}} = 4.32 \times 10^{-7} \text{ F/cm}^2.$$

$$K = \frac{\sqrt{\varepsilon_s q N_A}}{C_o} = \frac{\sqrt{11.9 \times 8.85 \times 10^{-14} \times 1.6 \times 10^{-19} \times 10^{17}}}{4.32 \times 10^{-7}} = 0.3.$$

$2\psi_B = 0.84$ V from Ex. 2.

Therefore, from Eq. 38,

$$V_{Dsat} \cong V_G - 2\psi_B + K^2\left(1 - \sqrt{1 + \frac{2V_G}{K^2}}\right)$$

$$= 3 - 0.84 + (0.3)^2 \left[1 - \sqrt{1 + 2 \times 3 \Big/ (0.3)^2} \right]$$

$$= 3 - 0.84 - 0.65 = 1.51 \text{ V}.$$

◀

The Subthreshold Region

When the gate voltage is below the threshold voltage and the semiconductor surface is only weakly inverted, the corresponding drain current is called the *subthreshold current*. The subthreshold region is particularly important when the MOSFET is used as a low-voltage, low-power device, such as a switch in digital logic and memory applications, because the subthreshold region describes how the switch turns on and off.

In the subthreshold region, the drain current is dominated by diffusion instead of drift and is derived in the same way as the collector current in a bipolar transistor with homogeneous base doping. If we consider the MOSFET as an *n-p-n* (source-substrate-drain) bipolar transistor, Fig. 16*b*, we have

$$I_D = -qAD_n \frac{\partial n}{\partial y} = -qAD_n \frac{n(0) - n(L)}{L}, \tag{41}$$

where A is the channel cross section of the current flow and $n(0)$ and $n(L)$ are the electron densities in the channel at the source and drain, respectively. The electron densities are given by Eq. 5a:

$$n(0) = n_i e^{q(\psi_s - \psi_B)/kT}, \tag{42a}$$

$$n(L) = n_i e^{q(\psi_s - \psi_B - V_D)/kT}, \tag{42b}$$

where ψ_s is the surface potential at the source. Substituting Eq. 42 into Eq. 41 gives

$$I_D = \frac{qAD_n n_i e^{-q\psi_B/kT}}{L} \left(1 - e^{-qV_D/kT} \right) e^{q\psi_s/kT}. \tag{43}$$

The surface potential ψ_s is approximately $V_G - V_T$. Therefore, the drain current will decrease exponentially when V_G becomes less than V_T:

$$\boxed{I_D \sim e^{q(V_G - V_T)/kT}.} \tag{44}$$

A typical measured curve for the subthreshold region is shown in Fig. 18. Note the exponential dependence of I_D on ($V_G - V_T$) for $V_G < V_T$. An important parameter in this region is the *subthreshold swing*, S, which is defined as $[\partial(\log I_D)/\partial V_G]^{-1}$. S is typically 70 ~ 100 mV/decade at room temperature. To reduce the subthreshold current to a negligible value, we must bias the MOSFET a half-volt or more below V_T.

6.2.2 Types of MOSFET

There are basically four types of MOSFETs, depending on the type of inversion layer. If, at zero gate bias, the channel conductance is very low and we must apply a positive voltage to the gate to form the *n*-channel, then the device is a normally off (enhancement)

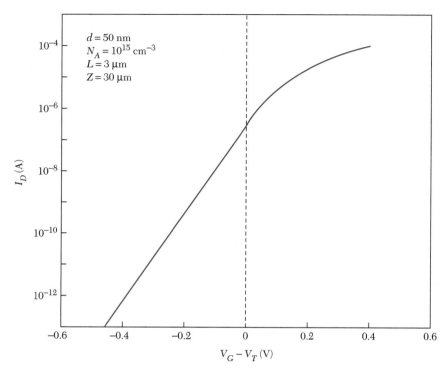

Fig. 18 Subthreshold characteristics of a MOSFET.

n-channel MOSFET. If an n-channel exists at zero bias and we must apply a negative voltage to the gate to deplete carriers in the channel to reduce the channel conductance, then the device is a normally on (depletion) n-channel MOSFET. Similarly, we have the p-channel normally off (enhancement) and normally on (depletion) MOSFETs.

The device cross sections, output characteristics (i.e., I_D versus V_D), and transfer characteristics (i.e., I_D versus V_G) of the four types are shown in Fig. 19. Note that for the normally off n-channel device, a positive gate bias larger than the threshold voltage V_T must be applied before a substantial drain current flows. For the normally on n-channel device, a large current can flow at $V_G = 0$, and the current can be increased or decreased by varying the gate voltage. This discussion can be readily extended to p-channel device by changing polarities.

6.2.3 Threshold Voltage Control

One of the most important parameters of the MOSFET is the threshold voltage. The ideal threshold voltage is given in Eq. 35. However, when we incorporate the effects of the fixed-oxide charge and the difference in work function, there is a flat-band voltage shift. Additionally, substrate bias can also influence the threshold voltage. When a reverse bias is applied between the substrate and the source, the depletion region is widened and the threshold voltage required to achieve inversion must be increased to accommodate the larger Q_{sc}. These factors in turn cause a change in the threshold voltage:

$$V_T \approx V_{FB} + 2\psi_B + \frac{\sqrt{2\varepsilon_s q N_A (2\psi_B + V_{BS})}}{C_o}, \tag{45}$$

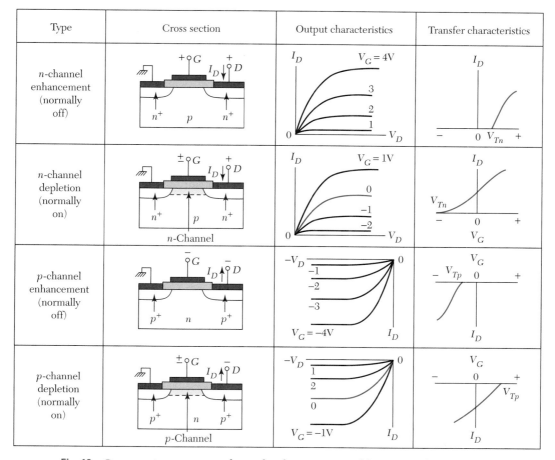

Type	Cross section	Output characteristics	Transfer characteristics
n-channel enhancement (normally off)			
n-channel depletion (normally on)			
p-channel enhancement (normally off)			
p-channel depletion (normally on)			

Fig. 19 Cross section, output, and transfer characteristics of four types of MOSFETs.

where V_{BS} is the reverse substrate-source bias. Figure 20 shows the calculated threshold of n-channel (V_{Tn}) and p-channel (V_{Tp}) MOSFETs with n^+-, p^+- polysilicon and mid-gap work function gate electrodes as a function of their substrate doping, assuming $d = 5$ nm, $V_{BS} = 0$, and $Q_f = 0$. Mid-gap gate materials are those with a work function of 4.61 eV, which equals the sum of the electron affinity $q\chi$ and $E_g/2$ of silicon (see Fig. 2).

Precise control of the threshold voltage of MOSFETs in an integrated circuit is essential for reliable circuit operation. Typically, the threshold voltage is adjusted through ion implantation into the channel region. For example, a boron implantation through a surface oxide is often used to adjust the threshold voltage of an n-channel MOSFET (with p-type substrate). Using this method, it is possible to obtain close control of threshold voltage because very precise quantities of impurity can be introduced. The negatively charged boron acceptors increase the doping level of the channel. As a result, V_T increases. Similarly, a shallow boron implant into a p-channel MOSFET can reduce V_T.

▷ **EXAMPLE 6**

For an n-channel n^+-polysilicon-SiO$_2$-Si MOSFET with $N_A = 10^{17}$ cm^{-3} and $Q_f/q = 5 \times 10^{11}$ cm^{-2}, calculate V_T for a gate oxide of 5 nm. What is the boron ion dose required to increase V_T to 0.6 V? Assume that the implanted acceptors form a sheet of negative charge at the Si-SiO$_2$ interface.

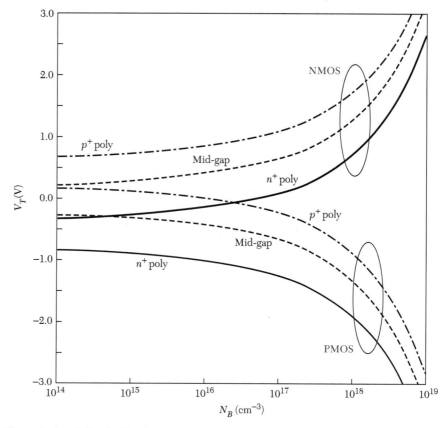

Fig. 20 Calculated threshold voltage of n-channel (V_{Tn}) and p-channel (V_{Tp}) MOSFETs as a function of impurity concentration, for devices with n^+-, p^+-polysilicon, and mid-gap work function gates assuming zero fixed charge. The thickness of the gate oxide is 5 nm. NMOS, n-channel MOSFET; PMOS, p-channel MOSFET.

SOLUTION From the examples in Section 6.1, we have $C_o = 6.9 \times 10^{-7}$ F/cm^2, $2\psi_B = 0.84$ V, and $V_{FB} = -1.1$ V. Therefore, from Eq. 45 (with $V_{BS} = 0$),

$$V_T = V_{FB} + 2\psi_B + \frac{\sqrt{2\varepsilon_s q N_A (2\psi_B)}}{C_o}$$

$$= -1.1 + 0.84 + \frac{\sqrt{2 \times 11.9 \times 8.85 \times 10^{-14} \times 1.6 \times 10^{-19} \times 10^{17} \times 0.84}}{6.9 \times 10^{-7}}$$

$$= -0.02 \text{ V}.$$

The boron charge causes a flat-band shift of qF_B/C_o. Thus,

$$0.6 = -0.02 + \frac{qF_B}{6.9 \times 10^{-7}},$$

$$F_B = \frac{0.62 \times 6.9 \times 10^{-7}}{1.6 \times 10^{-19}} = 2.67 \times 10^{12} \text{ cm}^{-2}.$$

◀

We can also control V_T by varying the oxide thickness. Threshold voltage becomes more positive for an n-channel MOSFET and more negative for a p-channel MOSFET as the oxide thickness is increased. This is simply due to the reduced field strength at a fixed gate voltage for a thicker oxide. Such an approach is used extensively for isolating transistors fabricated on a chip. Figure 21 shows the cross section of an isolation oxide (also called field oxide) between an n^+ diffusion and an n-well. Details about the field oxide formation and the well technology is given in Chapter 14. The n^+ diffusion region is the source or drain region of a normal n-channel MOSFET. The gate oxide of MOSFET is much thinner than the field oxide. When a conductor line is formed over the field oxide, a parasitic MOSFET, also called a field transistor, results with the n^+ diffusion and n-well regions as the source and drain, respectively. The V_T of the field oxide is typically an order of magnitude larger than that of the thin gate oxide. During circuit operation, the field transistor will not be turned on. Consequently, the field oxide offers good isolation between the n^+ diffusion and n-well regions.

▶ **EXAMPLE 7**

For an n-channel field transistor with $N_A = 10^{17}$ cm^{-3} and $Q_f/q = 5 \times 10^{11}$ cm^{-2}, calculate V_T for a gate oxide (i.e., the field oxide) of 500 nm.

SOLUTION $C_o = \varepsilon_{ox} / d = 6.9 \times 10^{-9}$ F/cm^2.

From Exs. 2 and 3, we have $2\psi_B = 0.84$ V, and $V_{FB} = -1.1$ V.

Therefore, from Eq.45 (with $V_{BS} = 0$)

$$V_T = V_{FB} + 2\psi_B + \frac{\sqrt{2\varepsilon_s q N_A \left(2\psi_B\right)}}{C_o}$$

$$= -1.1 + 0.84 + \frac{\sqrt{2 \times 11.9 \times 8.85 \times 10^{-14} \times 1.6 \times 10^{-19} \times 10^{17} \times 0.84}}{6.9 \times 10^{-9}}$$

$$= 24.12 \text{ V.}$$ ◀

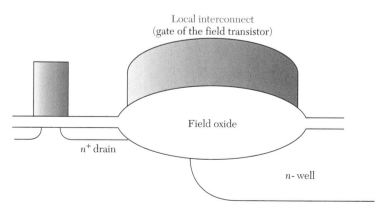

Local interconnect
(gate of the field transistor)

Field oxide

n^+ drain

n-well

Fig. 21   Cross section of a parasitic field transistor in an n-well structure.

Substrate bias can also be used to adjust the threshold voltage. According to Eq. 45, the change in threshold voltage due to the substrate bias is

$$\Delta V_T = \frac{\sqrt{2\varepsilon_s q N_A}}{C_o}\left(\sqrt{2\psi_B + V_{BS}} - \sqrt{2\psi_B}\right). \tag{46}$$

If we plot the drain current versus V_G, the intercept at the V_G-axis corresponds to the threshold voltage, Eq. 36. Such a plot is shown in Fig. 22 for three different substrate biases. As the magnitude of the substrate V_{BS} increases from 0 V to 2 V, the threshold voltage also increases from 0.56 V to 1.03 V. The substrate effect can be used to raise the threshold voltage of a marginal enhancement device ($V_T \sim 0$) to a larger value.

▶ **EXAMPLE 8**

For the MOSFET discussed in Ex. 6 with V_T of −0.02 V, if the substrate bias is increased from zero to 2 V, calculate the change in threshold voltage.

SOLUTION

From Eq. 46,

$$\Delta V_T = \frac{\sqrt{2\varepsilon_s q N_A}}{C_o}\left(\sqrt{2\psi_B + V_{BS}} - \sqrt{2\psi_B}\right)$$

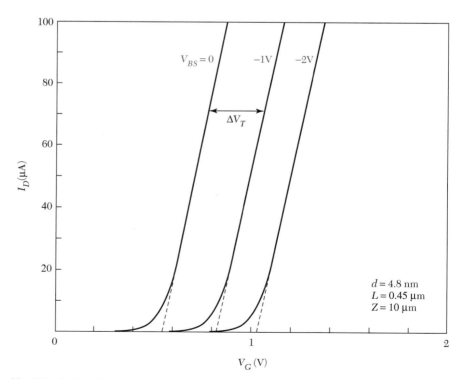

Fig. 22 Threshold voltage adjustment using substrate bias.

$$= \frac{\sqrt{2 \times 11.9 \times 8.85 \times 10^{-14} \times 1.6 \times 10^{-19} \times 10^{17}}}{6.9 \times 10^{-7}} \left(\sqrt{0.84 + 2} - \sqrt{0.84} \right),$$

$$= 0.27 \times (1.69 - 0.92) = 0.21 \text{ V.} \qquad \blacktriangleleft$$

Another way to control V_T is to adjust the work function difference by choosing an appropriate gate material. A number of conducting materials have been proposed, such as W, TiN, and a heavily doped polycrystalline silicon-germanium layer.[6] In deep submicron device fabrication, control of the threshold voltage and device performance becomes more difficult because of the geometric effects encountered in device scaling (see the discussion in the next section). The use of other gate materials to replace the conventional n^+ polysilicon could make device design more flexible in deep submicron fabrication.

6.3 MOSFET SCALING

Scaling down of MOSFET's dimensions is a continuous trend since its inception. Smaller device size enables higher device density in an integrated circuit. In addition, a smaller channel length improves the driving current ($I_D \sim 1/L$) and, thus, the operation performance. As a device's dimensions are reduced, however, influences from the side regions of the channel (i.e., source, drain, and isolation edge) become significant. Device characteristics, therefore, deviate from those derived from long-channel approximation.

6.3.1 Short-Channel Effects

The threshold voltage given in Eq. 45 is derived based on the gradual-channel approximation stated in Section 6.2.1. That is, the charges contained in the surface depletion region of the substrate are induced solely from the field created by the gate voltage. In other words, the third term in Eq. 45 is independent of the lateral fields from the source and drain. As channel length is reduced, however, the fields originating from the source/drain regions may influence the charge distribution and, thus, the device characteristics such as the threshold voltage control and device leakage.

Threshold Voltage Roll-Off in Linear Region
When the channel side effect becomes nonnegligible, the threshold voltage in the linear region usually becomes less positive as channel length decreases for n-channel MOSFETs and less negative as channel length decreases for p-channel MOSFETs. Figure 23 shows an example with $V_{DS} = 0.05$ V of this V_T roll-off phenomenon.[7] Roll-off can be explained by the charge-sharing model,[8] as illustrated in Fig. 24. This figure shows the cross section of an n-channel MOSFET. This device is operated in the linear region ($V_{DS} \leq 0.1$ V). Hence, the depletion region width for the drain junction is almost equal to that for the source junction. Since the channel depletion region overlaps the source and drain depletion regions, charges induced by the field created by the gate bias can be approximated by those within the trapezoidal region.

The threshold voltage shift ΔV_T is due to the reduction of charges in the depletion layer from the rectangular region $L \times W_m$ to the trapezoidal region $(L + L')W_m/2$. ΔV_T is given by (see Prob. 27)

$$\Delta V_T = - \frac{q N_A W_m r_j}{C_o L} \left(\sqrt{1 + \frac{2 W_m}{r_j}} - 1 \right), \qquad (47)$$

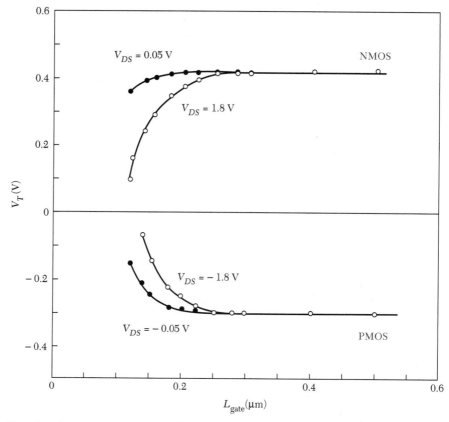

Fig. 23 Threshold voltage roll-off characteristics in a 0.15 μm complementary metal-oxide-semiconductor (CMOS) field-effect transistor technology.[7]

where N_A is the substrate doping concentration, W_m the depletion width, r_j the junction depth, L the channel length, and C_o the gate oxide capacitance per unit area.

For long-channel devices, the charge reduction is smaller, since Δ (Fig. 24) is much smaller than L. For short-channel devices, however, the charges needed to turn on the device are dramatically reduced, since Δ is comparable to L. As can be seen from Eq. 47, for a given set of N_A, W_m, r_j and C_o, the threshold voltage decreases with decreasing channel length.

Drain-Induced Barrier Lowering

When the drain voltage of a short-channel MOSFET increases from the linear region toward the saturation region, its threshold voltage roll-off becomes larger (see Fig. 23). This effect is called drain-induced barrier lowering (DIBL). The surface potentials between source and drain for several n-channel devices with different channel lengths are shown[9] in Fig. 25. The dotted lines are for $V_{DS} = 0$ and the solid lines for $V_{DS} > 0$. When the gate voltage is below V_T, the p-Si substrate forms a potential barrier between n^+ source and drain and limits the current flow from source to drain. For a device operated in the saturation region, the depletion-layer width of the drain junction is significantly wider than that of the source junction. In the long-channel case, the increase in depletion-layer region width of the drain junction will not affect the potential barrier height (see Fig. 25, the 1 μm case). Nevertheless, when the channel length is short enough, the increase in drain

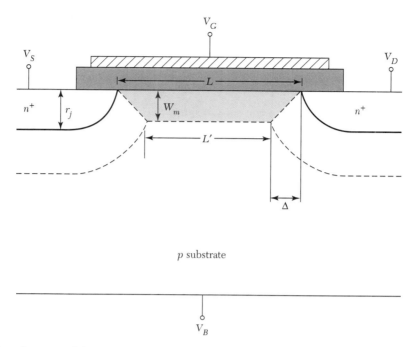

Fig. 24 Schematic of the charge sharing model.[8]

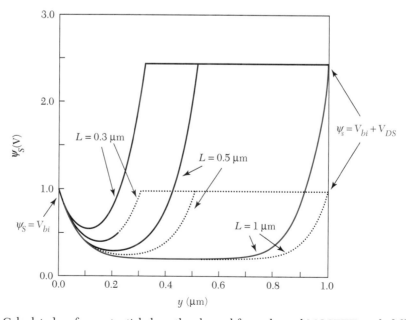

Fig. 25 Calculated surface potential along the channel for n-channel MOSFETs with different channel lengths.[9] The source-channel boundary is at $y = 0$. A low (0.05 V, dotted lines) and a high (1.5 V, solid lines) V_{DS} are applied. Oxide thickness d and substrate doping N_A are 10 nm and 10^{16} cm^{-3}, respectively. The substrate bias is 0 V.

voltage decreases the potential barrier height (the 0.3 and 0.5 μm cases in Fig. 25). This is ascribed to the field penetration at the surface region from the drain to the source when they are too close. Such a barrier-lowering effect leads to a substantial increase in electron injection from the source to the drain. As a result, subthreshold current increases. The threshold voltage, thus, decreases further with increasing drain bias in short-channel devices.

Figure 26 illustrates the subthreshold characteristics of a long and a short *n*-channel MOSFET at low and high drain bias conditions. The parallel shift in subthreshold current in the short-channel device (Fig. 26*b*) as the drain voltage increases indicates that a significant DIBL effect has been induced.

Bulk Punch-through

DIBL causes the formation of a leakage path at the SiO_2/Si interface. If the drain voltage is large enough, significant leakage current may also flow from drain to source via the bulk of the substrate for a short-channel MOSFET. This is also ascribed to the increase in the depletion-layer region width of the drain junction with increasing drain voltage. In a short-channel MOSFET, the sum of depletion-layer width for source and drain junctions is comparable to the channel length. The depletion region of the drain junction gradually merges with that of the source junction when the drain voltage is increased. As a result, a large leakage current may flow from the drain to the source through the bulk. Figure 27 shows the subthreshold characteristics of a short-channel ($L = 0.23$ μm) MOSFET. When the drain voltage is increased from 0.1 to 1 V, DIBL is induced with the parallel shift in the subthreshold characteristics similar to that shown in Fig. 26*b*. When the drain voltage is further increased to 4 V, the subthreshold swing is much larger than that for lower drain biases. Consequently, the device has a very high leakage current. This indicates that the bulk punch-through effect is very significant. The gate can no long turn the device completely off and loses control of the drain current. High-leakage current limits device operation for short-channel MOSFETs.

6.3.2 Scaling Rules

As device dimensions are reduced, the short-channel effects must be minimized to maintain normal device and circuit operation. Some guidelines are necessary in scaled-device design. One elegant approach for maintaining the long-channel behavior is to simply reduce all dimensions and voltages by a scaling factor $\kappa(>1)$, so that the internal electric fields are the same as those of a long-channel MOSFET. This approach is called *constant-field scaling*.[10]

Table 1 summarizes the scaling rules of constant-field scaling for various device parameters and circuit performance factors.[11] The circuit performance (speed and power consumption in the on-state) can be enhanced as the device dimensions are scaled down.§ In practical integrated circuit (IC) manufacturing, however, the electric fields inside the smaller devices are not kept constant but increase to some extent. This is mainly because the voltage factors (e.g., power supply, threshold voltage) cannot be scaled arbitrarily. If the threshold voltage is too small, the leakage level in the off-state ($V_G = 0$) will increase significantly because of the nonscalable subthreshold swing. Consequently, standby power consumption will also increase.[12] By applying the scaling rules, MOSFETs have been fabricated that have a channel length as short as 20 nm, a very high transconductance (>1000 mS/mm), and a reasonable subthreshold swing (~120 mV/decade).[13]

---

§A comparison of the cutoff frequency for different field-effect transistors (including MOSFET) is shown in Fig. 18 of Chapter 7.

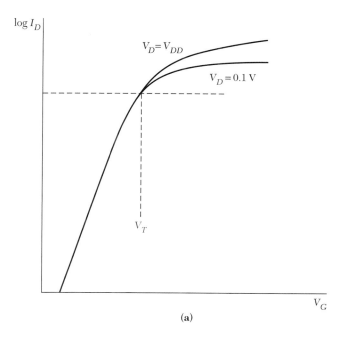

(a)

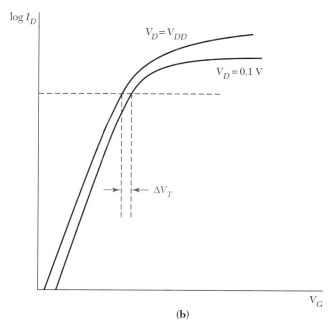

(b)

Fig. 26 Subthreshold characteristics of (*a*) a long-channel and (*b*) a short-channel MOSFET.

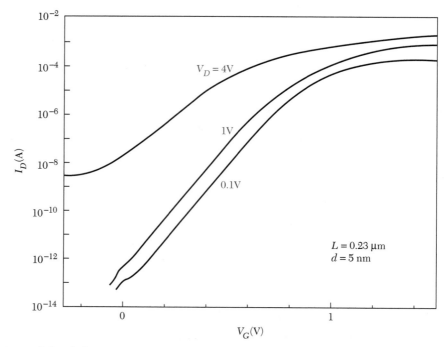

Fig. 27   Subthreshold characteristics of an n-channel MOSFET with V_{DS} = 0.1, 1 , and 4 V.

TABLE 1   Scaling of MOSFET Device and Circuit Parameters

Determinant	MOSFET device and circuit parameters	Multiplying factor ($\kappa > 1$)
Scaling assumptions	Device dimensions (d, L, W, r_j)	$1/\kappa$
	Doping concentration (N_A, N_D)	κ
	Voltage (V)	$1/\kappa$
Derived scaling behavior of device parameters	Electric field ($\mathscr{E}$)	1
	Carrier velocity (v)	1
	Depletion-layer width (W)	$1/\kappa$
	Capacitance ($C = \varepsilon A/d$)	$1/\kappa$
	Inversion-layer charge density (Q_n)	1
	Current, drift (I)	$1/\kappa$
	Channel resistance (R)	1
Derived scaling behavior of circuit parameters	Circuit delay time ($\tau \sim CV/I$)	$1/\kappa$
	Power dissipation per circuit ($P \sim VI$)	$1/\kappa^2$
	Power-delay product per circuit ($P\tau$)	$1/\kappa^3$
	Circuit density ($\sim 1/A$)	κ^2
	Power density (P/A)	1

► 6.4   CMOS AND BiCMOS

Complementary MOS (CMOS) refers to a complementary p-channel and n-channel MOS-FET pair. CMOS logic is the most popular technology utilized in present-day integrated circuit design. The main reasons for the success of CMOS are low power consumption and good noise immunity. In fact, currently only CMOS technology is used in advanced integrated-circuit manufacturing due to the low power-dissipation requirement.

6.4.1   The CMOS Inverter

A CMOS inverter, which is the basic element of CMOS logic circuits, is shown in Fig. 28. In a CMOS inverter, the gates of the p- and n-channel transistors are connected and serve as the input node to the inverter. The drains of the two transistors are also connected and serve as the output node to the inverter. The source and substrate contacts of n-channel MOSFETs are grounded, whereas those of p-channel MOSFETs are connected to the power supply (V_{DD}). Note that both p-channel and n-channel MOSFETs are enhancement-type transistors. When the input voltage is low (e.g., $V_{in} = 0$, $V_{GSn} = 0$ $< V_{Tn}$), the n-channel MOSFET is off.§ The p-channel MOSFET, however, is on, since $|V_{GSp}| \cong V_{DD} > |V_{Tp}|$ (V_{GSp} and V_{Tp} are negative). Consequently, the output node is charged to V_{DD} through the p-channel MOSFET. When the input voltage goes high so that the gate voltage equals V_{DD}, the n-channel MOSFET is turned on, since $V_{GSn} = V_{DD} > V_{Tn}$, and the p-channel MOSFET is turned off, since $|V_{GSp}| \cong 0 < |V_{Tp}|$. Therefore, the output node is discharged to ground through the n-channel MOSFET.

For more detailed understanding of the operation of the CMOS inverter, we can first plot the output characteristics of the transistors. This is given in Fig. 29, in which I_p and I_n are shown as a function of output voltage (V_{out}). I_p is the current of p-channel MOSFET in the direction from the source (connected to V_{DD}) to the drain (output node). I_n is the current of n-channel MOSFET in the direction from the drain (output node) to the source (connected to ground). It is noted that the increase in input voltage (V_{in}) tends to increase I_n but decrease I_p at a fixed V_{out}. In steady state, however, I_n should be equal to I_p. For a given V_{in}, we can determine the corresponding V_{out} from the intercept of $I_n(V_{in})$ and $I_p(V_{in})$, as shown in Fig. 29. The V_{in}-V_{out} curve, as shown in Fig. 30, is called the transfer curve of the CMOS inverter.[11]

An important characteristic of the CMOS inverter is that when the output is in a steady logic state, i.e., $V_{out} = 0$ or V_{DD}, only one transistor is on. The current flow from the power supply to ground is thus very low and is equal to the leakage current of the off device. In fact, there is significant current conduction only during the short transient period when the two devices are temporarily on. Therefore the power consumption is very low in the static state compared with other types of logic circuits, such as n-channel MOSFETs, bipolar, etc.

6.4.2   Latch-up

In order to fabricate both p-channel and n-channel MOSFETs in the same chip for CMOS applications, extra doping and diffusion steps are needed to form the "well" or "tub" in the substrate. The doping type in the well is different from that of the surrounding

---

§ V_{GSn} and V_{GSp} represent the voltage difference between the gate and the source for n- and p-channel MOSFETs, respectively.

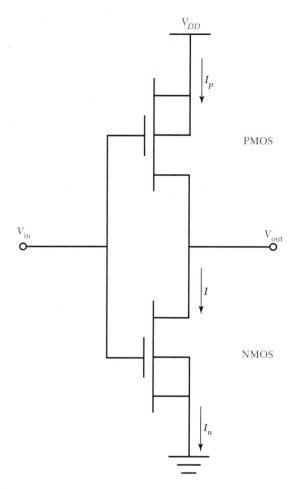

Fig. 28 The CMOS inverter.

substrate. Typical well types are the p-well, n-well, and twin well. Details of the well technology are given in Chapter 14. Figure 31 shows a cross-sectional view of a CMOS inverter fabricated using p-well technology. In this figure, the p-channel and n-channel MOSFETs are fabricated in the n-type Si substrate and the p-well region, respectively.

A major problem related to the well structure in CMOS circuits is the latch-up phenomenon. The cause of latch-up is the action of the parasitic p-n-p-n diode in the well structure. As shown in Fig. 31, the parasitic p-n-p-n diode consists of a lateral p-n-p and a vertical n-p-n bipolar transistors. The p-channel MOSFET's source, n-substrate, and p-well correspond to the emitter, base, and collector of the lateral p-n-p bipolar transistor, respectively. The n-channel MOSFET's source, p-well, and n-substrate are the emitter, base, and collector of the vertical n-p-n bipolar transistor, respectively. The equivalent circuit of the parasitic components is illustrated in Fig. 32. R_S and R_W are the series resistance in the substrate and the well, respectively. The base of each transistor is driven by the collector of the other to form a positive feedback loop. This configuration is similar to the thyristor discussed in Chapter 5. Latch-up is induced when the current gain product of the two bipolar transistors, $\alpha_{npn}\alpha_{pnp}$, is larger than 1. When latch-up occurs, a large current will flow from the power supply (V_{DD}) to the ground contact. This can interrupt

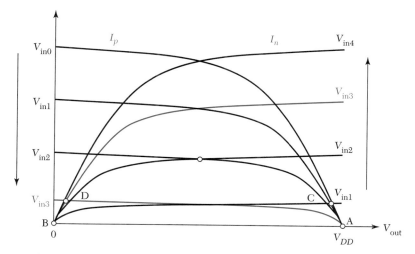

Fig. 29 I_p and I_n as a function of V_{out}. The intercepts of I_p and I_n (circled) represent the steady-state operation points of the CMOS inverter.[11] The curves are labeled by the input voltages: $0 = V_{in0} < V_{in1} < V_{in2} < V_{in3} < V_{in4} = V_{DD}$.

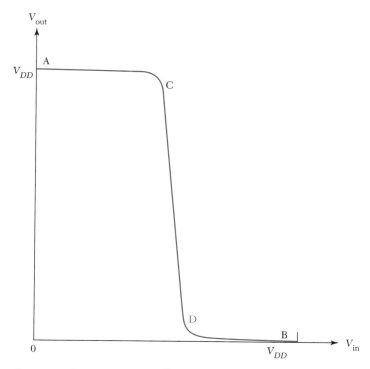

Fig. 30 Transfer curve of a CMOS inverter.[11] Points labeled A, B, C, and D correspond to those points labeled in Fig. 29.

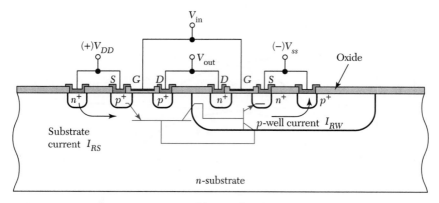

Fig. 31 Cross section of a CMOS inverter fabricated with p-well technology.

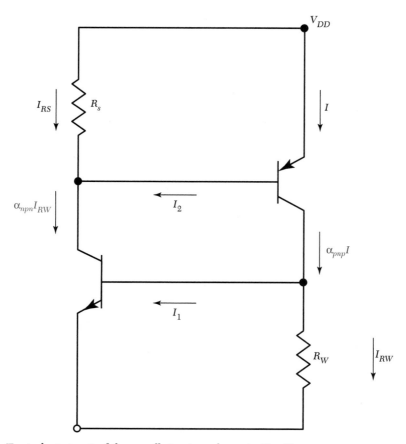

Fig. 32 Equivalent circuit of the p-well structure shown in Fig. 31.

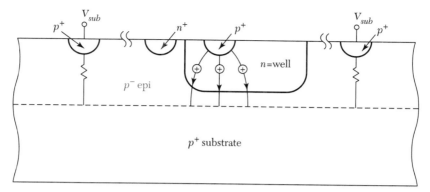

Fig. 33   Prevention of latch-up with a heavily doped substrate.[14]

normal circuit operation and even destroy the chip itself because of the high power dissipation.

To avoid latch-up, the current gains of the parasitic bipolar transistors must be reduced. One method is to use gold doping or neutron irradiation to lower the minority carrier lifetimes. However, this approach is difficult to control. Besides, it also causes an increase of the leakage current. A deeper well structure or high-energy implantation to form retrograde wells can also reduce the current gain of vertical bipolar transistor by raising the impurity concentration in the base. In the retrograde well, the peak of the well doping concentration is located within the substrate away from the surface.

Another way to reduce latch-up is to use a heavily doped substrate with devices fabricated on a lightly doped epitaxial layer, as shown[14] in Fig. 33. The heavily doped substrate provides a highly conductive path to collect the current. The current then is drained away through the surface contacts.

Latch-up can also be avoided with the trench isolation scheme. A process for forming trench isolation is discussed in Chapter 14. This approach can eliminate latch-up because the n-channel and p-channel MOSFETs are physically isolated by the trench.

6.4.3   BiCMOS

CMOS has the advantages of low power dissipation and high device density, which make it suitable for fabricating complex circuits. However, CMOS suffers from low drive capability compared with bipolar technology, limiting its circuit performance. BiCMOS is a technology that integrates both CMOS and bipolar device structures in the same chip. A BiCMOS circuit contains mostly CMOS devices, with a relatively small number of bipolar devices. The bipolar devices have better switching performance than their CMOS counterparts without consuming too much extra power. However, this performance enhancement is achieved at the expense of extra manufacturing complexity, longer fabrication time, and higher cost. The fabrication processes for BiCMOS are discussed in Chapter 14.

▶ 6.5 MOSFET ON INSULATOR

For certain applications, MOSFETs are fabricated on an insulating substrate rather than on a semiconductor substrate. The characteristics of these transistors are, however, similar to those of a MOSFET. Usually, we call such devices thin film transistors (TFT) if the channel layer is an amorphous or a polycrystalline silicon. If the channel layer is a monocrystalline silicon, we call it silicon-on-insulator (SOI).

6.5.1 The Thin Film Transistor (TFT)

Hydrogenated amorphous silicon (a-Si:H) and polysilicon are the two most popular materials for TFT fabrication. They are usually deposited on an insulating substrate such as a glass, quartz, or Si substrate with a thin SiO_2 capping layer.

The hydrogenated amorphous silicon TFT is an important device in electronic applications that require a large area, such as liquid crystal displays (LCD) and contact imaging sensors (CIS). The a-Si:H materials are usually deposited with a plasma-enhanced chemical vapor deposition (PECVD) system. Since the deposition temperature is low (typically 200°–400°C), inexpensive substrate materials such as glass can be used. The role played by the hydrogen atoms contained in the a-Si:H is to passivate dangling bonds in the amorphous silicon matrix and thus reduce the defect density. Without hydrogen passivation, the gate voltage can not adjust the Fermi level at the insulator and the a-Si interface, since the level is pinned by the large amount of defects.

The a-Si:H TFT is usually fabricated using the inverted staggered structure, as shown in Fig. 34. The inverted staggered structure is with a bottom-gate scheme. A metal gate can be used since the post process temperature is low (< 400°C). A dielectric layer such as silicon nitride or silicon dioxide, also deposited by PECVD, is often used as the gate dielectric. An undoped a-Si:H layer is subsequently deposited to form the channel. The source and drain of the TFT are formed with an in situ–doped n^+ a-Si:H layer complying with the requirement of low process temperature. A dielectric layer that serves as an etch-stop for patterning of n^+ a-Si:H is often used. Device characteristics of TFTs with the bottom-gate structure are usually better than those with the top-gate structure. This is because the a-Si:H channel could be damaged by plasma during PECVD gate-dielectric deposition of top-gated TFTs. In addition, the source/drain formation process is easier for the bottom-gate structure. A typical subthreshold characteristic of the a-Si:H TFT is shown in Fig. 35. Because of the amorphous matrix presents in the channel material, its carrier mobility is usually very low (< 1 cm²/V-s).

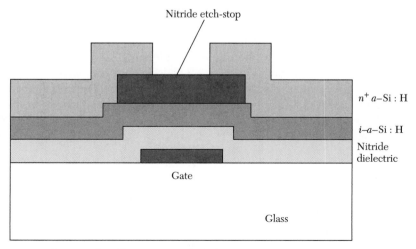

Fig. 34 A typical a-Si:H thin film transistor (TFT) structure.

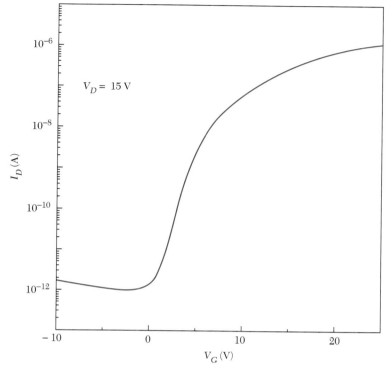

Fig. 35 Subthreshold characteristics of an a-Si:H TFT (L/Z = 10/60 μm/μm). The field-effect carrier mobility is 0.23 cm^2/ V-s.

The polysilicon TFT uses a thin polysilicon as the channel layer. Polysilicon is a material consisting of many Si grains. Within the grain are the monocrystalline Si lattices. The orientations of two side-by-side grains are, however, different from each other. The interface between the two grains is called the grain boundary. Polysilicon TFT exhibits much higher carrier mobility and thus better drive capability than a-Si:H TFT because of higher crystallinity. Carrier mobility of these devices typically ranges from 10 to several hundred cm^2/V-s, depending on the grain size and process conditions. Polysilicon is usually deposited with low-pressure CVD (LPCVD). The grain size of polysilicon is an important factor in determining the TFT performance, since the carrier mobility generally decreases with decreasing grain size. This is mainly because of the large number of defects contained in the grain boundaries that impede the transport of carriers.

The defects at the grain boundary could also affect the threshold voltage and subthreshold swing of the device. When gate voltage is applied to induce an inversion layer in the channel, these defects act as traps and impede the movement of the Fermi level in the forbidden gap. To alleviate these drawbacks, a hydrogenation step is often adopted after device fabrication. The hydrogenation treatment is usually done in a plasma reactor. Hydrogen atoms or ions generated in the plasma diffuse into the grain boundaries and passivate these defects. After hydrogenation, there is significant improvement in device performance .

Unlike a-Si:H TFT, polysilicon TFT is usually fabricated with the top-gate structure, as shown in Fig. 36. A self-aligned implant is used to form the source/drain. One main limitation of polysilicon TFT manufacturing is the high process temperature (> 600°C). Consequently, expensive substrates such as quartz are usually needed to tolerate the high

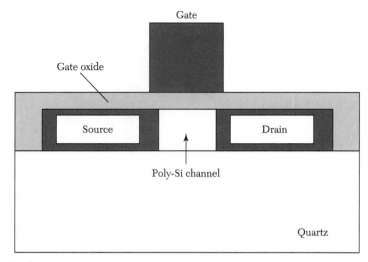

Fig. 36 A polysilicon TFT structure.

process temperatures. This makes polysilicon TFT less attractive than a-Si:H TFT in production for low-end applications because of higher cost. Laser crystallization of Si is a potential way to overcome the problem. In this method, an a-Si layer is deposited first on a glass substrate at low temperatures by PECVD or LPCVD. A high-power laser source is then used to irradiate the a-Si. The energy is absorbed by the a-Si and melting occurs locally in the a-Si layer. After cooling, the a-Si turns into polysilicon with very large grain size (≥ 100 μm). Very high carrier mobility, approaching that of crystalline Si MOSFETs, can be obtained using this method.

6.5.2 Silicon-on-Insulator (SOI) Devices

Many SOI devices have been proposed, including silicon-on-sapphire (SOS), silicon-on-spinel, silicon-on-nitride, and silicon-on-oxide.[15] Figure 37 shows a schematic diagram of an SOI CMOS built on silicon dioxide. Compared with CMOS built on a bulk Si substrate, also called bulk CMOS, SOI's isolation scheme is simplified and does not need complicated well structures. Device density can thus be increased. The latch-up phenomenon inherent in bulk CMOS circuits is also eliminated. The parasitic junction capac-

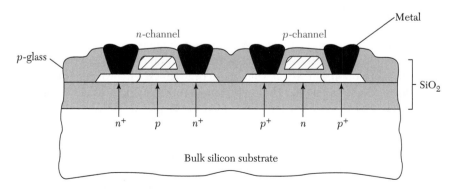

Fig. 37 Cross section of the silicon-on-insulator (SOI).

itance in the source and drain regions can be significantly reduced with the insulating substrate. Additionally, significant improvement over bulk CMOS in radiation-damage toleration is achieved in SOI. This is because of the small volume of Si available for electron-hole pair generation by radiation. This property is particularly important for space applications.

Depending on the thickness of the Si channel layer, SOI can be classified into partially depleted (PD) and fully depleted (FD) types. PD-SOI uses a thicker Si channel layer so that the depletion width of the channel does not exceed the thickness of Si layer. Device design and performance of a PD-SOI is similar to that of bulk CMOS. One major difference is the floating substrate used in SOI devices. During device operation, a high field near the drain could induce impact ionization there. Majority carriers, holes in the p-substrate for an n-channel MOSFET, generated by impact ionization will be stored in the substrate, since there is no substrate contact to drain away these charges. Therefore, the substrate potential will be changed and will result in a reduction of the threshold voltage. This, in turn, may cause an increase or a kink in the current-voltage characteristics. The kink phenomenon is shown[15] in Fig. 38. This float-body or kink effect is especially dramatic for n-channel devices, because of the higher impact-ionization rate of electrons. The kink effect can be eliminated by forming a substrate contact to the source of the transistor. This will, however, complicate the device layout and process flow.

FD-SOI uses a Si layer thin enough so that the channel of the transistor is completely depleted before threshold is reached. This allows the device to be operated at a lower field. In addition, the kink effect caused by high-field impact ionization can be eliminated. FD-SOI is very attractive for low-power applications. Nevertheless, the FD-SOI's characteristics are sensitive to the Si thickness variation. If an FD-SOI circuit is built on a wafer with nonuniform Si thickness, its operation will be unstable.

▶ **EXAMPLE 9**

Calculate the threshold voltage for an n-channel SOI device having $N_A = 10^{17}$ cm^{-3}, $d = 5$ nm, and $Q_f/q = 5 \times 10^{11}$ cm^{-2}. Si thickness, d_{Si}, for the device is 50 nm.

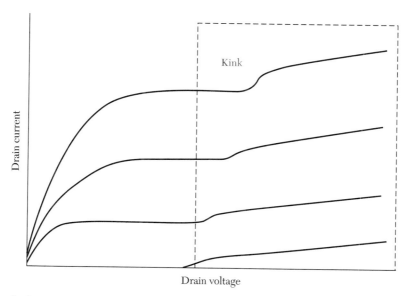

Fig. 38   The kink effect in the output characteristics of an n-channel SOI MOSFET.[15]

SOLUTION   From Ex. 1, the maximum depletion width, W_m, for a bulk NMOS device is 100 nm. Therefore, the SOI device is a fully depleted type. Since the width of the depletion region is now the Si thickness, W_m used in Eq. 17 and Eq. 45 for calculating the threshold voltage should be replaced by d_{Si}:

$$V_T = V_{FB} + 2\psi_B + \frac{qN_A d_{Si}}{C_o}.$$

From Exs. 2 and 3, we have $C_o = 6.9 \times 10^{-7}$ F/cm^2, $V_{FB} = -1.1$ V, and $2\psi_B = 0.84$ V.

Therefore,

$$V_T = -1.1 + 0.84 \; + \frac{1.6 \times 10^{-19} \times 10^{17} \times 5 \times 10^{-6}}{6.9 \times 10^{-7}} = -0.14 \text{ V.} \quad \blacktriangleleft$$

6.6 MOS MEMORY STRUCTURES

Semiconductor memories can be classified as volatile and nonvolatile memories. Volatile memories such as dynamic random access memories (DRAMs) and static random access memories (SRAMs) lose their stored information if the power supply is switched off. Nonvolatile memories, on the other hand, can retain the stored information. Currently, DRAM and SRAM are extensively used in personal computers and work stations, mainly because of DRAM's attributes of high density and low cost, and SRAM's attribute of high speed. The nonvolatile memory is used extensively in portable electronics systems such as the cellular phone, digital camera, and smart IC cards, mainly because of its attributes of low-power consumption and nonvolatility.

6.6.1 DRAM

Modern DRAM technology consists of a cell array using a storage cell structure shown[16] in Fig. 39. The cell includes an MOSFET and an MOS capacitor [i.e., one transistor/one capacitor (1T/1C) cell]. The MOSFET acts as a switch to control the writing, refreshing, and read-out actions of the cell. The capacitor is used for charge storage. During the write cycle, the MOSFET is turned on so that the logic state in the bit line is transferred to the storage capacitor. For practical applications, charges stored in the capacitor will be gradually lost because of the small but nonnegligible leakage current of the

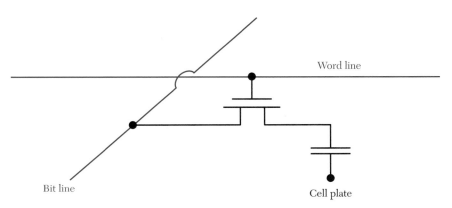

Fig. 39   Basic configuration of a dynamic random access memory (DRAM) cell.[16]

storage node. Consequently, the operation of DRAM is "dynamic," since the data need to be "refreshed" periodically within a fixed interval, typically 2–50 ms.

The 1T/1C DRAM cell has the advantages of very simple and small area construction. In order to increase the storage density of a chip, aggressive scaling of the cell size is necessary. However, this will degrade the storage capability of the capacitor, since the capacitor electrode area will be diminished as well. To solve this problem, three-dimension (3-D) capacitor structures are required. Some novel 3-D capacitor structures are discussed in Chapter 14. The utilization of high dielectric-constant materials to replace conventional oxide-nitride composite layers (dielectric constant: 4 ~ 6) as the capacitor dielectric materials can also be employed to increase the capacitance.

6.6.2 SRAM

SRAM is a matrix of static cells using a bistable flip-flop structure to store the logic state, as shown in Fig. 40. The flip-flop consists of two cross-coupled CMOS inverters (T1,T3 and T2,T4). The output of the inverter is connected to the input node of the other inverter. This configuration is called "latched." Two additional n-channel MOSFETs, T5 and T6, with their gates connected to the word line, are used to access the SRAM cell. The operation of the SRAM is static since the logic state is sustained as long as the power is applied. Therefore SRAM does not have to be refreshed. The two p-channel MOSFETs (T1 and T2) in the inverters are used as the load transistors. There is essentially no dc current flow through the cell, except during switching. In some situations, p-channel polysilicon TFTs or polysilicon resistors are used instead of bulk p-channel MOSFETs. These

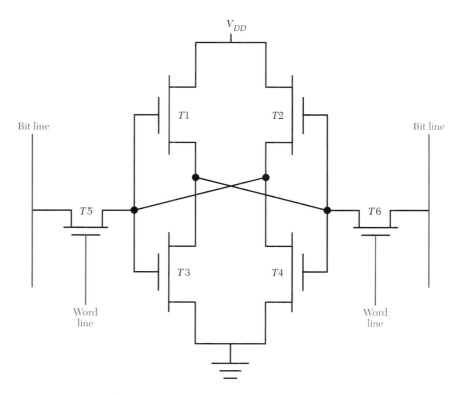

Fig. 40 Configuration of a CMOS SRAM cell. T1 and T2 are load transistors (p-channel). T3 and T4 are drive transistors (n-channel). T5 and T6 are access transistors (n-channel).

polysilicon load devices can be fabricated over the bulk n-channel MOSFETs. 3-D integration can effectively reduce the cell area and, therefore, increases the storage capacity of the chip.

6.6.3 Nonvolatile Memory

When the gate electrode of a conventional MOSFET is modified so that semipermanent charge storage inside the gate is possible, the new structure becomes a nonvolatile memory device. Since the first nonvolatile memory device proposed[17] in 1967, various device structures have been made. Nonvolatile memory devices have been used extensively in ICs such as the erasable-programmable read-only memory (EPROM), electrically erasable-programmable read-only memory (EEPROM), and flash memory.

Most of the devices used for present-day nonvolatile memories have a floating-gate structure (Fig. 41a). Charges are injected from the silicon substrate or the drain to the floating gate across the first insulator. They are then stored in the floating gate. This pro-

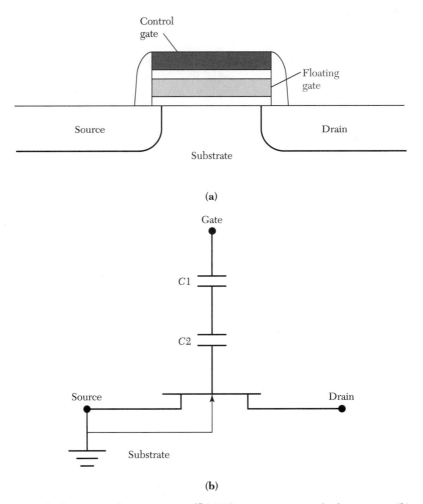

Fig. 41 Nonvolatile semiconductor memory.[17] (a) Floating-gate nonvolatile memory. (b) Equivalent circuit of the nonvolatile memory.

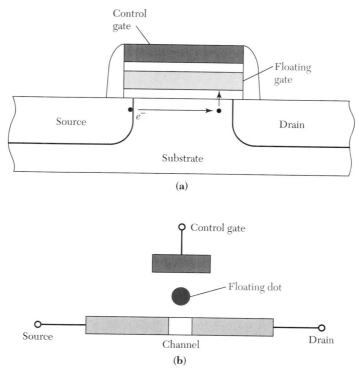

Fig. 42 (*a*) Illustration of hot electron injection in an *n*-channel, floating-gate nonvolatile memory. (*b*) Single-electron memory cell.[18]

cess is called "programming." The equivalent circuit for the floating-gate device can be represented by two capacitors in series for the gate structure, as illustrated in Fig. 41*b*. The stored charge gives rise to a threshold voltage shift, and the device is switched to a high-voltage state (e.g., logic 1). For a well-designed memory device, the charge retention time can be over 10 years. To erase the stored charge and return the device to a low-threshold voltage state (e.g., logic 0), an electrical bias or other means, such as UV light, can be used.

In floating-gate devices, the programming can be done by either hot carrier injection or a tunneling process. Figure 42*a* shows the hot electron injection scheme in an *n*-channel floating-gate device. The hot electrons are "hot" because they are heated to a high-energy state by the high field near the drain. Some of the hot electrons with energy higher than the barrier height of SiO$_2$/Si conduction band (~3.2 eV) can then surmount the barrier and are injected into the floating gate. Several types of floating-gate devices are differentiated by the erase mechanisms. In the EPROM, which has only a floating gate but no control gate, erasing is done by UV irradiation, since the UV light can excite the stored charges into the conduction band of gate oxide. The EPROM has the advantage of a small cell area due to the 1T/cell structure. Nevertheless, its erase scheme necessitates the use of an expensive package with a quartz window. In addition, the erasing time is long.

The EEPROM uses tunneling process to erase the stored charges. Unlike the EPROM device, in which all cells are erased during erasing, a cell in an EEPROM can be erased only when it is "selected." This function is accomplished through the selective transistor contained in each cell. Such "byte-erasable" characteristics make the use of EEPROM more flexible. However, the 2T (one selective transistor plus one storage transistor)/cell feature of EEPROM limits its storage capacity.

The storage cells for a flash memory are divided into several sectors (or blocks). The erasing scheme is performed on one selected sector with the tunneling process. During erasing, all cells in the selected sector are erased simultaneously. The erasing speed is much shorter than EPROM and the 1T/cell feature makes the storage capacity of the flash memory higher than EEPROM.

A related device structure is the single-electron memory cell (SEMC), which is a limiting case of the floating gate stucture.[18] By reducing the length of the floating gate to ultrasmall dimensions, say 10 nm, we obtain the SEMC. A cross-sectrional view of a SEMC is shown in Fig. 42b. The floating dot corresponds to the floating gate in Fig. 42a. Because of its small size the capacitance is also very small (~1 aF). When an electron tunnels into the floating dot, because of the small capacitance, a large tunneling barrier can arise to prevent the transfer of another electron. SEMC is an ultimate floating gate memory cell, since we need only one electron for information storage. A single-electron memory with densities as high as 256 terabits (256×10^{12} bits) that can be operated at room temperature has been projected.

6.7 THE POWER MOSFET

The input impedance of MOS devices is very high because of the insulating SiO_2 between the gate and semiconductor channel. This feature makes the MOSFET an attractive candidate in power-device applications. Because of the high-input impedance, the gate leakage is very low, and thus, the power MOSFET does not require complex input drive circuitry compared with bipolar devices. In addition, the switching speed of the power MOSFET is much faster than that of the power bipolar device. This is because the unipolar characteristics of MOS operation do not involve storage or recombination of minority carriers during turn-off.

Figure 43 shows three basic power MOSFET structures.[19] Unlike the MOSFET structure used in ULSI circuits, the power MOSFETs employ a vertical structure with the source and drain at the top and bottom surfaces of the wafer, respectively. This vertical scheme has the benefit of large channel width and reduced field crowding at the gate. These properties are important for power applications.

Figure 43a is the V-MOSFET in which the gate has a V-shaped groove. The V-shaped groove can be formed by preferential wet-etching using a KOH solution. When the gate voltage is larger than the threshold voltage, an inversion channel is induced at the surface along the edge of the V-shaped groove and forms the conductive path between source and drain. One main limitation of V-MOSFET development is related to process control. The high field at the tip of the V-shaped groove may lead to current crowding there and degrade device performance.

Figure 43b shows the cross-section of the U-MOSFET. It is similar to the V-MOSFET. The U-shaped trench is formed by reactive ion etching, and the electric fields at the bottom cornors are substantially lower than that at the tip of the V-shaped grove. Another power MOSFET is the D-MOSFET, shown in Fig. 43c. The gate is formed at the top surface and then serves as a mask for the subsequent double-diffusion process. The double diffusion process (this is the reason why it is call "D"-MOSFET) is used to form the p-base and n^+ source portions. The advantages of D-MOSFET are its short drift time across the p-base region and the avoidance of high-field corners.

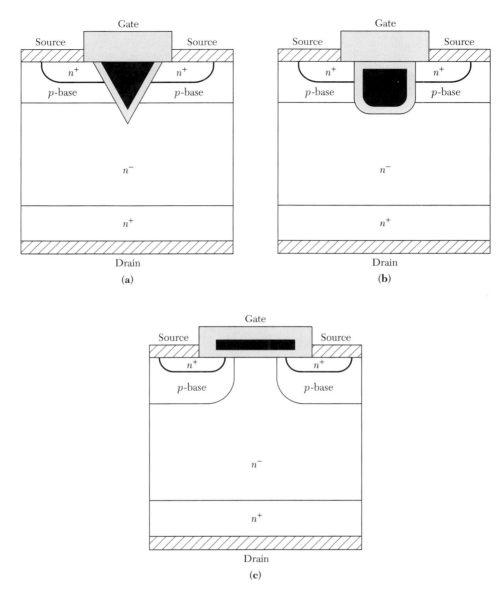

Fig. 43 (*a*) V-shaped MOS (VMOS), (*b*) U-shaped MOS (UMOS), and (*c*) double-diffused MOS (DMOS) power device structures.[19]

There is an n^- drift region in the drain for all the three power MOSFET structures. The doping concentration of n^- drift region is lower than p-base region. Therefore, when a positive voltage is applied to the drain and the drain/p-base junction is reverse biased, most of the depletion width will be developed across the n^- drift region. Therefore, the doping level and width of the n^- drift region are important parameters that determine the drain blocking voltage capability. On the other hand, there exists a parasitic n-p-n^--n^+ device in the power MOSFET structures. To prevent the action of the bipolar transistor during power MOSFET operation, the p-base and n^+ source (emitter) are shorted, as shown in Fig. 43. This can keep the p-base at a fixed potential.

▷ SUMMARY

In this chapter, we introduced the basic characteristics and the operational principles of the MOSFET. The core component of the MOSFET is the MOS diode. Charge distributions at the oxide/semiconductor interface (accumulation, depletion, and inversion) in an MOS device can be controlled by the gate voltage. The MOSFET is formed when a source and a drain are placed adjacent to the MOS diode. Output current (i.e., drain current) is controlled by varying the gate and drain voltages. The threshold voltage is the main parameter that determines the on–off characteristics of a MOSFET. Adjustment of the thresholds voltage can be done by choosing suitable substrate doping, oxide thickness, substrate bias, and gate materials (see Eq. 45).

The Si MOSFET is the most important device for advanced integral-circuit (IC) applications. Its success is mainly ascribed to the high-quality SiO_2 material and the stable Si/SiO_2 interface properties. To meet the stringent requirement of low power consumption in an IC chip, CMOS technology is currently the only viable solution and is widely implemented. Superior performance in power dissipation can be understood from the discussion of the CMOS inverter given in Section 6.4.

Scaling down of device dimensions is a continuing trend in CMOS technology to increase the device density, operating speed, and functionality of a chip. The short-channel effects, however, cause deviations in device operation and require attention in device scaling. Optimization of device structural parameters depends on the main requirement of the applications, such as minimized power consumption or faster operating speed.

TFT and SOI are MOSFET devices fabricated on insulating substrates, in contrast to the conventional MOSFETs fabricated on a bulk Si substrate. TFT uses an amorphous or a polycrystalline semiconductor as the active channel layer. The carrier mobility of the TFT is degraded by the presence of a large number of defects in the channel. However, TFT can be applied to a large-area substrate, which is difficult for bulk MOS technology, e.g., the switching element of pixels in a large-area flat-panel display. TFT can also be used as the load devices in the SRAM cell. SOI MOSFETs use a monocrystalline Si channel layer. Compared with bulk-MOS devices, SOI devices provide lower parasitic junction capacitance and improved resistance to radiation damage. SOI is also more attractive for low-power, high-speed applications.

MOSFETs have been employed for semiconductor memory applications, including DRAM, SRAM, and nonvolatile memory. These products share a significant portion of the IC market. Owing to the aggressive shrinkage of device size, the storage capacity of the MOS memories improves rapidly. For example, the DRAM density has been doubled every 18 months and the single-electron memory has been projected to reach a multiterabit level. Finally, we discussed three power MOSFETs. These devices use a vertical structure to allow higher operating voltage and current.

▷ REFERENCES

1. E. H. Nicollian and J. R. Brews, *MOS Physics and Technology,* Wiley, New York, 1982.

2. A. S. Grove, *Physics and Technology of Semiconductor Devices,* Wiley, New York, 1967.

3. B. E. Deal, "Standardized Terminology for Oxide Charge Associated with Thermally Oxidized Silicon," *IEEE Trans. Electron Devices,* **ED-27**, 606 (1980).

4. W. S. Boyle and G. E. Smith, "Charge Couple Semiconductor Devices," *Bell Syst. Tech. J.,* **49**, 587 (1970).

5. (a) D. Kahng and M. M. Atalla, "Silicon-Silicon Dioxide Field Induced Surface Devices," *IRE Solid State Device Res. Conf.,* Pittsburgh, PA, 1960. (b) D. Kahng, "A Historical Perspective on the

Development of MOS Transistors and Related Devices," *IEEE Trans. Electron Devices*, **ED-23**, 65 (1976).

6. Y. V. Ponomarev, et al., "Gate-Work function Engineering Using Poly-(Si,Ge) for High Performance 0.18 μm CMOS Technology," in *Tech. Dig. Int. Electron Devices Meet.* (IEDM), p.829 (1997).

7. H. Kawaguchi, et al., "A Robust 0.15 μm CMOS Technology with $CoSi_2$ Salicide and Shallow Trench Isolation," in *Tech. Dig. Symp. VLSI Technol.*, p. 125 (1997).

8. L. D. Yau, "A Simple Theory to Predict the Threshold Voltage in Short-Channel IGFETs," *Solid-State Electron.*, **17**, 1059 (1974).

9. Z. H. Liu, et al., "Threshold Voltage Model for Deep-Submicrometer MOSFETs," *IEEE Trans. Electron Devices*, **ED-40**, 86 (1993)

10. R. H. Dennard, et al., "Design of Ion Implanted MOSFET's with Very Small Physical Dimensions," *IEEE J. Solid State Circuits*, **SC-9**, 256 (1974).

11. Y. Taur and T. K. Ning, *Physics of Modern VLSI Devices,* Cambridge Univ. Press, London, 1998.

12. H-S. P. Wong, "MOSFET Fundamentals," in *ULSI Devices*, C. Y. Chang and S. M. Sze, Eds., Wiley Interscience, New York, 1999.

13. D. Hisamoto, et al., "A Folded-Channel MOSFET for Deep-Sub-Tenth Micron Era," *Tech. Dig. IEEE Int. Electron Devices Meet.*, San Francisco, 1998, p. 1032–1034.

14. R. R. Troutman, *Latch-up in CMOS Technology*, Kluwer, Boston,1986.

15. J. P. Colinge, *Silicon-on-Insulator Technology: Materials to VLSI*, Kluwer, Boston, 1991.

16. (a) R. H. Dennard, "Field-effect Transistor Memory," U.S. Patent 3,387,286. (b) R.H. Dennard, "Evolution of the MOSFET DRAM—A Personal View," *IEEE Trans. Electron Devices*, **ED31**, 1549 (1984).

17. D. Kahng and S. M. Sze, "A Floating Gate and Its Application to Memory Devices," *Bell System Tech. J.*, **46**, 1283 (1967).

18. S. M. Sze, "Evolution of Nonvolatile Semiconductor Memory: from Floating-Gate Concept to Single-Electron Memory Cell," in J. Xu, and A. Zaslavsky, Eds. *Future Trends in Microelectronics*, S. Luryi, Wiley Interscience, New York, 1999.

19. B. J. Baliga, *Power Semiconductor Devices*, PWS Publishers, Boston, 1996.

► PROBLEMS (* DENOTES DIFFICULT PROBLEMS)

FOR SECTION 6.1   THE MOS DIODE

1. Plot the band diagram of an ideal MOS diode with n-type substrate at $V_G = V_T$.

2. Plot the band diagram of an n^+-polysilicon-gated MOS diode with p-type substrate at $V_G = 0$.

3. Plot the band diagram of an n^+-polysilicon–gated MOS diode with p-type substrate at flat-band condition.

4. Plot (a) the charge distribution, (b) electric-field distribution, and (c) potential distribution of an ideal MOS diode with n-type substrate under inversion.

5. For a metal-SiO_2-Si capacitor having $N_A = 5 \times 10^{16}$ cm^{-3}, calculate the maximum width of surface depletion region.

6. For a metal-SiO_2-Si capacitor having $N_A = 5 \times 10^{16}$ cm^{-3} and $d = 8$ nm, calculate the minimum capacitance on the C-V curve.

*7. For an ideal Si-SiO_2 MOS diode with $d = 5$ nm, $N_A = 10^{17}$ cm^{-3}, find the applied voltage and the electric field at the interface required to make the silicon surface intrinsic.

8. For an ideal Si-SiO_2 MOS diode with $d = 10$ nm, $N_A = 5 \times 10^{16}$ cm^{-3}, find the applied voltage and the electric field at the interface required to bring about strong inversion.

*9. Assume that the oxide trapped charge Q_{ot} in an oxide layer has an uniform volume charge density, $\rho_{ot}(y)$, of $q \times 10^{17}$ cm^{-3}, where y is the distance from the location of the charge to the metal-oxide interface. The thickness of the oxide layer is 10 nm. Find the change in the flat-band voltage due to Q_{ot}.

10. Assume that the oxide trapped charge Q_{ot} in an oxide layer is a charge sheet with an area density of 5×10^{11} cm^{-2} locating solely at $y = 5$ nm. The thickness of the oxide layer is 10 nm. Find the change in the flat-band voltage due to Q_{ot}.

11. Assume that the oxide trapped charge Q_{ot} in an oxide layer has a triangular distribution: $\rho_{ot}(y) = q \times (5 \times 10^{23} \times y)$ cm^{-3}. The thickness of oxide layer is 10 nm. Find the change in the flat-band voltage due to Q_{ot}.

12. Assume that initially there is a sheet of mobile ions at the metal-SiO$_2$ interface. After a long period of electrical stressing under a high positive gate voltage and raised temperature condition, the mobile ions completely drift to the SiO$_2$-Si interface. This leads to a change of 0.3 V in the flat-band voltage. The thickness of oxide layer is 10 nm. Find the area density of Q_m.

FOR SECTION 6.2   MOSFET FUNDAMENTALS

13. Derive Eq. 34 from Eq. 33 in the text assuming $V_D \ll (V_G - V_T)$.

*14. Derive the *I-V* characteristics of a MOSFET with the drain and gate connected together and the source and substrate grounded. Can one obtain the threshold voltage from these characteristics?

15. Consider a long-channel MOSFET with $L = 1$ μm, $Z = 10$ μm, $N_A = 5 \times 10^{16}$ cm^{-3}, $\mu_n = 800$ cm^2/ V-s, $C_o = 3.45 \times 10^{-7}$ F/cm^2, and $V_T = 0.7$ V. Find V_{Dsat} and I_{Dsat} for $V_G = 5$ V.

16. Consider a submicron MOSFET with $L = 0.25$ μm, $Z = 5$ μm, $N_A = 10^{17}$ cm^{-3}, $\mu_n = 500$ cm^2/ V-s, $C_o = 3.45 \times 10^{-7}$ F/cm^2, and $V_T = 0.5$ V. Find the channel conductance for $V_G = 1$ V and $V_D = 0.1$ V.

17. For the device stated in Prob. 16, find the transconductance.

18. An *n*-channel, n^+-polysilicon-SiO$_2$-Si MOSFET has $N_A = 10^{17}$ cm^{-3}, $Q_f/q = 5 \times 10^{10}$ cm^{-2}, and $d = 10$ nm. Calculate the threshold voltage.

19. For the device stated in Prob. 18, boron ions are implanted to increase the threshold voltage to +0.7 V. Find the implant dose, assume that the implanted ions form a sheet of negative charges at the Si-SiO$_2$ interface.

20. A *p*-channel, n^+-polysilicon-SiO$_2$-Si MOSFET has $N_D = 10^{17}$ cm^{-3}, $Q_f/q = 5 \times 10^{10}$ cm^{-2}, and $d = 10$ nm. Calculate the threshold voltage.

21. For the device stated in Prob. 20, boron ions are implanted to decrease the value of threshold voltage to -0.7 V. Find the implant dose, assuming that the implanted ions form a sheet of negative charges at the Si-SiO$_2$ interface.

22. For the device stated in Prob. 20, if the n^+ poly-Si gate is replaced by p^+ poly-Si gate, what will the threshold voltage be?

23. A field transistor with a structure similar to Fig. 21 in the text has $N_A = 10^{17}$ cm^{-3}, $Q_f/q = 10^{11}$ cm^{-2}, and an n^+ polysilicon local interconnect as the gate electrode. If the requirement for sufficient isolation between device and well is $V_T > 20$ V, calculate the minimum field oxide thickness.

24. A MOSFET has a threshold voltage of $V_T = 0.5$ V, a subthreshold swing of 100 mV/decade, and a drain current of 0.1 μA at V_T. What is the subthreshold leakage current at $V_G = 0$?

25. For the device stated in Prob. 24, calculate the reverse substrate-source voltage required to reduce the leakage current by one order of magnitude. ($N_A = 5 \times 10^{17}$ cm^{-3}, $d = 5$ nm).

FOR SECTION 6.3   MOSFET SCALING

26. When the linear dimensions of MOSFET are scaled down by a factor of 10 based on the constant field scaling, what is the scaling factor for the corresponding switching energy?

*27. Based on the charge-sharing model, Fig. 24, show that the threshold voltage roll-off is given by Eq. 47.

FOR SECTION 6.4   CMOS AND BiCMOS

28. Describe the pros and cons of BiCMOS.

FOR SECTION 6.5   MOSFET ON INSULATOR

29. For an n-channel FD-SOI device having $N_A = 5 \times 10^{17}$ cm^{-3} and $d = 4$ nm, calculate the maximum allowable thickness for Si channel layer (d_{Si}).

30. For an n-channel SOI device with n$^+$-polysilicon gate having $N_A = 5 \times 10^{17}$ cm^{-3}, $d = 4$ nm, and $d_{Si} = 30$ nm, calculate the threshold voltage. Assume that Q_f, Q_{ot}, and Q_m are all zero.

31. For the device stated in Prob. 29, calculate the range of V_T distribution if the thickness variation of d_{Si} across the wafer is ±5 nm.

FOR SECTION 6.6   MOS MEMORY STRUCTURES

32. What is the capacitance of a DRAM capacitor if it is planar, 1 μm × 1 μm, with an oxide thickness of 10 nm? Calculate the capacitance if the same surface area is used for a trench that is 7 μm deep and with the same oxide thickness.

33. A DRAM must operate with a minimum refresh time of 4 ms. The storage capacitor in each cell has a capacitance of 50 fF and is fully charged to 5 V. Find the worst-case leakage current (i.e., during the refresh cycle 50% of the stored charge is lost) that the dynamic node can tolerate.

34. A floating-gate nonvolatile memory has an initial threshold voltage of –2 V, and a linear-region drain conductance of 10 μmhos at a gate voltage of –5 V. After a write operating, the drain conductance increases to 40 μmhos at the same gate voltage. Find the threshold voltage shift.

FOR SECTION 6.7   THE POWER MOSFET

35. A power MOSFET has an n^+-polysilicon gate and a p-base with $N_A = 10^{17}$ cm^{-3}. Gate oxide thickness $d = 100$ nm. Calculate the threshold voltage.

36. For the device stated in Prob. 35, calculate the effect of a positive fixed charge density of 5×10^{11} cm^{-3} on the threshold voltage.

MESFET and Related Devices

The metal-semiconductor field-effect transistor (MESFET) has current-voltage characteristics similar to that of a metal-oxide-semiconductor field-effect transistor (MOSFET). However, it uses a metal-semiconductor rectifying contact instead of a MOS structure for the gate electrode. In addition, the source and drain contacts of MESFET are ohmic,[§] whereas in a MOSFET they are p–n junctions.

Like other field-effect devices, a MESFET has a negative temperature coefficient at high current levels; that is, the current decreases as temperature increases. This characteristic leads to more uniform temperature distribution and the device is, therefore, thermally stable, even when the active area is large or when many devices are connected in parallel. Furthermore, because MESFETs can be made from compound semiconductors with high electron mobilities, such as GaAs and InP, they have higher switching speeds and higher cutoff frequencies than do silicon MOSFETs.

The basic building block of a MESFET is the metal-semiconductor contact. This contact is electrically similar to a one-sided abrupt p–n junction, yet it can be operated as a majority carrier device with inherently fast response. There are two types of metal-semiconductor contacts: the rectifying and the nonrectifying or ohmic types. In this chapter, we begin with the two types of contacts and then consider the basic characteristics and microwave performance of the MESFET. In the last section we discuss modulation-doped field-effect transistor (MODFET), which is similar to the device configuration of a MESFET but can offer even higher-speed performance.

Specifically, we cover the following topics:

- Rectifying metal-semiconductor contact and its current-voltage characteristics.
- Ohmic metal-semiconductor contact and its specific contact resistance.
- The MESFET and its high-frequency performance.

[§] The concept of rectifying was discussed in Chapter 4. The concept of ohmic contact is presented in Section 7.1.

- The MODFET and its two-dimensional electron gas.
- A comparison of three field-effect transistors—MOSFET, MESFET, and MODFET.

7.1 METAL-SEMICONDUCTOR CONTACTS

The first practical semiconductor device was the metal-semiconductor contact in the form of a point contact rectifier, that is, a metallic whisker pressed against a semiconductor. The device found many applications beginning in 1904. In 1938, Schottky, suggested that the rectifying behavior could arise from a potential barrier as a result of the stable space charges in the semiconductor. The model arising from this concept is known as the Schottky barrier. Metal-semiconductor contacts can also be nonrectifying; that is, the contact has a negligible resistance regardless of the polarity of the applied voltage. This type of contact is called an ohmic contact. All semiconductor devices as well as integrated circuits need ohmic contact to make connections to other devices in an electronic system. We consider the energy band diagram and the current-voltage characteristics of both the rectifying and ohmic metal-semiconductor contacts.

7.1.1 Basic Characteristics

The characteristics of point contact rectifiers were not reproducible from one device to another. They have been largely replaced by metal-semiconductor contacts fabricated by planar processes (see Chapters 10–14). A schematic diagram of such a device is shown in Fig. 1a. To fabricate the device, a window is opened in an oxide layer, and metal layer is deposited in a vacuum system. The metal layer covering the window is subsequently defined by a lithographic step. We consider a one-dimensional structure of the metal-semiconductor contact shown in Fig. 1b, which corresponds to the central section in Fig. 1a, between the dashed lines.

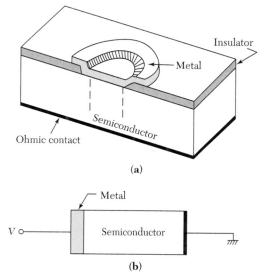

(a)

(b)

Fig. 1 (a) Perspective view of a metal-semiconductor contact fabricated by the planar process. (b) One-dimensional structure of a metal-semiconductor contact.

Figure 2*a* shows the energy band diagram of an isolated metal adjacent to an isolated n-type semiconductor. Note that the metal work function $q\phi_m$ is generally different from the semiconductor work function $q\phi_s$. The work function is defined as the energy difference between the Fermi level and the vacuum level. Also shown is the electron affinity $q\chi$, which is the energy difference between the conduction band edge and the vacuum level in the semiconductor. When the metal makes intimate contact with the semiconductor, the Fermi levels in the two materials must be equal at thermal equilibrium. In addition, the vacuum level must be continuous. These two requirements determine a unique energy band diagram for the ideal metal-semiconductor contact, as shown in Fig. 2*b*.

For this ideal case, the barrier height $q\phi_{Bn}$ is simply the difference between the metal work function and the semiconductor electron affinity[§]:

$$q\phi_{Bn} = q\phi_m - q\chi. \tag{1}$$

Similarly, for the case of an ideal contact between a metal and a *p*-type semiconductor, the barrier height $q\phi_{Bp}$ is given by

$$q\phi_{Bp} = E_g - (q\phi_m - q\chi), \tag{2}$$

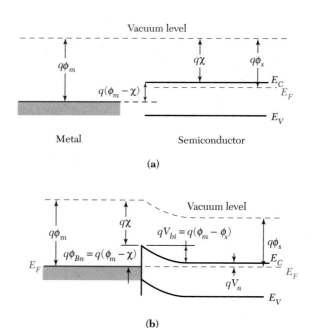

(a)

(b)

Fig. 2 (*a*) Energy band diagram of an isolated metal adjacent to an isolated n-type semiconductor under thermal nonequilibrium condition. (*b*) Energy band diagram of a metal-semiconductor contact in thermal equilibrium.

[§] Both $q\phi_{Bn}$ (in electron volts) and ϕ_{Bn} (in volts) are referred to as the barrier height.

where E_g is the bandgap of the semiconductor. Therefore, for a given semiconductor and for any metal, the sum of the barrier heights on n-type and p-type substrates is expected to be equal to the bandgap:

$$q\left(\phi_{Bn} + \phi_{Bp}\right) = E_g . \tag{3}$$

On the semiconductor side in Fig. 2b, V_{bi} is the built-in potential that is seen by electrons in the conduction band trying to move into the metal.

$$V_{bi} = \phi_{Bn} - V_n . \tag{4}$$

The qV_n is the distance between the bottom of the conduction band and the Fermi level. Similar results can be given for the p-type semiconductor.

Figure 3 shows the measured barrier heights for n-type silicon[2] and n-type gallium arsenide.[3] Note that $q\phi_{Bn}$ increases with increasing $q\phi_m$. However, the dependence is not as strong as predicted by Eq. 1. This is because in practical Schottky diodes, the disruption of the crystal lattice at the semiconductor surface produces a large number of surface energy states located in the forbidden bandgap. These surface states can act as donors or acceptors that influence the final determination of the barrier height. For silicon and gallium arsenide, Eq. 1 generally underestimated the n-type barrier height and Eq. 2 overestimates the p-type barrier height. The sum of $q\phi_{Bn}$ and $q\phi Bp$, however, is in agreement with Eq. 3.

Figure 4 shows the energy band diagrams for metals on both n-type and p-type semiconductors under different biasing conditions. Consider the n-type semiconductor first. When the bias voltage is zero, as shown in the left side of Fig. 4a, the band diagram is under a thermal equilibrium condition. The Fermi levels for both materials are equal. If we apply a positive voltage to the metal with respect to the n-type semiconductor, the semiconductor-to-metal barrier height decreases as shown on the left side of Fig. 4b. This is a forward bias. When a forward bias is applied, electrons can move easily from the semiconductor into the metal because the barrier has been reduced by a voltage V_F. For a

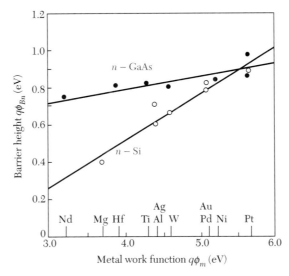

Fig. 3 Measured barrier height for metal-silicon and metal-gallium arsenide contacts.[2,3]

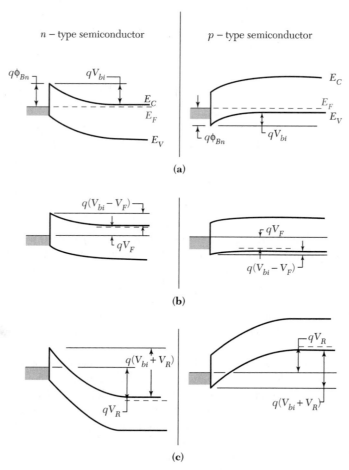

Fig. 4 Energy band diagrams of metal n-type and p-type semiconductors under different biasing conditions: (*a*) thermal equilibrium; (*b*) forward bias; and (*c*) reverse bias.

reverse bias (i.e., a negative voltage is applied to the metal), the barrier has been increased by a voltage V_R, as depicted on the left side of Fig. 4c. It is more difficult for electrons to flow from the semiconductor into the metal. We have similar results for p-type semiconductor, however, the polarities must be reversed. In the following derivations, we consider only the metal–n-type semiconductor contact. The results are equally applicable to a p-type semiconductor with an appropriate change of polarities.

The charge and field distributions for a metal-semiconductor contact are shown in Fig. 5a and 5b, respectively. The metal is assumed to be a perfect conductor; the charge transferred to it from the semiconductor exists in a very narrow region at the metal surface. The extent of the space charge in the semiconductor is W, i.e., $\rho_s = qN_D$ for x < W and $\rho_s = 0$ for x > W. Thus, the charge distribution is identical to that of a one-sided abrupt p^+-n junction.

The magnitude of the electric field is decreasing linearly with distance. The maximum electric field $\mathscr{E}_m$ is located at the interface. The electric field distribution is then given by

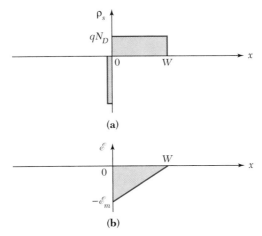

(a)

(b)

Fig. 5 (*a*) Charge distribution and (*b*) electric-field distribution in a metal-semiconductor contact.

$$\left| \mathscr{E}(x) \right| = \frac{qN_D}{\varepsilon_s}(W - x) = \mathscr{E}_m - \frac{qN_D}{\varepsilon_s}x, \tag{5}$$

$$\mathscr{E}_m = \frac{qN_D W}{\varepsilon_s}, \tag{6}$$

where ε_s is the dielectric permittivity of the semiconductor. The voltage across the space-charge region, which is represented by the area under the field curve in the Fig. 5*b*, is given by

$$V_{bi} - V = \frac{\mathscr{E}_m W}{2} = \frac{qN_D W^2}{2\varepsilon_s}. \tag{7}$$

The depletion-layer width W is expressed as

$$W = \sqrt{2\varepsilon_s(V_{bi} - V)/qN_D}, \tag{8}$$

and the space-charge density, Q_{SC}, in the semiconductor is given as

$$Q_{SC} = qN_D W = \sqrt{2q\varepsilon_s N_D(V_{bi} - V)} \quad \text{C/cm}^2, \tag{9}$$

where the voltage V equal to $+V_F$ for forward bias and to $-V_R$ for reverse bias. The depletion-layer capacitance C per unit area can be calculated by using Eq. 9 :

$$C = \left| \frac{\partial Q_{sc}}{\partial V} \right| = \sqrt{\frac{q\varepsilon_s N_D}{2(V_{bi} - V)}} = \frac{\varepsilon_s}{W} \quad \text{F/cm}^2 \tag{10}$$

and

$$\boxed{\frac{1}{C^2} = \frac{2(V_{bi} - V)}{q\varepsilon_s N_D} \quad \left(\text{F / cm}^2\right)^{-2}.} \tag{11}$$

We can differentiate $1/C^2$ with respect to V. Rearranging terms we obtain :

$$N_D = \frac{2}{q\varepsilon_s}\left[\frac{-1}{d(1/C^2)/dV}\right].\qquad(12)$$

Thus, measurements of the capacitance C per unit area as a function of voltage can provide the impurity distribution from Eq. 12. If N_D is constant throughout the depletion region, we should obtain a straight line by plotting $1/C^2$ versus V. Figure 6 is a plot of the measured capacitance versus voltage for tungsten-silicon and tungsten-gallium arsenide Schottky diodes.[4] From Eq. 11, the intercept at $1/C^2 = 0$ corresponds to the built-in potential V_{bi}. Once V_{bi} is determined, the barrier height ϕ_{Bn} can be calculated from Eq. 4.

▶ **EXAMPLE 1**

Find the donor concentration and the barrier height of the tungsten-silicon Schottky diode shown in Fig. 6.

SOLUTION The plot of $1/C^2$ versus V is a straight line, which implies that the donor concentration is constant throughout the depletion region. We find

$$\frac{d\left(1/C^2\right)}{dV} = \frac{6.2\times10^{15} - 1.8\times10^{15}}{-1-0} = -4.4\times10^{15}\ \left(\frac{cm^2/F}{V}\right)^2.$$

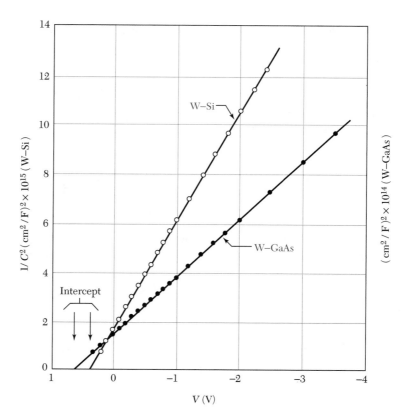

Fig. 6 $1/C^2$ versus applied voltage for W-Si and W-GaAs diodes.[4]

From Eq. 12,

$$N_D = \left[\frac{2}{1.6 \times 10^{-19} \times (11.9 \times 8.85 \times 10^{-14})}\right] \times \left(\frac{1}{4.4 \times 10^{15}}\right) = 2.7 \times 10^{15} \text{ cm}^{-3},$$

$$V_n = 0.0259 \times \ln\left(\frac{2.86 \times 10^{19}}{2.7 \times 10^{15}}\right) = 0.24 \text{ V}.$$

Since the intercept V_{bi} is 0.42 V, then the barrier height is ϕ_{Bn} = 0.42 + 0.24 = 0.66 V. ◀

7.1.2 The Schottky Barrier

A Schottky barrier refers to a metal-semiconductor contact having a large barrier height (i,e., ϕ_{Bn} or ϕ_{Bp} >> kT) and a low doping concentration that is less than the density of states in the conduction band or valence band.

The current transport in a Schottky barrier is due mainly to majority carrier, in contrast to a p–n junction, where current transport is due mainly to minority carriers. For Schottky diodes operated at moderate temperature (e.g., 300 K), the dominate transport mechanism is thermionic emission of majority carriers from the semiconductor over the potential barrier into the metal.

Figure 7 illustrates the thermionic emission process.[5] At thermal equilibrium (Fig. 7a), the current density is balanced by two equal and opposite flows of carriers, thus there is zero net currents. Electrons in the semiconductor tend to flow (or emit) into the metal, and there is an opposing balanced flow of electrons from metal into the semiconductor. These current components are proportional to the density of electrons at the boundary.

As discussed in Section 3.5 of Chapter 3, at the semiconductor surface an electron can be thermionically emitted into the metal if its energy is above the barrier height. Here the semiconductor work function $q\phi_s$ is replaced by $q\phi_{Bn}$, and

$$n_{th} = N_C \exp\left(-\frac{q\phi_{Bn}}{kT}\right), \tag{13}$$

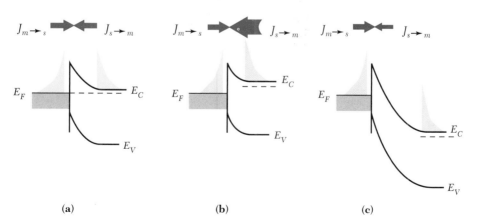

Fig. 7 Current transport by the thermionic emission process. (*a*) Thermal equilibrium; (*b*) forward bias; and (*c*) reverse bias.[5]

where N_C is the density of states in the conduction band. At thermal equilibrium we have

$$|J_{m \to s}| = |J_{s \to m}| \propto n_{th} \tag{14}$$

or

$$|J_{m \to s}| = |J_{s \to m}| = C_1 N_C \exp\left(-\frac{q\phi_{Bn}}{kT}\right), \tag{14a}$$

where $J_{m \to s}$ is the current from the metal to the semiconductor, $J_{s \to m}$ is the current from the semiconductor to the metal, and C_1 is a proportionality constant.

When a forward bias V_F is applied to the contact (Fig. 7b), the electrostatic potential difference across the barrier is reduced, and the electron density at the surface increases to

$$n_{th} = N_C \exp\left[-\frac{q(\phi_{Bn} - V_F)}{kT}\right]. \tag{15}$$

The current $J_{s \to m}$ that results from the electron flow out of the semiconductor is therefore altered by the same factor (Fig. 7b). The flux of electrons from the metal to the semiconductor, however, remains the same because the barrier ϕ_{Bn} remains at its equilibrium value. The net current under forward bias is then

$$J = J_{s \to m} - J_{m \to s}$$

$$= C_1 N_C \exp\left[-\frac{q(\phi_{Bn} - V_F)}{kT}\right] - C_1 N_C \exp\left(-\frac{q\phi_{Bn}}{kT}\right)$$

$$= C_1 N_C e^{-q\phi_{Bn}/kT}\left(e^{qV_F/kT} - 1\right). \tag{16}$$

Using the same argument for the reverse-bias condition(see Fig. 7c), the expression for the net current is identical to Eq. 16 except that V_F is replaced by $-V_R$.

The coefficient $C_1 N_C$ is found to be equal to A^*T^2, where A^* is called the *effective Richardson constant* (in units of A/K²-cm²), and T is the absolute temperature. The value of A^* depend on the effective mass and are equal to 110 and 32 for n- and p-type silicon, respectively, and 8 and 74 for n- and p-type gallium arsenide, respectively.[6]

The current-voltage characteristic of a metal-semiconductor contact under thermionic emission condition is then

$$\boxed{J = J_s\left(e^{qV/kT} - 1\right),} \tag{17}$$

$$\boxed{J_s = A^*T^2 e^{-q\phi_{Bn}/kT},} \tag{17a}$$

where J_s is the saturation current density and the applied voltage V is positive for forward bias and negative for reverse bias. Experimental forward I-V characteristics of two Schottky diodes[4] are shown in Fig. 8. By extrapolating the forward I-V curve to $V = 0$, we can find J_s. From J_s and Eq. 17a we can obtain the barrier height.

In addition to the majority carrier (electron) current, a minority-carrier (hole) current exists in a metal n-type semiconductor contact because of hole injection from the

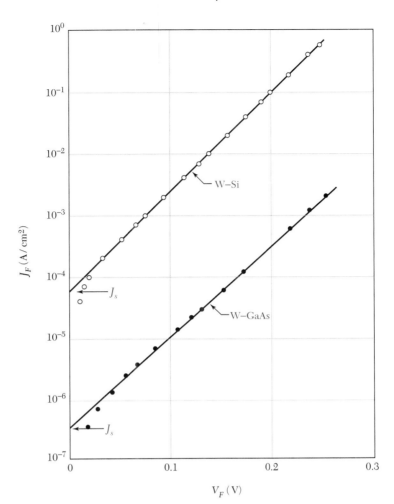

Fig. 8   Forward current density versus applied voltage of W-Si and W-GaAs diodes.[4]

metal to the semiconductor. The hole injection is the same as in a p^+-n junction, which is described in Chapter 4. The current density is given by

$$J_p = J_{po}\left(e^{qV/kT} - 1\right),\tag{18}$$

where

$$J_{po} = \frac{qD_p n_i^2}{L_p N_D}.\tag{18a}$$

Under normal operating conditions, the minority-carrier current is orders of magnitude smaller than the majority-carrier current. Therefore, a Schottky diode is a unipolar device (i.e., predominately only one type of carrier participates in the conduction process).

▷ EXAMPLE 2

For a tungsten-silicon Schottky diode with $N_D = 10^{16}$ cm^{-3}, find the barrier height and depletion-layer width from Fig. 8. Compare the saturation current J_s with J_{po}, assuming that the minority-carrier lifetime in Si is 10^{-6} s.

SOLUTION From Fig. 8, we have $J_s = 6.5 \times 10^{-5}$ A/cm^2. The barrier height can be obtained from Eq. 17a:

$$\phi_{Bn} = 0.0259 \times \ln\left(\frac{110 \times 300^2}{6.5 \times 10^{-5}}\right) = 0.67 \text{ V.}$$

This result is in the close agreement with the C-V measurement (see Fig. 6 and Ex. 1).

The built-in potential is given by $\phi_{Bn} - V_n$, where

$$V = 0.0259 \times \ln\left(\frac{N_C}{N_D}\right) = 0.0259 \ln\left(\frac{2.86 \times 10^{19}}{1 \times 10^{16}}\right) = 0.17 \text{ V.}$$

Therefore,

$V_{bi} = 0.67 - 0.17 = 0.50$ V.

The depletion-layer width at thermal equilibrium is given by Eq. 8 with $V = 0$:

$$W = \sqrt{\frac{2\varepsilon_s V_{bi}}{qN_D}} = 2.6 \times 10^{-5} \text{ cm.}$$

To calculate the minority-carrier current density J_{po}, we need to know D_p, which is 10 cm^2/s for $N_D = 10^{16}$ cm^{-3}, and L_p which is $\sqrt{D_p\tau_p} = \sqrt{10 \times 10^{-6}} = 3.1 \times 10^{-3}$ cm. Therefore,

$$J_{po} = \frac{qD_p n_i^2}{L_p N_D} = \frac{1.6 \times 10^{-19} \times 10 \times \left(9.65 \times 10^9\right)^2}{\left(3.1 \times 10^{-3}\right) \times 10^{16}} = 4.8 \times 10^{-12} \text{ A/cm}^2.$$

The ratio of the two current densities is

$$\frac{J_s}{J_{po}} = \frac{6.5 \times 10^{-5}}{4.8 \times 10^{-12}} = 1.3 \times 10^7.$$

From the comparison, we see that the majority-carrier current is over seven orders of magnitude greater than the minority-carrier current. ◀

7.1.3 The Ohmic Contact

An ohmic contact is defined as a metal-semiconductor contact that has a negligible contact resistance relative to the bulk or series resistance of the semiconductor. A satisfactory ohmic contact should not significantly degrade device performance and can pass the required current with a voltage drop that is small compared with the drop across the active region of the device.

A figure-of-merit for ohmic contacts is the specific contact resistance R_C, defined as

$$R_C \equiv \left(\frac{\partial J}{\partial V}\right)^{-1}_{V=0} \quad \Omega\text{-cm}^2. \tag{19}$$

For metal-semiconductor contacts with low doping concentrations, the thermionic-emission current dominates the current transport, as given by Eq. 17. Therefore,

$$R_C = \frac{k}{qA^*T}\exp\left(\frac{q\phi_{Bn}}{kT}\right). \tag{20}$$

Equation 20 shows that a metal-semiconductor contact with a low barrier height should be used to obtain a small R_C.

For contacts with high doping concentration, the barrier width becomes very narrow, and the tunneling current becomes dominate. The tunneling current, as described in the upper inset of Fig. 9, is proportional to the tunneling probability, which is given in Section 3.6 of Chapter 3:

$$I \sim \exp\left[-2W\sqrt{2m_n\left(q\phi_{Bn}-qV\right)/\hbar^2}\right], \tag{21}$$

where W is the depletion-layer, width which can be approximated as $\sqrt{(2\varepsilon_s/qN_D)(\phi_{Bn}-V)}$, m_n is the effective mass, and $\hbar$ is the reduced Planck constant. Substituting W into Eq. 21, we obtain

$$I \sim \exp\left[-\frac{C_2\left(\phi_{Bn}-V\right)}{\sqrt{N_D}}\right], \tag{22}$$

where C_2 equal $4\sqrt{m_n\varepsilon_s}/\hbar$. The specific contact resistance for high dopings is thus

$$R_C \sim \exp\left(\frac{C_2\phi_{Bn}}{\sqrt{N_D}}\right) = \exp\left(\frac{4\sqrt{m_n\varepsilon_s}\,\phi_{Bn}}{\sqrt{N_D}\,\hbar}\right). \tag{23}$$

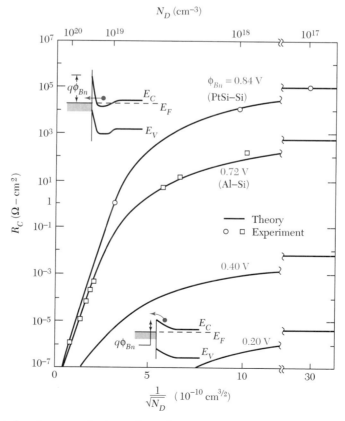

Fig. 9 Calculated and measured values of specific contact resistance. Upper inset shows the tunneling process. Lower inset shows thermionic emission over the low barrier.[6]

Equation 23 shows that in the tunneling range the specific contact resistance depends strongly on doping concentration and varies exponentially with the factor $\phi_{Bn}/\sqrt{N_D}$.

The calculated values of R_C are plotted[6] in Fig. 9 as a function of $1/\sqrt{N_D}$. For $N_D \geq 10^{19}$ cm^{-3}, R_C is dominated by the tunneling process and decreases rapidly with increased doping. On the other hand, for $N_D \leq 10^{17}$ cm^{-3}, the current is due to thermionic emission, and R_C is essentially independent of doping. Also shown in Fig. 9 are experimental data for platinum silicide-silicon (PtSi-Si) and aluminum-silicon (Al-Si) diodes. They are in close agreement with the calculated values. Figure 9 shows that a high doping concentration, a low barrier height, or both must be used to obtain a low value of R_C. These two approaches are used for all practical ohmic contacts.

▶ **EXAMPLE 3**

An ohmic contact has an area of 10^{-5} cm^2 and a specific contact resistance of 10^{-6} Ω–cm^2. The ohmic contact is formed in an n-type silicon. If $N_D = 5 \times 10^{19}$ cm^{-3}, and $\phi_{Bn} = 0.8$ V, and the electron effective mass is $0.26\ m_0$, find the voltage drop across the contact when a forward current of 1A flows through it.

SOLUTION The contact resistance for the ohmic contact is

$$\frac{R_C}{A} = 10^{-6}\,\Omega-\text{cm}^2/10^{-5}\ \text{cm}^2 = 10^{-1}\Omega,$$

$$C_2 = 2\sqrt{m_n \varepsilon_s}/\hbar = \frac{2\sqrt{0.26 \times 9.1 \times 10^{-31} \times \left(1.05 \times 10^{-12}\right)}}{1.05 \times 10^{-34}}$$

$$= 9.45 \times 10^{13} \left(\text{m}^{-3/2}/\text{V}\right).$$

From Eq. 22,

$$I = I_0 \exp\left[-\frac{C_2\left(\phi_{Bn} - V\right)}{\sqrt{N_D}}\right],$$

$$\left.\frac{\partial I}{\partial V}\right|_{V=0} = \frac{A}{R_C} = I_0 \left(\frac{C_2}{\sqrt{N_D}}\right)\exp\left(\frac{-C_2\phi_{Bn}}{\sqrt{N_D}}\right)$$

or,

$$I_0 = \frac{A}{R_C}\left(\frac{\sqrt{N_D}}{C_2}\right)\exp\left(\frac{C_2\phi_{Bn}}{\sqrt{N_D}}\right)$$

$$= 10 \times \left(\frac{\sqrt{5 \times 10^{19} \times 10^6}}{9.45 \times 10^{13}}\right)\exp\left(\frac{9.45 \times 10^{13} \times 0.8}{\sqrt{5 \times 10^{19} \times 10^6}}\right)$$

$$= 3.39 \times 10^4 \ \text{A}.$$

at $I = 1$A, we have

$$\phi_{Bn} - V = \frac{\sqrt{N_D}}{C_2}\ln\left(\frac{I_0}{I}\right) = 0.779\text{V}$$

or

$$V = 0.8 - 0.779 = 0.021 \text{ V} = 21 \text{ mV}.$$

Therefore, there is a negligibly small voltage drop across the ohmic contact. However, the voltage drop may become significant when the contact area is reduced to 10^{-8} cm^2 or smaller. ◁

7.2 MESFET

7.2.1 Device Structures

The metal-semiconductor field-effect transistor (MESFET) was proposed[7] in 1966. The MESFET has three metal-semiconductor contacts—one Schottky barrier for the gate electrode and two ohmic contacts for the source and drain electrodes. A perspective view of a MESFET is illustrated in Fig. 10a. The basic device parameters include L, the gate length, Z, the gate width, and a the thickness of the epitaxial layer. Most MESFETs are made of n-type III-V compound semiconductors, such as gallium arsenide, because of their high electron mobilities, which help to minimize series resistances, and because of their high saturation velocities, which result in the increase in the cutoff frequency.

Practical MESFETs are fabricated by using epitaxial layers on semiinsulating substrates to minimize parasitic capacitances. In Fig. 10a, the ohmic contacts are labeled source and drain, and the Schottky barrier is labeled as gate. A MESFET is often described in terms of the gate dimensions. If the gate length (L) is 0.5 μm and the gate width (Z)

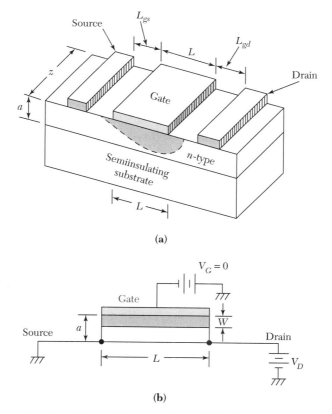

(a)

(b)

Fig. 10 (a) Perspective view of a metal-semiconductor field-effect transistor (MESFET). (b) Cross section of the gate region of a MESFET.

is 300 μm, the device is referred to as a 0.5×300 μm device. A microwave- or millimeter-wave device typically has a gate length in the range 0.1–1.0 μm. The thickness a of the epitaxial layer is typically one-third to one-fifth of the gate length. The spacing between the electrodes is one to four times that of the gate length. The current handling capability of a MESFET is directly proportional to the gate width Z because the cross-sectional area available for channel current is proportional to Z.

7.2.2 Principles of Operation

To understand the operation of a MESFET, we consider the section under the gate, Fig. 10b. The source is grounded, and the gate and drain voltage are measured with respect to the source. Under normal operating conditions, the gate voltage is zero or reverse biased and the drain voltage is zero or forward biased; that is, $V_G \le 0$ and $V_D \ge 0$. Since the channel is n-type material, the device is referred to as an n-channel MESFET. Most applications use the n-channel MESFET rather than the p-channel MESFET because of higher carrier mobility in n-channel devices.

The resistance of the channel is given by

$$R = \rho \frac{L}{A} = \frac{L}{q\mu_n N_D A} = \frac{L}{q\mu_n N_D Z(a - W)}, \tag{24}$$

where N_D is the donor concentration, A is the cross-section area for current flow and equals $Z(a - W)$, and W is the width of the depletion region of the Schottky barrier.

When no gate voltage is applied and V_D is small, as shown in Fig. 11a, a small drain current I_D flows in the channel. The magnitude of the current is given by V_D/R, where R is the channel resistance given in Eq. 24. Therefore, the current varies linearly with the drain voltage. Of course, for any given drain voltage, the voltage along the channel increases from zero at the source to V_D at the drain. Thus, the Schottky barrier becomes increasingly reverse biased as we proceed from the source to the drain. As V_D is increased, W increases, and the average cross-sectional area for current flow is reduced. The channel resistance R also increases. As a result, the current increases at a slower rate.

As the drain voltage is further increased, eventually the depletion region touches the semiinsulating substrate as shown in Fig. 11b. This happens when $W = a$ at the drain. We can obtain the corresponding value of the drain voltage, called the *saturation voltage*, V_{Dsat} from Eq. 7 where $V = -V_{Dsat}$:

$$V_{Dsat} = \frac{qN_D a^2}{2\varepsilon_s} - V_{bi} \quad \text{for} \quad V_G = 0. \tag{25}$$

At this drain voltage, the source and the drain are *pinched off* or completely separated by a reverse-biased depletion region. The location P in Fig. 11b is called the pinch-off point. At this point, a large drain current called the *saturation current* I_{Dsat} can flow across the depletion region. This is similar to the situation caused by injecting carriers into a reverse-biased depletion region such as the collector-base depletion region of a bipolar transistor.

Beyond the pinch-off point, as V_D is increased further, the depletion region near the drain will expand and point P will move toward the source, as indicated in Fig. 11c. However, the voltage at point P remains the same, V_{Dsat}. Thus, the number of electrons per unit time arriving from the source to point P, and hence the current flowing in the channel, remain the same because the potential drop in the channel from source to point

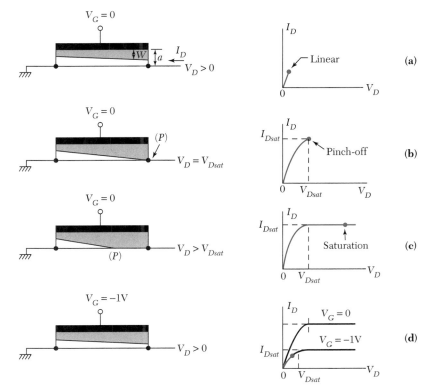

Fig. 11 Variation of the depletion-layer width and output characteristics of a MESFET under various biasing conditions. (*a*) $V_G = 0$ and a small V_D. (*b*) $V_G = 0$ and at pinch-off. (*c*) $V_G = 0$ at post pinch-off ($V_D > V_{Dsat}$). (*d*) $V_G = -1V$ and a small V_D.

P does not change. Therefore, for drain voltages larger than V_{Dsat}, the current remains essentially at the value I_{Dsat} and is independent of V_D.

When a gate voltage is applied to reverse bias the gate contact, the depletion-layer width W increases. For a small V_D, the channel again acts as a resistor but its resistance is higher because the cross-sectional area available for current flow is decreased. As indicated in Fig. 11*d*, the initial current is smaller for $V_G = -1$ V than for $V_G = 0$. When V_D is increased to a certain value, the depletion region again touches the semiinsulating substrate. The value of this V_D is given by

$$V_{Dsat} = \frac{qN_D a^2}{2\varepsilon_s} - V_{bi} - V_G.$$

(26)

For an *n*-channel MESFET, the gate voltage is negative with respect to the source, so we use the absolute value of V_G in Eq. 26 and in subsequent equations. We see from Eq. 26 that the application of a gate voltage V_G reduces the drain voltage required for the onset of pinch-off by an amount equal to V_G.

7.2.3 Current-Voltage Characteristics

We now consider a MESFET before the onset of pinch-off, as shown in Fig. 12a. The drain voltage variation along the channel is shown in Fig. 12b. The voltage drop across an elemental section dy of the channel is given by

$$dV = I_D dR = \frac{I_D dy}{q\mu_n N_D Z[a - W(y)]},$$ (27)

where we have used Eq. 24 for dR and we have replaced L by dy. The depletion-layer width at distance y from the source is given by

$$W(y) = \sqrt{\frac{2\varepsilon_s[V(y) + V_G + V_{bi}]}{qN_D}}.$$ (28)

The drain current I_D is a constant, independent of y. We can rewrite Eq. 27 as

$$I_D dy = q\mu_n N_D Z[a - W(y)] dV.$$ (29)

The differentiation of the drain voltage dV is obtained from Eq. 28:

$$dV = \frac{qN_D}{\varepsilon_s} W dW.$$ (30)

Substituting dV into Eq. 29 and integrating from $y = 0$ to $y = L$ yields

$$I_D = \frac{1}{L} \int_{W_1}^{W_2} q\mu_n N_D Z(a - W) \frac{qN_D}{\varepsilon_s} W dW$$

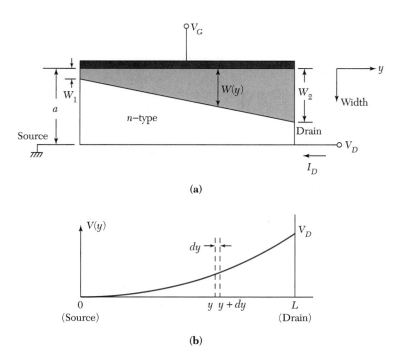

(a)

(b)

Fig. 12 (a) Expanded view of the channel region. (b) Drain voltage variation along the channel.

$$= \frac{Z\mu_n q^2 N_D^2}{2\varepsilon_s L}\left[a\left(W_2^2 - W_1^2\right) - \frac{2}{3}\left(W_2^3 - W_1^3\right)\right]$$

or

$$I = I_P\left[\frac{V_D}{V_P} - \frac{2}{3}\left(\frac{V_D + V_G + V_{bi}}{V_P}\right)^{3/2} + \frac{2}{3}\left(\frac{V_G + V_{bi}}{V_P}\right)^{3/2}\right],\tag{31}$$

where

$$I_P \equiv \frac{Z\mu_n q^2 N_D^2 a^3}{2\varepsilon_s L}\tag{31a}$$

and

$$V_P \equiv \frac{q N_D a^2}{2\varepsilon_s}.\tag{31b}$$

The voltage V_P is called the pinch-off voltage, that is, the total voltage $(V_D + V_G + V_{bi})$ at which $W_2 = a$.

In Fig. 13 we show the *I-V* characteristics of a MESFET having a pinch-off voltage of 3.2 V. The curves shown are calculated for $0 \leq V_D \leq V_{Dsat}$ using Eq. 31. Beyond V_{Dsat} the current is taken to be constant in accordance with our previous discussion. Note that there are three different regions in the current-voltage characteristics. When V_D is small,

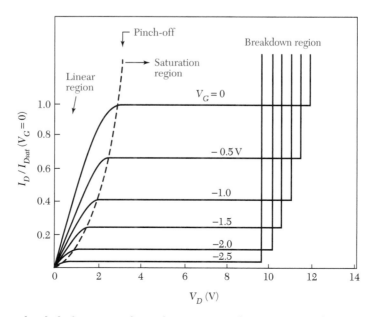

Fig. 13 Normalized ideal current-voltage characteristics of a MESFET with V_P = 3.2 V.

the cross-section area of the channel is essentially independent of V_D and the *I-V* characteristics are ohmic or linear. We refer to this region of operation as the linear region. For $V_D \geq V_{Dsat}$, the current saturates at I_{Dsat}. We refer to this region of operation as the saturation region. As the drain voltage is further increased, avalanche breakdown of the gate-to-channel diode occurs, and the drain current suddenly increases. This is the breakdown region.

In the linear region where $V_D << V_G$, Eq. 31 can be expended to give

$$I_D \cong \frac{I_P}{V_P}\left[1 - \sqrt{\left(\frac{V_G + V_{bi}}{V_P}\right)}\right]V \tag{32}$$

An important parameter of a MESFET is the transconductance g_m, which represents the change of drain current at a given drain voltage on a change in gate voltage. From Eq. 32, we obtain

$$g_m = \left.\frac{\partial I_D}{\partial V_G}\right|_{V_D} = \frac{I_P}{2V_P^2}\sqrt{\frac{V_P}{V_G + V_{bi}}}\, V_D. \tag{33}$$

In the saturation region, the drain current can be calculated from Eq. 31 by evaluating the current at the pinch-off point, that is, by setting $V_P = V_D + V_G + V_{bi}$:

$$I_{Dsat} = I_P\left[\frac{1}{3} - \left(\frac{V_G + V_{bi}}{V_P}\right) + \frac{2}{3}\left(\frac{V_G + V_{bi}}{V_P}\right)^{3/2}\right]. \tag{34}$$

The corresponding saturation voltage is given by

$$V_{Dsat} = V_P - V_G - V_{bi}. \tag{35}$$

The transconductance in the saturation region can be obtained from Eq. 34:

$$g_m = \frac{I_P}{V_P}\left(1 - \sqrt{\frac{V_G + V_{bi}}{V_P}}\right) = \frac{2Z\mu_n qN_D a}{L}\left(1 - \sqrt{\frac{V_G + V_{bi}}{V_P}}\right). \tag{36}$$

In the breakdown region, the breakdown voltage occurs at the drain end of the channel, where the reverse voltage is the highest:

$$V_B \text{ (breakdown voltage)} = V_D + |V_G|. \tag{37}$$

For example, in Fig. 13 the breakdown voltage is 12 V for $V_G = 0$. At $|V_G| = 1$, the breakdown voltage is still 12 V and the drain voltage at breakdown is $(V_B - |V_G|)$ or 11 V.

▶ **EXAMPLE 4**

Consider an *n*-channel GaAs MESFET at $T = 300$ K with a gold contact. Assume the barrier height is 0.89 V. The *n*-channel doping is 2×10^{15} cm^{-3} and the channel thickness is 0.6 μm. Calculate the pinch-off voltage and the built-in potential. The dielectric constant of GaAs is 12.4.

SOLUTION The pinch-off voltage is

$$V_P = \frac{qN_D}{2\varepsilon_s}a^2 = \frac{(1.6 \times 10^{-19})(2 \times 10^{15})}{2 \times 12.4 \times (8.85 \times 10^{-14})} \times (0.6 \times 10^{-4})^2 = 0.53 \text{ V}.$$

The difference between the conduction band and the Fermi level is given by

$$V_n = \frac{kT}{q}\ln\left(\frac{N_C}{N_D}\right) = 0.026\ln\left(\frac{4.7\times 10^{17}}{2\times 10^{15}}\right) = 0.14 \text{ V}.$$

The built-in potential is

$$V_{bi} = \phi_{Bn} - V_n = 0.89 - 0.14 = 0.75 \text{ V}.$$ ◀

So far we have considered only a normally on (or depletion mode) device; that is, the device has a conductive channel at $V_G = 0$. For high-speed, low-power applications, the normally off device is preferred. This device does not have a conductive channel at $V_G = 0$; that is, the built-in potential V_{bi} of the gate contact is sufficient to deplete the channel region. This is possible, for example, in a gallium arsenide MESFET with a very thin epitaxial layer on a semiinsulating substrate. For a normally off MESFET, a positive bias must be applied to the gate before channel current begins to flow. The required voltage, called the *threshold voltage V_T*, is given by

$$V_T = V_{bi} - V_P \tag{38a}$$

or

$$V_{bi} = V_T + V_P, \tag{38b}$$

where V_P is the pinch-off voltage defined in Eq. 31*b*. Near the threshold voltage, the drain current in the saturation region can be obtained by substituting V_{bi} of Eq. 38*b* in Eq. 34 and by using the Taylor series expansion assuming $(V_G - V_T)/V_p << 1$. We obtain

$$I_{Dsat} = I_P\left\{\frac{1}{3} - \left[1 - \left(\frac{V_G - V_T}{V_P}\right)\right] + \frac{2}{3}\left[1 - \left(\frac{V_G - V_T}{V_P}\right)\right]^{3/2}\right\}$$

or

$$\boxed{I_{Dsat} \approx \frac{Z\mu_n\varepsilon_s}{2aL}(V_G - V_T)^2.} \tag{39}$$

In deriving Eq. 39 we used a negative sign for V_G to account for its polarity.

The basic current-voltage characteristics of normally on and normally off devices are similar. Figure 14 compares these two modes of operation. The main difference is the shift of threshold voltage along the V_G axis. The normally off device (Fig. 14*b*) has no current conduction at $V_G = 0$, and the current varies as in Eq. 39 when $V_G > V_T$. Since the built-in potential of the gate is less than about 1 V, the forward bias on the gate is limited to about 0.5 V to avoid excessive gate current.

The transconductance for a normally off device can be obtained from Eq. 39:

$$g_m = \frac{dI_{Dsat}}{dV_G} = \frac{Z\mu_n\varepsilon_s}{aL}(V_G - V_T). \tag{40}$$

7.2.4 High-Frequency Performance

For high-frequency applications of MESFETs, an important figure of merit is the cut-off frequency f_T, which is the frequency at which the MESFET can no longer amplify

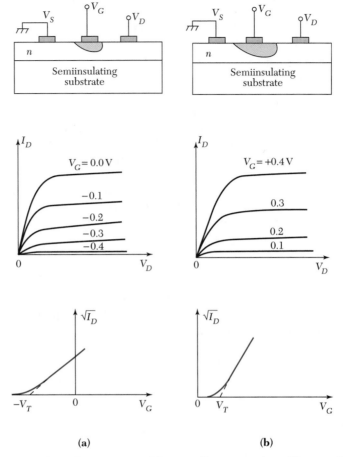

Fig. 14 Comparison of *I-V* characteristics. (*a*) Normally on MESFET. (*b*) Normally off MESFET.

the input signal. The small-signal input current is the product of the gate admittance and the small-signal gate voltage, assuming that the device has negligibly small series resistance:

$$\tilde{i}_{in} = 2\pi f C_G \tilde{v}_g. \tag{41}$$

where C_G is the gate capacitance equal to $ZL(\varepsilon_s/\overline{W})$ and $\overline{W}$ is the average depletion-layer width under the gate electrode. The small-signal output current is obtainable from the definition of the transconductance:

$$g_m = \frac{\partial I_C}{\partial V_G} = \frac{\tilde{i}_{out}}{\tilde{v}_g} \tag{42}$$

or

$$\tilde{i}_{out} = g_m \tilde{v}_g. \tag{42a}$$

Equating Eqs. 41 and 42*a*, we obtain the cutoff frequency

$$f_T = \frac{g_m}{2\pi C_G} < \frac{I_P/V_P}{2\pi ZL\left(\varepsilon_s/\overline{W}\right)} \approx \frac{\mu_n q N_D a^2}{2\pi \varepsilon_s L^2}, \tag{43}$$

where we have used Eq. 36 for g_m. From Eq. 43 we see that to improve high-frequency performance, we should use a MESFET having high carrier mobility and short channel length. This is the reason that n-channel MESFET, which has higher electron mobility, is preferred.

These derivations are based on the assumption that the carrier mobility in the channel is a constant independent of the applied field. However, for very-high-frequency operations, the longitudinal field, i.e., the electric field directed from the source to the drain, is sufficiently high that the carriers travel at their saturation velocity.

Under these conditions, the saturation channel current is given by

$$I_{Dsat} = (\text{ area for carrier transport })\times qnv_s,$$
$$= Z(a-W)qN_D v_s. \qquad (44)$$

The transconductance is then

$$g_m = \frac{\partial I_{Dsat}}{\partial V_G} = \frac{\partial I_{Dsat}}{\partial W}\cdot\frac{\partial W}{\partial V_G} = \left[qN_D v_s Z(-1)\right]\left(\frac{1}{-qN_D W / \varepsilon_s}\right) \qquad (45)$$

or

$$g_m = Zv_s\varepsilon_s / W. \qquad (45a)$$

In Eq. 45, we obtain $\partial W/\partial V_G$ from Eq. 28.

From Eq. 45a, we can obtain the cutoff frequency under saturation-velocity condition:

$$f_T = \frac{g_m}{2\pi C_G} = \frac{Zv_s\varepsilon_s / W}{2\pi ZL(\varepsilon_s / W)} = \frac{v_s}{2\pi L}. \qquad (46)$$

Therefore, to increase f_T, we must reduce the gate length L and to employ a semiconductor with a high velocity. Figure 15 shows the electron drift velocity versus electric field

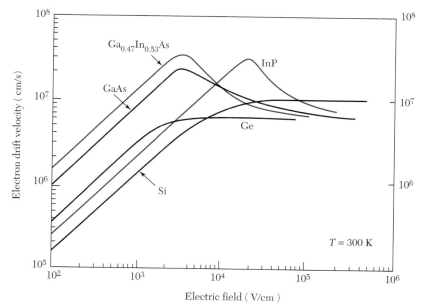

Fig. 15 The drift velocity versus the electric field for electrons in various semiconductor materials.[8]

for five semiconductors.[8] Note that GaAs has an average velocity[§] of 1.2×10^7 cm/s and a peak velocity of 2×10^7 cm/s, which are 20%–100% higher than the saturation velocity of silicon. Also note that $Ga_{0.47}In_{0.53}As$ and InP have higher average and peak velocities than GaAs. Consequently, the cutoff frequencies of these semiconductors will be higher than that from GaAs.

► 7.3 MODFET

7.3.1 MODFET Fundamentals

The modulation-doped field-effect transistor (MODFET) is a heterostructure field-effect device. Other names commonly applied to the device include high electron mobility transistor (HEMT), two-dimensional electron gas field-effect transistor (TEGFET), and selectively doped heterostructure transistor (SDHT). Frequently, it is referred to by a general name of heterojunction field-effect transistor (HFET).

Figure 16 shows a perspective view of a conventional MODFET. The special features of a MODFET are its heterojunction structure under the gate, and the modulation doped layers. For the device in Fig. 16, AlGaAs is the wide bandgap semiconductor, whereas GaAs is the narrow bandgap semiconductor. The two semiconductors are modulation doped, i.e., the AlGaAs is doped, except for a narrow region d_o, which is undoped, whereas the GaAs is undoped. Electrons in the AlGaAs will diffuse to the undoped GaAs, where a conduction channel can be formed at the surface of the GaAs.

Figure 17a shows the band diagram of a MODFET in thermal equilibrium condition. Similar to a standard Schottky barrier, $q\phi_{Bn}$ is the barrier height of the metal on the

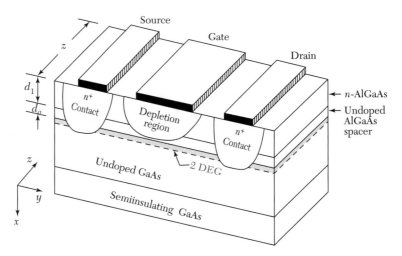

Fig. 16 Perspective view of a conventional modulation-doped field-effect transistor (MODFET) structure.

[§]The average velocity is defined as $\bar{v} \equiv \left[\dfrac{1}{L}\displaystyle\int_0^L \dfrac{dx}{v(x)}\right]^{-1}$. If $v(x)$ is a constant v_0, then $\bar{v} = v_0$.

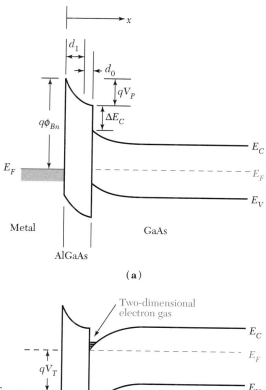

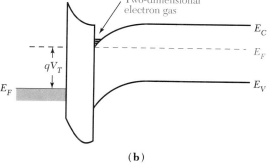

Fig. 17 Energy band diagrams for a normally-off MODFET at (a) thermal equilibrium, and (b) the onset of threshold. d_1 and d_0 are the doped and undoped regions, respectively.[9]

wide-bandgap semiconductor.[9] ΔE_C is the conduction band discontinuity for the heterojunction structure, and V_P is the pinch-off voltage given by

$$V_P = \frac{q}{\varepsilon_s} \int_0^d N_D(x)x\,dx = \frac{qN_D d_1^2}{2\varepsilon_s},\qquad (47)$$

where d_1 is the thickness of the doped region in AlGaAs and ε_s is the dielectric permittivity.

A key parameter for the operation of a MODFET is the threshold voltage V_T, which is the gate bias at which the channel starts to form between the source and drain. With reference to Fig. 17b, V_T corresponds to the situation when the bottom of the conduction band at the GaAs surface coincides with the Fermi level:

$$V_T = \phi_{Bn} - \frac{\Delta E_C}{q} - V_P.\qquad (48)$$

The threshold voltage V_T can be adjusted by using different values for ϕ_{Bn} and V_p. However, ΔE_C is fixed for a given set of semiconductors. Figure 17b has a positive V_T, and the MOD-FET is an enhancement-mode device (normally off), as opposed to a depletion-mode device with a negative V_T (normally on).

When the gate voltage is larger than V_T, a charge sheet $n_s(y)$ is capacitively induced by the gate at the heterojunction interface. The charge sheet is similar to the charge Q_n/q in the inversion layer of a MOSFET (see Section 6.1):

$$n_s(y) = \frac{C_i\left[V_G - V_T - V(y)\right]}{q},$$ (49)

where

$$C_i = \frac{\varepsilon_s}{d_1 + d_0 + \Delta d},$$ (49a)

d_1 and d_0 are the doped and undoped AlGaAs thickness (Fig. 16), and Δd is the channel thickness or the thickness of the inversion layer, estimated to be about 8 nm. $V(y)$ is the channel potential with respect to the source. It varies along the channel from 0 to the drain bias V_D, similar to that shown in Fig. 12b. The charge sheet is also called a *two-dimensional electron gas*. This is because that the electrons in the inversion layer are confined in the x-direction by ΔE_C on the left side and by the potential distribution of the conduction band on the right side (Fig. 17b). However, these electrons can make two-dimensional movements: in the y-direction from the source to the drain and in the z-direction parallel to the channel width (Fig. 16).

Equation 49 shows that a negative gate bias will reduce the two-dimensional electron gas. If, on the other hand, a positive V_G is applied, n_s will increase.

▶ EXAMPLE 5

Consider an AlGaAs/GaAs heterojunction with n-AlGaAs doped to 2×10^{18} cm^{-3} and a thickness of 40 nm. Assume the undoped spacer layer is 3 nm and the Schottky barrier height is 0.85 V and

$\dfrac{\Delta E_C}{q}$ = 0.23 V. The dielectric constant of the AlGaAs is 12.3. Calculate the two-dimensional

electron gas concentration for such heterojunction at $V_G = 0$.

SOLUTION

$$V_p = \frac{qN_D d_1^2}{2\varepsilon_s} = \frac{1.6 \times 10^{-19} \times 2 \times 10^{18} \times (40 \times 10^{-7})^2}{2 \times 12.3 \times 8.85 \times 10^{-14}} = 2.35 \text{ V}.$$

The threshold voltage is

$$V_T = \phi_{Bn} - \frac{\Delta E_C}{q} - V_p = 0.85 - 0.23 - 2.35 = -1.73 \text{ V}.$$

Therefore, the device is a normally on MODFET.

The two-dimensional electron gas at the source for $V_G = 0$ is

$$n_s = \frac{12.3 \times 8.85 \times 10^{-14}}{1.6 \times 10^{-19} \times (40 + 3 + 8) \times 10^{-7}} \times \left[0 - (-1.73)\right] = 2.29 \times 10^{12} \text{ cm}^{-2}.$$

◀

7.3.2 Current-Voltage Characteristics

The current-voltage characteristics of a MODFET can be obtained by using the gradual channel approximation similar to that of a MOSFET. The current at any point along the channel is given by

$$I = Zq\mu_n n_s \mathscr{E}_y$$

$$= Z\mu_n C_i \left[V_G - V_T - V(y) \right] \frac{dV(y)}{dy}. \tag{50}$$

Since the current is constant along the channel, integrating Eq. 50 from source to drain $(y = 0 \text{ to } y = L)$ gives

$$I = \frac{Z}{L} \mu_n C_i \left[\left(V_G - V_T \right) V_D - \frac{V_D^2}{2} \right]. \tag{51}$$

The output characteristics for an enhancement-mode MODFET are similar to that shown in Fig. 14b. In the linear region where $V_D << (V_G - V_T)$, Eq. 51 can be reduced to

$$I = \frac{Z}{L} \mu_n C_i \left(V_G - V_T \right) V_D. \tag{52}$$

For a large drain voltage, the charge sheet $n(y)$ at the drain is reduced to zero. This is the pinch-off condition previously discussed and shown in Fig. 11b. From Eq. 49, we obtain the saturation voltage V_{Dsat}, at which $n_s(y = L) = 0$,

$$V_{Dsat} = V_G - V_T, \tag{53}$$

and the saturation current is obtainable from Eqs. 51 and 53,

$$I = \frac{Z\mu_n C_i}{2L} \left(V_G - V_T \right)^2 = \frac{Z\mu_n \varepsilon_s}{2L(d_1 + d_0 + \Delta d)} (V_G - V_T)^2. \tag{54}$$

Note that this equation is very similar to Eq. 39. A similar expression can be obtained for the transconductance given in Eq. 40.

For high-speed operations, the longitudinal field (i.e., the field along the channel) is sufficiently high to cause carrier velocity saturation. The current in the velocity-saturation region is

$$I_{sat} = Z v_s q n_s$$

$$\cong Z v_s C_i (V_G - V_T). \tag{55}$$

The transconductance becomes

$$g_m = \frac{\partial I_{sat}}{\partial V_D} = Z v_s C_i. \tag{56}$$

Note that I_{sat} is independent of gate length and g_m is independent of both gate length and gate voltage in the velocity-saturation regime.

7.3.3 Cutoff Frequency

The speed of a MODFET is measured by the cutoff frequency

$$f_T = \frac{g_m}{2\pi\left(\text{total capacitance}\right)} = \frac{Zv_sC_i}{2\pi\left(ZLC_i + C_p\right)}$$

$$= \frac{v_s}{2\pi\left(L + C_p / ZC_i\right)}, \tag{57}$$

where C_p is the parasitic capacitance. To improve f_T, we should consider a semiconductor with large v_s, a gate structure with an ultrashort gate length, and a device configuration with minimum parasitic capacitance.

A comparison of the cutoff frequencies of various FETs is shown in Fig. 18. The cutoff frequency f_T is plotted against the channel or gate length.[8,10] Note that for a given length, a silicon n-type MOSFET has the lowest f_T; this is because of the relatively low

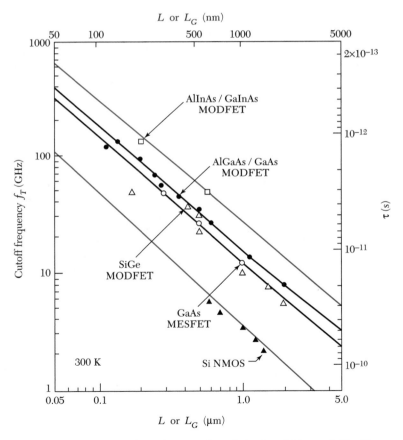

Fig. 18 Cutoff frequency versus channel or gate length for five different field-effect transistors.[8,10]

mobility and a low average velocity of electron in silicon. GaAs MESFET has an f_T about three times higher than does silicon MOSFET.

Also shown are three MODFETs. The conventional GaAs MODFET (i.e., AlGaAs-GaAs structure) has an f_T about 30% higher than that of the GaAs MESFET. For the pseudomorphic SiGe MODFET (i.e., Si-SiGe structure where the SiGe lattice is slightly shrinked to match the silicon lattice), the best devices have an f_T comparable to the GaAs MODFET. SiGe MODFETs are attractive because they can be processed in a silicon fabrication facility. For even higher cutoff frequencies, we have the $Al_{0.48}In_{0.52}As$-$Ga_{0.47}In_{0.53}As$ MODFET formed on a InP substrate. The superior performance is mainly due to the high electron mobility in $Ga_{0.47}In_{0.53}As$ and its high average and peak velocities. It is anticipated that at a gate length of 50 nm f_T can be as high as 600 GHz.

▷ SUMMARY

When a metal makes intimate contact with a semiconductor, it forms a *metal-semiconductor contact*. There are two types of contacts. The first type is the rectifying contact, also called the Schottky barrier contact, which has a relatively large barrier height and is formed on a semiconductor with a relatively low-doping concentration. The potential and field distribution in a Schottky barrier are identical to those of a one-sided abrupt $p–n$ junction. However, the current transport in a Schottky barrier is by thermionic emission and, therefore, has an inherent fast response.

The second type is the ohmic contact, which is formed on a degenerate semiconductor in which carrier transport is by the tunneling process. An ohmic contact can pass the required current with a very small voltage drop across it. All semiconductor devices and integrated circuits need ohmic contacts to make connections to other devices in an electronic system.

Metal-semiconductor contacts are building blocks for MESFET and MODFET devices. By employing a Schottky barrier as the gate electrode and two ohmic contacts as the source and drain electrodes, we form a MESFET. This three-terminal device is important for high-frequency application, especially for monolithic microwave integrated circuits (MMIC). Most MESFETs are made with n-type III-V compound semiconductors because of their high electron mobilities and high average drift velocities. Of particular importance is GaAs, because of its relatively mature technology and the availability of high-quality GaAs wafers.

The MODFET is a device with enhanced high-frequency performance. This device structure is similar to that of a MESFET except there is a heterojunction under the gate. A two-dimensional electron gas, i.e., a conductive channel, is formed at the heterojunction interface, and electrons with high mobility and high average drift velocity can be transported from the source through the channel to the drain.

The output characteristics of all field-effect transistors (FETs) are similar. They all have a linear region at low-drain biases. As the bias increases, the output current eventually saturates, and at a sufficiently high voltage, avalanche breakdown occurs at the drain. Depending on whether it requires a positive- or negative-threshold voltage, FET can be either normally off (enhancement mode) or normally on (depletion mode).

The cutoff frequency f_T is a figure of merit for the high-frequency performance of an FET. For a given length, the silicon MOSFET (n-type) has the lowest f_T, and the GaAs MESFET has an f_T about three times higher than that of silicon. The conventional GaAs MODFET and the pseudomorphic SiGe MODFET have an f_T about 30% higher than that of the GaAs MESFET. For even higher cutoff frequencies, we have the GaInAs MODFET, which has a projected f_T of 600 GHz at a gate length of 50 nm.

▶ REFERENCES

1. W. Schottky, "Halbleitertheorie der Sperrschicht," *Naturwissenschaften*, **26**, 843 (1938).

2. A. M. Cowley and S. M. Sze, "Surface States and Barrier Height of Metal Semiconductor System," *J. Appl. Phys.*, **36**, 3212 (1965).

3. G. Myburg, et al., "Summary of Schottky Barrier Height Data on Epitaxially Grown *n*- and *p*-GaAs," *Thin Solid Films*, **325**, 181 (1998).

4. C. R. Crowell, J. C. Sarace, and S. M. Sze, "Tungsten-Semiconductor Schottky-Barrier Diodes," *Trans. Met. Soc. AIME*, **23**, 478 (1965).

5. V. L. Rideout, "A Review of the Theory, Technology and Applications of Metal-Semiconductor Rectifiers," *Thin Solid Films*, **48**, 261 (1978).

6. S. M. Sze, *Physics of Semiconductor Devices*, 2nd ed., Wiley, New York, 1981.

7. C. A. Mead, "Schottky Barrier Gate Field-Effect Transistor," *Proc. IEEE*, **54**, 307 (1966).

8. S. M. Sze, Ed., *High Speed Semiconductor Device*, Wiley, New York, 1992.

9. K. K. Ng, *Complete Guide to Semiconductor Devices*, McGraw Hill, New York, 1995.

10. S. Luryi, J. Xu, and A. Zaslavsky, Eds., *Future Trends in Microelectronics*, Wiley, New York, 1999.

▶ PROBLEMS (* DENOTES DIFFICULT PROBLEMS)

FOR SECTION 7.1   METAL-SEMICONDUCTOR CONTACTS

1. Calculate the theoretical barrier height and built-in potential in a metal-semiconductor diode for zero applied bias. Assume the metal work function is 4.55 eV, the electron affinity is 4.01 eV, and $N_D = 2 \times 10^{16}$ cm^{-3} at 300 K.

2. (a) Find the donor concentration and barrier height of the W-GaAs Schottky barrier diode shown in Fig. 6. (b) Compare the barrier height with that obtained from the saturation current density of 5×10^{-7} A/cm^2 shown in Fig. 8. (c) For a reverse bias of –1 V, calculate the depletion-layer width W, the maximum field, and the capacitance.

3. Copper is deposited on a carefully prepared *n*-type silicon substrate to form an ideal Schottky diode. $\phi_m = 4.65$ eV, the electron affinity is 4.01 eV, $N_D = 3 \times 10^{16}$ /cm^3, and $T = 300$ K. Calculate the barrier height, the built-in potential, the depletion-layer width, and the maximum field at a zero bias.

*4. The capacitance of a Au-*n*-type GaAs Schottky barrier diode is given by the relation $1/C^2 = 1.57 \times 10^5 - 2.12 \times 10^5 \, V_a$, where C is expressed in μF and V_a is in volts. Taking the diode area to be 10^{-1} cm^2, calculate the built-in potential, the barrier height, the dopant concentration, and the work function.

5. Calculate the value of V_{bi} and ϕ_m in an ideal metal-Si Schottky barrier contact. Assume the barrier height is 0.8 eV, $N_D = 1.5 \times 10^{16}$ cm^{-3}, and $q\chi = 4.01$ eV.

6. In a metal-Si Schottky barrier contact, the barrier height is 0.75 eV and $A^* = 110$ A/cm^2 – K^2. Calculate the ratio of the injected hole current to the electron current at 300 K, assuming $D_p = 12$ cm^2 s^{-1}, $L_p = 1 \times 10^{-3}$ cm, and $N_D = 1.5 \times 10^{16}$ cm^{-3}.

FOR SECTION 7.2   MESFET

7. Given $\phi_{Bn} = 0.9$ eV and $N_D = 10^{17}$ cm^{-3}, find the minimum value of the thickness of the epitaxial layer for a GaAs MESFET to be a depletion mode device (i.e., $V_T < 0$).

8. Assume the doping in a GaAs MESFET is $N_D = 7 \times 10^{16}$ cm^{-3} and the dimensions are $a = 0.3$ μm, $L = 1.5$ μm, $Z = 5$ μm, $\mu_n = 4500$ cm^2/V-s, and $\phi_{Bn} = 0.89$ V. Calculate the ideal value of g_m for $V_G = 0$, and $V_D = 1$ V.

9. The *n*-channel GaAs MESFET shown in Fig. 10 has a barrier height $\phi_{Bn} = 0.9$ V, $N_D = 10^{17}$ cm^{-3}, $a = 0.2$ μm, $L = 1$ μm, and $Z = 10$ μm. (a) Is this an enhancement or depletion-

mode device? (b) Find the threshold voltage (the enhancement mode indicates $V_T > 0$; the depletion mode indicates $V_T < 0$).

10. An n-channel GaAs MESFET has a channel doping $N_D = 2 \times 10^{15}$ cm^{-3}, $\phi_{Bn} = 0.8$ V, $a = 0.5$ μm, $L = 1$ μm, $\mu_n = 4500$ cm^2/V-s, and $Z = 50$ μm. Find the pinch-off potential, threshold voltage, and the saturation current at $V_G = 0$.

11. The barrier height ϕ_{Bn} of two GaAs n-channel MESFETs are the same and equal to 0.85 V. The channel doping in device 1 is $N_D = 4.7 \times 10^{16}$ cm^{-3}, and that in device 2 is $N_D = 4.7 \times 10^{17}$ cm^{-3}. Determine the channel thickness required in each device such that the threshold voltage is zero for each device.

FOR SECTION 7.3   MODFET

12. For an abrupt AlGaAs/GaAs heterojunction with the n-AlGaAs layer doped to 3×10^{18} cm^{-3}, the Schottky barrier is 0.89 V and the heterojunction conduction-band edge discontinuity ΔE_C is 0.23 eV. Calculate the thickness of the doped AlGaAs layer d_1 so that the threshold voltage is -0.5 V. Assume the permittivity of the AlGaAs is 12.3.

*13. Find the thickness of the undoped spacer layer d_0, such that the two-dimensional electron gas concentration of an AlGaAs/GaAs heterojunction is 1.25×10^{12} cm^{-2} at zero gate bias. Assume that the n-AlGaAs is doped to 1×10^{18} cm^{-3} and has a thickness d_1 of 50 nm, the Schottky barrier height is 0.89 V, and $\Delta E_C/q = 0.23$ V. The permittivity of the AlGaAs is 12.3.

14. Consider a AlGaAs/GaAs HFET with a 50 nm n-AlGaAs and a 10 nm undoped AlGaAs spacer. Assume the threshold voltage is -1.3 V, N_D is 5×10^{17} cm^{-3}, $\Delta E_C = 0.25$ eV, the channel width is 8 nm, and the permittivity of AlGaAs is 12.3. Calculate the Schottky barrier height and the two-dimensional electron gas concentration at $V_G = 0$.

15. The AlGaAs/GaAs has a two-dimensional electron gas concentration of 1×10^{12} cm^{-2}, the spacer is 5 nm, the channel width is 8 nm, the pinch-off voltage is 1.5 V, $\Delta E_C/q = 0.23$ V, and the doping concentration of AlGaAs is 10^{18} cm^{-3}. The Schottky barrier height is 0.8 V. Find the thickness of the doped AlGaAs and the threshold voltage.

16. Consider an n-AlGaAs–intrinsic GaAs abrupt heterojunction. Assume that the AlGaAs is doped to $N_D = 3 \times 10^{18}$ cm^{-3} and has a thickness of 35 nm (there is no spacer). Let $\phi_{Bn} = 0.89$ V, and assume that $\Delta E_C = 0.24$ eV and the dielectric constant is 12.3. Calculate (a) V_P and (b) n_s for $V_G = 0$.

CHAPTER 8

Microwave Diodes, Quantum-Effect, and Hot-Electron Devices

Many semiconductor devices discussed in the previous chapters can be operated in the microwave region (0.1 ~ 3000 GHz). However, two-terminal devices generate the highest power levels per device area in system applications, especially at higher frequencies. Additionally, pulsed operation of these devices overcomes thermal limits and increases peak rf (radio frequency) power levels by more than an order of magnitude.[1] In this chapter, we introduce some basic microwave technology and consider some special two-terminal microwave devices, including the tunnel diode, the IMPATT diode, the transferred-electron device, and the resonant tunneling diode.

In the past two decades, we have witnessed considerable research and effort in the development of device structures that could exploit quantum-effect and hot-electron phenomena to enhance circuit performances. Speed is often cited as a primary benefit that quantum-effect devices (QEDs) and hot-electron devices (HEDs) can offer. The tunneling process, on which most QEDs rely, is an intrinsically fast process. In HEDs, carriers under ballistic transport can move at velocities considerably in excess of their equilibrium peak velocity. However, a more significant advantage of QEDs and HEDs is their higher functionality. These devices can perform relatively complex circuit functions with a greatly reduced device count, replacing large numbers of transistors or passive circuit components.[1] The basic device structures and operating principles of QEDs and HEDs are discussed in this chapter.

Specifically, we cover the following topics:

- Advantages of millimeter-wave devices over those operated at lower frequencies.
- The quantum tunneling phenomenon and its related devices—tunnel diode, resonant tunneling diode (RTD), and unipolar resonant tunneling transistor.

- The IMPATT diode—the most powerful semiconductor source of millimeter-wave power.

- The transferred-electron device and its transit-time domain mode.

- The real-space–transfer transistor and its advantages as a functional device.

► 8.1 BASIC MICROWAVE TECHNOLOGY

The microwave frequencies cover the range from about 0.1 GHz (10^8 Hz) to 3000 GHz with corresponding wavelengths from 300 cm to 0.01 cm. For frequencies from 30 to 300 GHz, we have the millimeter-wave band because the wavelength is between 10 and 1 mm. For even higher frequencies, we have the submillimeter-wave band. The microwave frequency range is usually grouped into different bands.[2] The bands and the corresponding frequency ranges as designated by the Institute of Electrical and Electronics Engineers (IEEE) are listed in Table 1. It is recommended that both the band and the corresponding frequency range be used when referring to microwave devices.

The development of microwave technology was driven by the demands of short-wavelength radio (and later radar) systems. The history of microwaves started with the first experiments of Heinrich Hertz around 1887. Hertz used a spark transmitter that produced signals in a very broad frequency band, and he selected from these a frequency around 420 MHz with an antenna that measured half a wavelength for his experiments. The rapid development of wireless communication products has led to an explosion in microwave technology. Since the introduction of cellular telephone service in the 1980s, there has been rapid growth of those systems as well as mobile paging devices and a variety of wireless "data communication" services, under the broad heading of personal communication services (PCS). In addition to these terrestrial communication systems, the field of satellite-based video, telephone, and data communication systems has also grown rapidly. These systems use the microwave frequencies from several hundred MHz to well over 60 GHz—the millimeter-wave region.[3]

Millimeter-wave technology offers many advantages for communications and radar systems, such as radio astronomy, clear-air turbulence detection, nuclear spectroscopy, air-traffic–control beacons, and weather radar. The advantages of millimeter waves over lower microwave and infrared systems include light-weight, small-size, broad bandwidths (several GHz), operation in adverse weather conditions, and narrow beamwidths with

TABLE 1   IEEE Microwave Frequency Bands

Designation	Frequency range (GHz)	Wavelength (cm)
VHF	0.1–0.3	300.00–100.00
UHF	0.3–1.0	100.00–30.00
L band	1.0–2.0	30.00–15.00
S band	2.0–4.0	15.00–7.50
C band	4.0–8.0	7.50–3.75
X band	8.0–13.0	3.75–2.31
Ku band	13.0–18.0	2.31–1.67
K band	18.0–28.0	1.67–1.07
Ka band	28.0–40.0	1.07–0.75
Millimeter	30.0–300.0	1.00–0.10
Submillimeter	300.0–3000.0	0.10–0.01

high resolution. The principal frequencies of interest in the millimeter-wave band are centered around 35, 60, 94, 140, and 220 GHz.[4] The reason for choosing these specific frequencies is mainly the atmospheric absorption of horizontally propagated millimeter waves, as shown in Fig. 1. The atmospheric "windows" where absorption is at a local minimum are found at about 35, 94, 140, and 220 GHz. The absorption peak due to O_2 at 60 GHz can be used for secure communication systems.

Ordinary electronic components behave differently at microwave frequencies than they do at lower frequencies. The distributed effects must be taken into account at microwave frequencies because at these frequencies the wavelength approximates the physical size of the components. At microwave frequencies, a thin film resistor, for example, looks like a complex RLC network with distributed L and C values and a different R value. These distributed components have immense significance at microwave frequencies, even though they can be ignored at lower frequencies. At microwave frequencies, capacitors and inductors are often realized by transmission-line segments. Transmission lines are often used for microwave circuit interconnections as well. Transmission lines are actually complex networks containing the equivalent of all of the three basic electrical components: resistor, capacitor, and inductor. Planar transmission lines are the mainstay of modern microwave circuit technology. Such lines consist of one or more flat conductors on a thin dielectric substrate with a ground surface.

Figure 2 shows several basic types of planar transmission lines: microstrip, coplanar waveguide (CPW) stripline, and suspended-substrate stripline (SSSL).[5] Microstrip is the most common type of transmission line. The coplanar waveguide is somewhat lossier, that is, the loss in signal propagation is larger, but it minimizes the parasitic inductance of ground connections. The characteristic impedance Z_0 of a transmission line is given by

$$Z_0 = \sqrt{\frac{R + j\omega L}{G + j\omega C}} \text{ ohms,} \tag{1}$$

where R is resistance per unit length in ohms, G is conductance per unit length in siemens, L is inductance per unit length in henrys, C is capacitance per unit length in farads, and

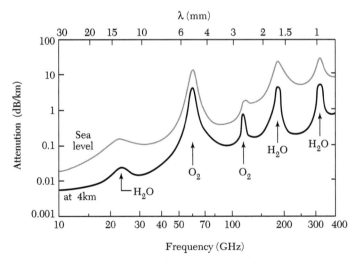

Fig. 1 Average atmosphere absorption of millimeter waves.[4] The upper curve is at the sea level; the lower curve is at 4 km above the sea level.

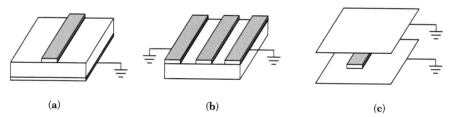

Fig. 2 Basic types of planar transmission lines: (*a*) microstrip, (*b*) coplanar waveguide stripline, and (*c*) suspended-substrate stripline.

ω is angular frequency in radians per second. In microwave circuits, the resistances are typically very low compared with the reactances, so Eq. 1 can be simplified to

$$Z_0 = \sqrt{\frac{L}{C}}. \tag{2}$$

The characteristic impedance for a specific type transmission line is a function of the conductor geometry (size, spacing) and the dielectric constant of the insulating material used between the conductors.

▷ **EXAMPLE 1**

Find the characteristic impedance of a nearly lossless transmission line (R is very small) that has a unit-length inductance of 10 nH and a unit-length capacitance of 4 pF.

SOLUTION

$$Z_0 = \sqrt{\frac{L}{C}} = \sqrt{\frac{10 \times 10^{-9}}{4 \times 10^{-12}}} = \sqrt{2.5 \times 10^3} = 50 \ \Omega. \qquad \blacktriangleleft$$

At lower microwave frequencies, resonant circuits can be made using inductive and capacitive elements. For millimeter-wave and higher frequency, however, the values of LC components at resonance are too small for practical implementation. Other methods for providing resonant circuits must be used. One popular solution is the resonant cavity, also called the tuned cavity.

The resonant cavity[2] is a metal-walled chamber made of low-resistivity material enclosing a good dielectric (for example, vacuum, dry air, or dry nitrogen). It is analogous to a section of waveguide with both ends shorted and has a means for injecting energy into or extracting energy from the cavity. As shown in Fig. 3, the cavity supports both TE (transverse electric) and TM (transverse magnetic) modes of propagation. The electromagnetic wave is confined by the wall of the cavity. Energy stored in the electric field represents the capacitance, whereas energy stored in the magnetic field represents the inductance. Thus, both elements of the LC tuned resonant tank circuit are present in the cavity. The resonant mode occur in a cavity at those frequencies for which the length d along the Z axis is a half-wavelength (see Fig. 3a). The alphanumeric system for designating modes in a cavity is $Tx_{m,n,p}$, where x is E for electric dominant modes and M for magnetic dominant modes, m is the number of half-wavelengths in the a dimension, n is the number of half-wavelengths in the b dimension, and p is the number of half-wavelengths in the

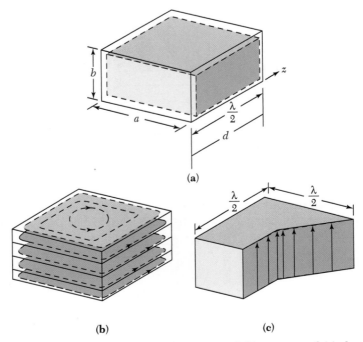

Fig. 3 Resonant cavity: (*a*) resonator shape, (*b*) magnetic field pattern, and (*c*) electric field pattern.[2]

d dimension. The general mode dependent equation for the resonant frequency of the cavity is

$$f_r = \frac{1}{2\sqrt{\mu\varepsilon}} \sqrt{\left(\frac{m}{a}\right)^2 + \left(\frac{n}{b}\right)^2 + \left(\frac{p}{d}\right)^2},\tag{3}$$

where μ and ε are the permeability and the permittivity of the material in the cavity. For vacuum, $\mu = \mu_0$ and $\varepsilon = \varepsilon_0$, and $\sqrt{\mu_0\varepsilon_0} = c^{-1}$, where c is the speed of light in vacuum. We can rewrite Eq. 3 as

$$f_r = \frac{c}{2} \sqrt{\left(\frac{m}{a}\right)^2 + \left(\frac{n}{b}\right)^2 + \left(\frac{p}{d}\right)^2}.\tag{4}$$

► **EXAMPLE 2**

For a cavity of the dimensions $a = 5$ cm (0.05 m), $b = 2.5$ cm (0.025 m), and $d = 10$ cm (0.1 m), find the resonant frequency in the dominant TE_{101} mode.

SOLUTION

$$f_r = \frac{c}{2} \sqrt{\left(\frac{m}{a}\right)^2 + \left(\frac{n}{b}\right)^2 + \left(\frac{p}{d}\right)^2}$$

$$= \frac{c}{2} \sqrt{\left(20\right)^2 + 0 + \left(10\right)^2}$$

$$= \frac{3 \times 10^8 \ \text{m/s}}{2} \times \sqrt{500} = 3.354 \ \text{GHz}.$$

◄

8.2 TUNNEL DIODE

The tunnel diode is associated with the quantum tunneling phenomena.[6] The tunneling time across the device is very short, permitting its use well into the millimeter-wave region. Because of its mature technology, the tunnel diode is used in special low-power microwave applications, such as local oscillators and frequency-locking circuits.

A tunnel diode consists of a simple *p–n* junction in which both the *p*- and *n*-sides are degenerate (i.e., very heavily doped with impurities). Figure 4 shows a typical static current-voltage characteristic of a tunnel diode under four different bias conditions. The *I-V* characteristic is the result of two current components: tunneling current and thermal current.

When there is no voltage applied to the diode, it is in thermal equilibrium ($V = 0$). Because of the high dopings, the depletion region is very narrow and the tunneling distance d is quite small (5–10 nm). The dopings also cause the Fermi levels to be located within the allowed bands. The amount of degeneracy, qV_p and qV_n, shown at the far left of Fig. 4, is typically 50–200 meV.

When a forward bias is applied, there exists a band of energy states that is occupied on the *n* side and a corresponding band of energy states that is available and unoccupied on the *p* side. The electrons can tunnel from the *n*-side to the *p*-side. When the applied

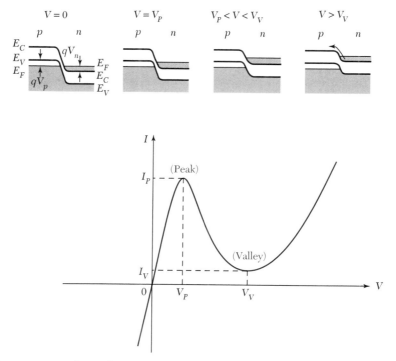

Fig. 4 Static current-voltage characteristics of a typical tunnel diode. I_P and V_P are the peak current and peak voltage, respectively. I_V and V_V are the valley current and valley voltage, respectively. The upper figures show the band diagrams of the device at different bias voltages.

bias equals approximately $(V_p + V_n)/3$, the tunneling current reaches its peak value I_P and the corresponding voltage is called the peak voltage V_p. When the forward voltage is further increased, there are fewer available unoccupied states on the p-side ($V_p < V < V_V$, where V_V is the valley voltage) and the current decreased. Eventually, the band is "uncrossed," and at this point the tunneling current can no longer flow. With still further increase of the voltage, the normal thermal current will flow (for $V > V_V$).

From this discussion we expect that in the forward direction the tunneling current increases from zero to a peak current I_P as the voltage increases. With a further increase in voltage, the current then decreases to zero when $V = V_n + V_p$, where V is the applied forward voltage. The decreasing portion after the peak current in Fig. 4 is the negative differential resistance region. The values of the peak current I_P and the valley current I_V determine the magnitude of the negative resistance. For this reason their ratio I_P/I_V is used as a figure of merit for tunnel diode.

An empirical form for the I-V characteristics is given by

$$
I = I_P \left(\frac{V}{V_P} \right) \exp\left(1 - \frac{V}{V_P} \right) + I_0 \exp\left(\frac{qV}{kT} \right),
\tag{5}
$$

where the first term is the tunnel current and I_P and V_P are the peak current and peak voltage, respectively, as shown in Fig. 4. The second term is the normal thermal current. The negative differential resistance can be obtained from the first term in Eq. 5:

$$
R = \left(\frac{dI}{dV} \right)^{-1} = -\left[\left(\frac{V}{V_P} - 1 \right) \frac{I_P}{V_P} \exp\left(1 - \frac{V}{V_P} \right) \right]^{-1}.
\tag{6}
$$

Figure 5 shows a comparison of the typical current-voltage characteristics of Ge, GaSb, and GaAs tunnel diodes at room temperature. The current ratios of I_P/I_V are 8:1 for Ge and 12:1 for GaSb and GaAs. Because of its smaller effective mass ($0.042\ m_0$) and small bandgap (0.72 eV), the GaSb tunnel diode has the largest negative resistance of the three devices.

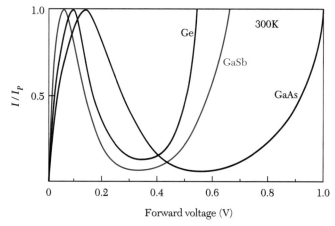

Fig. 5 Typical current-voltage characteristics of Ge, GaSb, and GaAs tunnel diodes at room temperature.

▷ 8.3 IMPATT DIODE

The name IMPATT stands for impact ionization avalanche transit-time. IMPATT diodes employ impact ionization and transit-time properties of semiconductor devices to produce a negative resistance at microwave frequencies. The IMPATT diode is one of the most powerful solid-state sources of microwave power. At present, the IMPATT diode can generate the highest cw (continuous wave) power output of all solid-state devices at millimeter-wave frequencies—above 30 GHz. It is extensively used in radar systems and alarm systems. There is one noteworthy difficulty in IMPATT applications: the noise is high because of random fluctuations of the avalanche multiplication processes.

8.3.1 Static Characteristics

The IMPATT diode family includes many different $p-n$ junction and metal-semiconductor devices. The first IMPATT oscillation was obtained from a simple silicon $p-n$ junction diode biased into reverse avalanche breakdown and mounted in a microwave cavity.[7] Figure 6a shows the doping profile and electric-field distribution at avalanche breakdown of a one-sided abrupt $p-n$ junction. Because of the strong dependence of the ionization rate on the electric field, most of the avalanche multiplication processes occur in a narrow region near the highest field between 0 and x_A (shaded area). x_A is the width of the avalanche region, the distance over which 95% of the contribution to the ionization integrand is obtained.

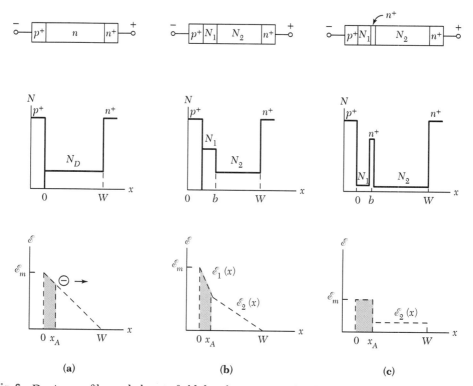

Fig. 6 Doping profiles and electric-field distributions at avalanche breakdown of three single-drift IMPATT diodes: (a) one-sided abrupt $p-n$ junction; (b) hi-lo structure; and (c) lo-hi-lo structure.

Figure 6b shows the hi-lo structure, in which a high doping N_1 region is followed by a lower doping N_2 region. With proper choices of the doping N_1 and its thickness b, the avalanche region can be confined within the N_1 region. Figure 6c is the lo-hi-lo structure in which a "clump" of donor atoms is located at $x = b$. Since a nearly uniform high-field region exists from $x = 0$ to $x = b$, the avalanche region x_A is equal to b, and the maximum field can be much lower than that for a hi-lo structure.

The breakdown voltage V_B (including the built-in potential V_{bi}) is given by the area underneath the electric-field versus distance plot (Fig. 6). For the one-sided abrupt junction (Fig. 6a), V_B is simply given by $\mathscr{E}_m W / 2$. For the hi-lo diode and the lo-hi-lo diode, the breakdown voltages are given, respectively, by

$$V_B (\text{hi - lo}) = \left(\mathscr{E}_m - \frac{qN_1 b}{2\varepsilon_s} \right) b - \frac{1}{2} \left(\mathscr{E}_m - \frac{qN_1 b}{2\varepsilon_s} \right) (W - b), \tag{7}$$

$$V_B (\text{lo - hi - lo}) = \mathscr{E}_m b + \left(\mathscr{E}_m - \frac{qQ}{\varepsilon_s} \right) (W - b), \tag{8}$$

where Q in Eq. 8 is the number of impurities/cm² in the clump. The maximum field at breakdown for a hi-lo diode with a given N_1 is the same as the value of the one-sided abrupt junction with the same N_1. The maximum field of a lo-hi-lo structure can be calculated from the ionization coefficient. These structures are single-drift IMPATT diodes because only one type of charge carriers, electrons, traverses the drift region. If, on the other hand, we form a p^+-p-n-n^+ structure, we have a double-drift IMPATT diode in which both electrons and holes participate in device operation over two separate drift regions, i.e., electrons move to the right side and holes move to the left side from the avalanche region. Similar approaches can be used to obtained breakdown voltages for various double-drift diodes.

8.3.2 Dynamic Characteristics

We now use the lo-hi-lo structure, shown in Fig. 6c, to discuss the injection delay and transit-time effect of the IMPATT diode. When a reverse direct current (dc) voltage V_B is applied to the diode so that the critical field for avalanche $\mathscr{E}_c$ is just reached (Fig. 7a), avalanche multiplication will begin. An alternating current (ac) voltage is superimposed onto this dc voltage at $t = 0$. This voltage is shown in Fig 7e. Holes generated in the avalanche region move to the p^+-region, and electrons enter the drift region. As the applied ac voltage goes positive, more electrons are generated in the avalanche region, as shown by the dotted line in Fig 7b. The electron pulse keeps increasing as long as the electrical field is above $\mathscr{E}_c$. Therefore, the electron pulse reaches its peak value not at $\pi/2$ when the voltage is maximum, but at π (Fig. 7c). The important consequence is that there is a $\pi/2$ phase delay inherent in the avalanche process itself, that is, the injected-carrier density (electron pulse) lags the ac voltage by 90°.

An additional delay is provided by the drift region. Once the applied voltage drops below V_B ($\pi \le \omega t \le 2\pi$), the injected electrons will drift toward the n^+-contact (Fig. 7d) with a saturation velocity, provided the field across the drift region is sufficiently high.

The situation described above is illustrated by the injected carriers in Fig. 7f. By comparing Figs. 7e and 7f, we note that the peak value of the ac field (or voltage) occurs at $\pi/2$, but the peak of the injected carrier density occurs at π. The injected carriers then traverse the drift region at saturation velocity, thereby introducing the transit-time delay.

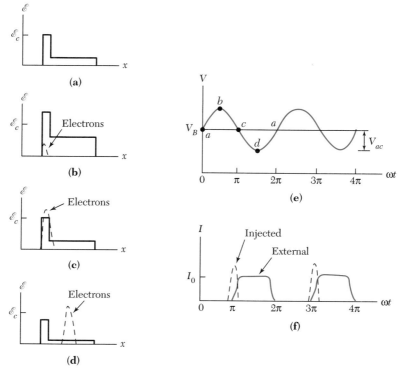

Fig. 7 Field distributions and generated-carrier densities of an IMPATT diode during an ac cycle at four intervals of time (*a–d*); (*e*) the ac voltage, and (*f*) the injected and external current.[7]

The induced external current is also shown in Fig. 7*f*. Comparing the ac voltage and the external current shows that the diode exhibits a negative resistance characteristic.

The injected carriers (electron pulse) will traverse the length W of the drift region during the negative half-cycle if we choose the transit time to be one-half the oscillation period, that is,

$$\frac{W - x_A}{v_s} = \frac{1}{2}\left(\frac{1}{f}\right) \tag{9}$$

or

$$f = \frac{v_s}{2\left(W - x_A\right)}, \tag{10}$$

where v_s is the saturation velocity, which is 10^7 cm/s for silicon at 300 K.

▷ **EXAMPLE 3**

Consider a lo-hi-lo silicon IMPATT diode (p^+-i-n^+-i-n^+) having $b = 1$ μm and $W = 6$ μm. If the field at breakdown is 3.3×10^5 V/cm, $Q = 2.0 \times 10^{12}$ charges/cm^2, find the dc breakdown voltage, the field in the drift region, and the operating frequency.

SOLUTION From Eq. 8 we can calculated the breakdown voltage:

$$V_B = 3.3 \times 10^5 \times 10^{-4} + \left(3.3 \times 10^5 - \frac{1.6 \times 10^{-19} \times 2.0 \times 10^{12}}{11.9 \times 8.85 \times 10^{-14}} \right) \times \left(5 \times 10^{-4} \right),$$

$$= 33 + 13 = 46 \text{ V.}$$

The field in the drift region is $\dfrac{13}{5 \times 10^{-4}} = 2.6 \times 10^4$ V / cm.

The drift field is high enough for the injected carriers to maintain their saturation velocity. Therefore

$$f = \frac{v_s}{2(W - x_A)} = \frac{10^7}{2 \times (6 - 1) \times 10^{-4}} = 10^{10} \text{ Hz} = 10 \text{ GHz.} \qquad \blacktriangleleft$$

We can also estimate the dc-to-ac power conversion efficiency of the IMPATT diode using Figs. 7e and 7f. The dc power input is the product of the average dc voltage and the average dc current, that is, $V_B(I_0/2)$. The ac power output can be estimated by assuming that the maximum ac voltage swing to be $1/2\ V_B$, that is, $V_{ac} = V_B/2$, and the external current is zero between $0 \le \omega t \le \pi$ and is I_0 between $\pi \le \omega t \le 2\pi$. Therefore, the microwave power-generating efficiency η is

$$\eta = \frac{\text{ac power output}}{\text{dc power input}} = \frac{\int_0^{2\pi} \left(V_{ac} \sin \omega t \right) I\, d(\omega t)}{\left(V_B \dfrac{I_0}{2} \right) 2\pi}$$

$$= \frac{\int_\pi^{2\pi} \left(\dfrac{V_B}{2} \sin \omega t \right) I_0\, d(\omega t)}{V_B I_0 \pi} = \frac{1}{\pi} = 32\%. \qquad (11)$$

State-of-the-art IMPATT diodes have cw power capabilities up to 3 W at 30 GHz with over 22% efficiency, up to 1 W at 100 GHz with 10% efficiency, and 50 mW at 250 GHz with 1% efficiency.[8] The substantial reduction in power and efficiency at higher frequencies is due to difficulties in device fabrication and circuit optimization. The reduction is also caused by the nonoptimal transit-time delays introduced by the finite time required to transfer energy to the carriers and the tunneling process for the very narrow depletion-layer width.

▶ ## 8.4 TRANSFERRED-ELECTRON DEVICES

The transferred-electron effect was first observed in 1963. In the first experiment,[9] a microwave output was generated when a dc electric field that exceeded a critical threshold value of several thousand volts per centimeter was applied across a short n-type sample of GaAs or InP. The transferred-electron device (TED) is an important microwave device. It is used extensively as a local oscillator and power amplifier covering the microwave frequency range from 1 to 150 GHz. The power output and efficiency of TEDs are generally lower than that of IMPATT diodes. However, TEDs have lower noise, lower operating voltages, and relatively easier circuit designs. The TEDs have matured to become important solid-state microwave sources used in detection systems, remote controls, and microwave test instruments.

8.4.1 Negative Differential Resistance

In Chapter 3 we considered the transferred-electron effect, that is, the transfer of conduction electrons from a high-mobility energy valley to low-mobility higher-energy satellite valleys. Based on the analysis in Chapter 3, the current density as shown in Fig. 8 takes on asymptotic values

$$J \cong qn\mu_1\mathscr{E} \quad \text{for} \quad 0 < \mathscr{E} < \mathscr{E}_a, \tag{12}$$

$$J \cong qn\mu_2\mathscr{E} \quad \text{for} \quad \mathscr{E} > \mathscr{E}_b. \tag{13}$$

If $\mu_1\mathscr{E}_a$ is larger than $\mu_2\mathscr{E}_b$, there is a region of negative differential resistance (NDR)[10] between $\mathscr{E}_a$ and $\mathscr{E}_b$ as shown in Fig. 8. Also shown are the threshold field $\mathscr{E}_T$ corresponding to the onset of the NDR, the threshold current density J_T, the valley field $\mathscr{E}_V$, and the valley current density J_V. The NDR region is between $\mathscr{E}_T$ and $\mathscr{E}_V$.

For the transferred-electron mechanism to give rise to the NDR, certain requirements must be met. (a) The lattice temperature must be low enough that in the absence of an electric field most of the electrons are in the lower valley (the conduction band minimum), that is, the energy separation between the two valleys $\Delta E > kT$. (b) In the lower valley the electrons must have high mobility and small effective mass, whereas in the upper satellite valleys the electrons must have low mobility and large effective mass. (c) The energy separation between the two valleys must be smaller than the semiconductor bandgap (i.e., $\Delta E < E_g$) so that avalanche breakdown does not begin before the transfer of electrons into the upper valleys.

Of the semiconductors satisfying these requirements, n-type gallium arsenide and n-type indium phosphide are the most widely studied and used. The measured room-temperature velocity field characteristics for these semiconductors are shown in Fig. 15 of Chapter 7. The threshold field $\mathscr{E}_T$ is 3.2 kV/cm for gallium arsenide and 10.5 kV/cm for indium phosphide. The peak velocity V_p is about 2.2×10^7 cm/s for gallium arsenide and 2.5×10^7 cm/s for indium phosphide. The maximum negative differential mobility

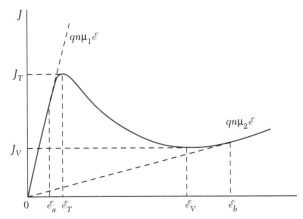

Fig. 8 The current versus electric-field characteristic of a two-valley semiconductor. $\mathscr{E}_T$ is the threshold field and $\mathscr{E}_V$ is the valley field.

(i.e., $dv/d\mathscr{E}$) is about -2400 cm²/V-s for gallium arsenide and -2000 cm²/V-s for indium phosphide.

A semiconductor exhibiting NDR is inherently unstable because a random fluctuation of carrier density at any point in the semiconductor produces a momentary space charge that will grow exponentially with time. The one-dimensional continuity equation is given by

$$\frac{\partial n}{\partial t} = \frac{1}{q}\frac{\partial J}{\partial x}. \tag{14}$$

If there is a small local fluctuation of the majority carriers from the uniform equilibrium concentration n_0, the locally created space charge density is $n - n_0$. Poisson's equation and the current density equation are

$$\frac{\partial \mathscr{E}}{\partial x} = \frac{-q(n - n_0)}{\varepsilon_s}, \tag{15}$$

$$J = qn_0\bar{\mu}\mathscr{E} + qD\frac{\partial n}{\partial x}, \tag{16}$$

where $\bar{\mu}$ is the average mobility (defined by Eq. 83 in Chapter 3), ε_s is the dielectric permittivity, and D is the diffusion constant. Differentiating Eq. 16 with respect to x and inserting Poisson's equation yields

$$\frac{1}{q}\frac{\partial J}{\partial x} = -\frac{n - n_0}{\varepsilon_s / qn_0\bar{\mu}} + D\frac{\partial^2 n}{\partial x^2}. \tag{17}$$

Substituting this expression into Eq. 14 gives

$$\frac{\partial n}{\partial t} = -\frac{n - n_0}{\varepsilon_s / qn_0\bar{\mu}} + D\frac{\partial^2 n}{\partial x^2}. \tag{18}$$

We can solve Eq. 18 by separation of variables, that is, let $n(x,t) = n_1(x)n_2(t)$. For the temporal response, the solution of Eq. 18 is

$$n - n_0 = (n - n_0)_{t=0}\exp\left(\frac{-t}{\tau_R}\right), \tag{19}$$

where τ_R is the *dielectric relaxation time* given by

$$\tau_R = \frac{\varepsilon_s}{qn_0\bar{\mu}}. \tag{20}$$

τ_R represents the time constant for the decay of the space charge to neutrality if the mobility $\bar{\mu}$ is positive. However, if the semiconductor exhibits NDR, any charge imbalance will grow with a time constant equal to $|\tau_R|$.

8.4.2 Device Operation

The TEDs require very pure and uniform materials with a minimum of deep impurity levels and traps. Modern TEDs almost always have epitaxial layers on n^+-substrates deposited by various epitaxial techniques. Typical donor concentrations range from 10^{14} to 10^{16} cm⁻³ and typical device lengths range from a few microns to several hundred microns. A TED having an epitaxial n-layer on n^+-substrate and an ohmic n^+-contact to the cathode electrode is shown in Fig. 9a. Also shown are the energy band diagram at

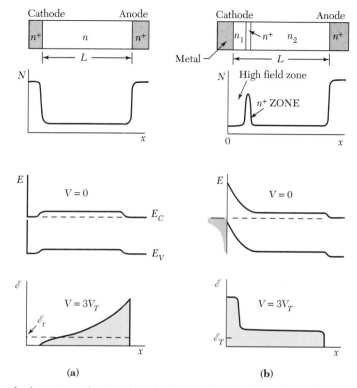

Fig. 9 Two cathode contacts for transferred-electron devices (TEDs). (*a*) Ohmic contact and (*b*) two-zone Schottky barrier contact.

thermal equilibrium and the electric-field distribution when a voltage $V = 3V_T$ is applied to the device where V_T is the product of the threshold field $\mathscr{E}_T$ and the device length L. For such an ohmic contact there is always a low-field region near the cathode, and the field is nonuniform across the device length.

To improve device performance, we use the two-zone cathode contact instead of the n^+-ohmic contact. The two-zone cathode contact consists of a high-field zone and an n^+-zone (Fig. 9*b*). This configuration is similar to that of a lo-hi-lo IMPATT diode. Electrons are "heated" in the high-field zone and subsequently injected into the active region, which has a uniform field. This structure has been used successfully over a wide temperature range with high efficiency and high power output.

The operational characteristics of a TED depends on five factors: doping concentration and doping uniformity in the device, length of the active region, cathode contact characteristics, type of circuit, and operating bias voltage.

We have shown that for a device with NDR, the initial space charge will grow exponentially with time (Eq. 19) and that the time constant is given by Eq. 20:

$$\left|\tau_R\right| = \frac{\varepsilon_s}{q n_0 \left|\mu_-\right|}, \tag{21}$$

where μ_- is the negative differential mobility. If Eq. 19 remains valid throughout the entire transit time of the space-charge layer, the maximum growth factor would be $\exp(L/v|\tau_R|)$, where L is the length of the active region and v is the average drift velocity of the space-charge layer. For large space-charge growth, this growth factor must be greater than unity, making $L/v|\tau_R| > 1$, or

$$n_0 L > \frac{\varepsilon_s v}{q|\mu_-|} \approx 10^{12}\,\text{cm}^{-2} \tag{22}$$

for GaAs and InP.

An important mode of operation for the TED is the transit-time domain mode. When there are positive and negative charges separated by a small distance, we have a dipole formation (also called a domain)[10] as shown in Figs. 10a and 10b. From Poisson's equa-

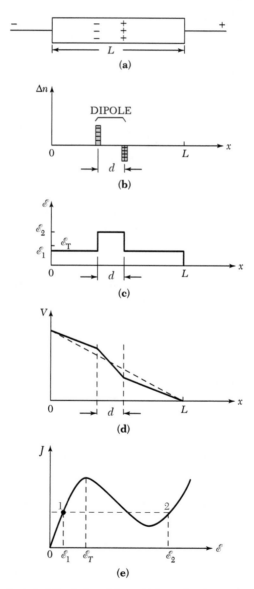

Fig. 10 Formation of a domain (dipole layer) in a medium that has a negative differential resistivity.[10]

tion, we find that the electric field inside the dipole would be greater than the field on either side of it (Fig. 10c). The corresponding voltage variation across the device can be obtained by integrating the Poisson's equation once more, as shown in Fig. 10d. Because of the NDR, the current in the low-field region would be greater than that in the high-field region. The two field values tend toward the equilibrium level outside the NDR region, where the high and low currents are the same, Fig. 10e. The dipole has now reached a stable configuration. The dipole layer moves through the active region and disappears at the anode, at which time the field begins to rise uniformly across the device through the threshold (i.e., $\mathscr{E} > \mathscr{E}_T$), thus forming a new dipole, and the process repeats itself. The time required for the domain to travel from the cathode to anode is L/v, where L is the active device length and v is the average velocity. The corresponding frequency for the transit time domain mode is $f = v/L$.

Figure 11 shows a simulated time-dependent behavior of a domain in a gallium arsenide TED 100 μm long with a doping of 5×10^{14} cm^{-3} ($n_0 L = 5 \times 10^{12}$ cm^{-2}).[11] The time between successive vertical displays of $\mathscr{E}(x,t)$ is $16\tau_R$, where τ_R is the low-field dielectric relaxation time from Eq. 20 ($\tau_R = 1.5$ ps for this device). Each time a domain is absorbed at the anode, the current in the external circuit increases and a new domain is nucleated at the cathode contact where the largest doping fluctuation and space-charge perturbation usually exist.

State-of-the-art TED diodes have cw power capabilities up to 0.5 W at 30 GHz with 15% efficiency, up to 0.2 W at 100 GHz with 7% efficiency, and 70 mW at 150 GHz with 1% efficiency. The power output of TED is lower than that of IMPATT; however, TED has much lower noise (e.g., 20 dB less at 135 GHz).[8]

▷ **EXAMPLE 4**

A GaAs TED is 10 μm long and is operated in the transit-time domain mode. Find the minimum electron density n_0 required and the time between current pulses.

SOLUTION For transit-time domain mode, we require $n_0 L \geq 10^{12}$ cm^{-2}:

$n_0 \geq 10^{12} / L = 10^{12} / 10 \times 10^{-4} = 10^{15}$ cm^{-3}.

The time between current pulses is the time required for the domain to travel from the cathode to the anode:

$$t = L/v = 10 \times 10^{-4} / 10^7 = 10^{-10}\text{s} = 0.1 \text{ ns.} \qquad \blacktriangleleft$$

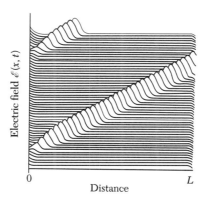

Fig. 11 Numerical simulation of the time-dependant behavior of a cathode-nucleated TED for the transit-time domain mode.[11]

▷ 8.5 QUANTUM-EFFECT DEVICES

A quantum-effect device (QED) uses quantum mechanical tunneling to provide controlled carrier transport. In such a device, the active layer thickness is very small, on the order of 10 nm. These small dimensions give rise to a quantum size effect that can alter the band structures and enhance device transport properties. The basic QED is the resonant tunneling diode (RTD), discussed in Section 8.5.1. Many novel current-voltage characteristics can be obtained by combining a RTD with conventional devices considered in previous chapters. QEDs are of particular importance because they can serve as *functional devices*, that is, they can performance a give circuit function with a greatly reduced number of components.

8.5.1   Resonant Tunneling Diode

Figure 12 shows the band diagram of a RTD. It has a semiconductor double-barrier structure containing four heterojunctions, a GaAs/AlAs/GaAs/AlAs/GaAs structure, and one quantum well in the conduction band. There are three important device parameters for a RTD—the energy barrier height E_0, which is the conduction band discontinuity, the energy barrier thickness L_B, and the quantum well thickness L_W.

We now concentrate on the conduction band of a RTD as shown[12] in Fig.13a. If the well thickness L_W is sufficiently small (on the order of 10 nm or less), a set of discrete energy levels will exist inside the well (such as E_1, E_2, E_3, and E_4 in Fig.13a). If the barrier thickness L_B is also very small, resonant tunneling will occur. When an incident electron has an energy E that exactly equals one of the discrete energy levels inside the well, it will tunnel through the double barrier with a unity (100%) transmission coefficient.

The transmission coefficient decreases rapidly as the energy E deviates from the discrete energy levels. For example, an electron with an energy 10 meV higher or lower than the level E_1 will result in 10^5 times reduction in the transmission coefficients, as depicted in Fig.13b. The transmission coefficient can be calculated by solving the one-dimensional Schroedinger equations in the five regions in Fig. 13a (I, II, III, IV, V). Since

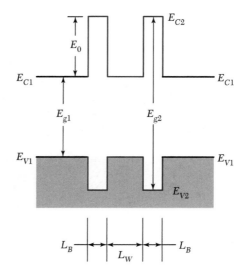

Fig. 12   Band diagram of a resonant-tunneling diode.

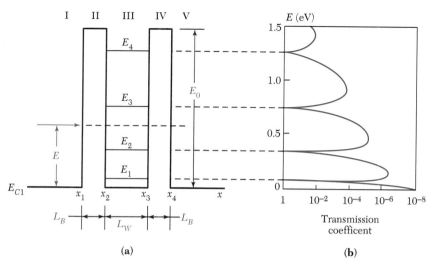

Fig. 13 (*a*) Schematic illustration of AlAs/GaAs/AlAs double-barrier structure with a 2.5 nm barrier and a 7 nm well. (*b*) Transmission coefficient versus electron energy for the structure.[12]

the wavefunctions and their first derivatives at each potential discontinuity must be continuous, we can obtain the transmission coefficient T_t. Appendix J shows the calculation of the transmission coefficient for the RTD.

The energy levels, E_n, at which the transmission coefficient exhibits its first and second resonant peaks in GaAs/AlAs RTD are shown in Fig. 14*a* as a function of barrier thickness L_B with the well thickness L_W as a parameter.[13] It is apparent that E_n is essentially independent of L_B but is dependent on L_W. The calculated width of the peak ΔE_n (i.e., the full width at the half-maximum point of the transmission coefficient where T_t = 0.5) is shown in Fig. 14*b* as function of L_B and L_W. For a given L_W, the width ΔE_n decreases exponentially with L_B.

The cross section of a RTD is shown[13] in Fig. 15. The alternating GaAs/AlAs layers are grown sequentially by molecular beam epitaxy (MBE) on an n^+ GaAs substrate (the MBE process is discussed in Chapter 10). The barrier thicknesses are 1.7 nm and the well thickness is 4.5 nm. The active regions are defined with ohmic contacts. The top contact is used as a mask to isolate the region under the contact by etching mesas.

The measured current-voltage characteristic of the above RTD is shown in Fig. 16. Also shown are the band diagrams for varies dc biases. Note that the *I-V* curve is similar to that of a tunnal diode (Fig. 4). At thermal equilibrium, $V = 0$, the energy diagram is similar to that in Fig. 13*a* (here only the lowest energy level E_1 is shown). As we increase the applied voltage, the electrons in the occupied energy states near the Fermi level to the left side of the first barrier tunnel into the quantum well. The electrons subsequently tunnel through the second barrier into the unoccupied states in the right side. Resonance occurs when the energy of the injected electrons becomes approximately equal to the energy level E_1, where the transmission probability is maximum. This is illustrated by the energy diagram for $V = V_1 = V_P$, where the conduction band edge on the left side is lined up with E_1. The magnitude of the peak voltage must be at least $2E_1/q$ but is usually larger because of additional voltage drops in the accumulation and depletion regions:

$$V_P > \frac{2E_1}{q} .$$ (23)

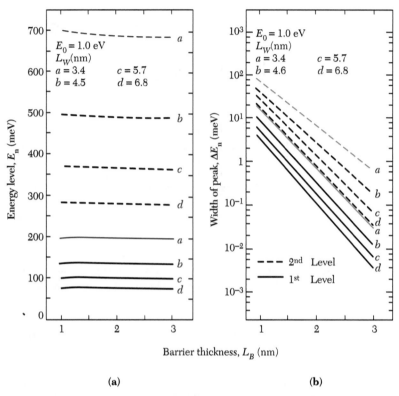

(a) **(b)**

Barrier thickness, L_B (nm)

Fig. 14 (a) Calculated energy of electrons at which the transmission coefficient shows the resonant peak in an AlAs/GaAs/AlAs structure as a function of barrier thickness for various well thicknesses. (b) Full width at half maximum of the transmission coefficient versus barrier thickness for the first and second resonant peak. [13]

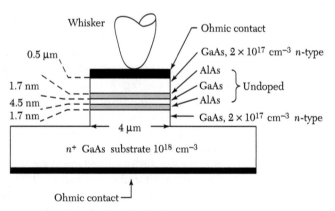

Fig. 15 A mesa-type resonant tunneling diode.[13]

When the voltage is further increased, that is, at $V = V_2$, the conduction band edge is above E_1 and the number of electrons that can tunnel decreases, resulting in a small current. The valley current I_V is due mainly to the excess current components, such as electrons that tunnel via an upper valley in the barrier. At room temperature and higher,

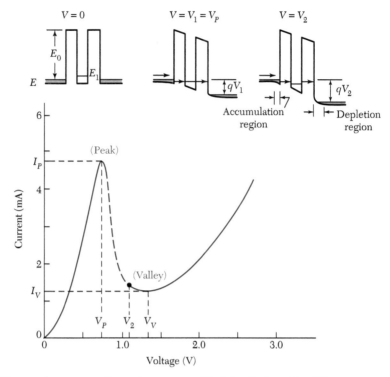

Fig. 16   Measured current–voltage characteristics[13] of the diode in Fig. 15.

there are other components due to tunneling current associated with either lattice vibrations or impurity atoms. To minimize the valley current, we must improve the quality of the heterojunction interfaces and eliminate impurities in the barrier and well regions. For even higher applied voltages, $V > V_V$, we have the thermionic current component I_{th}, due to electrons injected through higher discrete energy levels in the well or thermionically injected over the barriers. The current I_{th} increases monotonically with increasing voltage similar to that of a tunnel diode. To reduce I_{th}, we should increase the barrier height and design a diode that operates at relatively low bias voltages.

RTDs can be operated at very high frequencies because of their smaller parasitics. In the RTD the main contribution to the capacitance is from the depletion region (refer to the band diagram for $V = V_2$ in Fig. 16). Since the doping density there can be much lower than in a degenerate p–n junction, the depletion capacitance is much smaller. The cutoff frequency for a RTD can reach the THz (10^{12} Hz) range. It can be used in ultra fast pulse-forming circuits, in THz radiation detection systems, and in oscillators to generate THz signals.

> **EXAMPLE 5**

Find the ground energy level and the corresponding width of the peak for the RTD in Fig. 15. Compare the peak voltage V_P in Fig. 16 with $2E_1/q$.

SOLUTION   From Fig. 14 we find that the ground energy level for L_W = 4.5 nm is at 140 meV, and the width of the peak ΔE_1 is about 1 meV. In Fig. 16, V_P is 700 mV, which is larger than 280 mV ($2E_1/q$). The difference (420 mV) is due to voltage drops at the accumulation and the depletion regions.   ◂

8.5.2 Unipolar Resonant Tunneling Transistor

A schematic band diagram of a unipolar resonant tunneling transistor [14] is shown in Fig. 17. The structure consists of a resonant tunneling (RT) double barrier placed between the GaAs emitter and base layers. The RT structure was made of a GaAs quantum well 5.6 nm thick sandwiched between two $Al_{0.33}Ga_{0.67}As$ barriers 5 nm thick. High-energy electrons can be injected from the emitter through the RT into the base region. The electrons are then transported through the n^+-base region 100 nm thick before being collected at the 300 nm thick $Al_{0.2}Ga_{0.8}As$ collector barrier. The barriers and the quantum well are undoped whereas the emitter, base, and collector layers are n-type, doped to 1 $\times 10^{18}$ cm^{-3}.

The operation of the device in the common-emitter configuration with a fixed collector-emitter voltage V_{CE} is shown in the band diagrams of Fig.17. When the base-emitter voltage V_{BE} is zero (Fig. 17a), there is no electron injection; hence, the emitter and collector currents are zero even with a positive V_{CE}. A peak in the emitter and collector currents occurs when V_{BE} is equal to $2E_1/q$, where E_1 is the energy of the first resonant level in the quantum well (Fig. 17b). With a further increase in V_{BE}, RT is quenched (Fig. 17c) with a corresponding drop in the collector current. The current-voltage characteristics is shown in Fig. 17d. Note that a current peak occurs around $V_{BE} = 0.4$ V. If we connect two inputs A and B to the base terminal (Fig. 17e), the device can performance

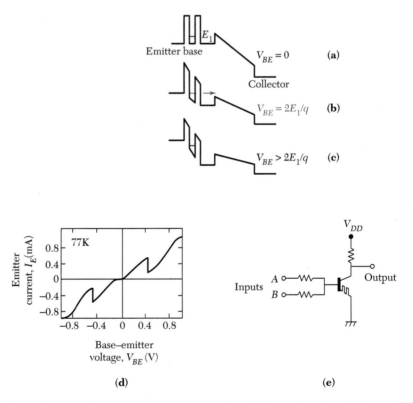

Fig. 17 Band diagrams of a unipolar resonant tunneling (RT) transistor[14] at (a) $V_{BE} = 0$, (b) $V_{BE} = 2E_1/q$ (maximum RT current), (c) $V_{BE} > 2E_1/q$ (RT quenched), (d) base-emitter current–voltage characteristics measured at 77 K, and (e) an exclusive NOR circuit.

an exclusive NOR logic function[§], that is, the output voltage will be high, if A and B are both high or both low; otherwise the output voltage will be low. To perform the same function, we need eight conventional MESFETs. Therefore, many quantum-effect devices are useful as functional devices.

8.6 HOT-ELECTRON DEVICES

Hot electrons are electrons with kinetic energies substantially above kT, where k is the Boltzmann's constant and T is the lattice temperature. As the dimensions of semiconductor devices shrink and the internal fields rise, a large fraction of carriers in the active regions of the device during its operation is in states of high kinetic energy. At a given point in time and space the velocity distribution of carriers may be narrowly peaked, in which case one speaks about "ballistic" electron packets. At other times and locations, the electron ensemble can have a broad velocity distribution, similar to a conventional Maxwellian distribution but with an effective electron temperature T_e larger than the lattice temperature T.

Over the years, many hot electron devices have been studied. We now consider two important devices—the hot-electron heterojunction bipolar transistor and the real-space–transfer transistor.

8.6.1 Hot-Electron HBT

Hot-electron injection is enabled in heterojunction bipolar transistors (HBTs) by designing structures with a wider-bandgap emitter, shown[15] in Fig. 18 for an AlInAs/GaInAs HBT lattice matched to InP. There are several advantages of the hot-electron effect. Electrons are injected by thermionic emission over the emitter-base barrier at an energy $\Delta E_C = 0.5$ eV above the conduction band-edge in the p-GaInAs base. Here the purpose

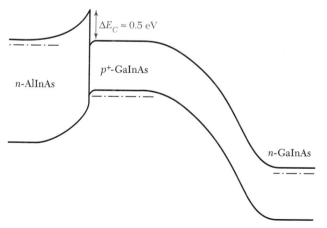

Fig. 18 Energy band diagram of a hot electron heterojunction bipolar transistor.[15]

[§]Exclusive NOR logic function: when the two inputs are both high or both low, the output is high. Otherwise, the output is low.

of the ballistic injection is to shorten the base traversal time by replacing the relatively slow diffusion motion by faster ballistic propagation.

8.6.2 Real-Space–Transfer Transistor

The original real-space–transfer (RST) structure, illustrated in Fig. 19a, is a heterostructure with alternate doped wide-gap AlGaAs and undoped narrow-gap GaAs layers. In thermal equilibrium the mobile electrons reside in the undoped GaAs quantum wells and are spatially separated from their parent donors in AlGaAs layers. [16] If the power input into the structure exceeds the rate of energy loss by the system to the lattice, then the carriers "heat up" and undergo partial transfer into the wide-gap layer where they may have a different mobility (Fig. 19b). If the mobility in layer 2 is much lower, negative differential resistance will occur in the two-terminal circuit (Fig. 19c). There is a strong analogy to the transferred-electron effect, based on the momentum-space intervalley transfer, therefore the name real-space transfer.

Figure 20 shows a schematic cross-section and the corresponding band diagram of a three-terminal RST transistor (RSTT) implemented in a GaInAs/AlInAs material system.[17,18] The source and drain contacts are to an undoped high-mobility $Ga_{0.47}In_{0.53}As$ ($E_g = 0.75$ eV with $\mu_n = 13800$ cm^2/V-s) channel and the collector contact is to a doped $Ga_{0.47}In_{0.53}As$ conducting layer. This layer is separated from the channel by a larger bandgap material. (i.e., an $Al_{0.48}In_{0.52}As$ with $E_g = 1.45$ eV). At $V_D = 0$ an electron density is induced in the source-drain channel by a sufficiently positive collector bias V_C with respect to the

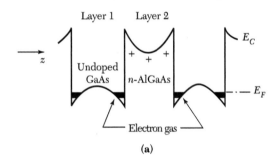

(a)

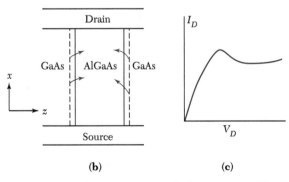

(b)   (c)

Fig. 19 (a) A heterostructure with alternate GaAs and AlGaAs layers. (b) Electrons, heated by an applied electric field, transfer into the wide-gap layers. (c) If the mobility in layer 2 is lower, the transfer results in a negative differential conductivity.[16]

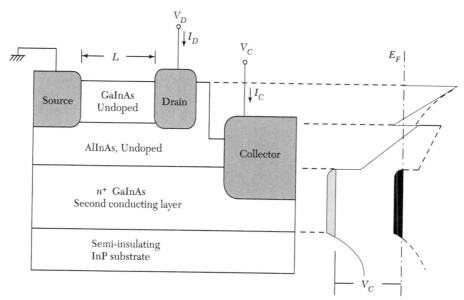

Fig. 20 Cross section and band diagram of a real-space–transfer transistor in a GaInAs/AlInAs material system.[17,18]

ground source, but no collector current I_C flows because of the AlInAs barrier. As V_D increases, however, a drain current I_D begins to flow and the channel electrons heat up to some effective temperature T_e (V_D). This electron temperature determines the RST current injected over the collector barrier. The injected electrons are swept into the collector by the V_C–induced electric field, giving rise to I_C. Transistor action results from control of the electron temperature T_e in the source-drain channel, which modulates I_C flowing into the collector electrode.

The drain current I_D and the collector current I_C versus drain voltage V_D at a fixed V_C = 3.9 V are shown[19] in Fig. 21. On the I_D-V_D characteristics, the RSTT shows pronounced negative differential resistance, with a peak-to-valley ratio that reaches 7000 at 300 K. In the I_C-V_D characteristics, the collector current increases approximately linearly and eventually reaches a saturation value similar to a field-effect transistor.

The RSTT can be used as a conventional high-speed transistor with high transconductance, $g_m \equiv \partial I_C / \partial V_D$ (at fixed V_C), and a high cutoff frequency f_T. In addition, the RSTT is another useful functional device for logic circuits. This is because the source and drain contacts of an RSTT are symmetrical. A single device, such as that shown in Fig. 20, can perform an exclusive OR (XOR) logic function[§] because the collector current I_C flows if the source and drain are at different logic values, regardless of which is "high." With additional input terminals a single RSTT can perform more complex functions. For example, three input terminals permit an implementation of an OR-NAND gate.[°] Figure 22*a* shows such an implementation using a Si/SiGe material system.[20] Depending on whether

[§]Exclusive OR logic function: when one of the two inputs, but not both are high, the output is high.

[°]OR logic function: irrespective of the number of inputs, the output is high if any of the inputs is high. NAND logic function: irrespective of the number of inputs, the output is high at all times except when all inputs are high.

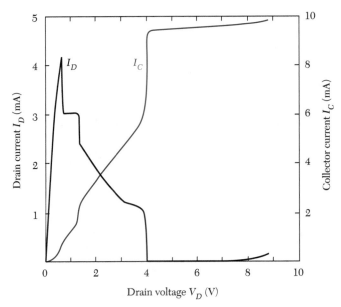

Fig. 21 Experimental real-space–transfer transistor characteristics[19] at T = 300 K. Drain current I_D and collector current I_C versus drain voltage V_D at a fixed collector voltage V_C = 3.9 V.

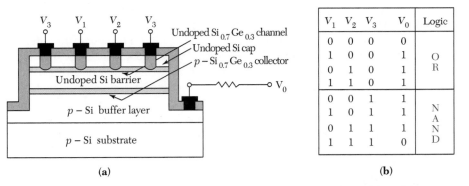

V_1	V_2	V_3	V_0	Logic
0	0	0	0	
1	0	0	1	O
0	1	0	1	R
1	1	0	1	
0	0	1	1	N
1	0	1	1	A
0	1	1	1	N
1	1	1	0	D

(a)   **(b)**

Fig. 22 (*a*) A Si/SiGe RSTT OR–NAND gate with three inputs. The channel length between different inputs is 1 μm, and the device width is 50 μm. (*b*) Truth table for OR–NAND logic operation.[20]

the control input is high or low, the output current behaves as either a NAND or an OR function of the other two inputs. A truth table for the output and the logic functions is shown in Fig. 22*b*.

▶ SUMMARY

Diodes associated with tunneling phenomenon (such as the tunnel diode), the avalanche breakdown (IMPATT diode), and momentum-space transfer of electrons (TED) are devices that are used at microwave frequencies. Those two-terminal devices have relatively simple construction and have much less parasitic resistance and capacitance than do their three-terminal counterparts. These microwave diodes can operate in the millimeter-wave band (30–300 GHz) and some devices can operate in the submillimeter-wave band (>300 GHz). Among microwave devices, the IMPATT diode is the most extensively used semiconductor device for millimeter-wave power applications. However, the TED is often used in local oscillators and amplifiers because it has lower noise and can be operated at lower voltages than the IMPATT diode.

In this chapter, we also considered the quantum-effect and hot-electron devices. Quantum effects become important when the device dimensions are reduced to about 10 nm. A key quantum-effect device is the resonant tunneling diode (RTD), which is a heterostructure having a double barrier and a quantum well. If the incoming carrier energy is equal to a discrete energy level in the quantum well, the tunneling probability through the double barrier becomes 100%. This effect is called resonant tunneling. Microwave detectors that operate up to the THz (10^{12} Hz) range have been made using a RTD. By combining RTD with conventional devices we can obtain many novel characteristics. One example is the unipolar resonant tunneling transistor, which can perform a given logic function with a greatly reduced number of components.

HEDs can be classified into two groups—ballistic devices and the RST devices, depending on the type of a hot-electron ensemble employed in the operation. Ballistic devices, such as the hot-electron heterojunction bipolar transistor, have the potential for ultrahigh-speed operation. In ballistic devices, high–kinetic-energy electrons are injected over the emitter-base barrier by thermionic emission. This "ballistic propagation" greatly reduces the transit time through the base. In RST devices, electrons in a narrow-gap semiconductor can gain energy from the input power and undergo transfer into a wide-gap semiconductor, giving rise to a NDR characteristic. These devices such as the RSTT have high transconductance and a high cutoff frequency. RSTTs are also used in logic circuits, where they permit a lower component count for a given function than do other devices.

▷ REFERENCES

1. S. M. Sze, Ed., *Modern Semiconductor Device Physics*, Wiley, New York, 1998.

2. J. J. Carr, *Microwave and Wireless Communications Technology*, Butterworth-Heinemann, Newton, MA, 1997.

3. L. E. Larson, *RF and Microwave Circuit Design for Wireless Communications*, Artech House, Norwood, MA, 1996.

4. G. R. Thorn, "Advanced Applications and Solid-State Power Sources for Millimeterwave Systems," *Proc. Soc. Photo-Optic. Inst. Opt. Eng.* (SPIE), **544**, 2 (1985).

5. B. C. Wadell, *Transmission-Line Design Handbook*, Artech House, Norwood, MA, 1991.

6. (a) L. Esaki, "New Phenomenon in Narrow Ge *p–n* Junction," *Phys. Rev.*, **109**, 603 (1958); (b) L. Esaki, "Discovery of the Tunnel Diode," *IEEE Trans. Electron Devices*, **ED-23**, 644 (1976).

7. (a) B. C. DeLoach, Jr., "The IMPATT Story," *IEEE Trans. Electron Devices*, **ED-23**, 57 (1976); (b) R. L. Johnston, B. C. DeLoach, Jr., and B. G. Cohen, "A Silicon Diode Oscillator," *Bell Syst. Tech. J.*, **44**, 369 (1965).

8. H. Eisele and G. I. Haddad, "Active Microwave Diodes," in S. M. Sze, Ed., *Modern Semiconductor Device Physics*, Wiley, New York, 1998.

9. J. B. Gunn, "Microwave Oscillation of Current in III-V Semiconductors," *Solid State Comm.*, **1**, 88 (1963).

10. H. Kroemer, "Negative Conductance in Semiconductor," *IEEE Spectr.*, **5**, 47 (1968).

11. M. Shaw, H. L. Grubin, and P. R. Solomon, *The Gunn–Hilsum Effect*, Academic, New York, 1979.

12. M. Tsuchiya, H. Sakaki, and J. Yashino, "Room Temperature Observation of Differential Negative Resistance in AlAs/GaAs/AlAs Resonant Tunneling Diode," *Jpn. J. Appl. Phys.* **24**, L466 (1985).

13. E. R. Brown, et al., "High Speed Resonant Tunneling Diodes," *Proc. Soc. Photo-Opt. Inst. Eng.* (SPIE), **943**, 2 (1988).

14. N. Yokoyama, et al., "A New Functional Resonant Tunneling Hot Electron Transistor," *Jpn. J. Appl. Phys.*, **24**, L853 (1985).

15. B. Jalali et al, "Near-Ideal Lateral Scaling in Abrupt AlInAs/InGaAs Heterostructure Bipolar Transistor Prepared by Molecular Beam Epitaxy," *Appl. Phys. Lett.*, **54**, 2333 (1989).

16. K. Hess, et al., "Negative Differential Resistance Through Real-Space-Electron Transfer," *Appl. Phys. Lett.*, **35**, 469 (1979).

17. S. Luryi, "Hot Electron Transistors," in S. M. Sze, Ed., *High Speed Semiconductor Devices*, Wiley, New York, 1990.

18. S. Luryi and A. Zashavsky, "Quantum-Effect and Hot-Electron Devices," in S. M. Sze, Ed., *Modern Semiconductor Device Physics*, Wiley, New York, 1998.

19. P. M. Mensz, et al, "High Transconductance and Large Peak-to-Valley Ratio of Negative Differential Conductance in Three Terminal InGaAs/InAlAs Real-Space-Transfer Devices," *Appl. Phys. Lett.*, **57**, 2558 (1990).

20. M. Mastrapasqua, et al., "Charge Injection Transistor and Logic Elements in Si/SiGe Heterostructures," in S. Luryi, J. Xu, and A. Zaslavsky, Eds., *Future Trends in Microelectrons*, Kluwer, Dordrecht, 1996.

▶ PROBLEMS (* DENOTES DIFFICULT PROBLEMS)

FOR SECTION 8.1   BASIC MICROWAVE TECHNOLOGY

1. For a nearly lossless transmission line (R is very small) with a 75 Ω characteristic impedance, if this transmission line has a unit length capacitance of 2 pF, what is the unit length inductance of this transmission line?

2. For a cavity of the dimensions a = 10 cm (0.1 m), b = 5 cm (0.05 m), and d = 25 cm (0.25 m), find the resonant frequency in the dominant TE_{101} mode.

FOR SECTION 8.2   TUNNEL DIODE

3. Find the depletion-layer capacitance and depletion-layer width at 0.25 V forward bias for a GaAs tunnel diode doped to 10^{19} cm^{-3} on both sides, using the abrupt junction approximation and assuming $V_n = V_p = 0.03$ V.

4. The current-voltage characteristic of a GaSb tunnel diode can be expressed by the empirical form of Eq. 5. with I_P = 10 mA, V_P = 0.1 V, and I_0 = 0.1 nA. Find the largest negative differential resistance and the corresponding voltage.

FOR SECTION 8.3   IMPATT DIODE

5. The variation of electric field in the depletion region due to avalanche-generated space charge gives rise to an incremental resistance for abrupt p^+-n diode. The incremental resistance is called the space-charge resistance, R_{SC}, and is given by $(1/I)\int_0^W \Delta \mathscr{E} dx$, where $\Delta \mathscr{E}$ is given by

$$\Delta \mathscr{E}(W) = \frac{\int_0^W \rho_S dx}{\varepsilon_S} = \frac{IW}{A \varepsilon_S v_S}.$$

(a) Find R_{SC} for a p^+-n Si IMPATT diode with $N_D = 10^{15}$ cm^{-3}, W = 12 μm, and A = 5×10^{-4} cm^2. (b) Find the total applied dc voltage for a current density of 10^3 A/cm^2.

6. A GaAs IMPATT diode is operated at 10 GHz with a dc bias of 100 V and an average biasing current ($I_0/2$) of 100 mA. (a) If the power-generating efficiency is 25% and the thermal resistance of the diode is 10° C/W, find the junction temperature rise above the

room temperature. (b) If the breakdown voltage increases with temperature at a rate of 60 mV/°C, find the breakdown voltage of the diode at room temperature.

7. Consider a GaAs single drift lo-hi-lo IMPATT diode shown in Fig. 6c with an avalanche region width (where the electric field is constant) of 0.4 μm and a total depletion width of 3 μm. The n^+ clump has a charge Q of 1.5×10^{12}/cm². (a) Find the breakdown voltage of the diode and the maximum field at breakdown. (b) Is the field in the drift region high enough to maintain the velocity saturation of electrons? (c) Find the operating frequency.

*8. A silicon n^+-p-π-p^+ IMPATT diode has a p-layer 3 μm thick and a π-layer (low-doping p-layer) 9 μm thick. The biasing voltage must be high enough to cause avalanche break-down in the p-region and velocity saturation in the π region. (a) Find the minimum required biasing voltage and the doping concentration of the p-region. (b) Estimate the transit time of the device.

FOR SECTION 8.4   TRANSFERRED-ELECTRON DEVICES

9. An InP TED is 1 μm long with a cross-section area of 10^{-4} cm² and is operated in the transit-time mode. (a) Find the minimum electron density n_0 required for transit-time mode. (b) Find the time between current pulses. (c) Calculate the power dissipated in the device if it is biased at one-half the threshold.

*10. (a) Find the effective density of states in the upper valley N_{CU} of the GaAs conduction band. The upper-valley effective mass is 1.2 m_0. (b) The ratio of electron concentrations between the upper and lower valleys is given by $(N_{CU}/N_{CL}) \exp(-\Delta E/kT_e)$, where N_{CL} is the effective density of states in the lower valley, $\Delta E = 0.31$ eV is the energy difference, and T_e is the effective electron temperature. Find the ratio at $T_e = 300$ K. (c) When elec-trons gain kinetic energies from the electric field, T_e increases. Find the concentration ratio for $T_e = 1500$ K.

FOR SECTION 8.5   QUANTUM-EFFECT DEVICES

11. Modecular beam epitaxy interfaces are typically abrupt to within one or two monolayers (one monolayer = 0.28 nm in GaInAs) because of terrace formation in the growth plane. Estimate the energy level broadening for the ground and first excited-electron states of a 15 nm GaInAs quantum well bound by thick AlInAs barriers. (Hint: assume the case of two-monolayer thickness fluctuation and an infinity deep quantum well. The electron effective mass in GaInAs is 0.0427 m_0.)

12. Find the first excited level and the corresponding width of the peak ΔE_2 for a RTD with AlAs (2 nm)/GaAs (6.78 nm)/AlAs (2 nm). If we want to maintain the same energy level but increase the width ΔE_2 by a factor of 10, what should be the thicknesses of AlAs and GaAs?

Photonic Devices

Photonic devices are devices in which the basic particle of light—the photon—plays a major role. In this chapter, we consider four groups of photonic devices: *light-emitting diodes* (LEDs) and *lasers* (*l*ight *a*mplification by *s*timulated *e*mission of *r*adiation), which convert electrical energy to optical energy; *photodetectors*, which electrically detect optical signals; and *solar cells*, which convert optical energy into electrical energy.

Specifically, we cover the following topics:

- Basic interactions between a photon and an electron.
- Generation of photons by spontaneous emission for the conventional and organic LEDs.
- Generation of photons by stimulated emission for heterostructure lasers.
- Absorption of photons to create electron-hole pairs for photodetectors.
- Absorption of photons to convert them to electrical energy for solar cells.

▷ 9.1 RADIATIVE TRANSITIONS AND OPTICAL ABSORPTION

Figure 1 shows the electromagnetic spectrum of the optical region. The detectable range of light by the human eye extends only from approximately 0.4 μm to 0.7 μm. Figure 1 also shows the major color bands from violet to red in the expanded scale. The ultraviolet region includes wavelengths from 0.01 μm to 0.4 μm, and the infrared region extends from 0.7 μm to 1,000 μm. In this chapter, we are primarily interested in the wavelength range from near-ultraviolet (~0.3 μm) to near-infrared (~1.5 μm).

Figure 1 also shows the photon energy on a separate horizontal scale. To convert the wavelength to photon energy, we use the relationship

$$\lambda = \frac{c}{\nu} = \frac{hc}{h\nu} = \frac{1.24}{h\nu(\text{eV})} \quad \mu\text{m},$$

(1)

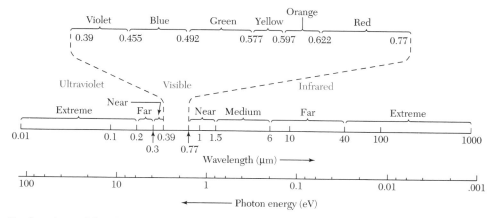

Fig. 1   Chart of the electromagnetic spectrum from the ultraviolet region to the infrared region.

where c is the speed of light in vacuum, v is the frequency of light, h is Planck's constant, and hv is the energy of a photon, and is measured in electron volts. For example, a 0.5 μm green light corresponds to a photon energy of 2.48 eV.

9.1.1   Radiative Transitions

There are basically three processes for interaction between a photon and an electron in a solid: absorption, spontaneous emission, and stimulated emission. We use a simple system to demonstrate these processes.[1] Consider two energy levels E_1 and E_2 of an atom, where E_1 corresponds to the ground state and E_2 corresponds to the excited state (Fig. 2). Any transition between these states involves the emission or absorption of a photon with frequency v_{12} given by $hv_{12} = E_2 - E_1$. At room temperature, most of the atoms in a solid are at the ground state. This situation is disturbed when a photon of energy exactly equal to hv_{12} impinges on the system. An atom in state E_1 absorbs the photon and thereby goes to the excited state E_2. The change in the energy state is the *absorption* process, shown in Fig. 2a. The excited state of the atom is unstable. After a short time, without any external stimulus, it makes a transition to the ground state, giving off a photon of energy hv_{12}. This process is called *spontaneous emission* (Fig. 2b). When a photon of energy hv_{12} impinges on an atom while it is in the excited state (Fig. 2c), the atom can be stimulated to make a transition to the ground state and gives off a photon of energy hv_{12}, which is in phase with the incident radiation. This process is called *stimulated emission*. The radiation from stimulated emission is monochromatic because each photon has precisely an energy hv_{12} and is coherent because all photons emitted are in phase.

The dominant operating process for LEDs is spontaneous emission, for the laser diodes (LDs) it is stimulated emission, and for the photodetectors and the solar cell it is absorption.

Let us assume that the instantaneous populations of E_1 and E_2 are n_1 and n_2, respectively. Under a thermal equilibrium condition and for $(E_2 - E_1) > 3\,kT$, the population is given by Boltzmann distribution:

$$\frac{n_2}{n_1} = e^{-(E_2 - E_1)/kT} = e^{-hv_{12}/kT}. \tag{2}$$

The negative exponent indicates that n_2 is less than n_1 in thermal equilibrium; that is, most electrons are at the lower energy level.

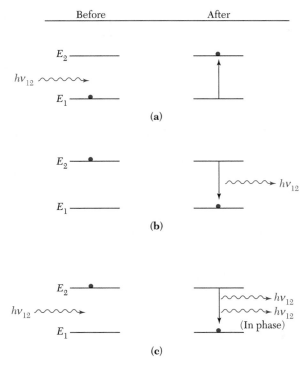

Fig. 2 The three basic transition processes between two energy levels. [1] Black dots indicated the state of the atom. The initial state is at the left; the final state, after the transition, is at the right. (*a*) Absorption. (*b*) Spontaneous emission. (*c*) Stimulated emission.

In steady state, the stimulated-emission rate (i.e., the number of stimulated-emission transitions per unit time) and the spontaneous-emission rate must be balanced by the rate of absorption to maintain the population n_1 and n_2 constant. The stimulated-emission rate is proportional to the photon-field energy density $\rho\,(h\nu_{12})$, which is the total energy in the radiation field per unit volume per unit frequency. Therefore, the stimulated-emission rate can be written as $B_{21}n_2\rho(h\nu_{12})$, where n_2 is the number of electrons in the upper level and B_{21} is a proportionality constant. The spontaneous-emission rate is proportional only to the population of the upper level and can be written as $A_{21}n_2$, where A_{21} is a constant. The absorption rate is proportional to the electron population at the lower level and to $\rho\,(h\nu_{12})$; this rate can be written as $B_{12}n_1\rho(h\nu_{12})$, where B_{12} is a proportionality constant. Therefore, we have at steady state

stimulated-emission rate + spontaneous-emission rate = absorption rate,

or

$$B_{21}n_2\rho\,(h\nu_{12}) + A_{21}n_2 = B_{12}n_1\rho\,(h\nu_{12}). \tag{3}$$

From Eq. 3 we observe that

$$\frac{\text{stimulated} - \text{emission rate}}{\text{spontaneous} - \text{emission rate}} = \frac{B_{21}}{A_{21}}\,\rho\!\left(h\nu_{12}\right). \tag{4}$$

To enhance stimulated emission over spontaneous emission, we must have a very large photon-field energy density $\rho\,(h\nu_{12})$. To achieve this density, an optical resonant cavity is used to increase the photon field. We also observe from Eq. 3 that

$$\frac{\text{stimulated} - \text{emission rate}}{\text{absorption rate}} = \frac{B_{21}}{B_{12}}\left(\frac{n_2}{n_1}\right). \tag{5}$$

If the stimulated emission of photon is to dominate over the absorption of photons, we must have higher electron density in the upper level than in the lower level. This condition is called *population inversion*, since under an equilibrium condition the reverse is true. In Section 9.3 on semiconductor lasers, we consider various ways to have a large photon-field energy density and achieve population inversion, so that the stimulated emission becomes dominant over both spontaneous emission and absorption.

9.1.2 Optical Absorption

Figure 3 shows the basic transitions in a semiconductor. When the semiconductor is illuminated, photons are absorbed to create electron-hole pairs, as shown in Fig. 3a, if the photon energy is equal to the bandgap energy, that is, $h\nu$ equals E_g. If $h\nu$ is greater than E_g, an electron-hole pair is generated and, in addition, the excess energy $(h\nu - E_g)$ is dissipated as heat, as shown in Fig. 3b. Both processes, (Figs. *a* and *b*) are called *intrinsic transitions* (or band-to-band transitions). On the other hand, for $h\nu$ less than E_g, a photon will be absorbed only if there are available energy states in the forbidden bandgap due to chemical impurities or physical defects, as shown in Fig. 3c. That process is called *extrinsic transition*. This discussion also is generally true for the reverse situation. For example, an electron at the conduction band edge combining with a hole at the valence band edge will result in the emission of a photon with energy equal to that of the bandgap.

Assume that a semiconductor is illuminated from a light source with $h\nu$ greater than E_g and a photon flux of Φ_0 in units of photons per square centimeter per second. As the photon flux travels through the semiconductor, the fraction of the photons absorbed is proportional to the intensity of the flux. Therefore, the number of photons absorbed within

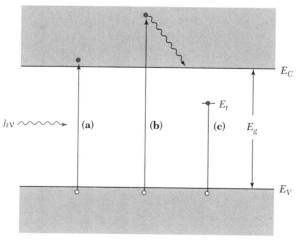

Fig. 3   Optical absorption for (*a*) $h\nu = E_g$, (*b*) $h\nu > E_g$, and (*c*) $h\nu < E_g$.

an incremental distance Δx (Fig. 4a) is given by $\alpha \Phi(x)\Delta x$, where α is a proportionality constant defined as the absorption coefficient. From the continuity of photon flux as shown in Fig. 4a, we obtain

$$\Phi(x + \Delta x) - \Phi(x) = \frac{d\Phi(x)}{dx}\Delta x = -\alpha\Phi(x)\Delta x$$

or

$$\frac{d\Phi(x)}{dx} = -\alpha\Phi(x). \tag{6}$$

The negative sign indicates a decreasing intensity of the photon flux due to absorption. The solution of Eq. 6 with the boundary condition $\Phi(x) = \Phi_0$, at $x = 0$ is

$$\Phi(x) = \Phi_0\, e^{-\alpha x}. \tag{7}$$

The fraction of photon flux that exits from the other end of the semiconductor at $x = W$ (Fig. 4b) is

$$\boxed{\Phi(W) = \Phi_0 e^{-aW}.} \tag{8}$$

The absorption coefficient α is a function of $h\nu$. Figure 5 shows the measured optical absorption coefficient for some important semiconductors that are used for photonic devices.[2] Also shown is the absorption coefficient for amorphous silicon (dashed curve), which is an important material for solar cells. The absorption coefficient decreases rapidly at the cutoff wavelength λ_c that is,

$$\lambda_c = \frac{1.24}{E_g}\ \mu m \tag{9}$$

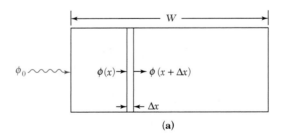

(a)

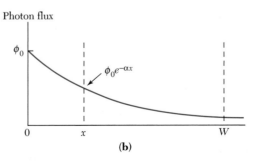

(b)

Fig. 4 Optical absorption. (a) Semiconductor under illumination. (b) Exponential decay of photon flux.

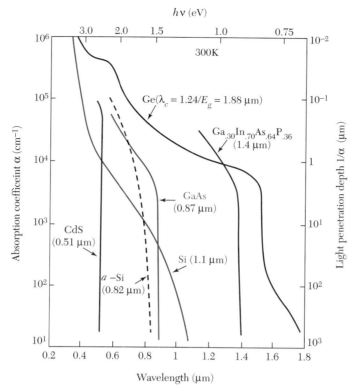

Fig. 5 Optical absorption coefficients for various semiconductor materials.[2] The value in the parenthesis is the cutoff wavelength.

because the optical band-to-band absorption becomes negligible for $h\nu < E_g$, or $\lambda > \lambda_c$.

▶ **EXAMPLE 1**

A single-crystal silicon sample 0.25 μm-thick is illuminated with a monochromatic light (single frequency) having an $h\nu$ of 3 eV. The incident power is 10 mW. Find the total energy absorbed by the semiconductor per second, the rate of excess thermal energy dissipated to the lattice, and the number of photons per second given off from recombination by intrinsic transitions.

SOLUTION   From Fig. 5 the absorption coefficient α is 4×10^4 cm^{-1}. The energy absorbed per second is

$$\Phi_0(1 - e^{-\alpha W}) = 10^{-2}\left[1 - \exp\left(-4 \times 10^4 \times 0.25 \times 10^{-4}\right)\right]$$

$$= 0.0063 \text{ J/s} = 6.3 \text{ mW}.$$

The portion of each photon's energy that is converted to heat is

$$\frac{h\nu - E_g}{h\nu} = \frac{3 - 1.12}{3} = 62\%.$$

Therefore, the amount of energy dissipated per second to the lattice is

$$62\% \times 6.3 = 3.9 \text{ mW}.$$

Since the recombination radiation accounts for 2.4 mW (i.e., 6.3 mW –3.9 mW) at 1.12 eV/ photon, the number of photons per second from recombination is

$$\frac{2.4 \times 10^{-3}}{1.6 \times 10^{-19} \times 1.12} = 1.3 \times 10^{16} \text{ photons / s.}$$

◀

9.2 LIGHT-EMITTING DIODES

Light-emitting diodes (LEDs) are *p-n* junctions that can emit spontaneous radiation in ultraviolet, visible, or infrared regions. The visible LED has a multitude of applications as an information link between electronic instruments and their users. The infrared LED is useful in opto-isolators and for optical-fiber communication.

9.2.1 Visible LEDs

Figure 6 shows the relative eye response as a function of wavelength (or the corresponding photon energy). The maximum sensitivity of the eye is at 0.555 μm. The eye response falls to nearly zero at the extremes of the visible spectrum at about 0.4 and 0.7 μm. For normal vision at the peak response of the eye, 1 W of radiant energy is equivalent to 683 lumen.

Since the eye is only sensitive to light with a photon energy $h\nu$ equal to or greater than 1.8 eV (≤0.7 μm), semiconductors of interest must have an energy bandgap larger than this limit. Figure 6 also shows the bandgaps of various semiconductors. Table 1 lists the semiconductors used to produce light in the visible and infrared parts of the spectrum. Among all the semiconductors shown, the most important materials for visible LEDs are the alloy $GaAs_{1-y}P_y$ and $Ga_xIn_{1-x}N$ III-V compound systems. An alloy III-V compound is formed when more than one group III element is distributed randomly on group III

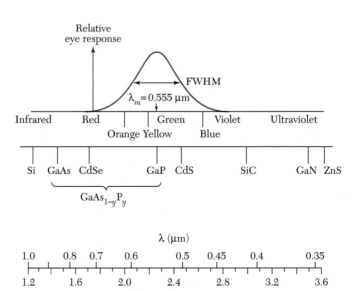

Fig. 6 Semiconductors of interest as visible LEDs. Figure includes relative response of the human eye.

TABLE 1   Common III-V materials used to produce LEDs and their emission wavelengths.

Material	Wavelength (nm)
InAsSbP/InAs	4200
InAs	3800
GaInAsP/GaSb	2000
GaSb	1800
$Ga_xIn_{1-x}As_{1-y}P_y$	1100-1600
$Ga_{0.47}In_{0.53}As$	1550
$Ga_{0.27}In_{0.73}As_{0.63}P_{0.37}$	1300
GaAs:Er,InP:Er	1540
Si:C	1300
GaAs:Yb,InP:Yb	1000
$Al_xGa_{1-x}As$:Si	650-940
GaAs:Si	940
$Al_{0.11}Ga_{0.89}As$:Si	830
$Al_{0.4}Ga_{0.6}As$:Si	650
$GaAs_{0.6}P_{0.4}$	660
$GaAs_{0.4}P_{0.6}$	620
$GaAs_{0.15}P_{0.85}$	590
$(Al_xGa_{1-x})_{0.5}In_{0.5}P$	655
GaP	690
GaP:N	550-570
$Ga_xIn_{1-x}N$	340,430,590
SiC	400-460
BN	260,310,490

lattice sites (e.g., gallium sites) or more than one group V element are distributed randomly on group V lattice sites (e.g., arsenic sites). The notation used is $A_xB_{1-x}C$ or $AC_{1-y}D_y$ for *ternary* (three elements) compounds and $A_xB_{1-x}C_yD_{1-y}$ for *quaternary* (four elements) compounds, where A and B are the group III elements, C and D are the group V elements, and x and y are the mole fractions, that is, the ratios of the number of atoms of a given species to the total number of group III or group V atoms in the alloy compound.

Figure 7a shows the energy gap for $GaAs_{1-y}P_y$ as a function of the mole fraction y. For $0 < y < 0.45$, the bandgap is direct and increases from $E_g = 1.424$ eV at $y = 0$ to $E_g = 1.977$ eV at $y = 0.45$. For $y > 0.45$, the bandgap is indirect. Figure 7b shows the corresponding energy-momentum plots for selected alloy compositions.[3] As indicated, the conduction band has two minima. The one along $p = 0$ is the direct minimum, and the one along $p = p_{max}$ is the indirect minimum. Electrons in the direct minimum of the conduction band and holes at the top of the valence band have equal momenta ($p = 0$). But electrons in the indirect minimum of the conduction band and holes at the top of the valence band have different momenta. The radiative transition mechanisms are found predominantly in direct-bandgap semiconductors, such as gallium arsenide and $GaAs_{1-y}P_y$ ($y < 0.45$), since the momentum is conserved. The photon energy is approximately equal to the bandgap energy of the semiconductor.

However, for $GaAs_{1-y}P_y$ with y greater than 0.45 and gallium phosphide, which are indirect-bandgap semiconductors, the probability for radiative transitions is very small, since lattice interactions or other scattering agents must participate in the process to

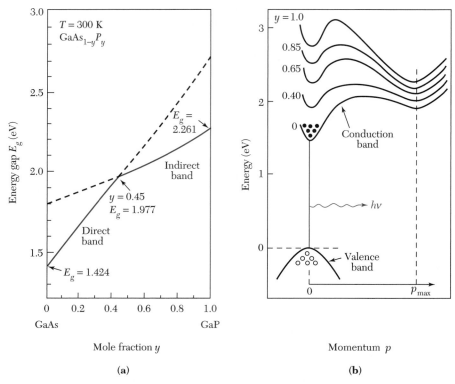

Fig. 7 (a) Compositional dependence for the direct- and indirect-energy bandgap for GaAs$_{1-y}$P$_y$. (b) The alloy compositions shown correspond to red (y = 0.4), orange (0.65), yellow (0.85), and green light (1.0). [3]

conserve momentum. Therefore, for indirect-bandgap semiconductors, special recombination centers are incorporated to enhance the radiative processes. Incorporating nitrogen into the crystal lattice can form efficient radiative recombination centers in GaAs$_{1-y}$P$_y$. When nitrogen is introduced, it replaces phosphorous atoms in the lattice sites. The outer electronic structure of nitrogen is similar to that of phosphorus (both are group V elements in the periodic table), but the electronic core structures of these atoms are different. This difference results in the creation of an electron trap level close to the bottom of the conduction band. A recombination center is thus produced, and it is called an *isoelectronic center*. This recombination center can greatly enhance the probability of radiative transition in indirect-bandgap semiconductors.

Figure 8 shows the quantum efficiency, the number of photons generated per electron-hole pair, versus alloy composition for GaAs$_{1-y}$P$_y$. with and without the isoelectronic impurity nitrogen.[4] The efficiency without nitrogen drops sharply in the composition range $0.4 < y < 0.5$ because the bandgap changes from direct to indirect at y = 0.45. The efficiency with nitrogen is considerably higher for $y > 0.5$ but nevertheless decreases steadily with an increasing y because of the increasing separation between the direct and indirect bandgap (Fig. 7*b*).

Figure 9 shows the basic visible LED structures in the flat-diode configuration.[5] Figure 9*a* shows the cross section of a direct-bandgap LED, which emits red light, fabricated on gallium arsenide substrate. Figure 9*b* is a indirect bandgap LED, which emits orange, yellow, or green light fabricated on gallium phosphide substrates. A graded-alloy

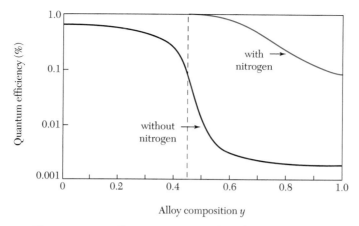

Fig. 8 Quantum efficiency versus alloy composition with and without isoelectronic impurity nitrogen. [4]

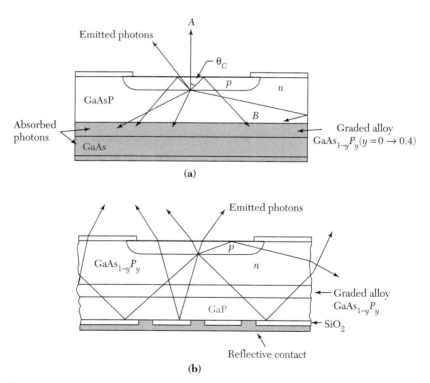

Fig. 9 Basic structure of a flat-diode LED and the effects of (a) an opaque substrate $(GaAs_{1-y}P_y)$ and (b) a transparent substrate (GaP) on photons emitted at the p-n junction .[5]

$GaAs_{1-y}P_y$ layer is grown epitaxially to minimize the nonradiative centers at the interface that result from lattice mismatch.

For high-brightness blue LEDs (0.455–0.492 μm), II-VI compounds such as ZnSe, III-V nitride semiconductors such as GaN, and IV-IV compounds such as SiC have been investigated. However, their short lifetimes prevent II-IV–based devices from being commercialized at present; and the indirect bandgap of SiC results in the low brightness of SiC blue LEDs.

The most promising candidates are GaN (E_g = 3.44 eV) and related III-V nitride semiconductors such as AlGaInN, which have direct bandgaps ranging from 1.95 eV to 6.2 eV with corresponding wavelengths from 0.2 μm to 0.63 μm.[6] Although there are no lattice-matched substrates for the growth of GaN, high-quality GaN has been grown on sapphire (Al_2O_3) using a low-temperature grown AlN as a buffer layer. Figure 10 shows a III-V nitride LED grown on sapphire substrate. Because the sapphire substrate is an insulator, both the n- and p-type ohmic contacts are formed on the top surface. The blue light originates from the radiative recombination in the $Ga_xIn_{1-x}N$ region which is sandwiched between two larger bandgap semiconductors: a p-type $Al_xGa_{1-x}N$ layer and an n-type GaN layer.

There are three loss mechanisms that reduce the quantity of emitted photons: absorption within the LED material, reflection loss when light passes from a semiconductor to air due to differences in refractive index, and total internal reflection of light at angles greater than the critical angle θ_c (Fig. 9a) defined by Snell's law:

$$\sin\theta_c = \frac{\bar{n}_1}{\bar{n}_2}, \tag{10}$$

where the light passes from a medium with a refractive index of $\bar{n}_2$, such as GaAs with $\bar{n}_2$ = 3.66 at $\lambda \cong 0.8$ μm, to a medium of $\bar{n}_1$, such as air with $\bar{n}_1$ = 1. For gallium arsenide, the critical angle is about 16°; and for gallium phosphide with $\bar{n}_2$ = 3.45 at $\lambda \cong 0.8$ μm, the critical angle is about 17°.

The forward current-voltage characteristics of a LED is similar to that of the GaAs p–n junction discussed in Chapter 4. At low forward voltages, the diode current is dominated by the nonradiative recombination current due mainly to surface recombination near the perimeter of the LED chip. At higher forward voltages, the diode current is dominated by the radiative diffusion current. At even higher voltages, the series resistance will limit the diode current. The total diode current can be written as

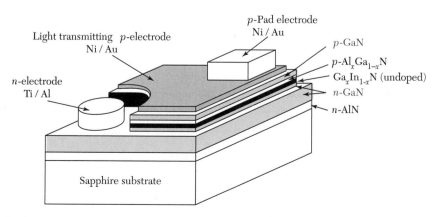

Fig. 10 III-V nitride LED grown on sapphire substrate.[6]

$$I = I_d \exp\left[\frac{q(V - IR_s)}{kT}\right] + I_r \exp\left[\frac{q(V - IR_s)}{2kT}\right], \tag{11}$$

where R_s is the device series resistance and I_d and I_r are the saturation currents due to diffusion and recombination, respectively. To increase the output power of the LED, we must reduce I_r and R_s.

The emission spectra of LEDs are similar to the eye response curve shown in Fig. 6. The spectral width is given by the full width at half maximum (FWHM) intensity. The spectral width generally varies as λ_m^2, where λ_m, is the wavelength at the maximum intensity.[7] Thus, the FWHM becomes larger as the wavelength is increased from visible to infrared. For example, at $\lambda_m = 0.55\ \mu$m (green), FWHM is about 20 nm, but at 1.3 μm (infrared), FWHM is over 120 nm.

Visible LEDs can be used for full-color displays, full-color indicators, and lamps with high efficiency and high reliability. Figure 11 shows the diagrams of two LED lamps.[8] An LED lamp contains an LED chip and a plastic lens, which is usually colored to serve as an optical filter and to enhance contrast. The lamp in Fig. 11a uses a conventional diode header. Figure 11b shows a package that is suited for a transparent semiconductor, such as gallium phosphide, which emits light through all five facets (four sides and the top) of the LED chip.

Figure 12 shows the basic formats for LED displays. The seven segments (Fig. 12a) display numbers from 0 to 9. The 5×7 matrix array (Fig. 12b) displays alphanumerics (A–Z and 0–9). The displays can be made by monolithic processes similar to those used

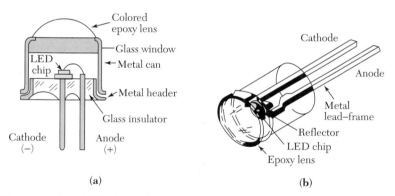

Fig. 11   Diagrams of two LED lamps.[8]

Fig. 12   LED display formats for numeric and alphanumeric: (a) 7-segment (numeric); (b) 5×7 array (alphanumeric).[8]

to make silicon integrated circuits (described in Chapters 10 through 14) or by using an individual LED chip mounted on a reflector to form a bar segment.

There has been interest in the development of white LEDs for general illumination because LEDs are three times as efficient as incandescent lamps and can last 10 times longer. A white LED requires LEDs of red, green, and blue colors. A wide-spread use of white LED will be realized when the costs of these LEDs, especially the blue LEDs, become competitive with the conventional light sources.

Orangic LED

In the previous discussions, we considered devices made only from *inorganic* semiconductor materials, such as GaAsP and GaN. In recent years, certain *organic semiconductors* have been studied for electroluminescent applications. The orangic light-emitting diode (OLED) is particularly useful for multicolor, large-area flat-panel display because of its attributes of low-power consumption and excellent emissive quality with a wide viewing angle.[9]

Figure 13a shows the molecular structures of two representative organic semiconductors.[10] They are the tris (8-hydroxy-quinolinato) aluminum (AlQ_3), which contains six benzene rings connected to a central aluminum atom, and the aromatic diamine, which also contains six benzene rings but with a different molecular arrangement. A basic OLED has a number of layers on a transparent substrate (e.g., glass). Onto the substrate we deposit, in sequence, a transparent conductive anode [e.g., ITO (indium tin oxide)], the diamine as a hole transport layer, the AlQ_3 as the electron transport layer, and the cathode contact (e.g., Mg alloy with 10% Ag). A cross sectional view is shown in Fig. 13b.

Figure 13c shows the band diagram of the OLED. It is basically a heterojunction formed between AlQ_3 and diamine. Under proper biasing conditions, electrons are injected from the cathode and move toward the heterojunction interface whereas holes are injected from the anode and also move toward the interface. Because of the energy barriers ΔE_C and ΔE_V, these carriers will accumulate at the interface to enhance the chance of radiative recombination.

From Fig. 13c, we can specify the design criteria of an OLED: (a) ultrathin layers for low biasing voltage—for example, the total thickness of the organic semiconductor

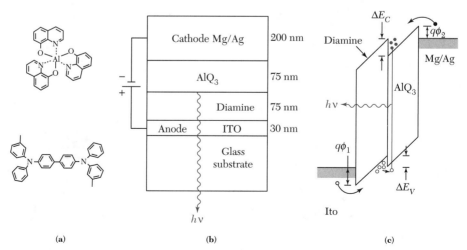

Fig. 13 (*a*) Organic semiconductors. (*b*) OLED cross sectional view. (*c*) Band diagram of an OLED.

layers shown is only 150 nm; (b) low injection barriers—the barrier height $q\phi_1$ for hole injection and the barrier height $q\phi_2$ for electron injection must be low enough to allow large carrier injections for high-current density operation, and (c) proper bandgaps for the required color. For AlQ_3, the emitted light is green color. By choosing different organic semiconductors with different bandgaps, various colors including red, yellow, and blue can be obtained.

9.2.2 Infrared LED

Infrared LEDs include gallium arsenide LEDs, which emit light near 0.9 µm, and many III-V compounds, such as the quaternary $Ga_xIn_{1-x}As_yP_{1-y}$ LEDs, which emit light from 1.1 to 1.6 µm.

An important application of infrared LEDs is in opto-isolators, where an input or control signal is decoupled from the output. Figure 14 shows an opto-isolator having an infrared LED as the light source and a photodiode as the detector. When an input signal is applied to the LED, light is generated and subsequently detected by the photodiode. The light is then converted back to an electrical signal as a current that flows through a load resistor. Opto-isolators transmit signals at the speed of light and are electrically isolated because there is no electrical feedback from the output to the input.

Another important application of infrared LEDs is for transmission of an optical signal through an optical fiber, as in a communication system. An optical fiber is a waveguide at optical frequencies. The fiber is usually drawn from a preform of glass to a diameter of about 100 µm. It is flexible and can guide optical signals over distances of many kilometers to a receiver, similar to the way a coaxial cable transmits electrical signals.

Two types of optical fibers are shown in Fig. 15. One type of fiber has a cladding layer of relatively pure fused silica (SiO_2) surrounding a core of doped glass (e.g., germanium doped glass) that has a higher refractive index than the cladding layer.[11] This type of fiber is called a *step index fiber*. The light is transmitted along the length of the fiber by internal reflection at the step in the refractive index. The critical angle for internal reflection is about 79° for $\overline{n}_1 = 1.457$ (cladding layer) and $\overline{n}_2 = 1.480$ (core, 20% Ge-doped) as calculated from Eq. 10. Note that different rays will propagate with different path lengths (Fig. 15*a*). A light pulse reaching the end of a step index fiber will result in a pulse spread. In a *graded-index fiber* (Fig. 15*b*), the index decreases from the core center by a parabolic law. Now, rays traversing toward the cladding have a high velocity (due to lower refractive index) than rays along the center of the core. The pulse spread is significantly reduced. As the light is transmitted along the optical fiber, the light signal will be attenuated. However, due to the transparency of ultrapure silica used for the fiber material in the wavelength region from 0.8 to 1.6 µm, the attenuation is quite low and

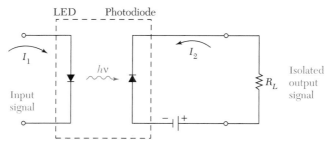

Fig. 14 An opto-isolator in which an input signal is decoupled from the output signal.

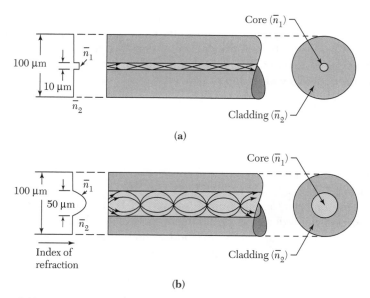

Fig. 15 Optical fibers. (*a*) Step-index fiber having a core with slightly larger reflective index. (*b*) Graded-index fiber having a parabolic grading of the reflective index in the core.[11]

is proportional to λ^{-4}. Typical attenuations are about 3 dB/km at a wavelength of 0.8 μm, 0.6 dB/km at 1.3 μm, and 0.2 dB/km at 1.55 μm.

A simple point-to-point optical-fiber communication system is shown in Fig. 16, where the electrical input signals are converted to optical signals using an optical source (LED or laser). The optical signals are coupled into the fiber and transmitted to the photodetector, where they are converted back to electrical signals.

The surface-emitting infrared InGaAsP LED used for optical-fiber communication is shown[12] in Fig. 17. The light is emitted from the central surface area and coupled into the optical fiber. The use of heterojunctions (e.g., InGaAsP-InP) can increase the efficiency that results from the confinement of the carrier by the layers of the higher-bandgap semiconductor InP, surrounding the radiative-recombination region InGaAsP. We con-

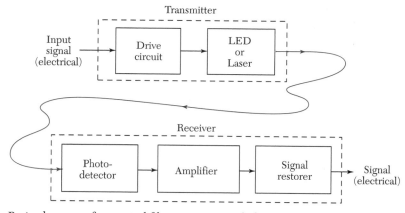

Fig. 16 Basic elements of an optical fiber transmission link.

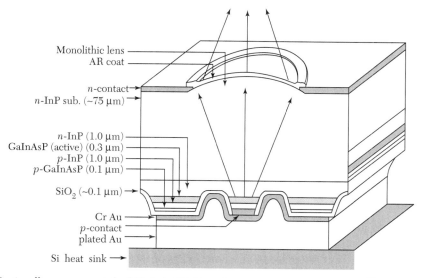

Monolithic lens
AR coat
n-contact
n-InP sub. (~75 µm)

n-InP (1.0 µm)
GaInAsP (active) (0.3 µm)
p-InP (1.0 µm)
p-GaInAsP (0.1 µm)

SiO$_2$ (~0.1 µm)

Cr Au
p-contact
plated Au

Si heat sink

Fig. 17 Small-area mesa-etched GaInAsP/InP surface-emitting LED structure.[12]

sider the carrier confinement in Section 9.3. The heterojunction can also serve as an optical window to the emitted radiation because the higher-bandgap–confining layers do not absorb radiation from the lower-bandgap–emitting region.

The electrical input signal is generally modulated at high frequencies. This signal causes direct modulation of the injected current in an LED. Parasitic elements such as the depletion-layer capacitance and series resistance can cause a delay of carrier injection into the junction and a delay in the light output. The ultimate limit on how fast one can vary the light output depends on the carrier lifetime, which is determined by various recombination processes, such as the surface recombination discussed in Chapter 3. If the current is modulated at an angular frequency ω, the light output $P(\omega)$ is given by

$$P(\omega) = \frac{P(0)}{\sqrt{1 + (\omega\tau)^2}},\qquad(12)$$

where $P(0)$ is the light output at $\omega = 0$, and τ is the carrier lifetime. The *modulation bandwidth* Δf is defined as the frequency at which the light output is reduced to $1/\sqrt{2}$ that at $\omega = 0$, that is,

$$\Delta f \equiv \frac{\Delta\omega}{2\pi} = \frac{1}{2\pi\tau}.\qquad(13)$$

▷ **EXAMPLE 2**

Calculate the modulation bandwidth of a GaAs-based LED with $\tau = 500$ ps.

SOLUTION From Eq. 13,

$$\Delta f = \frac{1}{2\pi \cdot 500 \cdot 10^{-12}} = 318 \text{ MHz.}$$

◀

▶ ## 9.3 SEMICONDUCTOR LASER

Semiconductor lasers are similar to the solid-state ruby laser and helium-neon gas laser in that the emitted radiation is highly monochromatic and produces a highly directional beam of light. However, the semiconductor laser differs from other lasers in that it is small (on the order of 0.1 mm long) and is easily modulated at high frequencies simply by modulating the biasing current. Because of these unique properties, the semiconductor laser is one of the most important light sources for optical-fiber communication. It is also used in video recording, optical reading, and high-speed laser printing. In addition, semiconductor lasers have significant applications in many areas of basic research and technology, such as high-resolution gas spectroscopy and atmospheric pollution monitoring.

9.3.1 Semiconductor Materials

All lasing semiconductors have direct bandgaps. This is expected because the momentum is conserved and the radiative-transition probability in a direct-bandgap semiconductor is high. At present, the laser emission wavelengths cover the range from 0.3 to over 30 μm. Gallium arsenide was the first material to emit laser radiation, and its related III-V compound alloys are the most extensively studied and developed.

The three most important III-V compound alloy systems are $Ga_xIn_{1-x}As_yP_{1-y}$, $Ga_xIn_{1-x}As_ySb_{1-y}$, and $Al_xGa_{1-x}As_ySb_{1-y}$ solid solutions. Figure 18 shows the bandgaps plotted against the lattice constant for various III-V binary semiconductors and their intermediate ternary and quaternary compounds.[13] To achieve a heterostructure with negligible interface traps, the lattices between the two semiconductors must be matched closely.

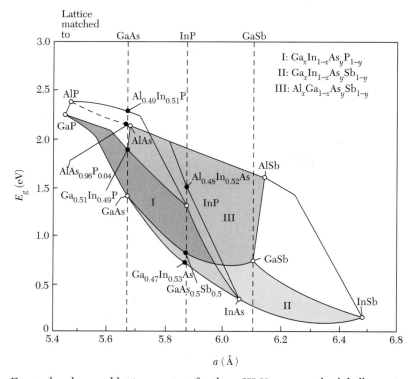

Fig. 18 Energy bandgap and lattice constant for three III-V compound solid alloy system. [13]

If we use GaAs (a = 5.6533 Å) as the substrate, the ternary compound $Al_xGa_{1-x}As$ can have a lattice mismatch less than 0.1%. Similarly, with InP (a = 5.8687 Å) as the substrate, the quaternary compound $Ga_xIn_{1-x}As_yP_{1-y}$ also can have a nearly perfect lattice match, as indicated by the center vertical line in Fig. 18.

Figure 19a shows the bandgap of ternary $Al_xGa_{1-x}As$ as a function of aluminum composition.[1] The alloy has a direct bandgap up to x = 0.45, then becomes an indirect-bandgap semiconductor. Figure 19b shows the compositional dependence of the refractive index. For example, for x = 0.3, the bandgap of $Al_xGa_{1-x}As$ is 1.789 eV, which is 0.365 eV larger than that of GaAs; its refractive index is 3.385, which is 6% smaller than that of GaAs. These properties are important for continuous operation of semiconductor lasers at and above room temperatures.

9.3.2 Laser Operation

Figure 20 shows schematic representations[14] of the band diagram under forward-bias , the refractive index profile, and the optical-field distribution of light generated at the junction of a homojunction laser (Fig. 20a) and a double-heterostructure (DH) laser (Fig. 20b).

Population Inversion

As discussed in Section 1, to enhance the stimulated emission for laser operation we need population inversion. To achieve population inversion in a semiconductor laser, we consider a p-n junction or a heterojunction formed between degenerate semiconductors. This means that the doping levels on both sides of the junction are high enough that the Fermi level E_{FV} is below the valence band edge on the p-side and E_{FC} is above the conduction band edge on the n-side. When a sufficiently large bias is applied (the band diagrams in Fig. 20), high injection occurs, that is, large concentrations of electrons and holes are injected into the transition region. As a result, the region d (Fig. 20) contains a large concentration of electrons in the conduction band and a large concentration of holes in the valence band; this is the required condition for population inversion. For band-to-band

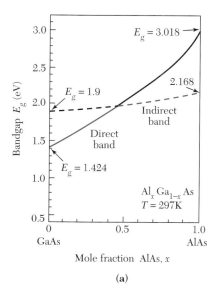

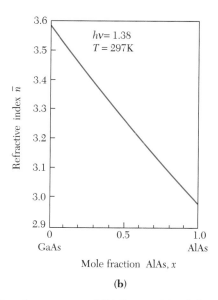

(a)   (b)

Fig. 19 (*a*) Compositional dependence of the $Al_xGa_{1-x}As$ energy gap.[1] (*b*) Compositional dependence of the refractive index at 1.38 eV.

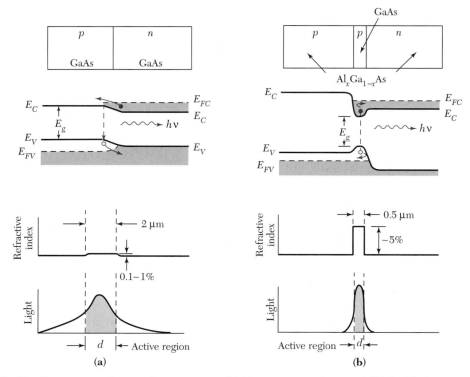

Fig. 20 Comparison of some characteristics of (*a*) homojunction laser and (*b*) double-hetero-junction (DH) laser. Second from the top row shows energy band diagrams under forward bias. The refractive index change for a homojunction laser is less than 1%. The refractive index change for DH laser is about 5%. The confinement of light is shown in the bottom row. [14]

transition, the minimum energy required is the bandgap energy E_g. Therefore, from the band diagram in Fig. 20, we can write the condition necessary for population inversion: $(E_{FC} - E_{FV}) > E_g$.

Carrier and Optical Confinement

As can be seen in the DH laser, the carriers are confined on both sides of the active region by the heterojunction barriers, whereas in the homojunction laser the carriers can move away from the active region, where radiative recombination occurs.

In the DH laser the optical field is also confined within the active region by the abrupt reduction of the refractive index outside the active region. The optical confinement can be explained by Fig. 21, which shows a three-layer dielectric wave guide with refractive indices $\bar{n}_1, \bar{n}_2$ and $\bar{n}_3$, where an active layer is sandwiched between two confining layers (Fig. 21*a*). Under the condition $\bar{n}_2 > \bar{n}_1 \geq \bar{n}_3$, the ray angle θ_{12} at the layer 1/layer 2 inter-face in Fig. 21*b* exceeds the critical angle given by Eq. 10. A similar situation occurs for θ_{23} at the layer 2/layer 3 interface. Therefore, when the refractive index in the active layer is larger than the index of its surrounding layers, the propagation of the optical radiation is guided (confined) in a direction parallel to the layer interfaces. We can define a *confinement factor* Γ, which is the ratio of the light intensity within the active layer to the

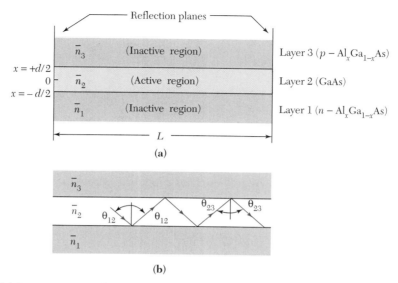

Fig. 21 (*a*) Representation of a three-layer dielectric waveguide. (*b*) Ray trajectories of the guided wave.

sum of light intensity both within and outside the active layer. The confinement factor is given as

$$\Gamma \cong 1 - \exp\left(-C\Delta\bar{n}d\right),$$

(14)

where C is a constant, $\Delta\bar{n}$ is the difference in the refractive index, and d is the thickness of the active layer. It is clear that the larger $\Delta\bar{n}$ and d are, the higher Γ will be.

Optical Cavity and Feedback

We have considered the condition necessary to produce laser action: population inversion. Photons released by stimulated emission are likely to cause further stimulations as long as there is population inversion. This is the phenomenon of optical gain. The gain obtained in a single travel of an optical wave down a laser cavity is small. To increase gain, multiple passes of a wave must occur. This is achieved using mirrors placed at either end of the cavity, shown as the reflection planes at the left side and right side in Fig. 21*a*. For a semiconductor laser, the cleaved ends of the crystal forming the device can act as the mirrors. For a GaAs device, cleaving along (110) plane creates two parallel identical mirrors. Sometimes the back mirror of the laser is metallized to enhance the reflectivity. The reflectivity R at each mirror can be calculated as

$$R = \left(\frac{\bar{n} - 1}{\bar{n} + 1}\right)^2,$$

(15)

where $\bar{n}$ is the refractive index in the semiconductor corresponding to the wavelength λ ($\bar{n}$ is generally a function of λ).

▷ **EXAMPLE 3**

Calculate the R for GaAs ($\bar{n} = 3.6$).

SOLUTION   From Eq. 15,

$$R = \left(\frac{3.6 - 1}{3.6 + 1}\right)^2 = 0.32,$$

that is, 32% of the light will be reflected at the cleaved surface.   ◀

If an integral number of half-wavelength fit between the two end planes, reinforced and coherent light will be reflected back and forth within the cavity. Therefore, for stimulated emission, the length L of the cavity must satisfy the condition

$$m\left(\frac{\lambda}{2\bar{n}}\right) = L \qquad (16)$$

or

$$m\lambda = 2\bar{n}L, \qquad (16a)$$

where m is an integral number. Obviously, many values of λ can satisfy this condition (Fig. 22a), but only those within the spontaneous emission spectrum will be produced (Fig. 22b). In addition, optical losses in the path traveled by the wave mean that only the strongest lines will survive, leading to a set of lasing modes, as given in Fig. 22c. The separation $\Delta\lambda$ between the allowed modes in the longitudinal direction is the difference in the wavelengths corresponding to m and $m+1$. Differentiating Eq. 16a with respect to λ, we obtain

$$\Delta\lambda = \frac{\lambda^2 \Delta m}{2\bar{n}L[1 - (\lambda / \bar{n})(d\bar{n}/d\lambda)]}. \qquad (17)$$

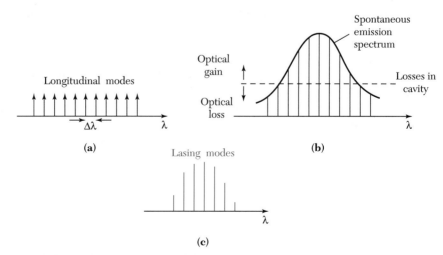

Fig. 22   (a) Resonant modes of a laser cavity. (b) Spontaneous emission spectrum. (c) Optical-gain wavelengths.

Although $\overline{n}$ is a function of λ, over the very small change in wavelength between adjacent modes $d\overline{n}/d\lambda$ is very small, and hence to a good approximation the mode spacing $\Delta\lambda$ is given by

$$\left|\Delta\lambda\right| \cong \frac{\lambda^2}{2\overline{n}L}. \tag{18}$$

▷ **EXAMPLE 4**

Calculate the mode spacing for a typical GaAs laser with λ = 0.94 μm, $\overline{n}$ = 3.6, and L = 300 μm.

SOLUTION   From Eq. 18,

$$\Delta\lambda \cong \frac{\left(0.94 \times 10^{-6}\right)^2}{2 \times 3.6 \times 300 \times 10^{-6}} = 4 \times 10^{-10} \text{ m } = 4 \text{ Å.}$$ ◀

9.3.3 Basic Laser Structure

Figure 23 shows three laser structures.[14,15] The first structure (Fig. 23*a*), is a basic *p-n* junction laser and is called a *homojunction laser* because it has the same semiconductor material (e.g., GaAs) on both sides of the junction. A pair of parallel planes (or facets) are cleaved or polished perpendicular to the ⟨110⟩-axis. Under appropriate biasing conditions laser light will be emitted from these planes (only the front emission is shown in Fig. 23). The two remaining sides of the diode are roughened to eliminate lasing in the directions other than the main ones. This structure is called a *Fabry–Perot cavity*, with a typical cavity length L of about 300 μm. The Fabry–Perot cavity configuration is used extensively for modern semiconductor lasers.

Figure 23*b* shows a *double-heterostructure* (DH) *laser*, in which a thin layer of a semiconductor (e.g., GaAs) is sandwiched between layers of a different semiconductor (e.g., $\text{Al}_x\text{Ga}_{1-x}\text{As}$). The laser structures shown in Figs. 23*a* and *b* are broad-area lasers because the entire area along the junction plane can emit radiation. Figure 23*c* shows a DH laser with a stripe geometry. The oxide layer isolates all but the stripe contact; consequently the lasing area is restricted to a narrow region under the contact. The stripe widths S are typically 5–30 μm. The advantages of the stripe geometry are reduced operating current, elimination of multiple-emission areas along the junction, and improved reliability that is the result of removing most of the junction perimeter.

Threshold Current Density

One of the most important parameters for laser operation is the threshold current density J_{th}, that is, the minimum current density required for lasing to occur. Figure 24 compares J_{th} versus operating temperature for a homojunction laser and a DH laser.[14] Note that as the temperature increases, J_{th} for the DH laser increases much more slowly than J_{th} for the homojunction laser. Because of the low values of J_{th} for DH lasers at 300 K, DH lasers can be operated continuously at room temperature. This characteristic has led to the increased use of semiconductor lasers, especially in optical-fiber communication systems.

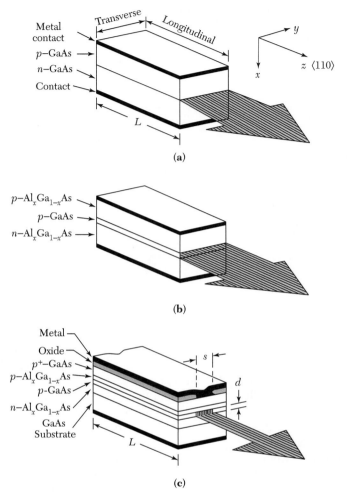

Fig. 23 Semiconductor laser structure in the Fabry–Perot-cavity configuration. (*a*) Homojunction laser. (*b*) Double-heterojunction (DH) laser. (*c*) Stripe-geometry DH laser. [14,15]

In a semiconductor laser, the gain g, the incremental optical energy flux per unit length, depends on the current density. The gain g can be expressed as a function of a nominal current density J_{nom}, which is defined for unity quantum efficiency (i.e., number of carriers generated per photon, $\eta = 1$) as the current density required to uniformly excite 1 μm thick an active layer. The actual current density is then given by

$$J(\mathrm{A/cm^2}) = \frac{J_{nom}d}{\eta}, \tag{19}$$

where d is the thickness of the active layer in μm. Figure 25 shows the calculated gain for a typical gallium arsenide DH laser.[16] The gain increases linearly with J_{nom} for $50 \leq g \leq 400$ cm^{-1}. The linear dashed line can be written as

$$g = \left(g_0 / J_0\right)\left(J_{nom} - J_0\right), \tag{20}$$

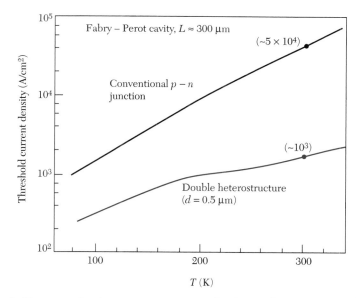

Fig. 24 Threshold current density versus temperature for the two laser structures shown[14] in Fig. 20.

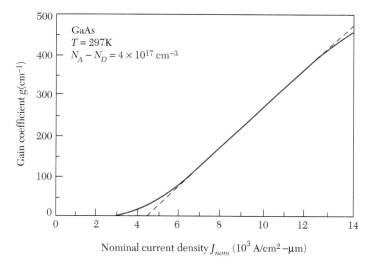

Fig. 25 Variation of gain coefficient versus nominal current density. Dashed line represents a linear dependence.[16]

where $g_0/J_0 = 5 \times 10^{-2}$ cm-μm/A and $J_0 = 4.5 \times 10^3$ A/ cm-μm.

As discussed previously, at low currents there is spontaneous emission in all directions. As the current increases, the gain increases (Fig. 25) until the threshold for lasing is reached, that is, until the gain satisfies the condition that a light wave makes a complete traversal of the cavity without attenuation:

$$R \exp\left[\left(\Gamma g - \alpha\right)L\right] = 1 \tag{21}$$

or

$$\Gamma g\left(\text{threshold gain}\right) = \alpha + \frac{1}{L}\ln\left(\frac{1}{R}\right), \tag{22}$$

where Γ is the confinement factor, α is the loss per unit length from absorption and other scattering mechanism, L is the length of the cavity shown in Fig. 23, and R is the reflectance of the ends of the cavity assuming that R for both ends is equal. Equations 19, 20, and 22 may be combined to give the threshold current density as

$$J_{th}(\text{A}/\text{cm}^2) = \frac{J_0 d}{\eta} + \left(\frac{J_0 d}{g_0 \eta \Gamma}\right)\left[\alpha + \frac{1}{L}\ln\left(\frac{1}{R}\right)\right]. \tag{23}$$

The term $(J_0 d/g_0 \eta \Gamma)$ is often called $1/\beta$, where β is known as the gain factor. To reduce J_{th}, we can increase η, Γ, L, and R and reduce d and α.

▶ **EXAMPLE 5**

Find the threshold current for a laser diode using the following data: front and rear mirror reflectivities are 0.44 and 0.99, respectively. The cavity length and width are 300 µm and 5 µm, respectively, $\alpha = 100$ cm^{-1}, $\beta = 0.1$ cm^{-3}A^{-1}, $g_0 = 100$ cm^{-1}, and $\Gamma = 0.9$.

SOLUTION With a known gain factor, the term $J_0 d/\eta$ in Eq. 23 can be expressed as $g_0 \Gamma/\beta$.

Due to different reflectivities of the two mirrors, Eq. 23 is modified to

$$J_{th}(\text{A}/\text{cm}^2) = \frac{g_0 \Gamma}{\beta} + \frac{1}{\beta}\left[\alpha + \frac{1}{2L}\ln\left(\frac{1}{R_1 R_2}\right)\right] \tag{23a}$$

Thus, $\quad J_{th} = \dfrac{100 \times 0.9}{0.1} + 10 \times \left[100 + \dfrac{1}{2 \times 300 \times 10^{-4}}\ln\left(\dfrac{1}{0.44 \times 0.99}\right)\right] = 2036 \text{ A}/\text{cm}^2,$

and so $I_{th} = 2036 \times 300 \times 10^{-4} \times 5 \times 10^{-4} = 30$ mA. ◀

Temperature Effect

Figure 26 shows the temperature dependence of the threshold current I_{th} for a cw (continuous wave) stripe geometry Al$_x$Ga$_{1-x}$As-GaAs DH laser.[17] Figure 26a shows cw light outputs versus injection current at various temperatures between 25° and 115°C. Note the excellent linearity in the light-current characteristics. The threshold current at a given temperature is the extrapolated value for zero output power. Figure 26b shows a plot of threshold currents as a function of temperature. The threshold current increases exponentially with temperature as

$$I_{th} \sim \exp\left(\frac{T}{T_0}\right), \tag{24}$$

where T is the temperature in °C and T_0 is 110°C for this laser.

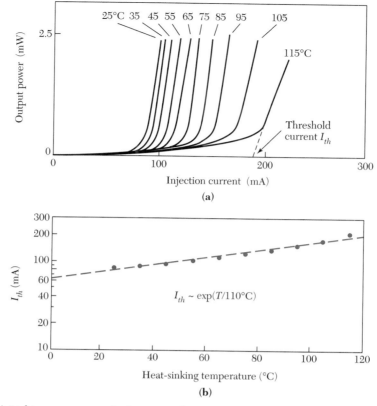

Fig. 26 (a) Light output versus diode current for a GaAs/AlGaAs heterostructure laser. (b) Temperature dependence of the continuous wave (cw)-current threshold. [17]

▷ **EXAMPLE 6**

Calculate the temperature at which the room-temperature value of the threshold current doubles for the laser shown in Fig. 26.

SOLUTION

$$\frac{J_{th}}{2J_{th}} = \frac{\exp\left(27/110\right)}{\exp\left(T/110\right)},$$

therefore $T = 27 + 110 \times \ln 2 = 27 + 76 = 103°C$. ◀

Modulation Frequency and Longitudinal Modes

For optical fiber communications, the optical source must be able to be modulated at high frequencies. Unlike LEDs, whose output power decreases with increasing modulation bandwidth (Eq. 12), the output power of typical GaAs or GaInAsP laser remains at a constant level (e.g., 10 mW per facet) well into GHz range.

For a stripe-geometry GaInAs-AlGaAs DH laser at a current above the threshold, many emission lines exist that are approximately evenly spaced with a separation of $\Delta\lambda$

(e.g., $\Delta\lambda = 4$ Å in Ex. 4). These emission lines belong to the longitudinal modes given in Eq. 17. Because of these longitudinal modes, the stripe geometry laser is not a spectrally pure light source. For optical-fiber communication systems, an ideal light source is one that has a single frequency. This is because light pulses of different frequencies travel through optical fiber at different speeds, thus causing pulse spread.

9.3.4 Distributed Feedback Lasers

Because of the multimodes in stripe-geometry lasers, these devices are only useful for telecommunication systems operated at relatively low rates (i.e., below 1 Gbit/s). For advanced optical fiber systems, *single-frequency* lasers are necessary. The fundamental approach is to take a laser cavity that allows only one mode to resonate and to provide a constructive interference mechanism that picks out a single frequency. Two laser configurations use this approach—the distributed Bragg reflector laser and the distributed feedback laser.[18]

Figure 27*a* shows the cross section of a distributed Bragg reflector (DBR) laser. The region which conducts electric current is called the pumped region. A wavelength-selection grating is placed outside the pumped region. Because of efficient coupling between the active region and the passive grating structure, the reflection is enhanced at the wavelength λ_B, known as the Bragg wavelength, which is related to the period of the grating Λ by

$$\lambda_B = \frac{2\overline{n}\Lambda}{l},\tag{25}$$

where $\overline{n}$ is the effective refractive index of the mode and l is the integer order of the grating. The mode at the Bragg wavelength that has the lowest loss, and thus, the lowest threshold gain will have the predominant output.

Figure 27*b* shows the distributed feedback (DFB) laser, which has a corrugated grating structure within the active region. The grating region has a periodically varying index of refraction that enhances the wavelength closest to the Bragg wavelength, thus achieving single-frequency operation. Because of the small temperature dependence of the refractive index, the lasing wavelength of the DFB laser has a very small temperature coefficient

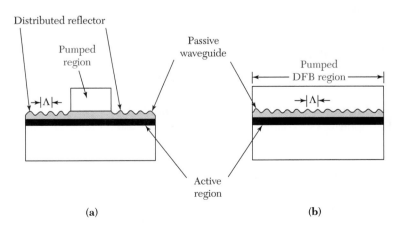

Fig. 27 Two methods of obtaining a single-frequency laser. (*a*) Distributed Bragg reflector (DBR) laser, and (*b*) a distributed feedback (DFB) laser.

(~0.5 Å/°C), while the temperature coefficient for a corresponding stripe-geometry laser is much larger (~3 Å/°C), because it follows the temperature dependence of the bandgap. DBR and DFB lasers are also useful as optical sources in integrated optics, which uses miniature optical waveguide components and circuits made by planar technology on rigid substrates.

9.3.5 Quantum-Well Lasers

The structure of a quantum-well (QW) laser[18,19] is similar to that of a DH laser except the thickness of the active layer in a QW laser is very small, about 10–20 nm. Figure 28a shows the band diagram of a QW laser where the central GaAs region ($L_y \cong 20$ nm) is sandwiched between two larger bandgap AlGaAs layers. The length L_y is comparable to the de Broglie wavelength ($\lambda = h/p$, where h is the Planck constant and p is the momentum of the charge carrier), and the carriers are confined in a finite potential well in the y-direction.

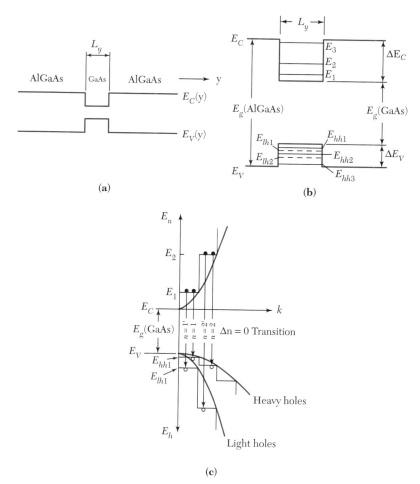

Fig. 28 The quantum-well (QW) laser: (*a*) single GaAs QW surrounded by AlGaAs, (*b*) discrete energy levels within the well, and (*c*) density of states for electrons and holes within the well.

The energy of the charge particle can be separated into a confinement component in the y-direction and two unconfined components in the x- and z-direction of the QW layer.

$$E(n, k_x, k_z) = E_n + \frac{\hbar^2}{2m^\circ}(k_x^2 + k_z^2), \tag{26}$$

where E_n is the nth eigenvalue of the confined particle, m° is the effective mass, and k_x and k_z are the wave number in the x- and z-direction, respectively. Figure 28b shows the energy levels in the quantum well. The values of E_n are shown as E_1, E_2, E_3 for electrons, E_{hh1}, E_{hh2}, E_{hh3}, for heavy holes,[§] and E_{lh1}, E_{lh2} for light holes.[18] The usual parabolic forms for the conduction and valence band density of states have been replaced by a "staircase" representation of discrete levels (Fig. 28c), each corresponding to a constant density of states per unit area given by

$$\frac{dN}{dE} = \frac{m^\circ}{\pi\hbar^2}. \tag{27}$$

Since the density of states is constant, rather than gradually increasing from zero, as in a conventional laser, there is a group of electrons of nearly the same energy available to recombine with a group of holes of nearly the same energy, for example, the level E_1 in the conduction with the level E_{hh1} in the valence band. QW lasers offer significant improvement in laser performance, such as reduction in the threshold current, high output power, and high speed, compared with the conventional DH lasers. QW lasers made in GaAs/ AlGaAs material systems have threshold current densities as low as 65 A/cm^2 and submilliampere threshold currents. These lasers operate at emission wavelengths around 0.9 μm.

For longer-wavelength operation, GaInAs/GaInAsP multiple–quantum-well (MQW) lasers for the 1.3 μm and 1.5 μm wavelength regions have been developed. Figure 29a shows a schematic diagram of a separate-confinement-heterostructure (SCH) MQW laser where four QWs of GaInAs with GaInAsP barrier layers are sandwiched between the InP cladding layers to form a waveguide with a step index change.[20] These alloy compositions are chosen so that they are lattice matched to the InP substrate. The active region is composed of four 8 nm thick, undoped GaInAs QWs (with E_g of 0.75 eV) separated by 30 nm thick undoped GaInAsP layers (with E_g of 0.95 eV). Figure 29b shows the corresponding band diagram of the active region. The n- and p-cladding InP layers are doped with sulfur (10^{18} cm^{-3}) and zinc (10^{17}cm^{-3}), respectively.

A graded-index SCH (GRIN-SCH) is shown in Fig. 29c, in which a GRIN of the waveguide is accomplished by several small stepwise increases of the bandgap energies of multiple cladding layers. The GRIN-SCH structure confines both the carriers and the optical field more effectively than the SCH structure and, consequently, leads to an even lower-threshold current density. With the MQW structure, a variety of advanced lasers and photonic integrated circuits becomes possible for future system applications.

[§] In GaAs, the effective mass for heavy holes is 0.62 m_0, whereas that for light holes is 0.074 m_0.

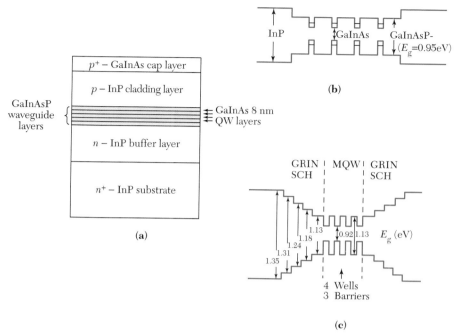

Fig. 29 (*a*) Schematic of the cross section of an GaInAs/GaInAsP multiple-quantum-well laser structure. (*b*) Schematic of the bandgaps of the SCH-MQW layers shown in (*a*). (*c*) GRIN-SCH-MQW structure with thin layers of increasing bandgaps to approximate the graded-index change. [20]

9.4 PHOTODETECTOR

Photodetectors are semiconductor devices that can convert optical signals into electrical signals. The operation of a photodetector involves three steps: carrier generation by incident light, carrier transport and/or multiplication by whatever current gain mechanism may be present, and interaction of current with the external circuit to provide the output signal.

Photodetectors have a broad range of applications, including infrared sensors in opto-isolators and detectors for optical-fiber communications. For these applications, the photodetectors must have high sensitivity at the operating wavelengths, high response speed, and low noise. In addition, the photodetector should be compact, use low biasing voltages or currents, and be reliable under the required operating conditions.

9.4.1 Photoconductor

A photoconductor consists simply of a slab of semiconductor with ohmic contacts at both ends of the slab (Fig. 30). When incident light falls on the surface of the photoconductor, electron-hole pairs are generated either by band-to-band transition (intrinsic) or by transitions involving forbidden-gap energy levels (extrinsic), resulting in an increase in conductivity.

For the intrinsic photoconductor, the conductivity is given by

$$\sigma = q\left(\mu_n n + \mu_p p\right), \tag{28}$$

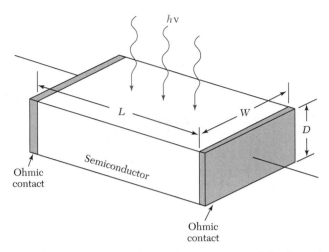

Fig. 30   Schematic diagram of a photoconductor that consists of a slab of semiconductor and two contacts at the ends.

and the increase in conductivity under illumination is mainly due to the increase in the number of carriers. The long-wavelength cutoff for an intrinsic photoconductor is given by Eq. 9. For the extrinsic photoconductor, photoexcitation may occur between the band edge and an energy level in the energy gap. In this case, the long-wavelength cutoff is determined by the depth of the forbidden-gap energy level.

Consider the operation of a photoconductor under illumination. At time zero, the number of carriers generated in a unit volume by a given photon flux is n_0. At a later time t, the number of carriers $n(t)$ in the same volume decays by recombination as

$$n = n_0 \exp\left(\frac{-t}{\tau}\right),$$ (29)

where τ is the carrier lifetime. From Eq. 29 the recombination rate is

$$\left|\frac{dn}{dt}\right| = \frac{1}{\tau}n_0 \exp\left(-\frac{t}{\tau}\right) = \frac{n}{\tau}.$$ (30)

If we assume a steady flow of photon flux impinging uniformly on the surface of a photoconductor (Fig. 30) with an area $A = WL$, the total number of photons arriving at the surface is $(P_{opt}/h\nu)$ per unit time, where P_{opt} is the incident optical power and $h\nu$ is the photon energy. At steady state, the carrier-generation rate G must be equal to the recombination rate n/τ. If the detector thickness D is much larger than the light penetration depth $1/\alpha$ the total steady-state carrier-generation rate per unit volume is

$$G = \frac{n}{\tau} = \frac{\eta\left(P_{opt}/h\nu\right)}{WLD},$$ (31)

where η is the quantum efficiency, the number of carriers generated per photon, and n is the carrier density, the number of carriers per unit volume. The photocurrent flowing between the electrodes is

$$I_p = \left(\sigma\mathscr{E}\right)WD = \left(q\mu_n n\mathscr{E}\right)WD = \left(qn\nu_d\right)WD,$$ (32)

where $\mathscr{E}$ is the electric field inside the photocoductor and v_d is the carrier drift velocity. Substituting n in Eq. 31 into Eq. 32 gives

$$I_p = q \left(\eta \frac{P_{opt}}{h\nu} \right) \cdot \left(\frac{\mu_n \tau \mathscr{E}}{L} \right). \tag{33}$$

If we define the primary photocurrent as

$$I_{ph} \equiv q \left(\eta \frac{P_{opt}}{h\nu} \right), \tag{34}$$

the photocurrent gain from Eq. 33 is

$$\boxed{\text{Gain} \equiv \frac{I_p}{I_{ph}} = \frac{\mu_n \tau \mathscr{E}}{L} = \frac{\tau}{t_r},} \tag{35}$$

where $t_r \equiv L/v_d = L/\mu_n \mathscr{E}$ is the carrier transit time. The gain depends on the ratio of carrier lifetime to the transit time.

▶ **EXAMPLE 7**

Calculate the photocurrent and gain when 5×10^{12} photons/s arriving at the surface of a photoconductor of $\eta = 0.8$. The minority carrier lifetime is 0.5 ns, and the device has $\mu_n = 2500$ cm^2/V-s , $\mathscr{E} = 5000$ V/cm, and $L = 10$ μm.

SOLUTION From Eq. 33,

$$I_p = q \left(0.8 \times 5 \times 10^{12} \text{ photons / s} \right) \cdot \left(\frac{2500 \text{ cm}^2/\text{V-s} \cdot 5 \times 10^{-10}\text{s} \cdot 5000 \text{ V/cm}}{10 \times 10^{-4}\text{cm}} \right),$$

$$= 4 \times 10^{-6} \text{A} = 4 \text{ μA,}$$

and from Eq. 35,

$$\text{Gain} = \frac{\mu_n \tau \mathscr{E}}{L} = \frac{2500 \cdot 5 \times 10^{-10} \cdot 5000}{10 \times 10^{-4}} = 6.25. \qquad ◀$$

For a sample with long minority-carrier lifetime and short electrode spacing, the gain can be substantially greater than unity. Gains as high as 10^6 can be obtained from some photoconductors. The response time of a photoconductor is determined by the transit time t_r. To achieve short transit time, small electrode spacing and a high electric field must be used. The response times of photoconductors cover a wide range, from 10^{-3} to 10^{-10} seconds. They are extensively used for infrared detection especially for wavelengths greater than a few microns.

9.4.2 Photodiode

A photodiode is basically a p-n junction or a metal-semiconductor contact operated under reverse bias. When an optical signal impinges on the photodiode, the depletion region serves to separate the photogenerated electron-hole pairs and an electric current flows in the external circuit. For high-frequency operation, the depletion region must be kept thin to reduce the transit time. On the other hand, to increase the quantum efficiency, the depletion layer must be sufficiently thick to allow a large fraction of the incident light to be absorbed. Thus, there is a trade-off between the response speed and quantum efficiency.

Quantum Efficiency

The quantum efficiency as mentioned previously is the number of electron-hole pairs generated for each incident photon:

$$\eta = \left(\frac{I_p}{q} \right) \cdot \left(\frac{P_{opt}}{h\nu} \right)^{-1}, \tag{36}$$

where I_p is the photogenerated current from the absorption of incident optical power P_{opt} at a wavelength λ (corresponding to a photon energy $h\nu$). One of the key factors that determine η is the absorption coefficient μ α (Fig. 5). Since α is a strong function of the wavelength, the wavelength range in which appreciable photocurrent can be generated is limited. The long-wavelength cutoff λ_c is established by the bandgap, Eq. 9, and is about 1.8 μm for germanium and 1.1 μm for silicon. For wavelengths longer than λ_c, the values of α are too small to give appreciable band-to-band absorption. The short-wavelength cutoff of the photoresponse comes about because the values of α are very large (~10^5 cm^{-1}), and hence the radiation is mostly absorbed very near the surface where recombination time is short. Therefore, the photocarriers can recombine before they can be collected in the p–n junction.

Figure 31 shows typical plots of quantum efficiency versus wavelength for some high-speed photodiodes.[21,22] Note that in the ultraviolet and visible region, metal-semiconductor photodiodes show good quantum efficiencies. In the near-infrared region, silicon photodiodes (with an antireflection coating) can reach 100% quantum efficiency near the 0.8- to 0.9-μm region. In the 1.0- to 1.6-μm region, germanium photodiodes and Group

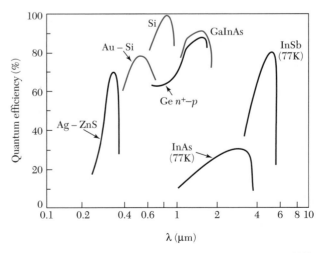

Fig. 31   Quantum efficiency versus wavelength for various photodetectors.[21,22]

III-V photodiodes (e.g., GaInAs) have shown high quantum efficiencies. For even longer wavelengths, photodiodes are cooled (e.g., to 77 K) for high-efficiency operation.

Response Speed

The response speed is limited by three factors: (1) diffusion of carriers, (2) drift time in the depletion region, and (3) capacitance of the depletion region. Carriers generated outside the depletion region must diffuse to the junction resulting in considerable time delay. To minimize the diffusion effect, the junction should be formed very close to the surface. The greatest amount of light will be absorbed when the depletion region is wide. However, the depletion layer must not be too wide or transit time effects will limit the frequency response. It also should not be too thin, or excessive capacitance C will result in a large RC time constant, where R is the load resistance. The optimal compromise is the width at which the depletion layer transit time is approximately one half the modulation period. For example, for a modulation frequency of 2 GHz, the optimal depletion layer thickness in silicon (with a saturation velocity of 10^7 cm/s) is about 25 μm.

p-i-n Photodiode

The *p-i-n* photodiode is one of the most common photodetectors because the depletion region thickness (the intrinsic layer) can be tailored to optimize the quantum efficiency and frequency response. Figure 32*a* shows a cross section of a *p-i-n* photodiode that has an antireflection coating to increase quantum efficiency.

Figures 32*b* and 32*c* show the energy band diagram of the *p-i-n* diode under reverse-bias condition and its optical absorption characteristics. Light absorption in the semiconductor produces electron-hole pairs. Pairs produced in the depletion region or within a diffusion length of it will eventually be separated by the electric field as shown in Fig. 32*b*, whereby a current flows in the external circuit as carriers drift across the depletion layer.

▷ **EXAMPLE 8**

On reaching the surface of the semiconductor, the incident optical power P_0 will have its level reduced to $P_0(1 - R)$ on entering the material, where R is the reflection coefficient. On passing though the semiconductor the light will be absorbed, and so at any depth x the amount of residual optical power $P(x)$ is given by $P(x) = P_0(1 - R) \exp(-\alpha x)$. For $\alpha = 10^4$ cm^{-1} and $R = 0.1$, calculate the depth at which half the incident optical power has been absorbed in a material.

SOLUTION

$$x = \frac{-1}{\alpha} \ln\left[\frac{P(x)}{P_0(1-R)} \right] = -10^{-4} \cdot \ln\left(\frac{1}{2 \times 0.9} \right) \text{ cm}$$

$$= 0.59 \text{ μm}.$$

◄

Metal-Semiconductor Photodiode

The construction of a high-speed metal-semiconductor photodiode is shown in Fig. 33. To avoid large reflection and absorption losses when the diode is illuminated through the metal contact, the metal film must be very thin (~10 nm) and an antireflection coating must be used. Metal-semiconductor photodiodes are particularly useful in the ultraviolet- and visible-light regions. In these regions the absorption coefficients, α, in most of the common semiconductors are very high, of the order of 10^4 cm^{-1} or more, which corresponds to an effective absorption length $1/\alpha$ of 1.0 μm or less. It is possible to choose

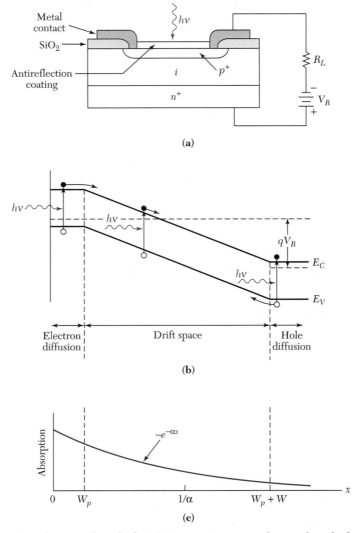

Fig. 32 Operation of a *p-i-n* photodiode (*a*) Cross-section view of *p-i-n* photodiode. (*b*) Energy band diagram under reverse bias. (*c*) Carrier absorption characteristics.

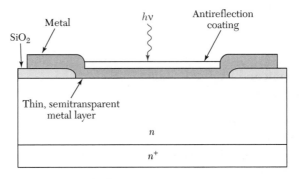

Fig. 33 Metal-semiconductor photodiode.

a metal and an antireflection coating so that a large fraction of the incident radiation will be absorbed near the surface of the semiconductor. As an example, for a gold-silicon photodetector having 10 nm gold and 50 nm zinc sulfide as the antireflection coating, more than 95% of the incident light with $\lambda = 0.6328$ μm (helium-neon laser wavelength, red light) will be transmitted into the silicon substrate.

Heterojunction Photodiode

Another photodiode structure is a heterojunction device formed by depositing a large-bandgap semiconductor epitaxially on a smaller-handgap semiconductor. One advantage of a heterojunction photodiode is that the quantum efficiency does not depend critically on the distance of the junction from the surface, because the large-bandgap material can be used as a window for the transmission of optical power. In addition, the heterojunction can provide unique material combinations so that the quantum efficiency and response speed can be optimized for a given optical-signal wavelength.

To obtain a heterojunction with low leakage current, the lattice constants of the two semiconductors must be closely matched. Ternary III-V compounds $Al_xGa_{1-x}As$ epitaxially grown on GaAs can form heterojunctions with perfectly matched lattices. These heterojunction are important for photonic devices operated in the wavelength range from 0.65 to 0.85 μm. At longer wavelengths (1 to 1.6 μm), ternary compounds such as $Ga_{0.47}In_{0.53}As$ (with $E_g = 0.75$ eV) and quaternary compounds such as $Ga_{0.27}In_{0.73}As_{0.63}P_{0.37}$ (with $E_g = 0.95$ eV) can be used. These compounds have a nearly perfect lattice match to an InP substrate. The quantum efficiency is greater than 70% over the wavelength range from 1 to 1.6 Mm as shown in Fig. 31 (GaInAs curve).

9.4.3 Avalanche Photodiode

An avalanche photodiode (APD) is operated under a reverse-bias voltage that is sufficient to enable avalanche multiplication. The multiplication results in internal current gain and the device can respond to light modulated at frequencies as high as microwave frequencies.

One important consideration in the design of an APD is the need to minimize avalanche noise. The avalanche noise comes about from the random nature of the avalanche multiplication process in which every electron-hole pair generated at a given distance in the depletion region does not experience the same multiplication. The avalanche noise depends on the ratio of the ionization coefficients α_p/α_n; the smaller the ratio, the smaller is the avalanche noise. This is because when $\alpha_p = \alpha_n$, each incident photocarrier results in three carriers in the multiplicating region: the primary carrier and its secondary hole and electron. A fluctuation that changes the number of carriers by one represents a large percentage change, and the noise will be large. On the other hand, if one of the ionization coefficients approaches zero (e.g., $\alpha_p \rightarrow 0$), each incident photocarrier can result in a large number of carriers in the multiplication region. In this case, a fluctuation of one carrier is a relatively insignificant perturbation. To minimize the avalanche noise, we should use semiconductors with a large difference in α_p and α_n. The noise factor is given by

$$F = M\left(\frac{\alpha_p}{\alpha_n}\right) + \left(2 - \frac{1}{M}\right)\left(1 - \frac{\alpha_p}{\alpha_n}\right) \tag{37}$$

where M is the multiplication factor. We can see from Eq. 37 that when $\alpha_p = \alpha_n$, the noise factor has a maximum value of M; while for $\alpha_p/\alpha_n = 0$ and for a large M, the minimum noise factor is 2.

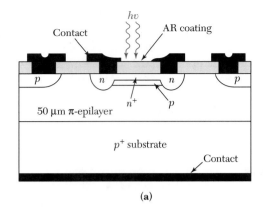

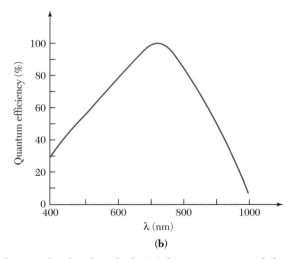

Fig. 34 A typical silicon avalanche photodiode: (*a*) device structure and (*b*) quantum efficiency.

Figure 34*a* shows the structure of a typical silicon APD having a n^+-p-π-p^+ doping profile (π is a lightly doped *p*-region). The quantum efficiency is near 100 % at a wavelength of about 0.75 μm for a device having a SiO_2-Si_3N_4 antireflection coating (Fig. 34*b*). Because the ratio of α_p/α_n is about 0.04, the noise factor obtained from Eq. 37 is 2.3 for $M = 10$.

▷ 9.5 SOLAR CELL

Solar cells are useful for both space and terrestrial applications. Solar cells furnish the long-duration power supply for satellites. The solar cell is an important candidate for an alternative terrestrial energy source because it can convert sunlight directly to electricity with good conversion efficiency, can provide nearly permanent power at low operating cost, and is virtually nonpolluting.[23,24]

9.5.1 Solar Radiation

The radiative energy output from the sun derives from a nuclear fusion reaction. In every second, about 6×10^{11} kg hydrogen is converted to helium, with a net mass loss of about

4×10^3 kg. The mass loss is converted through the Einstein relation ($E = mc^2$) to 4×10^{20} J. This energy is emitted primarily as electromagnetic radiation in the ultraviolet to infrared region (0.2 to 3 μm). The total mass of the sun is now about 2×10^{30} kg, and a reasonably stable life with a nearly constant radiative-energy output of over 10 billion (10^{10}) years is projected.

The intensity of solar radiation outside the earth's atmosphere, at the average distance of its orbit around the sun, is defined as the solar constant and has a value of 1367 W/m². Terrestrially, the sunlight is attenuated by clouds and by atmospheric scattering and absorption. The attenuation depends primary on the length of the light's path through the atmosphere, or the mass of air through which it passes. The "*air mass*" is defined as $1/\cos\phi$, where ϕ is the angle between the vertical and the sun's position.

▶ **EXAMPLE 9**

The air mass can most easily be estimated from the length of the shadow, s, of a vertical structure of height, h, as $\sqrt{1 + (s/h)^2}$. If $s = 1.118$ m and $h = 1.00$ m, find the air mass.

SOLUTION

$$\sqrt{1 + (1.118 / 1.0)^2} = \sqrt{2.25} = 1.5.$$

We have an air mass 1.5 (AM 1.5). The corresponding $\cos\phi$ is $1/1.5 = 0.667$ and the angle ϕ between the vertical and the sun's position is $\cos^{-1}(0.667) = 48°$. The maximum sunlight intensity occurs when the sun is straight overhead (i.e., AM 1.0 with $\phi = 0°$) ◂

Figure 35 shows two curves related to solar spectral irradiance (power per unit area per unit wavelength).[25] The upper curve, which represents the solar spectrum outside the Earth's atmosphere, is the air mass zero condition (AM0). The AM0 spectrum is relevant for satellite and space vehicle applications. Terrestrial solar-cell performance is

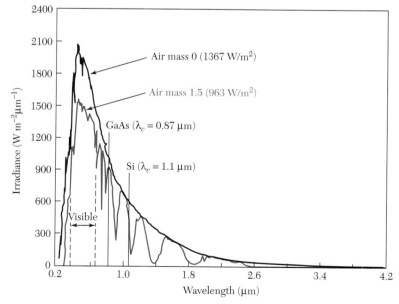

Fig. 35 Solar spectral irradiance [25] at air mass 0 and air mass 1.5 and the cutoff wavelength of GaAs and Si.

specified with reference to the air mass 1.5 (AM1.5) spectrum. This spectrum represents the sunlight at the Earth's surface when the sun is at an angle of 48° from the vertical. At this angle the incident power is about 963 W/m².

9.5.2 *p–n* Junction Solar Cell

A schematic representation of a *p–n* junction solar cell is shown in Fig. 36. It consists of a shallow *p–n* junction formed on the surface, a front ohmic contact stripe and fingers, a back ohmic contact that covers the entire back surface, and an antireflection coating on the front surface.

When the cell is exposed to the solar spectrum, a photon that has an energy less than the bandgap E_g makes no contribution to the cell output. A photon that has energy greater than E_g contributes an energy E_g to the cell output. Energy greater than E_g is wasted as heat. To derive the conversion efficiency, we shall consider the energy band diagram of a *p–n* junction under solar radiation shown in Fig. 37a. The equivalent circuit is shown in Fig. 37b, where a constant-current source is in parallel with the junction. The source I_L results from the excitation of excess carriers by solar radiation, I_s is the diode saturation current, and R_L is the load resistance.

The ideal *I-V* characteristics of such a device are given by

$$I = I_s\left(e^{qV/kT} - 1\right) - I_L \tag{38}$$

and

$$J_s = \frac{I_s}{A} = qN_CN_V\left(\frac{1}{N_A}\sqrt{\frac{D_n}{\tau_n}} + \frac{1}{N_D}\sqrt{\frac{D_p}{\tau_p}}\right) \cdot e^{-E_g/kT}, \tag{38a}$$

where A is the device area. A plot of Eq. 38 is given in Fig. 38a for I_L = 100 mA, I_s = 1 nA, cell area A = 4 cm², and T = 300 K. The curve passes through the fourth quadrant

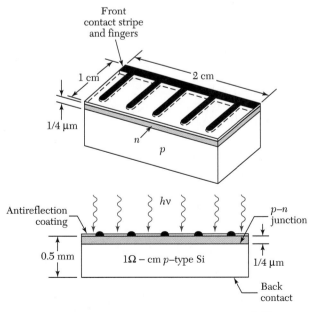

Fig. 36 Schematic representation of a silicon *p–n* junction solar cell. [23]

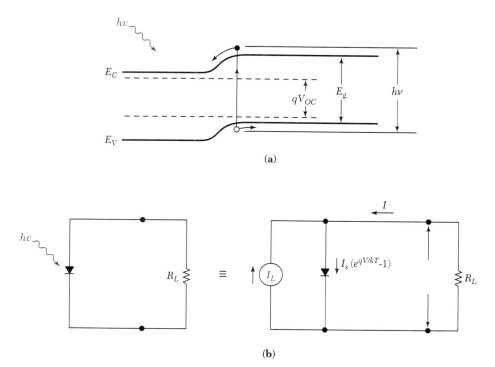

Fig. 37 (*a*) Energy band diagram of a *p–n* junction solar cell under solar irradiation. (*b*) Idealized equivalent circuit of a solar cell.

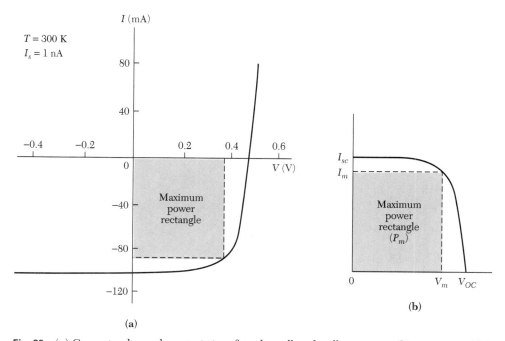

Fig. 38 (*a*) Current-voltage characteristics of a solar cell under illumination. (*b*) Inversion of (*a*) about the voltage axis.

and ,therefore, power can be extracted from the device. The *I-V* curve is more generally represented by Fig. 38*b*, which is an inversion of Fig. 38*a* about the voltage axis. By choosing a proper load, close to 80% of the product $I_{SC}V_{OC}$ can be extracted, where I_{SC} is the short-circuit current equal to I_L and V_{OC} is the open-circuit voltage of the cell; the shaded area in the figure is the maximum-power rectangle. Also defined in Fig. 38*b* are the quantities I_m and V_m that correspond to the current and voltage, respectively, for the maximum power output $P_m (= I_m \times V_m)$.

From Eq. 38 we obtain for the open-circuit voltage ($I = 0$)

$$V_{OC} = \frac{kT}{q}\ln\left(\frac{I_L}{I_s} + 1\right) \cong \frac{kT}{q}\ln\left(\frac{I_L}{I_s}\right). \tag{39}$$

Hence, for a given I_L, V_{OC} increases logarithmically with decreasing saturation current I_s. The output power is given by

$$P = IV = I_s V\left(e^{qV/kT} - 1\right) - I_L V. \tag{40}$$

The condition for maximum power is obtained when $dP/dV = 0$, or

$$V_m = \frac{kT}{q}\ln\left[\frac{1+(I_L/I_s)}{1+(qV_m/kT)}\right] \cong V_{OC} - \frac{kT}{q}\ln\left(1 + \frac{qV_m}{kT}\right), \tag{41a}$$

$$I_m = I_s\left(\frac{qV_m}{kT}\right)e^{qV_m/kT} \cong I_L\left(1 - \frac{1}{qV_m/kT}\right). \tag{41b}$$

The maximum output power P_m is then

$$P_m = I_m V_m \cong I_L\left[V_{OC} - \frac{kT}{q}\ln\left(1 + \frac{qV_m}{kT}\right) - \frac{kT}{q}\right]. \tag{42}$$

► **EXAMPLE 10**

Calculate the open-circuit voltage and the output power at a voltage of 0.35 V for a solar cell shown in Fig. 38*a*.

SOLUTION   From Eq. 39

$$V_{oc} = (0.026 \text{ V})\ln\left(\frac{100\times10^{-3}\,\text{A}}{1\times10^{-9}\,\text{A}}\right) = 0.48 \text{ V}.$$

The output power at 0.35 V is given by Eq. 40 (note that I_S and I_L are reverse currents so we need negative signs for them):

$$P = \left(-10^{-9}\,\text{A}\right)\cdot(0.35 \text{ V})\left(e^{0.35/0.026} - 1\right) - \left(-0.1 \text{ A}\right)\cdot(0.35 \text{ V}) = 3.48\times10^{-2} \text{ W}. \qquad ◄$$

9.5.3 Conversion Efficiency

Ideal efficiency

The power conversion efficiency of a solar cell is given by

$$\eta = \frac{I_m V_m}{P_{in}} = \frac{I_L \left[V_{OC} - \frac{kT}{q} \ln\left(1 + \frac{qV_m}{kT}\right) - \frac{kT}{q} \right]}{P_{in}} \tag{43}$$

or

$$\boxed{\eta = \frac{FF \cdot I_L V_{OC}}{P_{in}},} \tag{43a}$$

where P_{in} is the incident power and FF is the fill factor defined as

$$FF \equiv \frac{I_m V_m}{I_L V_{OC}} = 1 - \frac{kT}{qV_{OC}} \ln\left(1 + \frac{qV_m}{kT}\right) - \frac{kT}{qV_{OC}}. \tag{44}$$

The fill factor is the ratio of the maximum power rectangle (Fig. 38b) to the rectangle of $I_{SC} \times V_{OC}$. To maximize the efficiency, we should maximize all three items in the numerator of Eq. 43a.

The ideal efficiency can be obtained from the ideal I-V characteristics defined by Eq. 38. For a given semiconductor, the saturation current density is obtained from Eq. 38a. For a given air mass condition (e.g., AM 1.5), the short-circuit current I_L is the product q and the number of the available photons with energy $h\nu \geq E_g$ in the solar spectrum. Once I_s and I_L are known, the output power P and the maximum power P_m can be obtained from Eqs. 40 to 42. The input power P_{in} is the integration of all the photons in the solar spectrum (Fig. 35). Under AM 1.5 condition, the efficiency P_m/P_{in} has a broad maximum[24,26] of about 29% and does not depend critically on E_g. Therefore, semiconductors with bandgap between 1 and 2 eV can all be considered solar cell materials.

Spectrum Splitting

The simplest way to exceed the efficiency limit is by "spectrum splitting". By splitting sunlight into narrow wavelength bands and directing each band to a cell that has a bandgap optimally chosen to convert just this band, as shown in Fig. 39a, efficiency above 60% is possible in principle.[27] Fortunately, simply stacking cell on top of one another with the highest bandgap cell uppermost, as in Fig. 39b, automatically achieves an identical spectral-splitting effect, making this "tandem" cell approach a reasonably practical way of increasing cell efficiency.

Series Resistance and Recombination Current

Many factors degrade the ideal efficiency. One of the major factors is the series resistance R_s from the ohmic loss in the front surface. The equivalent circuit is shown in Fig.40. From the ideal diode current given by Eq. 38, the I-V characteristics are found to be

$$\ln\left(\frac{I + I_L}{I_s} + 1\right) = \frac{q}{kT}\left(V - IR_s\right). \tag{45}$$

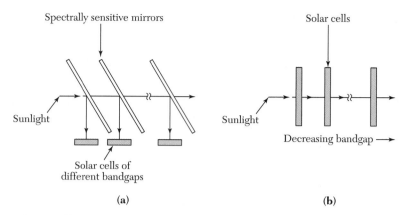

Fig. 39 Multigap cell concepts. (*a*) Spectrum splitting approach. (*b*) Tandem-cell approach.[27]

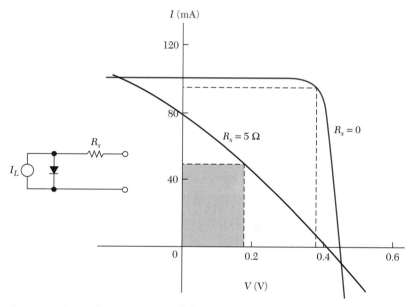

Fig. 40 Current-voltage characteristics and the equivalent circuit of solar cells that have resistances.

Plots of this equation are shown in Fig. 40, with $R_s = 0$ and 5 Ω and where the other parameter I_s, I_L, and T are the same as those in Fig. 38. It can he seen that a series resistance of only 5Ω reduces the available power to less than 30% of the maximum power with $R_s = 0$. The output current and output power are

$$I = I_s \left\{ \exp \left[\frac{q(V - IR_s)}{kT} \right] - 1 \right\} - I_L, \tag{46}$$

$$P = I \left[\frac{kT}{q} \ln \left(\frac{I + I_L}{I_s} \right) + IR_s \right]. \tag{47}$$

The series resistance depends on the junction depth, the impurity concentrations of *p*-type and *n*-type regions, and the arrangement of the front-surface ohmic contacts. For a typical silicon solar cell with the geometry shown in Fig. 36, the series resistance is about 0.7 Ω for n^+-p cells and 0.4 Ω for p^+-n cells. The difference in resistance is mainly the result of the lower resistivity in n-type substrates.

Another factor is the recombination current in the depletion region. For single-level centers, the recombination current can be expressed as

$$I_{rec} = I'_s \left[\exp\left(\frac{qV}{2kT} \right) - 1 \right]$$ (48)

and

$$\frac{I'_s}{A} = \frac{qn_i W}{\sqrt{\tau_p \tau_n}},$$ (48a)

where I'_s is the saturation current. The energy conversion equation can be put into a closed form yielding equations similar to Eqs. 39 through 42 with the exception that I_s is replaced by I'_s and the exponential factor is divided by 2. The efficiency for the recombination current case is found to be much less than the ideal current due to the degradation of both V_{OC} and the fill factor. For silicon solar cells at 300 K, the recombination current can cause a 25% reduction in efficiency.

9.5.4 Silicon and Compound-Semiconductor Solar Cells

Silicon is the most important semiconductor for solar cells. It is nontoxic and is second only to oxygen in the earth's crust. Therefore, silicon poses minimal environmental or resource depletion risks if used on a large scale. It also has a well established technological base because of its use in microelectronics.

III-V compound semiconductors and their alloy systems provide wide choices of bandgaps with closely matched lattice constants. These compounds are ideal for producing tandem solar cells. For example, AlGa/GaAs, GaInP/GaAs, and GaInAs/InP material systems have been developed for solar cells in satellite and space vehicle applications.

PERL Cell

The silicon *p*assivated *e*mitter *r*ear *l*ocally-diffused (PERL) cell[28] is shown in Fig.41. The cell has inverted pyramids on the top that are formed by using anisotropic etches to expose the slowly etching (111) crystallographic planes. The pyramids reduce reflections of light incident on the top surface, since light incident perpendicularly to the cell will strike one of the inclined (111) planes obliquely, and will be refracted obliquely into the cell. The rear contact is separated from the silicon by an intervening oxide layer. This gives much better rear reflection than an aluminum layer. To date, the PERL cell shows the highest conversion efficiency of 24%.

Amorphous Si Solar Cell

The basic cell structure for a series interconnected *a*-Si solar cells is shown[29] in Fig. 42. A layer of SiO_2 followed by a transparent conducting layer of a large bandgap, degenerately doped semiconductor such as SnO_2 is deposited onto a glass substrate and patterned using a laser. The substrate is then coated by a *p-i-n* junction stack of amorphous silicon by the decomposition of silane in a radio-frequency plasma-discharge system. After deposition, the α-Si layers are patterned in a second laser system. A layer of aluminum is sputtered onto the rest of the silicon and this layer is also patterned by laser. This technique forms a series of interconnected cells, as shown in Fig. 42. The cell has the lowest manufacturing cost and has a modest efficiency of 5%.

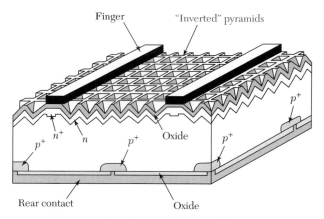

Fig. 41 Passivated emitter near locally diffused (PERL) cell.[24]

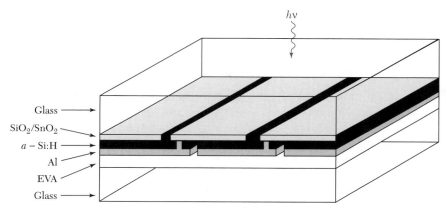

Fig. 42 Series-interconnected *a*-Si solar cells deposited on a glass substrate with a rear glass cover bonded using ethylene vinyl acetate (EVA). [29]

Tandem Solar Cell

Figure 43 shows the structure of a monolithic tandem solar cell.[24] A *p*-type germanium is used as the substrate, which has a lattice constant very close to that of GaAs and $Ga_{0.51}In_{0.49}P$. A *p*-GaInP layer is then grown to reduce the minority-carrier concentration near the rear contact. The bottom junction is the GaAs *p-n* junction (E_g = 1.42 eV), and the top junction is the GaInP junction (E_g = 1.9 eV). A tunneling p^+-n^+ GaAs junction is placed between the top and bottom junctions to connect the cells. Tandem solar cells with efficiency as high as 30% have been obtained.

9.5.5 Optical Concentration

Sunlight can be focused by using mirrors and lenses. Optical concentration offers an attractive and flexible approach to reducing high cell costs by substituting a concentrator area for much of the cell area. It also offers other advantages, such as a 20% increase in efficiency for a concentration of 1000 suns (an intensity of 963×10^3 W/m²).

Figure 44 shows the measured results of a typical silicon solar cell mounted in a concentrated system.[30] Note that device performances improve as the concentration increases

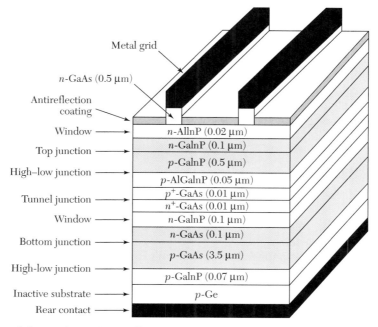

Fig. 43 Monolithic tandem solar cell.[24]

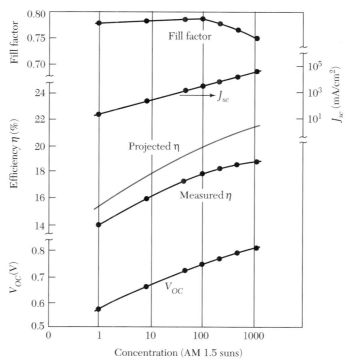

Fig. 44 Efficiency, open-circuit voltage, short-circuit current, and fill factor versus solar concentration. [30]

from one sun toward 1000 suns. The short-circuit current density increases linearly with concentration. The open-circuit voltage increases at a rate of 0.1 V per decade, while the fill factor varies slightly. The efficiency, which is the product of the foregoing three factors divided by the input power, increases at a rate of about 2% per decade. With a proper antireflection coating, we project an efficiency increase of 30% at 1000 suns. Therefore, one cell operated under 1000-sun concentration can produce the same power output as 1300 cells under one sun. Potentially, the optical concentration approach can replace expensive solar cells with less expensive concentrator materials and a related tracking and heat removal system to minimize the overall system cost.

▶ SUMMARY

The operation of the four groups of photonic devices — LEDs, laser diodes, photodetectors, and solar cells depends upon the emission or absorption of photons. Photons are emitted due to the recombination of charge carriers, and photons are absorbed to create charge carriers.

LEDs are p–n junctions that can emit spontaneous radiation due to recombinations of electrons and holes in a forward-biased junction. Visible LEDs can emit radiation with photon energies in the range of 1.8 to 2.8 eV, corresponding to wavelengths from 0.7 to 0.4 um. They are used extensively for display and various electronic instruments. By combining LEDs of different colors (i.e., red, green, and blue), we can form white LEDs which are useful for general illumination. Organic semiconductors can also be used for display applications. OLED is particularly useful for multi-color large-area flat-panel displays. Infrared LEDs can emit radiation with $h\nu < 1.8$ eV. They are used for opto-isolator and short-distance optical-fiber communication.

The laser diode is also a p–n junction operated under forward bias condition. However, the diode structure must provide the confinement of the carriers and the optical field so that the stimulated emission condition can be established. Laser diodes have evolved from the homojunction, to the distributed feedback configuration, and to quantum-well structures. The main objectives are to lower the threshold current density for lasing and to have all emitted photons at a single frequency. The laser diode is a key device for long distance, optical fiber communication systems. It is also extensively used for video recording, high-speed printing, and optical reading.

Photodetectors include photoconductors, photodiodes and avalanche photodiodes. They can convert optical signals into electrical signals. When photons are absorbed, electron-holes pairs are generated in the device, which are subsequently separated by an electrical field to produce a photo-current flowing between the electrodes. Photodetectors are used for optical sensing and detection in opto-isolators and optical-fiber communication systems.

A solar cell is similar to a photodiode and has the same operational principle. However, the solar cell differs from a photodiode in that it is a large-area device and covers a wide range of optical spectrum (solar radiation). Solar cells furnish the long-duration power supply for satellites. The solar cell is a major candidate for a terrestrial energy source because it can convert sunlight directly to electricity with good efficiency and is environmentally benign. Currently, the important solar cells are the highly efficient silicon PERL cell (24%), the GaInP/GaAs tandem cell (30%), and the low cost, series-interconnected a-Si solar cell (5 %).

REFERENCE

1. H. C. Casey, Jr. and M. B. Panish, *Heterostructure Lasers*, Academic, New York, 1978.

2. H. Melchior, "Demodulation and Photodetection Techniques," in F. T. Arecchi and E. O. Schulz-Dubois, Eds., *Laser Handbook*, Vol. 1, North-Holland, Amsterdam, 1972.

3. M. G. Craford, "Recent Developments in LED Technology," *IEEE Trans. Electron Devices*, **ED-24**, 935 (1977).

4. W. O. Groves, A. H. Herzog, and M. G. Craford, "The Effect of Nitrogen Doping on GaAsP Electroluminescent Diodes," *Appl. Phys. Lett.*, **19**, 184 (1971).

5. S. Gage, et al., *Optoelectronic Application Manual*, McGraw-Hill, New York, 1977.

6. S. Nakamura and G. Fasol, *The Blue Laser Diode*, Wiley, New York, 1997.

7. R. H. Saul, T. P. Lee, and C. A. Burrus, "Light-Emitting Diode Device Design," in R. K. Willardon and A. C. Bear, Eds., *Semiconductor and Semimetals*, Academic, New York, 1984.

8. A. A. Bergh and P. J. Dean, *Light Emitting Diodes*, Clarendon, Oxford, 1976.

9. N. Bailey, " The Future of Organic Light-Emitting Diodes," *Inf. Disp.*, **16**, 12 (2000).

10. C. H. Chen, J. Shi, and E. W. Tang, " Recent Development, in Molecular Organic Electroluminescent Materials," *Macromal. Symp.* **125**, 1(1997).

11. S. E. Miller and A. G. Chynoweth, Eds., *Optical Fiber Communications*, Academic, New York, 1979.

12. W. T. Tsang, "High Speed Photonic Devices," in S. M. Sze, Ed., *High Speed Semiconductor Devices*, Wiley, New York, 1990.

13. O. Madelung, Ed., *Semiconductor-Group IV Elements and III-V Compounds*, Springer-Verlag, Berlin, 1991.

14. M. B. Panish, I. Hayashi, and S. Sumski, "Double-Heterostructure Injection Lasers with Room Temperature Threshold As Low As 2300 A/cm^2," *Appl. Phys. Lett.*, **16**, 326(1970).

15. T. E. Bell, "Single-Frequency Semiconductor Lases," *IEEE Spectrum*, **20**, 38(1983).

16. F. Stern, "Calculated Spectral Dependence of Gain in Excited GaAs," *J. Appl. Phys.*, **47**, 5328(1976).

17. W. T. Tsang, R. A. Logan, and J. P. Van der Ziel, "Low-Current-Threshold Stripe-Buried-Heterostructure Laser with Self-Aligned Current Injection Stripes," *Appl. Phys. Lett.*, **34**, 644 (1979).

18. N. Holonyak, et al., "Quantum Well Heterostructure Laser," *IEEE J. Quant. Electron.*, **QE-16**, 170 (1980).

19. T. P. Lee, "High Speed Photonic Devices," in S. M. Sze, Ed., *Modern Semiconductor Device Physics*, Wiley Interscience, New York, 1998.

20. K. Kasukawa, Y. Imajo, and T. Makino, "1.3 µm GaInAsP/InP Buried Heterostructure Graded Index Separate Confinement Multiple Quantums Well Lasers Epitaxially Grown by MOCVD," *Electron. Lett.*, **25**,104(1989).

21. S. M. Sze, *Physics of Semiconductor Devices*, 2nd ed., Wiley, New York, 1981, Ch. 12-14.

22. S. R. Forrest, "Photodiodes for Long-Wavelength Communication systems," *Laser Focus*, **18**, 81 (1982).

23. D. M. Chapin, C. S. Fuller, and G. L. Pearson, "A New Silicon p-n Junction Photocell for Converting Solar Radiation into Electrical Power," *J. Appl. Phys.*, **25**, 676 (1954).

24. M. S. Green, "Solar Cells" in S. M. Sze, Ed., *Modern Semiconductor Device Physics*, Wiley Interscience, New York, 1998.

25. R. Hulstrom, R. Bird, and C. Riordan, "Spectral Solar Irradiance Data Sets for Selected Terrestrial Conditions," *Solar Cells*, **15**, 365 (1985).

26. C. H. Henry, "Limiting Efficiency of Ideal Single and Multiple Energy Gap Terrestrial Solar Cells," *J. Appl. Phys.*, **51**, 4494 (1980).

27. A. Luque, Ed, *Physical Limitation to Photovoltaic Energy Conversion*, IOP Press, Philadelphia, 1990.

28. M. A. Green, *Silicon Solar Cells: Advanced Principles and Practice*, Bridge Printery, Sydney, 1995.

29. J. Macneil, et. al. "Recent Improvements in Very Large Area *α-Si* PV Module Manufacturing," in *Proc., 10th Euro. Photovolt. Sol. Energy Conf.*, Lisbon, 1188, 1991.

30. R. I. Frank, J. L. Goodich, and R. Kaplow, "A Novel Silicon High-Intensity Photovoltaic Cell," *Conf. Rec. 14th IEEE Photovolt. Conf.*, IEEE, New York, p. 1350 (1980).

▶ PROBLEMS (* DENOTES DIFFICULT PROBLEMS)

FOR SECTION 9.1   RADIATIVE TRANSITIONS AND OPTICAL ABSORPTION

*1. A GaAs sample is illuminated with a light having a wavelength of 0.6 μm. The incident power is 15 mW. If one-third of the incident power is reflected and another third exits from the other end of the sample, what is the thickness of the sample? Find the thermal energy dissipated per second to the lattice.

FOR SECTION 9.2   LIGHT EMITTING DIODE

2. The efficiency for electrical-to-optical conversion in a LED is given by $4\bar{n}_1\bar{n}_2(1 - \cos\theta_c)/(\bar{n}_1 + \bar{n}_2)^2$, where $\bar{n}_1$ and $\bar{n}_2$ are the refractive index of air and the semiconductor, respectively, and θ_c is the critical angle. Find the efficiency of an $Al_{0.3}Ga_{0.7}As$ LED operated at 0.898 μm.

FOR SECTION 9.3   SEMICONDUCTOR LASER

3. An InGaAsP Fabry–Perot laser operating at a wavelength of 1.33 μm has a cavity length of 300 μm. The index of refraction of InGaAsP is 3.39. (a) What is the mirror loss expressed in cm^{-1}? (b) If one of the laser facets is coated to produce 90% reflectivity, how much threshold current reduction (as a percentage) can be expected, assuming α = 10 cm^{-1}.

4. Calculate the confinement factor for a GaAs laser with an active region thickness 1 μm, refractive index 3.6, and critical angle at the active-to-nonactive boundary of 84°. Assume the C constant to be 8×10^7 m^{-1}. Repeat the calculation for a GaAs/AlGaAs DH laser, where all the factors remain unchanged except for the critical angle, which is now 78°.

5. Derive Eq. 17 for the separation $\Delta\lambda$ between the allowed modes in the longitudinal direction. For a GaAs laser diode operated at λ = 0.89 μm, with $\bar{n}_1$ = 3.58, L = 300 μm, and $d\bar{n}_1/d\lambda$ = 2.5 μm^{-1}, find $\Delta\lambda$.

6. Calculate the gain coefficient for the two cases of Prob. 4 if the cavity length is 100 μm, the absorption coefficient is 10^4 m^{-1}, and the end mirrors are cleaved. How much shorter can the cavity be and still produce the same gain if one end-mirror is metallized to produce a reflectivity of 0.99?

7. Calculate the threshold current for the two cases of Prob. 4 if one end-mirror's reflectivity is 0.99. The cavity width is 5 μm and the gain factor for case 1 is 0.1 cm^{-3}A^{-1}.

*8. For a DFB laser with a cavity length of 300 μm, a material reflective index of 3.4 and an oscillating wavelength of 1.33 μm, find the Bragg wavelength and grating periodicity. The oscillating wavelength λ_0 is given by $\lambda_0 = \lambda_B \pm \dfrac{\left(m+\dfrac{1}{2}\right)\lambda_B^{\,2}}{2\pi L}$, where m is an integer.

9. For high-temperature laser operation, it is important to have a low-temperature coefficient of the threshold current $\xi = (dI_{th}/dT)/dI_{th}$. What is the coefficient ξ for the laser shown in Fig. 26? If T_0 = 50°C, is this laser better or worse for high-temperature operation?

FOR SECTION 9.4   PHOTODETECTOR

10. A photoconductor with dimensions L = 6 mm, W = 2 mm, and D = 1 mm (Fig. 30) is placed under uniform radiation. The absorption of the light increases the current by 2.83 mA. A voltage of 10 V is applied across the device. As the radiation is suddenly cut off, the current falls, initially at a rate of 23.6 A/s. The electron and hole mobility are 3600 and 1700 cm²/V-s, respectively. Find (a) the equilibrium density of electron-hole pairs generated under radiation, (b) the minority-carrier lifetime, and (c) the excess density of electrons and holes remaining 1 ms after the radiation is cut off.

11. Calculate the gain and current generated when 1 μW of optical power with $h\nu$ = 3 eV is shone onto a photoconductor of η = 0.85 and a minority carrier lifetime of 0.6 ns. The material has an electron mobility of 3000 cm²/V-s, the electric field is 5000 V/cm, and L = 10 μm.

*12. Show that the quantum efficiency η of a p-i-n photodetector is related to the responsivity $R(= I_p/P_{opt})$ at a wavelength λ (μm) by the equation $R = \dfrac{\eta\lambda}{1.24}$.

*13. A silicon n^+-p-π-p^+ avalanche photodiode operated at 0.8 μm has a p-layer of 3 μm and a n-layer 9 μm thick. The biasing voltage must be high enough to cause avalanche breakdown in the p-region and velocity saturation in the π-region. Find the minimum required biasing voltage and the corresponding doping concentration of the p-region. Estimate the transit time of the device.

FOR SECTION 9.5   SOLAR CELL

14. A p-n junction photodiode can be operated under photovoltaic conditions similar to that of a solar cell. The current-voltage characteristics of a photodiode under illumination are also similar (Fig. 37). State three major differences between a photodiode and a solar cell.

*15. Consider a silicon p-n junction solar cell of area 2 cm². If the dopings of the solar cell are N_A = 1.7× l0¹⁶ cm⁻³ and N_D = 5×10¹⁹ cm⁻³, and given τ_n = 10 μs, τ_p = 0.5 μs, D_n = 9.3 cm²/s, D_p = 2.5 cm²/s, and I_L = 95 mA, (a) calculate and plot the I-V characteristics of the solar cell, (b) calculate the open-circuit voltage, and (c) determine the maximum output power of the solar cell, all at room temperature.

*16. At 300 K, an ideal solar cell has a short circuit current of 3 A and an open-circuit voltage of 0.6 V. Calculate and sketch its power output as a function of operating voltage and find its fill factor from this power output.

17. For the solar cell shown in Fig. 40, find the relative maximum power output for a R_s of 0 and 5 Ω.

18. For the solar cell operated under solar-concentration conditions (Fig. 44 with the measured η), how many such solar cells operated under one-sun conditions are needed to produce the same power output as one cell operated under a 10-sun, 100-sun, or 1000-sun concentration?

Crystal Growth and Epitaxy

As discussed in Chapter 2, the two most important semiconductors for discrete devices and integrated circuits are silicon and gallium arsenide. In this chapter we describe the common techniques for growing single crystals of these two semiconductors. The basic process flow from starting materials to polished wafers is shown in Fig. 1. The starting

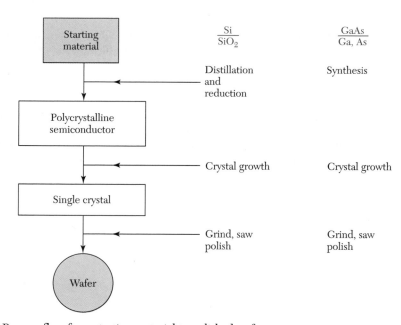

Fig. 1   Process flow from starting material to polished wafer

materials, silicon dioxide for a silicon wafer and gallium and arsenic for a gallium arsenide wafer, are chemically processed to form a high-purity polycrystalline semiconductor from which single crystals are grown. The single-crystal ingots are shaped to define the diameter of the material and sawed into wafers. These wafers are etched and polished to provide smooth, specular surfaces on which devices will be made.

A technology closely related to crystal growth involves the growth of single-crystal semiconductor layers on a single-crystal semiconductor substrate. This is called *epitaxy*, from the Greek words epi (meaning "on") and taxis (meaning "arrangement"). The epitaxial layer and the substrate materials may be the same, giving rise to *homoepitaxial*. For example, an *n*-type silicon can be grown epitaxially on an n^+-silicon substrate. On the other hand, if the epitaxial layer and the substrate are chemically and often crystallographically different, we have *heteroepitaxy*, such as the epitaxial growth of $Al_xGa_{1-x}As$ on GaAs.

Specifically, we cover the following topics:

- Basic techniques to grow silicon and GaAs single-crystal ingots.
- Wafer-shaping steps from ingots to polished wafers.
- Wafer characterization in term of its electrical and mechanical properties.
- Basic techniques of epitaxy, that is, to grow a single-crystal layer on a single-crystal substrate.
- Structures and defects of lattice-matched and strained-layer epitaxial growth.

10.1 SILICON CRYSTAL GROWTH FROM THE MELT

The basic technique for silicon crystal growth from the melt, which is material in liquid form, is the Czochralski technique. A substantial percentage (> 90%) of the silicon crystals for the semiconductor industry is prepared by the Czochralski technique, and virtually all the silicon used for fabricating integrated circuits is prepared by this technique.

10.1.1   Starting Material

The starting material for silicon is a relatively pure form of sand (SiO_2) called quartzite. This is placed in a furnace with various forms of carbon (coal, coke, and wood chips). Although a number of reactions take place in the furnace, the overall reaction is

$$SiC \text{ (solid)} + SiO_2 \text{ (solid)} \rightarrow Si \text{ (solid)} + SiO \text{ (gas)} + CO \text{ (gas)}. \tag{1}$$

This process produces metallurgical-grade silicon with a purity of about 98%. Next, the silicon is pulverized and treated with hydrogen chloride (HCl) to form trichlorosilane ($SiHCl_3$):

$$Si \text{ (solid)} + 3HCl \text{ (gas)} \xrightarrow{300°C} SiHCl_3 \text{ (gas)} + H_2 \text{ (gas)}. \tag{2}$$

The trichlorosilane is a liquid at room temperature (boiling point 32°C). Fractional distillation of the liquid removes the unwanted impurities. The purified $SiHCl_3$ is then used in a hydrogen reduction reaction to prepare the electronic-grade silicon (EGS):

$$SiHCl_3 \text{ (gas)} + H_2 \text{ (gas)} \rightarrow Si \text{ (solid)} + 3HCl \text{ (gas)}. \tag{3}$$

This reaction takes place in a reactor containing a resistance-heated silicon rod, which serves as the nucleation point for the deposition of silicon. The EGS, a polycrystalline material of high purity, is the raw material used to prepare device-quality, single-crystal silicon. Pure EGS generally has impurity concentrations in the parts-per-billion range.[1]

10.1.2 The Czochralski Technique

The Czochralski technique uses an apparatus called a crystal puller, shown in Chapter 2. A simplified version is shown in Fig. 2. The puller has three main components: (a) a furnace, which includes a fused-silicon (SiO_2) crucible, a graphite susceptor, a rotation mechanism (clockwise as shown), a heating element, and a power supply; (b) a crystal-pulling mechanism, which includes a seed holder and a rotation mechanism (counter–clockwise); and (c) an ambient control, which includes a gas source (such as argon), a flow control, and a exhaust system. In addition, the puller has an overall microprocessor-based control system to control process parameters such as temperature, crystal diameter, pull rate, and rotation speeds, as well as to permit programmed process steps. Also, various sensors and feedback loops allow the control system to respond automatically, reducing operator intervention.

In the crystal-growing process, polycrystalline silicon (EGS) is placed in the crucible and the furnace is heated above the melting temperature of silicon. A suitably oriented seed crystal (e.g., <111>) is suspended over the crucible in a seed holder. The seed is inserted into the melt. Part of it melts, but the tip of the remaining seed crystal still touches the liquid surface. It is then slowly withdrawn. Progressive freezing at the solid-liquid

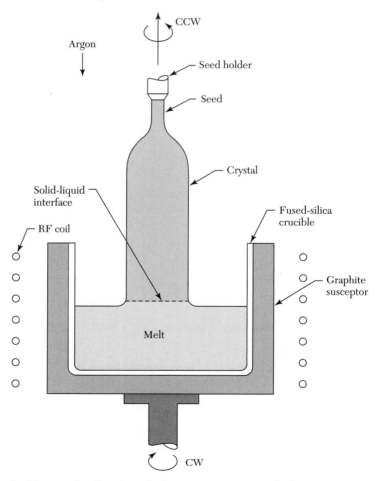

Fig. 2 Czochralski crystal puller. CW, clockwise; CCW, counter clockwise.

interface yields a large, single crystal. A typical pull rate is a few millimeters per minute. For large-diameter silicon ingots, an external magnetic field is applied to the basic Czochialski puller. The purpose of the external magnetic field is to control the concentration of defects, impurities, and oxygen contents.[2] Figure 3 shows a 300 mm (12 in.) and a 400 mm (16 in.) Czochralski grown silicon ingots.

10.1.3 Distribution of Dopant

In crystal growth, a known amount of dopant is added to the melt to obtain the desired doping concentration in the grown crystal. For silicon, boron and phosphorus are the most common dopants for *p*- and *n*-type materials, respectively.

As a crystal is pulled from the melt, the doping concentration incorporated into the crystal (solid) is usually different from the doping concentration of the melt (liquid) at the interface. The ratio of these two concentrations is defined as the *equilibrium segregation coefficient* k_0:

$$k_0 \equiv \frac{C_s}{C_l},$$ (4)

where C_s and C_l are, respectively, the equilibrium concentrations of the dopant in the solid and liquid near the interface. Table 1 lists values of k_0 for the commonly used dopants for silicon. Note that most values are below 1, which means that during growth the dopants

Fig. 3 300 mm (12 in.) and 400 mm (16 in.) Czochralski-grown silicon ingots. (Photo courtesy of Shin-Etsu Handotai Co., Tokyo.)

TABLE 1 Equilibrium Segregation Coefficients for Dopants in Si

Dopant	k_0	Type	Dopant	k_0	Type
B	8×10^{-1}	p	As	3.0×10^{-1}	n
Al	2×10^{-3}	p	Sb	2.3×10^{-2}	n
Ga	8×10^{-3}	p	Te	2.0×10^{-4}	n
In	4×10^{-4}	p	Li	1.0×10^{-2}	n
O	1.25	n	Cu	4.0×10^{-4}	—[a]
C	7×10^{-2}	n	Au	2.5×10^{-5}	—[a]
P	0.35	n			

[a]Deep-lying impurity level.

are rejected into the melt. Consequently, the melt becomes progressively enriched with the dopant as the crystal grows.

Consider a crystal being grown from a melt having an initial weight M_0 with an initial doping concentration C_0 in the melt (i.e., the weight of the dopant per 1 g of melt). At a given point of growth when a crystal of weight M has been grown, the amount of dopant remaining in the melt (by weight) is S. For an incremental amount of the crystal with weight dM, the corresponding reduction of the dopant ($-dS$) from the melt is $C_s \, dM$, where C_s is the doping concentration in the crystal (by weight):

$$-dS = C_s \, dM. \tag{5}$$

Now, the remaining weight of the melt is $M_0 - M$, and the doping concentration in the liquid (by weight), C_l, is given by

$$C_l = \frac{S}{M_0 - M}. \tag{6}$$

Combining Eqs. 5 and 6 and substituting $C_s/C_l = k_0$ yields

$$\frac{dS}{S} = -k_0 \left(\frac{dM}{M_0 - M} \right). \tag{7}$$

Given the initial weight of the dopant, $C_0 M_0$, we can integrate Eq. 7 :

$$\int_{C_0 M_0}^{S} \frac{dS}{S} = k_0 \int_0^M \frac{-dM}{M_0 - M}. \tag{8}$$

Solving Eq. 8 and combining with Eq. 6 gives

$$\boxed{C_s = k_0 C_0 \left(1 - \frac{M}{M_0} \right)^{k_0 - 1}.} \tag{9}$$

Figure 4 illustrates the doping distribution as a function of the fraction solidified (M/M_0) for several segregation coefficients.[3,4] As crystal growth progresses, the composition initially at $k_0 C_0$ will increase continually for $k_0 < 1$ and decrease continually for $k_0 > 1$. When $k_0 \cong 1$, a uniform impurity distribution can be obtained.

▶ **EXAMPLE 1**

A silicon ingot, which should contain 10^{16} boron atoms/cm³, is to be grown by the Czochralski technique. What concentration of boron atoms should be in the melt to give the required concentra-

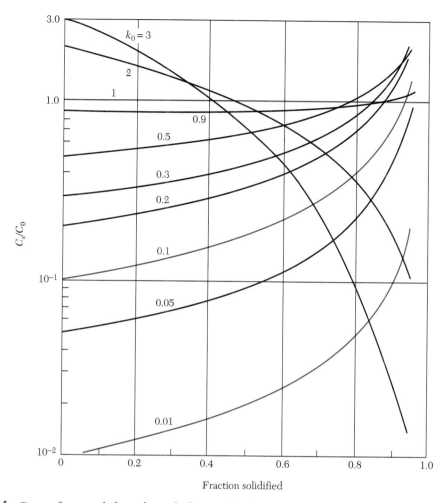

Fig. 4 Curves for growth from the melt showing the doping concentration in a solid as a function of the fraction solidified.[4]

tion in the ingot? If the initial load of silicon in the crucible is 60 kg, how many grams of boron (atomic weight 10.8) should be added? The density of molten silicon is 2.53 g/cm^3.

SOLUTION Table 1 shows that the segregation coefficient k_0 for boron is 0.8. We assume that $C_s = k_0 C_l$ throughout the growth. Thus, the initial concentration of boron in the melt should be

$$\frac{10^{16}}{0.8} = 1.25 \times 10^{16} \text{ boron atoms / cm}^3.$$

Since the amount of boron concentration is so small, the volume of melt can be calculated from the weight of silicon. Therefore, the volume of 60 kg of silicon is

$$\frac{60 \times 10^3}{2.53} = 2.37 \times 10^4 \text{ cm}^3.$$

The total number of boron atoms in the melt is

1.25×10^{16} atoms/cm$^3 \times 2.37 \times 10^4$ cm^3 = 2.96×10^{20} boron atoms, so that

$$\frac{2.96 \times 10^{20} \text{ atoms} \times 10.8 \text{ g / mol}}{6.02 \times 10^{23} \text{ atoms / mol}} = 5.31 \times 10^{-3} \text{ g of boron,}$$

$$= 5.31 \text{ mg of boron.}$$

Note the small amount of boron needed to dope such a large load of silicon. ◀

10.1.4 Effective Segregation Coefficient

While the crystal is growing, dopants are constantly being rejected into the melt (for k_0 < 1). If the rejection rate is higher than the rate of which the dopant can be transported away by diffusion or stirring, then a concentration gradient will develop at the interface, as illustrated in Fig. 5. The segregation coefficient (given in Section 10.1.3) is $k_0 = C_s/C_l(0)$. We can define an effective segregation coefficient k_e, which is the ratio of C_s and the impurity concentration far away from the interface:

$$k_e \equiv \frac{C_s}{C_l}. \tag{10}$$

Consider a small, virtually stagnant layer of melt with width δ in which the only flow is that required to replace the crystal being withdrawn from the melt. Outside this stagnant layer, the doping concentration has a constant value C_l. Inside the layer, the doping concentration can be described by the continuity equation (Eq. 58) derived in Chapter 3. At steady state, the only significant terms are the second and third terms on the right-hand side (we replace n_p by C and $\mu_n \mathscr{E}$ by v):

$$0 = v\frac{dC}{dx} + D\frac{d^2C}{dx^2}, \tag{11}$$

where D is the dopant diffusion coefficient in the melt, v is the crystal growth velocity, and C is the doping concentration in the melt.

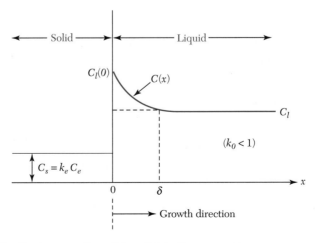

Fig. 5 Doping distribution near the solid-melt interface.

The solution of Eq. 11 is

$$C = A_1 e^{-vx/D} + A_2 \qquad (12)$$

where A_1 and A_2 are constants to be determined by the boundary conditions. The first boundary condition is that $C = C_l(0)$ at $x = 0$. The second boundary condition is the conservation of the total number of dopants; that is, the sum of the dopant fluxes at the interface must be zero. By considering the diffusion of dopant atoms in the melt (neglecting diffusion in the solid), we have

$$D\left(\frac{dC}{dx}\right)_{x=0} + \left[C_l(0) - C_s\right]v = 0. \qquad (13)$$

Substituting these boundary conditions into Eq. 12 and noting that $C = C_l$ at $x = \delta$ gives

$$e^{-v\delta/D} = \frac{C_l - C_s}{C_l(0) - C_s}. \qquad (14)$$

Therefore,

$$\boxed{k_e \equiv \frac{C_s}{C_l} = \frac{k_0}{k_0 + (1 - k_0)e^{-v\delta/D}}.} \qquad (15)$$

The doping distribution in the crystal is given by the same expression as in Eq. 9, except that k_0 is replaced by k_e. Values of k_e are larger than those of k_0 and can approach 1 for large values of the growth parameter $v\delta/D$. Uniform doping distribution ($k_e \to 1$) in the crystal can be obtained by employing a high pull rate and a low rotation speed (since δ is inversely proportional to the rotation speed). Another approach to achieve uniform doping is to add ultrapure polycrystalline silicon continuously to the melt so that the initial doping concentration is maintained.

10.2 SILICON FLOAT–ZONE PROCESS

The float–zone process can be used to grow silicon that has lower contaminations than that normally obtained from the Czochralski technique. A schematic setup of the float zone process is shown in Fig. 6a. A high-purity polycrystalline rod with a seed crystal at the bottom is held in a vertical position and rotated. The rod is enclosed in a quartz envelope within which an inert atmosphere (argon) is maintained. During the operation, a small zone (a few centimeters in length) of the crystal is kept molten by a radio-frequency heater, which is moved from the seed upward so that this *floating zone* traverses the length of the rod. The molten silicon is retained by surface tension between the melting and growing solid-silicon faces. As the floating zone moves upward, a single-crystal silicon freezes at the zone's retreating end and grows as an extension of the seed crystal. Materials with higher resistivities can be obtained from the float-zone process than from the Czochralski process because it can be used to purify the crystal more easily. Furthermore, since no crucible is used in the float-zone process, there is no contamination from the crucible (as with Czochralski growth). At the present time, float-zone crystals are used mainly for high-power, high-voltage devices, where high-resistivity materials are required.

To evaluate the doping distribution of a float-zone process, consider a simplified model, as shown in Fig. 6b. The initial, uniform doping concentration in the rod is C_0 (by weight).

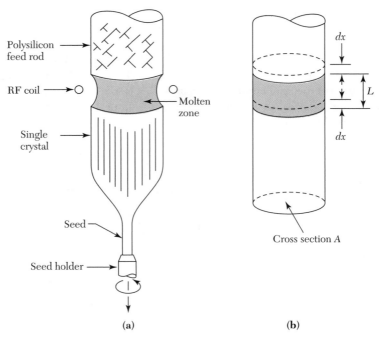

Fig. 6 Float-zone process. (*a*) Schematic setup. (*b*) Simple model for doping evaluation.

L is the length of the molten zone at a distance x along the rod, A the cross-sectional area of the rod, ρ_d the specific density of silicon, and S the amount of dopant present in the molten zone. As the zone traverses a distance dx, the amount of dopant added to it at its advancing end is $C_0\rho_d A\,dx$, whereas the amount of dopant removed from it at the retreating end is $k_e(S\,dx/L)$, where k_e is the effective segregation coefficient. Thus,

$$dS = C_0\rho_d A dx - \frac{k_e S}{L}\,dx = \left(C_0\rho_d A - \frac{k_e S}{L}\right)dx, \tag{16}$$

so that

$$\int_0^x dx = \int_{S_0}^{S} \frac{dS}{C_0\rho_d A - (k_e S / L)}, \tag{16a}$$

where $S_0 = C_0\rho_d AL$ is the amount of dopant in the zone when it was first formed at the front end of the rod. From Eq. 16a we obtain

$$\exp\left(\frac{k_e x}{L}\right) = \frac{C_0\rho_d A - (k_e S_0 / L)}{C_0\rho_d A - (k_e S / L)} \tag{17}$$

or

$$S = \frac{C_0 A \rho_d L}{k_e}\left[1 - (1 - k_e)^{-k_e x/L}\right]. \tag{17a}$$

Since C_s (the doping concentration in the crystal at the retreating end) is given by $C_s = k_e(S/A\rho_d L)$, then

$$\boxed{C_s = C_0[1 - (1 - k_e)^{-k_e x/L}].} \tag{18}$$

Figure 7 shows the doping concentration versus the solidified zone length for various values of k_e.

These two crystal growth techniques can also be used to remove impurities. A comparison of Fig. 7 with Fig. 4 shows that a single pass in the float-zone process does not produce as much purification as a single Czochralski growth. For example, for $k_0 = k_e = 0.1$, C_s/C_0 is smaller over most of the solidified ingot made by the Czochralski growth. However, multiple float-zone passes can be performed on a rod much more easily than a crystal can be grown, the end region cropped off, and regrown from the melt. Figure 8 shows the impurity distribution for an element with $k_e = 0.1$ after a number of successive passes of the zone along the length of the rod. [4] Note that there is a substantial reduction of impurity concentration in the rod after each pass. Therefore, the float-zone process is ideally suited for crystal purification. This process is also called the zone-refining technique, which can provide a very-high-purity level of the raw material.

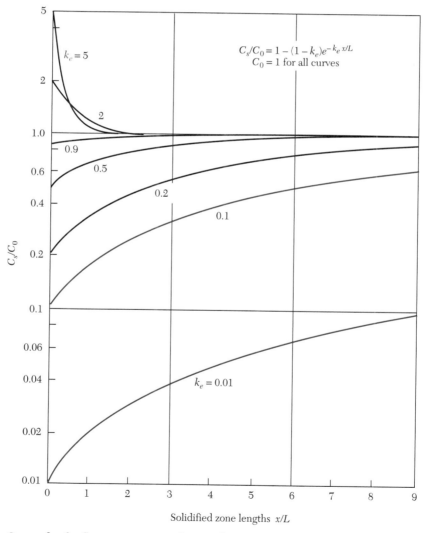

Fig. 7 Curves for the float-zone process showing doping concentration in the solid as a function of solidified zone lengths. [4]

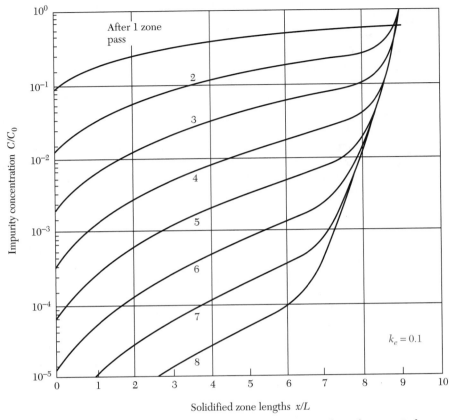

Fig. 8 Relative impurity concentration versus zone length for a number of passes. L denotes the zone length.[4]

If it is desirable to dope the rod rather than purify it, consider the case in which all the dopants are introduced in the first zone ($S_0 = C_1 A\rho_d L$) and the initial concentration C_0 is negligibly small. Equation 17 gives

$$S_0 = S \exp\left(\frac{k_e x}{L}\right). \tag{19}$$

Since $C_s = k_e(S/A\rho_d L)$, we obtained from Eq. 19

$$C_s = k_e C_1 e^{-k_e x/L}. \tag{20}$$

Therefore, if $k_e x/L$ is small, C_s will remain nearly constant with distance except at the end that is last to solidify.

For certain switching devices, such as high-voltage thyristors, discussed in Chapter 4, large chip areas are used, frequently an entire wafer for a single device. This size imposes stringent requirements on the uniformity of the starting material. To obtain homogeneous distribution of dopants, we use a float-zone silicon slice that has an average doping concentration well below the required amount. The slice is then irradiated with thermal neutrons. This process, called *neutron irradiation*, gives rise to fractional transmutation of silicon into phosphorus and dopes the silicon *n*-type:

$$\text{Si}_{14}^{30} + \text{neutron} \longrightarrow \text{Si}_{14}^{31} + \gamma\,\text{ray} \xrightarrow{\ 2.62\,\text{hr}\ } \text{P}_{15}^{31} + \beta\,\text{ray}. \tag{21}$$

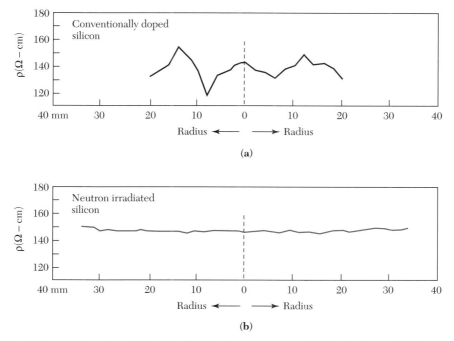

Fig. 9 (*a*) Typical lateral resistivity distribution in a conventionally doped silicon. (*b*) Silicon doped by neutron irradiation.[5]

The half-life of the intermediate element Si_{14}^{31} is 2.62 hours. Because the penetration depth of neutrons in silicon is about 100 cm, doping is very uniform throughout the slice. Figure 9 compares the lateral resistivity distributions in conventionally doped silicon and in silicon doped by neutron irradiation.[5] Note that the resistivity variations for the neutron-irradiated silicon are much smaller than that for the conventionally doped silicon.

10.3 GaAs CRYSTAL-GROWTH TECHNIQUES

10.3.1 Starting Materials

The starting materials for the synthesis of polycrystalline gallium arsenide are the elemental, chemically pure gallium and arsenic. Because gallium arsenide is a combination of two materials, its behavior is different from that of a single material such as silicon. The behavior of a combination can be described by a *phase diagram*. A phase is a state (e.g., solid, liquid, or gaseous) in which a material may exist. A phase diagram shows the relationship between the two components, gallium and arsenic, as a function of temperature.

Figure 10 shows the phase diagram of the gallium-arsenic system. The abscissa represents various compositions of the two components in terms of atomic percent (lower scale) or weight percent (upper scale).[6,7] Consider a melt that is initially of composition x (e.g., 85 atomic percent arsenic shown in Fig. 10). When the temperature is lowered, its composition will remain fixed until the *liquidus* line is reached. At the point (T_1, x), material of 50 atomic percent arsenic (i.e., gallium arsenide) will begin to solidify.

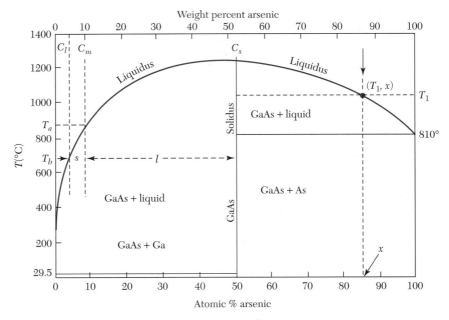

Fig. 10 Phase diagram for the gallium-arsenic system.[6]

▶ **EXAMPLE 2**

In Fig.10, consider a melt of initial composition C_m (weight percent scale) that is cooled from T_a (on the liquidus line) to T_b. Find the fraction of the melt that will be solidified.

SOLUTION At T_b, M_l is the weight of the liquid, M_s the weight of the solid (i.e., GaAs), and C_l and C_s the concentrations of dopant in the liquid and the solid, respectively. Therefore, the weight of arsenic in the liquid and solid are $M_l C_l$ and $M_s C_s$, respectively. Because the total arsenic weight is $(M_l + M_s)C_m$, we have

$$M_l C_l + M_s C_s = (M_l + M_s)C_m$$

or

$$\frac{M_s}{M_l} = \frac{\text{weight of GaAs at } T_b}{\text{weight of liquid at } T_b} = \frac{C_m - C_l}{C_s - C_m} = \frac{s}{l},$$

where s and l are the lengths of the two lines measured from C_m to the liquidus and solidus line, respectively. As can be seen from Fig. 10, about 10% of the melt is solidified.   ◀

Unlike silicon, which has a relatively low vapor pressure at its melting point ($\sim 10^{-6}$ atm at 1412°C), arsenic has much higher vapor pressures at the melting point of gallium arsenide (1240°C). In its vapor phase, arsenic has As_2 and As_4 as its major species. Figure 11 shows the vapor pressures of gallium and arsenic along the liquidus curve.[8] Also shown for comparison is the vapor pressure of silicon. The vapor pressure curves for gallium arsenide are double valued. The dashed curves are for arsenic-rich gallium arsenide melt (right side of liquidus line in Fig. 10), and the solid curves are for gallium-rich gallium arsenide melt (left side of liquidus line in Fig. 10). Because there is a larger amount of arsenic in an arsenic-rich melt than in a gallium-rich melt, more arsenic (As_2 and As_4) will be vaporized from the arsenic-rich melt, thus resulting in a higher vapor pressure. A similar argument can explain the higher vapor pressure of gallium in a gallium-rich melt. Note that long before the melting point is reached, the surface layers of liquid gal-

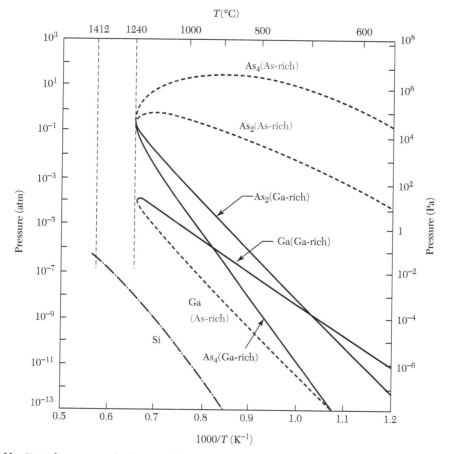

Fig. 11 Partial pressure of gallium and arsenic over gallium arsenide as a function of temperature.[8] Also shown is the partial pressure of silicon.

lium arsenide may decompose into gallium and arsenic. Since the vapor pressure of gallium and arsenic is different, there is a preferential loss of the more volatile arsenic species, and the liquid becomes gallium rich.

To synthesize gallium arsenide, an evacuated, sealed quartz tube system with a two-temperature furnace is commonly used. The high-purity arsenic is placed in a graphite boat and heated to 610°–620°C, whereas the high-purity gallium is placed in another graphite boat and heated to slightly above the gallium arsenide melting temperature (1240°–1260°C). Under these conditions, an overpressure of arsenic is established (a) to cause the transport of arsenic vapor to the gallium melt, converting it into gallium arsenide, and (b) to prevent decomposition of the gallium arsenide while it is being formed in the furnace. When the melt cools, a high-purity polycrystalline gallium arsenide results. This serves as the raw material to grow single-crystal gallium arsenide. [7]

10.3.2 Crystal-Growth Techniques

There are two techniques for GaAs crystal growth: the Czochralski technique and the Bridgman technique. Most gallium arsenide is grown by the Bridgman technique. However, the Czochralski technique is more popular for the growth of larger-diameter GaAs ingots.

For Czochralski growth of gallium arsenide, the basic puller is identical to that for silicon. However, to prevent decomposition of the melt during crystal growth, a liquid encapsulation method is employed. The liquid encapsulant is a molten boron trioxide (B_2O_3) layer about 1 cm thick. Molten boron trioxide is inert to the gallium arsenide surface and serves as a cap to cover the melt. This cap prevents decomposition of the gallium arsenide as long as the pressure on its surface is higher than 1 atm (760 Torr). Because boron trioxide can dissolve silicon dioxide, the fused-silica crucible is replaced with a graphite crucible.

To obtain the desired doping concentration in the grown crystal of GaAs, cadmium and zinc are commonly used for p-type materials, whereas selenium, silicon, and tellurium are used for n-type materials. For semiinsulating GaAs, the material is undoped. The equilibrium segregation coefficients for dopants in GaAs are listed in Table 2. Similar to those in Si, most of the segregation coefficients are less than 1. The expressions derived previously for Si are equally applicable to GaAs (Eqs. 4 to 15).

Figure 12 shows a Bridgman system in which a two-zone furnace is used for growing single-crystal gallium arsenide. The left-hand zone is held at a temperature (~610°C) to maintain the required overpressure of arsenic, whereas the right-hand zone is held just above the melting point of gallium arsenide (1240°C). The sealed tube is made of quartz and the boat is made of graphite. In operation, the boat is loaded with a charge of polycrystalline gallium arsenide, with the arsenic kept at the other end of the tube.

As the furnace is moved toward the right, the melt cools at one end. Usually, there is a seed placed at the left end of the boat to establish a specific crystal orientation. The gradual freezing (solidification) of the melt allows a single crystal to propagate at the liquid-solid interface. Eventually, a single crystal of gallium arsenide is grown. The impurity distribution can be described essentially by Eqs. 9 and 15, where the growth rate is given by the traversing speed of the furnace.

TABLE 2   Equilibrium Segregation Coefficients for Dopants in GaAs

Dopant	k_0	Type
Be	3	p
Mg	0.1	p
Zn	4×10^{-1}	p
C	0.8	n/p
Si	1.85×10^{-1}	n/p
Ge	2.8×10^{-2}	n/p
S	0.5	n
Se	5.0×10^{-1}	n
Sn	5.2×10^{-2}	n
Te	6.8×10^{-2}	n
Cr	1.03×10^{-4}	Semiinsulating
Fe	1.0×10^{-3}	Semiinsulating

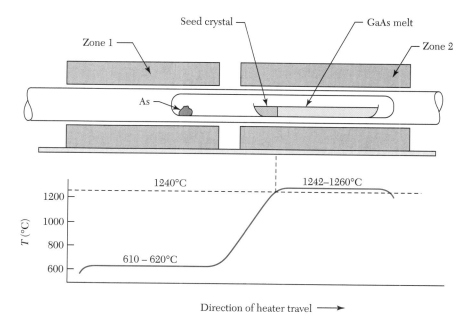

Fig. 12 Bridgman technique for growing single-crystal gallium arsenide and a temperature profile of the furnace.

10.4 MATERIAL CHARACTERIZATION

10.4.1 Wafer Shaping

After a crystal is grown, the first shaping operation is to remove the seed and the other end of the ingot, which is last to solidify.[1] The next operation is to grind the surface so that the diameter of the material is defined. After that, one or more flat regions are ground along the length of the ingot. These regions, or *flats*, mark the specific crystal orientation of the ingot and the conductivity type of the material. The largest flat, the *primary flat*, allows a mechanical locator in automatic processing equipment to position the wafer and to orient the devices relative to the crystal. Other smaller flats, called *secondary flats*, are ground to identify the orientation and conductivity type of the crystal, as shown in Fig. 13. For crystals with diameters equal or larger than 200 mm, no flats are ground. Instead, a small groove is ground along the length of the ingot.

The ingot is ready to be sliced by diamond saw into wafers. Slicing determines four wafer parameters: *surface orientation* (e.g., <111> or <100>), *thickness* (e.g., 0.5–0.7 mm, depending on wafer diameter); *taper*, which is the wafer thickness variations from one end to another; and *bow*, which is the surface curvature of the wafer, measured from the center of the wafer to its edge.

After slicing, both sides of the wafer are lapped using a mixture of Al_2O_3 and glycerine to produce a typical flatness uniformity within 2 μm. The lapping operation usually leaves the surface and edges of the wafer damaged and contaminated. The damaged and contaminated regions can be removed by chemical etching (see Chapter 12). The final step of wafer shaping is polishing. Its purpose is to provide a smooth, specular surface where device features can be defined by lithographic processes (see Chapter 12). Figure 14 shows 200 mm (8 in.) and 400 mm (16 in.) polished silicon wafers in cassettes.

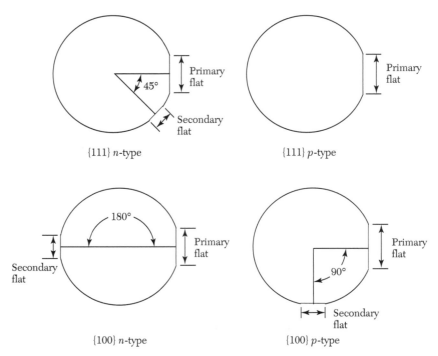

Fig. 13 Identifying flats on a semiconductor wafer.

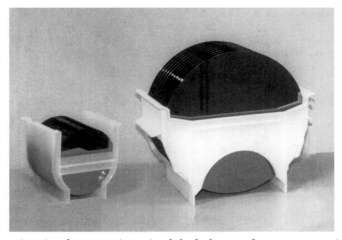

Fig. 14 200 mm (8 in.) and 400 mm (16 in.) polished silicon wafers in cassettes. (Photo courtesy of Shin-Etsu Handotai Co., Tokyo.)

Table 3 shows the specifications for 125, 150, 200, and 300 mm diameter polished silicon wafers from the Semiconductor Equipment and Materials Institute (SEMI). As mentioned previously, for large crystals (≥ 200 mm diameter) no flats are ground; instead, a groove is made on the edge of the wafer for positioning and orientation purpose.

Gallium arsenide is a more fragile material than silicon. Although the basic shaping operation of gallium arsenide is essentially the same as that for silicon, greater care must be exercised in gallium arsenide wafer preparation. The state of gallium arsenide tech-

TABLE 3   Specification for Polished Monocyrstalline Silicon Wafers

Parameter	125 mm	150 mm	200 mm	300 mm
Diameter (mm)	125±1	150±1	200±1	300±1
Thickness (mm)	0.6–0.65	0.65–0.7	0.715–0.735	0.755–0.775
Primary flat length (mm)	40–45	55–60	NA[a]	NA
Secondary flat length (mm)	25–30	35–40	NA	NA
Bow (μm)	70	60	30	< 30
Total thickness variation (μm)	65	50	10	< 10
Surface orientation	$(100) \pm 1°$	Same	Same	Same
	$(111) \pm 1°$	Same	Same	Same

[a]NA: not available.

nology is relatively primitive compared with that of silicon. However, the technology of group III-V compounds has advanced partly because of the advances in silicon technology.

10.4.2   Crystal Characterization

Crystal Defects

A real crystal (such as a silicon wafer) differs from the ideal crystal in important ways. It is finite; thus, surface atoms are incompletely bonded. Furthermore, it has defects, which strongly influence the electrical, mechanical, and optical properties of the semiconductor. There are four categories of defects: point defects, line defects, area defects, and volume defects.

Figure 15 shows several forms of *point defects*.[1,9] Any foreign atom incorporated into the lattice at either a substitutional site [i.e., at a regular lattice site (Fig. 15*a*)] or interstitial site [i.e., between regular lattice sites (Fig.15*b*)] is a point defect. A missing atom in the lattice creates a vacancy, also considered a point defect (Fig. 15*c*). A host atom that is situated between regular lattice sites and adjacent to a vacancy is called a *Frenkel defect* (Fig. 15*d*). Point defects are particularly important subjects in the kinetics of diffusion and oxidation processes. These topics are considered in Chapters 11 and 13.

The next class of defects is the *line defect*, also called a dislocation.[10] There are two types of dislocations: the edge and screw types. Figure 16*a* is a schematic representation of an edge dislocation in a cubic lattice. There is an extra plane of atoms *AB* inserted into the lattice. The line of the dislocation would be perpendicular to the plane of the page. The screw dislocation may be considered as being produced by cutting the crystal partway through and pushing the upper part one lattice spacing over, as show in Fig. 16*b*. Line defects in devices are undesirable because they act as precipitation sites for metallic impurities, which may degrade device performance.

Area defects represent a large area discontinuity in the lattice. Typical defects are twins and grain boundaries. Twinning represents a change in the crystal orientation across a plane. A grain boundary is a transition between crystals having no particular orientational relationship to one another. Such defects appear during crystal growth. Another area defect is the stacking fault.[9] In this defect, the stacking sequence of atomic layer is interrupted. In Fig. 17 the sequence of atoms in a stack is *ABCABC* . . . . When a part of layer *C* is missing, this is called the intrinsic stacking fault, Fig. 17*a*. If an extra plane *A* is inserted between layers *B* and *C*, this is an extrinsic stacking fault (Fig. 17*b*). Such defects may appear during crystal growth. Crystals having these area defects are not usable for integrated-circuit manufacture and are discarded.

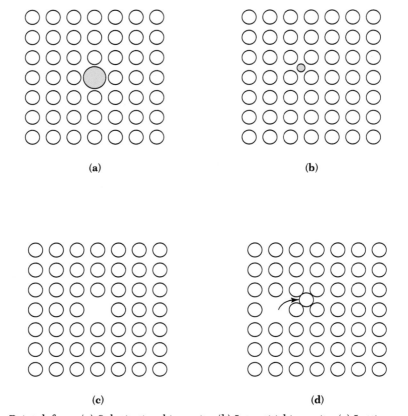

Fig. 15 Point defects. (*a*) Substitutional impurity. (*b*) Interstitial impurity. (*c*) Lattice vacancy. (*d*) Frenkel-type defect.[9]

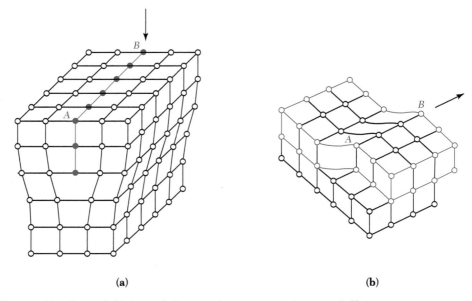

Fig. 16 (*a*) Edge and (*b*) screw dislocation formation in cubic crystals.[10]

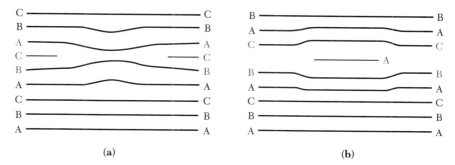

Fig. 17 Stacking fault in semiconductor. (*a*) Intrinsic stacking fault. (*b*) Extrinsic stacking fault.[9]

Precipitates of impurities or dopant atoms make up the fourth class of defects, the volume defects. These defects arise because of the inherent solubility of the impurity in the host lattice. There is a specific concentration of impurity that the host lattice can accept in a solid solution of itself and the impurity. Figure 18 shows solubility versus temperature for a variety of elements in silicon.[11] The solubility of most impurities decreases with decreasing temperature. Thus, at a given temperature, if an impurity is introduced to the maximum concentration allowed by its solubility and the crystal is then cooled to a lower temperature, the crystal can only achieve an equilibrium state by precipitating the impurity atoms in excess of the solubility level. However, the volume mismatch between the host lattice and the precipitates results in dislocations.

Material Properties

Table 4 compares silicon characteristics and the requirements for ultralarge-scale integration§ (ULSI).[12,13] The semiconductor material properties listed in Table 4 can be measured by various methods. The resistivity is measured by the four-point probe method discussed in Section 3.1, and the minority-carrier lifetime can be measured by the photoconductivity method considered in Section 3.3. The trace impurities such as oxygen and carbon in silicon can be analyzed by the secondary-ion–mass spectroscope (SIMS) techniques to be described in Chapter. 13. Note that although the current capabilities can meet most of the wafer specifications listed in Table 3, many improvements are needed to satisfy the stringent requirements for ULSI technology.[13]

The oxygen and carbon concentrations are substantially higher in Czochralski crystals than in float-zone crystals because of to the dissolution of oxygen from the silica crucible and transport of carbon to the melt from the graphite susceptor during crystal growth. Typical carbon concentrations range from 10^{16} to about 10^{17} atoms/cm^3 and carbon atoms in silicon occupy substitutional lattice sites. The presence of carbon is undesirable because it aids the formation of defects. Typical oxygen concentrations range from 10^{17} to 10^{18} atoms/cm^3. Oxygen, however, has both deleterious and beneficial effects. It can act as a donor, distorting the resistivity of the crystal caused by intentional doping. On the other hand, oxygen in an interstitial lattice site can increase the yield strength of silicon.

In addition, the precipitates of oxygen due to the solubility effect can be used for *gettering*. Gettering is a general term meaning a process that removes harmful impurities or defects from the region in a wafer where devices are fabricated. When the wafer is subjected to high-temperature treatment (e.g., 1050°C in N_2), oxygen evaporates from

§ The number of components in an ultralarge-scale integrated circuit is more than 10^7.

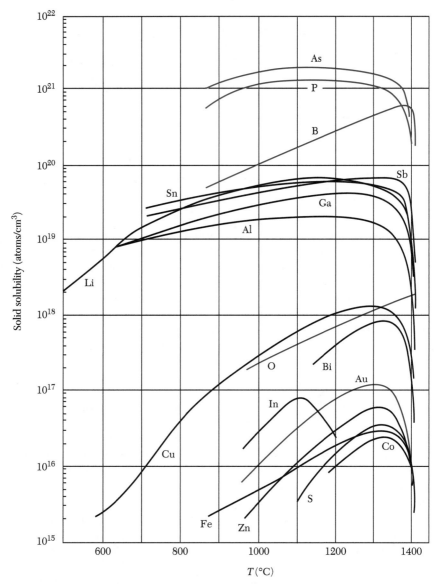

Fig. 18 Solid solubilities of impurity elements in silicon.[11]

the surface. This lowers the oxygen content near the surface. The treatment creates a defect-free (or *denuded*) zone for device fabrication, as shown in the inset[1] of Fig. 19. Additional thermal cycles can be used to promote the formation of oxygen precipitates in the interior of the wafer for gettering of impurities. The depth of the defect-free zone depends on the time and temperature of the thermal cycle and on the diffusivity of oxygen in silicon. Measured results for the denuded zone are shown[1] in Fig. 19. It is possible to obtain Czochralski crystals of silicon that are virtually free of dislocations.

Commercial melt-grown materials of gallium arsenide are heavily contaminated by the crucible. However, for photonic applications, most requirements call for heavily doped materials (between 10^{17} and 10^{18} cm^{-3}). For integrated circuits or for discrete MESFET (metal-semiconductor field-effect transistor) devices, undoped gallium arsenide can be

TABLE 4   Comparison of Silicon Material Characteristics and Requirements for ULSI

Property[a]	Characteristics		Requirements for ULSI
	Czochralski	Float zone	
Resistivity (phosphorus) n-type (ohm-cm)	1–50	1–300 and up	5-50 and up
Resistivity (antimony) n-type (ohm-cm)	0.005–10	—	0.001–0.02
Resistivity (boron) p-type (ohm-cm)	0.005–50	1–300	5–50 and up
Resistivity gradient (four-point probe) (%)	5–10	20	< 1
Minority carrier lifetime (μs)	30–300	50–500	300–1000
Oxygen (ppma)	5–25	Not detected	Uniform and controlled
Carbon (ppma)	1–5	0.1–1	< 0.1
Dislocation (before processing) (per cm^2)	≤ 500	≤ 500	≤ 1
Diameter (mm)	Up to 200	Up to 100	Up to 300
Slice bow (μm)	≤ 25	≤ 25	< 5
Slice taper (μm)	≤ 15	≤ 15	< 5
Surface flatness (μm)	≤ 5	≤ 5	< 1
Heavy-metal impurities (ppba)	≤ 1	≤ 0.01	< 0.001

[a] ppma, parts per million atoms; ppba, parts per billion atoms.

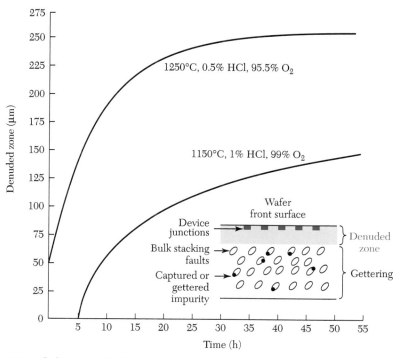

Fig. 19   Denuded zone width for two sets of processing conditions. Inset shows a schematic of the denuded zone and gettering sites in a wafer cross section.[1]

used as the starting material with a resistivity of 10^9 Ω-cm. Oxygen is an undesirable impurity in GaAs because it can form a deep donor level, which contributes to a trapping charge in the bulk of the substrate and increases its resistivity. Oxygen contamination can be minimized by using graphite crucibles for melt growth. The dislocation content for Czochralski-grown gallium arsenide crystals is about two orders of magnitude higher than that for silicon. For Bridgman GaAs crystals, the dislocation density is about an order of magnitude lower than that for Czochralski-grown GaAs crystals.

10.5 EPITAXIAL-GROWTH TECHNIQUES

In an epitaxial process, the substrate wafer acts as the seed crystal. Epitaxial processes are differentiated from the melt-growth processes described in previous sections in that the epitaxial layer can be grown at a temperature substantially below the melting point, typically 30–50% lower. The common techniques for epitaxial growth are chemical-vapor deposition (CVD) and molecular-beam epitaxy (MBE).

10.5.1 Chemical-Vapor Deposition

CVD is also known as vapor-phase epitaxy (VPE). CVD is a process whereby an epitaxial layer is formed by a chemical reaction between gaseous compounds. CVD can be performed at atmospheric pressure (APCVD) or at low pressure (LPCVD).

Figure 20 shows three common susceptors for epitaxial growth. Note that the geometric shape of the susceptor provides the name for the reactor: horizontal, pancake, and barrel susceptors—all made from graphite blocks. Susceptors in the epitaxial reactors are

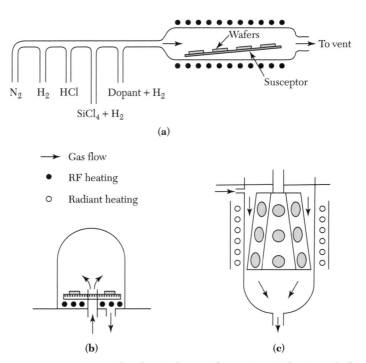

Fig. 20 Three common susceptors for chemical vapor deposition: (*a*) horizontal, (*b*) pancake, and (*c*) barrel susceptor.

analogous to the crucible in the crystal growing furnaces. Not only do they mechanically support the wafer, but in induction-heated reactors they also serve as the source of thermal energy for the reaction. The mechanism of CVD involves a number of steps: (a) the reactants such as the gases and dopants are transported to the substrate region, (b) they are transferred to the substrate surface where they are adsorbed, (c) a chemical reaction occurs, catalyzed at the surface, followed by growth of the epitaxial layer, (d) the gaseous products are desorbed into the main gas stream, and (e) the reaction products are transported out of the reaction chamber.

CVD for Silicon

Four silicon sources have been used for VPE growth. They are silicon tetrachloride ($SiCl_4$), dichlorosilane (SiH_2Cl_2), trichlorosilane ($SiHCl_3$), and silane (SiH_4). Silicon tetrachloride has been the most studied and has the widest industrial use. The typical reaction temperatures is 1200°C. Other silicon sources are used because of lower reaction temperatures. The substitution of a hydrogen atom for each chlorine atom from silicon tetrachloride permits about a 50°C reduction in the reaction temperature. The overall reaction of silicon tetrachloride that results in the growth of silicon layers is

$$SiCl_4 \text{ (gas)} + 2H_2 \text{ (gas)} \leftrightarrows Si \text{ (solid)} + 4HCl \text{ (gas)}. \qquad (22)$$

An additional competing reaction is taking place along with that given in Eq. 22:

$$SiCl_4 \text{ (gas)} + Si \text{ (solid)} \leftrightarrows 2SiCl_2 \text{ (gas)}. \qquad (23)$$

As a result, if the silicon tetrachloride concentration is too high, etching rather than growth of silicon will take place. Figure 21 shows the effect of the concentration of silicon tetrachloride in the gas on the reaction, where the *mole fraction* is defined as the ratio of the number of molecules of a given species to the total number of molecules.[14] Note that

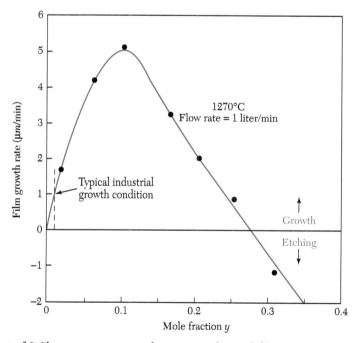

Fig. 21 Effect of $SiCl_4$ concentration on silicon epitaxial growth.[14]

initially the growth rate increases linearly with an increasing concentration of silicon tetrachloride. As the concentration of silicon tetrachloride is increased, a maximum growth rate is reached. Beyond that, the growth rate starts to decrease and eventually etching of the silicon will occur. Silicon is usually grown in the low-concentration region, as indicated in Fig. 21.

The reaction of Eq. 22 is reversible, that is, it can take place in either direction. If the carrier gas entering the reactor contains hydrochloric acid, removal or etching will take place. Actually, this etching operation is used for in-situ cleaning of the silicon wafer prior to epitaxial growth.

The dopant is introduced at the same time as the silicon tetrachloride during epitaxial growth (Fig. 20a). Gaseous diborane (B_2H_6) is used as the p-type dopant, whereas phosphine (PH_3) and arsine (AsH_3) are used as n-type dopants. Gas mixtures are ordinarily used with hydrogen as the diluent to allow reasonable control of flow rates for the desired doping concentration. The dopant chemistry for arsine is illustrated in Fig. 22, which shows arsine being adsorbed on the surface, decomposing, and being incorporated into the growing layer. Figure 22 also shows the growth mechanisms at the surface, which are based on the surface adsorption of host atoms (silicon) as well as the dopant atom (e.g., arsenic) and the movement of these atoms toward the ledge sites.[15] To give these adsorbed atoms sufficient mobility for finding their proper positions within the crystal lattice, epitaxial growth needs relatively high temperatures.

CVD for GaAs

For gallium arsenide, the basic setup is similar to that shown in Fig. 20a. Since gallium arsenide decomposes into gallium and arsenic upon evaporation, its direct transport in the vapor phase is not possible. One approach is the use of As_4 for the arsenic component and gallium chloride ($GaCl_3$) for the gallium component. The overall reaction leading to epitaxial growth of gallium arsenide is

$$As_4 + 4GaCl_3 + 6H_2 \rightarrow 4GaAs + 12HCl. \qquad (24)$$

The As_4 is generated by thermal decomposition of arsine (AsH_3):

$$4As_4H_3 \rightarrow As_4 + 6H_2, \qquad (24a)$$

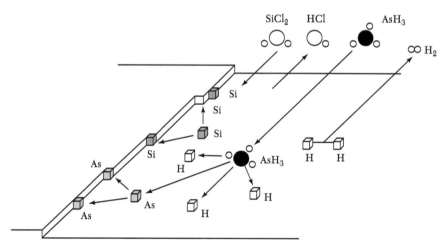

Fig. 22 Schematic representation of arsenic doping and the growing processes.[15]

and the gallium chloride is generated by the reaction

$$6HCl + 2Ga \rightarrow 2GaCl_3 + 3H_2. \tag{24b}$$

The reactants are introduced into a reactor with a carrier gas (e.g., H_2). The gallium arsenide wafers are typically held within the 650°–850°C temperature range. There must be sufficient arsenic overpressure to prevent thermal decomposition of the substrate and the growing layer.

Metalorganic CVD

Metalorganic CVD (MOCVD) is also a VPE process based on pyrolytic reactions. Unlike the conventional CVD, MOCVD is distinguished by the chemical nature of the precursor. It is important for those elements that do not form stable hydrides or halides but that form stable metalorganic compounds with reasonable vapor pressure. MOCVD has been extensibly applied in the heteroepitaxial growth of III-V and II-VI compounds.

To grow GaAs, we can use metalorganic compounds such as trimethylgallium $Ga(CH_3)_3$ for the gallium component and aresine AsH_3 for the arsenic component. Both chemicals can be transported in vapor form into the reactor. The overall reaction is

$$AsH_3 + Ga(CH_3)_3 \rightarrow GaAs + 3CH_4. \tag{25}$$

For Al-containing compounds, such as AlAs, we can use trimethylaluminum $Al(CH_3)_3$. During epitaxy, the GaAs is doped by introducing dopants in vapor form. Diethylzine $Zn(C_2H_5)_2$ and diethylcadmium $Cd(C_2H_5)_2$ are typical p-type dopants and silane SiH_4 is an n-type dopant for III-V compounds. The hydrides of sulfur and selenium or tetramethyltin are also used for n-type dopants; and chromyl chloride is used to dope chromium into GaAs to form semiinsulating layers. Since these compounds are highly poisonous and often spontaneously inflammable in air, rigorous safety precautions are necessary in the MOCVD process.

A schematic of an MOCVD reactor is shown[16] in Fig. 23. Typically, the metalorganic compound is transported to the quartz reaction vessel by hydrogen carrier gas, where it is mixed with AsH_3 in the case of GaAs growth. The chemical reaction is induced by heating the gases to 600°–800°C above a substrate placed on a graphite susceptor using radio-frequency heating. A pyrolytic reaction forms the GaAs layer. The advantages of using metalorganics are that they are volatile at moderately low temperatures and there are no troublesome liquid Ga or In sources in the reactor.

10.5.2 Molecular-Beam Epitaxy

MBE[17] is an epitaxial process involving the reaction of one or more thermal beams of atoms or molecules with a crystalline surface under ultrahigh-vacuum conditions ($\sim 10^{-8}$ Pa).[§] MBE can achieve precise control in both chemical compositions and doping profiles. Single-crystal multilayer structures with dimensions on the order of atomic layers can be made using MBE. Thus, the MBE method enables the precise fabrication of semiconductor heterostructures having thin layers from a fraction of a micron down to a monolayer. In general, MBE growth rates are quite low, and for GaAs, a value of 1 μm/hr is typical.

[§] The international unit for pressure is the Pascal (Pa); 1 Pa =1 N/m². However, various other units have been used. The conversion of these units are as follows: 1 atm = 760 mm Hg = 760 Torr = 1.013×10^5 Pa.

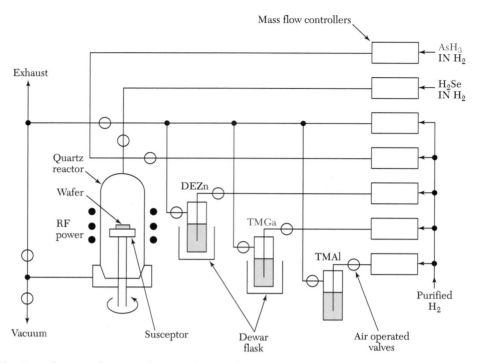

Fig. 23   Schematic diagram of a vertical atmospheric-pressure metalorganic chemical-vapor deposition (MOCVD) reactor.[16] DEZn is diethylozinc $Zn(C_2H_5)_2$, TMGa is trimethylgallium $Ga(CH_3)_3$, and TMAl is trimethylaluminum $Al(CH_3)_3$.

A schematic of MBE system is shown in Fig. 24 for gallium arsenide and related III-V compounds such as $Al_xGa_{1-x}As$. The system represents the ultimate in film deposition control, cleanliness, and in-situ chemical characterization capability. Separate effusion ovens made of pyrolytic boron nitride are used for Ga, As, and the dopants. All the effusion ovens are housed in an ultrahigh-vacuum chamber ($\sim 10^{-8}$ Pa). The temperature of each oven is adjusted to give the desired evaporation rate. The substrate holder rotates continuously to achieve uniform epitaxial layers (e.g., ±1% in doping variations and ±0.5% in thickness variations).

To grow GaAs, an overpressure of As is maintained, since the sticking coefficient of Ga to GaAs is unity, whereas that for As is zero, unless there is a previously deposited Ga layer. For a silicon MBE system, an electron gun is used to evaporate silicon. One or more effusion ovens are used for the dopants. Effusion ovens behave like small-area sources and exhibit a $\cos\theta$ emission, where θ is the angle between the direction of the source and the normal to the substrate surface.

MBE uses an evaporation method in a vacuum systems. An important parameter for vacuum technology is the molecular impingement rate, that is, how many molecules impinge on a unit area of the substrate per unit time. The impingement rate ϕ is a function of the molecular weight, temperature, and pressure. The rate is derived in Appendix K and can be expressed as[18]

$$\phi = P(2\pi mkT)^{-1/2} \tag{26}$$

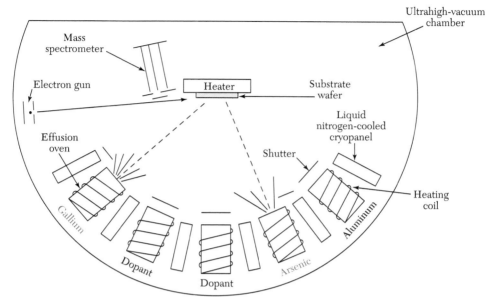

Fig. 24 Arrangement of the sources and substrate in a conventional molecular-beam epitaxy (MBE) system. (Courtesy of M. B. Panish, Bell Laboratories, Lucent Technologies.)

or

$$\phi = 2.64 \times 10^{20} \left(\frac{P}{\sqrt{MT}} \right) \text{ molecules / cm}^2 \text{- s,} \qquad (26a)$$

where P is the pressure in Pa, m is the mass of a molecule in kg , k is Boltzmams constant in J/K, T is the temperature in Kelvin, and M is the molecular weight. Therefore, at 300 K and 10^{-4} Pa pressure, the impingement rate is 2.7×10^{14} molecules/cm²-s for oxygen ($M = 32$).

▷ **EXAMPLE 3**

At 300 K, the molecular diameter of oxygen is 3.64 Å, and the number of molecules per unit area N_s is 7.54×10^{14} cm⁻². Find the time required to form a monolayer of oxygen at pressures of 1, 10^{-4}, and 10^{-8} Pa.

SOLUTION The time required to form a monolayer (assuming 100% sticking) is obtained from the impingement rate:

$$t = \frac{N_s}{\phi} = \frac{N_s \sqrt{MT}}{2.64 \times 10^{20} P}.$$

therefore,

$$t = 2.8 \times 10^{-4} \approx 0.28 \text{ ms} \qquad \text{at } 1 \quad \text{Pa,}$$
$$= 2.8 \text{ s} \qquad \text{at } 10^{-4} \text{Pa,}$$
$$= 7.7 \text{ hr} \qquad \text{at } 10^{-8} \text{Pa.}$$

To avoid contamination of the epitaxial layer, it is of paramount importance to maintain ultrahigh-vacuum conditions ($\sim 10^{-8}$ Pa) for the MBE process. ◀

During molecular motion, they will collide with other molecules. The average distance traversed by all the molecules between successive collisions with each other is defined as the mean free path. It can be derived from a simple collision theory. A molecule having a diameter d and a velocity v will move a distance $v\delta t$ in the time δt. The molecule suffers a collision with another molecule if its center is anywhere within the distance d of the center of another molecule. Therefore, it sweeps out (without collision) a cylinder of diameter $2d$. The volume of the cylinder is

$$\delta V = \frac{\pi}{4}(2d)^2 \, v\delta t = \pi d^2 v\delta t. \qquad (27)$$

Since there are n molecules/cm^3, the volume associated with one molecule is on the average $1/n$ cm^3. When the volume δV is equal to $1/n$, it must contain on the average one other molecule; thus, a collision would have occurred. Setting $\tau = \delta t$ as the average time between collision,

$$\frac{1}{n} = \pi d^2 v\tau, \qquad (28)$$

and the mean free path λ is then

$$\lambda = v\tau = \frac{1}{\pi n d^2} = \frac{kT}{\pi P d^2}. \qquad (29)$$

A more rigorous derivation gives

$$\lambda = \frac{kT}{\sqrt{2\pi P d^2}} \qquad (30)$$

and

$$\boxed{\lambda = \frac{0.66}{P\,(\text{in Pa})} \text{ cm}} \qquad (31)$$

for air molecules (equivalent molecular diameter of 3.7Å) at room temperature. Therefore, at a system pressure of 10^{-8} Pa, λ would be 660 km.

▶ **EXAMPLE 4**

Assume an effusion oven geometry of area $A = 5$ cm^2 and a distance L between the top of the oven and the gallium arsenide substrate of 10 cm. Calculate the MBE growth rate for the effusion oven filled with gallium arsenide at 900°C. The surface density of gallium atom is 6×10^{14}cm^{-2}, and the average thickness of a monolayer is 2.8 Å.

SOLUTION On heating gallium arsenide, the volatile arsenic vaporizes first, leaving a gallium-rich solution. Therefore, only the pressures marked Ga-rich in Fig.11 are of interest. The pressure at 900°C is 5.5×10^{-2} Pa for gallium and 1.1 Pa for arsenic (As$_2$). The arrival rate can be obtained from the impingement rate (Eq. 26a) by multiplying it by $A/\pi L^2$:

$$\text{Arrival rate} = 2.64 \times 10^{20} \left(\frac{P}{\sqrt{MT}} \right) \left(\frac{A}{\pi L^2} \right) \text{molecules / cm}^2\text{- s.}$$

The molecular weight M is 69.72 for Ga and 74.92 × 2 for As$_2$. Substituting values of P, M, and T (1173 K) into the above equation gives

$$\text{Arrival rate} = 8.2 \times 10^{14} / \text{cm}^2\text{-s} \qquad \text{for Ga,}$$
$$= 1.1 \times 10^{16} / \text{cm}^2\text{-s} \qquad \text{for As}_2.$$

The growth rate of gallium arsenide is found to be governed by the arrival rate of gallium. The growth rate is

$$\frac{8.2 \times 10^{14} \times 2.8}{6 \times 10^{14}} \approx 3.8 \text{ Å / s} = 23 \text{ nm / min.}$$

Note that the growth rate is relatively low compared with that of VPE. ◀

There are two ways to clean a surface in situ for MBE. High-temperature baking can decompose native oxide and remove other adsorbed species by evaporation or diffusion into the wafer. Another approach is to use a low-energy ion beam of an inert gas to sputter-clean the surface, followed by a low-temperature annealing to reorder the surface lattice structure.

MBE can use a wide variety of dopants (compared with CVD and MOCVD), and the doping profile can be exactly controlled. However, the doping process is similar to the vapor-phase growth process: a flux of evaporated dopant atoms arrives at a favorable lattice site and is incorporated along the growing interface. Fine control of the doping profile is achieved by adjusting the dopant flux relative to the flux of silicon atoms (for silicon epitaxial films) or gallium atoms (for gallium arsenide epitaxial films). It is also possible to dope the epitaxial film using a low-current, low-energy ion beam to implant the dopant (see Chapter 13).

The substrate temperatures for MBE range from 400°–900°C; and the growth rates range from 0.001 to 0.3 μm/min. Because of the low-temperature process and low-growth rate, many unique doping profiles and alloy compositions not obtainable from conventional CVD can be produced in MBE. Many novel structures have been made using MBE. These include the *superlattice*, which is a periodic structure consisting of alternating ultra-thin layers with its period less than the electron mean free path (e.g., GaAs/Al$_x$Ga$_{1-x}$As, with each layer 10 nm or less in thickness), and the heterojunction field-effect transistors discussed in Chapter 7.

A further development in MBE has replaced the group III elemental sources by metalorganic compounds such as trimethygallium (TMG) or triethylegallium (TEG). This approach is called metalorganic molecular-beam epitaxy (MOMBE) and is also referred to as chemical-beam epitaxy (CBE). Although closely related to MOCVD, it is considered a special form of MBE. The metalorganics are sufficiently volatile that they can be admitted directly into the MBE growth chamber as a beam and are not decomposed before forming the beam. The dopants are generally elemental sources, typically Be for *p*-type and Si or Sn for *n*-type GaAs epitaxial layers.

10.6 STRUCTURES AND DEFECTS IN EPITAXIAL LAYERS

10.6.1 Lattice-Matched and Strained-Layer Epitaxy

For conventional homoepitaxial growth, a single-crystal semiconductor layer is grown on a single-crystal semiconductor substrate. The semiconductor layer and the substrate are the same material having the same lattice constant. Therefore, homoepitaxy is, by definition, a lattice-matched epitaxial process. The homoepitaxial process offers one important means of controlling the doping profiles so that device and circuit performance can be optimized. For example, an *n*-type silicon layer with a relatively low doping concentration can be grown epitaxially on an n^+-silicon substrate. This structure substantically reduces the series resistance associated with the substrate.

For heteroepitaxy, the epitaxial layer and the substrate are two different semiconductors, and the epitaxial layer must be grown in such a way that an idealized interfacial structure is maintained. This implies that atomic bonding across the interface must be continuous without interruption. Therefore, the two semiconductor must either have the same lattice spacing or be able to deform to adopt a common spacing. These two cases are referred to as lattice-matched epitaxy and strained-layer epitaxy.

Figure 25a shows a lattice-matched epitaxy where the substrate and the film have the same lattice constant. An important example is the epitaxial growth of $Al_xGa_{1-x}As$ on a GaAs substrate where for any x between 0 and 1, the lattice constant of $Al_xGa_{1-x}As$ differs from that of GaAs by less than 0.13%.

For the lattice-mismatched case, if the epitaxial layer has a larger lattice constant and is flexible, it will be compressed in the plane of growth to conform to the substrate spacing. Elastic forces then compel it to dilate in a direction perpendicular to the interface. This type of structure is called strained-layer epitaxy and is illustrated[19] in Fig. 25b. On the other hand, if the epitaxial layer has a smaller lattice constant, it will be dilated in the plane of growth and compressed in a direction perpendicular to the interface. In the above strained-layer epitaxy, as the strained-layer thickness increases, the total number of atoms under strain or the distorted atomic bonds grows, and at some point misfit dislocations are nucleated to relieve the homogeneous strain energy. This thickness is referred to as the *critical layer thickness* for the system. Figure 25c shows the case in which there are edge dislocations at the interface.

The critical layer thicknesses for two material systems are shown[20] in Fig. 26. The upper curve is for the strained-layer epitaxy of a Ge_xSi_{1-x} layer on a silicon substrate, and

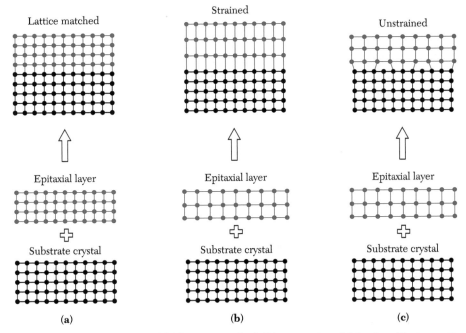

Fig. 25 Schematic illustration of (*a*) lattice-matched, (*b*) strained, and (*c*) relaxed heteroepitaxial structures.[19] Homoepitaxy is structurally identical to the lattice-matched heteroepitaxy.

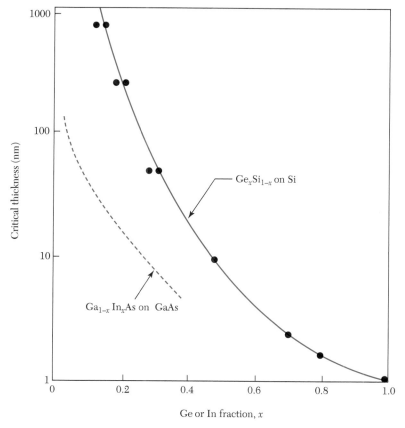

Fig. 26 Experimentally determined critical layer thickness for defect-free, strained-layer epitaxy[20] of Ge_xSi_{1-x} on Si, and $Ga_{1-x}In_xAs$ on GaAs.

the lower curve is for a $Ga_{1-x}In_xAs$ layer on a GaAs substrate. For example, for $Ge_{0.3}Si_{0.7}$ on silicon, the maximum epitaxial thickness is about 70 nm. For thicker films, edge dislocations will occur.

A related heteroepitaxial structure is the strained-layer superlattice (SLS). A superlattice is an artificial one-dimensional periodic structure constituted by different materials with a period of about 10 nm. Figure 27 shows[17] a SLS having two semiconductors with different equilibrium lattice constants $a_1 > a_2$, grown in a structure with a common inplane lattice constant b, where $a_1 > b > a_2$. For sufficiently thin layers, the lattice mismatch is accommodated by uniform strains in the layers. Under these condition, no misfit dislocations are generated at the interfaces, so high-quality crystalline materials can be obtained. These artificially structured materials can be grown by MBE. These materials provide a new area in semiconductor research and permit new solid-state devices, especially for high-speed and photonic applications.

10.6.2 Defects in Epitaxial Layers

Defects in epitaxial layers will degrade device properties. For example, defects can result in reduced mobility or increased leakage current. The defects in epitaxial layers can be categorized into five groups.

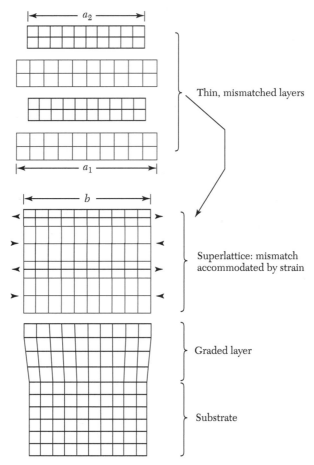

Fig. 27 Illustration of the elements and formation of an strained-layer superlattice.[17] Arrows show the direction of the strain.

1. Defects from the substrates. These defects may propagate from the substrate into the epitaxial layer. To avoid these defects, dislocation-free semiconductor substrates are required.

2. Defects from the interface. The oxide precipitates or any contamination at the interface of the epitaxial layer and substrate may cause the formation of misoriented clusters or nuclei containing stacking faults. These clusters and stacking faults may coalesce with normal nuclei and grow into the film in the shape of an inverted pyramid. To avoid these defects, the surface of the substrate must be thoroughly cleaned. In addition, an in-situ etch back may be used such as the reversable reaction of Eq. 22.

3. Precipitates or dislocation loops. Their formation is due to supersaturation of impurities or dopants. Epitaxial layers containing very high intentional or unintentional dopants or impurity concentrations are susceptible to such defects.

4. Low-angle grain boundaries and twins. Any misoriented areas of an epitaxial film during growth may meet and coalesce to form these defects.

5. Edge dislocations. These are formed in the heteroepitaxy of two lattice-mismatched semiconductors. If both lattices are rigid, they will retain their fundamental lattice spacings, and the interface will contain rows of misbonded atoms described as misfit or edge dislocations. The edge dislocations can also form in a strained layer when the layer thickness becomes larger than the critical layer thickness.

▶ SUMMARY

Several techniques are available to grow single crystals of silicon and gallium arsenide. For silicon crystals, we use sand (SiO_2) to produce polycrystalline silicon, which then serves as the raw material in a Czochralski puller. A seed crystal with the desired orientation is used to grow a large ingot from the melt. Over 90% of silicon crystals are prepared by this technique. During crystal growth, the dopant in the crystal will redistribute. A key parameter is the segregation coefficient, i.e., the ratio of the dopant concentration in the solid to that in the melt. Since most of the coefficients are less than 1, the melt becomes progressively enriched with the dopant as the crystal grows.

Another growth technique for silicon is the float-zone process. It offers lower contamination than that normally obtained from the Czochralski technique. Float-zone crystals are used mainly for high-power, high-voltage devices where high-resistivity materials are required.

To make GaAs, we use chemically pure gallium and arsenic as the starting materials that are synthesized to form polycrystalline GaAs. Single crystals of GaAs can be grown by the Czochralski technique. However, a liquid encapsulant (e.g., B_2O_3) is required to prevent decomposition of GaAs at the growth temperature. Another technique is the Bridgman process, which uses a two-zone furnace for gradual solidification of the melt.

After a crystal is grown, it usually goes through wafer-shaping operations to give an end product of highly polished wafers with a specified diameter, thickness, and surface orientation. For example, 200 mm silicon wafers for a MOSFET (metal-oxide-semiconductor field-effect transistor) fabrication line should have a diameter of 200 ± 1 mm, a thickness of 0.725 ± 0.01 mm, and a surface orientation of $(100) \pm 1°$. Wafers with diameters larger than 200 mm are being manufactured for future integrated circuits. Their specifications are listed in Table 3.

A real crystal has defects that influence the electrical, mechanical, and optical properties of the semiconductor. These defects are point defects, line defects, area defects, and volume defects. We also discussed means to minimize such defects. For the more demanding ULSI applications, the dislocation density must be less than 1 per square centimeter. Other important requirements are listed in Table 4.

A technology closely related to crystal growth is the epitaxial process. In this process, the substrate wafer is the seed. High-quality, single-crystal films can be grown at a temperature 30%–50% lower than the melting point. The common techniques for epitaxial growth are chemical-vapor deposition (CVD), metalorganic CVD (MOCVD), and molecular-beam epitaxy (MBE). CVD and MOCVD are chemical deposition processes. Gases and dopants are transported in vapor form to the substrate, where a chemical reaction occurs that results in the deposition of the epitaxial layer. Inorganic compounds are used for CVD, whereas metalorganic compounds are used for MOCVD. MBE, on the other hand, is a physical deposition process. It is done by the evaporation of a species in an ultrahigh vacuum system. Because it is a low-temperature process that has a low growth rate, MBE can grow single-crystal, multilayer structures with dimensions on the order of atomic layers.

In addition to conventional homoepitaxy, such as n-type silicon on an n^+-silicon substrate, we have also considered heteroepitaxy that includes lattice-matched and strained-layer structures. For strained-layer epitaxy, there is a critical layer thickness above which edge dislocations will nucleate to relieve the strain energy.

Besides the edge dislocations in an epitaxial layer, we have defects from the substrate, defects from the interface, precipitates, and low-angle grain boundaries and twins. These defects degrade device performance. Various means have been presented to minimize or even to eliminate these defects so that a defect-free semiconductor layer can be grown either homoepitaxially or heteroepitaxially.

▶ REFERENCES

1. C. W. Pearce, "Crystal Growth and Wafer Preparation" and "Epitaxy," in S. M. Sze, Ed., *VLSI Technology*, McGraw-Hill, New York, 1983.

2. T. Abe, "Silicon Crystals for Giga-Bit Scale Integration," in T. S. Moss, Ed., *Handbook on Semiconductors*, Vol. 3, Elsevier Science B. V., Amsterdam/New York, 1994.

3. W. R. Runyan, *Silicon Semiconductor Technology*, McGraw-Hill, New York, 1965.

4. W. G. Pfann, *Zone Melting*, 2nd Ed., Wiley, New York, 1966.

5. E. W. Hass and M. S. Schnoller, "Phosphorus Doping of Silicon by Means of Neutron Irradiation," *IEEE Trans. Electron Devices*, **ED-23**, 803 (1976).

6. M. Hansen, *Constitution of Binary Alloys*, McGraw-Hill, New York, 1958.

7. S. K. Ghandhi, *VLSI Fabrication Principles*, Wiley, New York, 1983.

8. J. R. Arthur, "Vapor Pressures and Phase Equilibria in the GaAs System," *J. Phys. Chem. Solids*, 28, 2257 (1967).

9. B. El-Kareh, *Fundamentals of Semiconductor Processing Technology*, Kluwer Academic, Boston, 1995.

10. C. A. Wert and R. M. Thomson, *Physics of Solids*, McGraw-Hill, New York, 1964.

11. (a) F. A. Trumbore, "Solid Solubilities of Impurity Elements in Germanium and Silicon," *Bell Syst. Tech. J.*, **39**, 205 (1960); (b) R. Hull, *Properties of Crystalline Silicon*, INSPEC, London, 1999.

12. Y. Matsushita, "Trend of Silicon Substrate Technologies for 0.25 μm Devices," *Proc. VLSI Technol. Workshop*, Honolulu, (1996).

13. *The International Technology Roadmap for Semiconductors*, Semiconductor Industry Association, San Jose, CA, 1999.

14. A. S. Grove, *Physics and Technology of Semiconductor Devices*, Wiley, New York, 1967.

15. R. Reif, T. I. Kamins, and K. C. Saraswat, "A Model for Dopant Incorporation into Growing Silicon Epitaxial Films," *J. Electrochem. Soc.*, **126**, 644, 653 (1979).

16. R. D Dupuis, *Science*, **226**, 623 (1984).

17. M. A. Herman and H. Sitter, *Molecular beam Epitaxy*, Springer-Verlag, Berlin, 1996.

18. A. Roth, *Vacuum Technology*, North-Holland, Amsterdam, 1976.

19. M. Ohring, *The Materials Science of Thin Films*, Academic, New York, 1992.

20. J. C. Bean, "The Growth of Novel Silicon Materials," *Physics Today*, **39**, 10, 36 (1986).

▶ PROBLEMS (* DENOTES DIFFICULT PROBLEMS)

FOR SECTION 10.1 SILICON CRYSTAL GROWTH FROM THE MELT

1. Plot the doping distribution of arsenic at distances of 10, 20, 30, 40, and 45 cm from the seed in a silicon ingot 50 cm long that has been pulled from a melt with an initial doping concentration of 10^{17} cm^{-3}.

2. In silicon, the lattice constant is 5.43 Å. Assume a hard-sphere model. (a) Calculate the radius of a silicon atom. (b) Determine the density of silicon atoms in atoms/cm^3. (c) Use the Avogadro constant to find the density of silicon.

3. Assuming that a 10 kg pure silicon charge is used, what is the amount of boron that must be added to get the boron-doped silicon having a resistivity of 0.01 Ω-cm when one half of the ingot is grown?

4. A silicon wafer 1-mm thick having a diameter of 200 mm contains 5.41 mg of boron uniformly distributed in substitutional sites. Find (a) the boron concentration in atoms/cm^3 and, (b) the average distance between boron atoms.

5. The seed crystal used in Crochralski process is usually necked down to a small diameter (5.5 mm) as a means to initiate dislocation-free growth. If the critical yield strength of silicon is 2×10^6 g/cm^2, calculate the maximum length of a silicon ingot 200 mm diameter that can be supported by such a seed.

6. Plot the curve of C_s/C_0 value for $k_0 = 0.05$ in Czochralski technique.

7. A Czochralski grown crystal is doped with boron. Why is the boron concentration larger at the tail end of the crystal than at the seed end?

8. Why is the impurity concentration larger in the center of the wafer than at its perimeter?

FOR SECTION 10.2   SILICON FLOAT-ZONE PROCESS

9. We use the float-zone process to purify a silicon ingot that contains a uniform gallium concentration of 5×10^{16} cm^{-3}. One pass is made with a molten zone 2 cm long. Over what distance is the resulting gallium concentration below 5×10^{15} cm^{-3}?

10. From Eq. 18 find the C_s/C_0 value at $x/L = 1$ and 2 with $k_e = 0.3$.

11. If p^+-n abrupt-junction diodes are fabricated using the silicon materials shown in Fig. 9, find the percentage change of breakdown voltages for the conventionally doped silicon and the neutron-irradiated silicon.

FOR SECTION 10.3   GaAs CRYSTAL-GROWTH TECHNIQUES

12. From Fig. 10, if C_m is 20%, what fraction of the liquid at T_b will remain?

13. From Fig. 11, explain why the GaAs liquid always becomes gallium rich?

FOR SECTION 10.4   MATERIAL CHARACTERIZATION

14. The equilibrium density of vacancy n_s is given by $N\exp(-E_s/kT)$, where N is the density of semiconductor atoms and E_s is the energy of formation. Calculate n_s in silicon at 27°C, 900°C, and 1200°C. Assume $E_s = 2.3$ eV.

15. Assume the energy of formation (E_f) of a Frenkel-type defect to be 1.1 eV and estimate the defect density at 27°C and 900°C. The equilibrium density of Frenkel type defects is given by $n_f = \sqrt{NN'}\, e^{-E_f/2kT}$, where N is the atomic density of silicon (cm^{-3}), and N' is the density of available interstitial sites (cm^{-3}) and it is represented by

$N' = 1 \times 10^{27} e^{-3.8(\text{eV})/kT}$ cm^{-3}.

16. How many chips of area 400 mm^2 can be placed on a wafer 300 mm in diameter? Explain your assumptions on the chip shape and unused wafer perimeter?

FOR SECTION 10.5   EPITAXIAL-GROWTH TECHNIQUES

*17. Find the average molecular velocity of air at 300 K (the molecular weight for air is 29).

18. The distance between source and wafer in a deposition chamber is 15 cm. Estimate the pressure at which this distance become 10% of the mean free path of source molecules.

*19. Find the number of atoms per unit area, N_s, needed to form a monolayer under close-packing condition (i.e., each atom is in contact with its six neighboring atoms), assuming the diameter d of the atom is 4.68 Å.

*20. Assume an effusion oven geometry of $A = 5$ cm^2 and $L = 12$ cm. (a) Calculate the arrival rate of gallium and the MBE growth rate for the effusion oven filled with gallium arsenide at 970°C. (b) For a tin effusion oven operated at 700°C under the same geometry, calculate the doping concentration (assuming tin atoms are fully incorporated in the gallium arsenide grown at the aforementioned rate). The molecular weight for tin is 118.69, and the pressure at 700°C for tin is 2.66×10^{-6} Pa.

FOR SECTION 10.6   STRUCTURES AND DEFECTS IN EPITAXIAL LAYERS

21. Find the maximum percentage of In, i.e., the x value for Ga_xIn_{1-x} As film grown on GaAs substrate without formation of misfit disloation, if the final film thickness is 10 nm.

22. The lattice misfit, f, of a film is defined as $f \equiv [a_0(s) - a_0(f)]/a_0(f) = \Delta a_0/a_0$ where $a_0(s)$ and $a_0(f)$ are the unstrained lattice constants of the substrate and the film, respectively. Find the f values for InAs-GaAs and Ge-Si systems.

Film Formation

To fabricate discrete devices and integrated circuits, we use many different kinds of thin films. We can classify thin films into four groups: thermal oxides, dielectric layers, polycrystalline silicon, and metal films. Figure 1 shows a schematic view of a conventional silicon n-channel MOSFET (metal-oxide-semiconductor field-effect transitor) that uses all four groups of films. The first important thin film from the thermal oxide group is the gate-oxide layer under which a conducting channel can be formed between the source and the drain. A related layer is the field oxide, which provides isolation from other devices. Both gate and field oxides generally are grown by a thermal oxidation process because only thermal oxidation can provide the highest-quality oxides having the lowest interface trap densities.

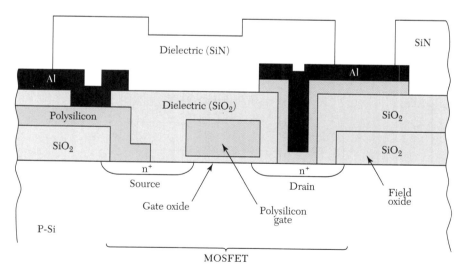

Fig. 1 Schematic cross section of a metal-oxide-semiconductor field-effect transitor (MOSFET).

Dielectric layers such as silicon dioxide and silicon nitride are used for insulation between conducting layers, for diffusion and ion implantation masks, for capping doped films to prevent the loss of dopants, and for passivation to protect devices from impurities, moisture, and scratches. Polycrystalline silicon, usually referred to as polysilicon, is used as a gate electrode material in MOS devices, a conductive material for multilevel metallization, and a contact material for devices with shallow junctions. Metal films such as aluminum and silicides are used to form low-resistance interconnections, ohmic contacts, and rectifying metal-semiconductor barriers.

Specifically, we cover the following topics:

- The thermal oxidation process to form silicon dioxide (SiO_2).

- Deposition techniques to form low–dielectric-constant and high–dielectric-constant films as well as polysilicon films.

- Deposition techniques to form aluminum and copper interconnections as well as the related global planarization process.

- Characteristics of these thin films and their compatibility with integrated-circuit processing.

▶ 11.1 THERMAL OXIDATION

Semiconductors can be oxidized by various methods. These include thermal oxidation, electrochemical anodization, and plasma reaction. Among these methods, thermal oxidation is by far the most important for silicon devices. It is the key process in modern silicon integrated-circuit technology. For gallium arsenide, however, thermal oxidation results in generally nonstoichiometric films. The oxides provide poor electrical insulation and semiconductor surface protection; hence, these oxides are rarely used in gallium arsenide technology. Consequently, in this section we concentrate on the thermal oxidation of silicon.

The basic thermal oxidation setup is shown[1] in Fig. 2. The reactor consists of a resistance-heated furnace, a cylindrical fused-quartz tube containing the silicon wafers held vertically in a slotted quartz boat, and a source of either pure dry oxygen or pure water vapor. The loading end of the furnace tube protudes into a vertical flow hood where a filtered flow of air is maintained. Flow is directed as shown by the arrow in Fig. 2. The hood reduces dust and particulate matters in the air surrounding the wafers and mini-

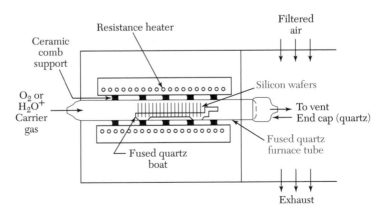

Fig. 2 Schematic cross section of a resistance-heated oxidation furnace.

mizes contamination during wafer loading. The oxidation temperature is generally in the range of 900°–1200°C and the typical gas flow rate is about 1 liter/min. The oxidation system uses microprocessors to regulate the gas flow sequence, to control the automatic insertion and removal of silicon wafers, to ramp the temperature up (i.e., to increase the furnace temperature linearly) from a low temperature to the oxidation temperature so that the wafers will not warp due to sudden temperature change, to maintain the oxidation temperature to within ±1°C, and to ramp the temperature down when oxidation is completed.

11.1.1 Kinetics of Growth

The following chemical reactions describe the thermal oxidation of silicon in oxygen or water vapor:

$$\text{Si (solid)} + \text{O}_2 \text{ (gas)} \rightarrow \text{SiO}_2 \text{ (solid)}, \tag{1}$$

$$\text{Si (solid)} + 2\text{H}_2\text{O (gas)} \rightarrow \text{SiO}_2 \text{ (solid)} + 2\text{H}_2 \text{ (gas)}. \tag{2}$$

The silicon-silicon dioxide interface moves into the silicon during the oxidation process. This creates a fresh interface region, with surface contamination on the original silicon ending up on the oxide surface. The densities and molecular weights of silicon and silicon dioxide are used in the following example to show that growing an oxide of thickness x consumes a layer of silicon $0.44x$ thick (Fig. 3).

▶ **EXAMPLE 1**

If a silicon oxide layer of thickness x is grown by thermal oxidation, what is the thickness of silicon being consumed? The molecular weight of Si is 28.9 g/mol, and the density of Si is 2.33 g/cm³. The corresponding values for SiO₂ are 60.08 g/mol and 2.21 g/cm³.

SOLUTION   The volume of 1 mol of silicon is

$$\frac{\text{Molecular weight of Si}}{\text{Density of Si}} = \frac{28.9 \text{ g/mole}}{2.33 \text{ g/cm}^3} = 12.06 \text{ cm}^3/\text{mol}.$$

The volume of 1 mol of silicon dioxide is

$$\frac{\text{Molecular weight of SiO}_2}{\text{Density of SiO}_2} = \frac{60.08 \text{ g/mol}}{2.2 \text{ g/cm}^3} = 27.18 \text{ cm}^3/\text{mol}.$$

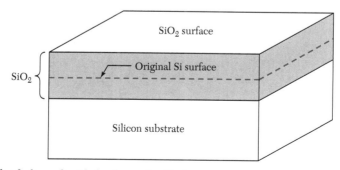

Fig. 3   Growth of silicon dioxide by thermal oxidation.

Since 1 mol of silicon is converted to 1 mol of silicon dioxide,

$$\frac{\text{Thickness of Si} \times \text{area}}{\text{Thickness of SiO}_2 \times \text{area}} = \frac{\text{volume of 1 mol of Si}}{\text{volume of 1 mol of SiO}_2},$$

$$\frac{\text{Thickness of Si}}{\text{Thickness of SiO}_2} = \frac{12.06}{27.18} = 0.44,$$

Thickness of silicon = 0.44 (thickness of SiO_2).

For example, to grow a silicon dioxide layer of 100 nm, a layer of 44 nm of silicon is consumed. ◀

The basic structural unit of thermally grown silicon dioxide is a silicon atom surrounded tetrahedrally by four oxygen atoms, as illustrated[1] in Fig. 4a. The silicon-to-oxygen internuclear distance is 1.6 Å, and the oxygen-to-oxygen internuclear distance is 2.27 Å. These tetrahedra are joined together at their corners by oxygen bridges in a variety of ways to form the various phases or structures of silicon dioxide (also called silica). Silica has several crystalline structures (e.g., quartz) and an amorphous structure. When silicon is thermally oxidized, the silicon dioxide structure is amorphous. Typically amorphous silica has a density of 2.21 g/cm³, compared with 2.65 g/cm³ for quartz.

The basic difference between the crystalline and amorphous structures is that the former is a periodic structure, extending over many molecules, whereas the latter has no periodic structure at all. Figure 4b is a two-dimensional schematic diagram of a quartz crystalline structure made up of rings with six silicon atoms. Figure 4c is a two-dimensional schematic diagram of an amorphous structure for comparison. In the amorphous

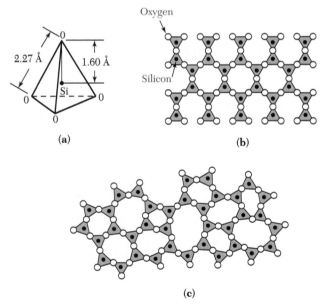

(a)   (b)

(c)

Fig. 4 (a) Basic structural unit of silicon dioxide. (b) Two dimensional representation of a quartz crystal lattice. (c) Two-dimensional representation of the amorphous structure of silicon dioxide.[1]

structure there is still a tendency to form characteristic rings with six silicon atoms. Note that the amorphous structure in Fig. 4c is quite open because only 43% of the space is occupied by silicon dioxide molecules. The relatively open structure accounts for the lower density and allows a variety of impurities (such as sodium) to enter and diffuse readily through the silicon dioxide layer.

The kinetics of thermal oxidation of silicon can be studied based on a simple model illustrated[2] in Fig. 5. A silicon slice contacts the oxidizing species (oxygen or water vapor), resulting in a surface concentration of C_0 molecules/cm^3 for these species. The magnitude of C_0 equals the equilibrium bulk concentration of the species at the oxidation temperature. The equilibrium concentration generally is proportional to the partial pressure of the oxidant adjacent to the oxide surface. At 1000°C and at a pressure of 1 atm, the concentration C_0 is 5.2×10^{16} molecules/cm^3 for dry oxygen and 3×10^{19} molecules/cm^3 for water vapor.

The oxidizing species diffuses through the silicon dioxide layer, resulting in a concentration C_s at the surface of silicon. The flux F_1 can be written as

$$F_1 = D \frac{dC}{dx} \cong \frac{D(C_0 - C_s)}{x},$$ (3)

where D is the diffusion coefficient of the oxidizing species and x is the thickness of the oxide layer already present.

At the silicon surface, the oxidizing species reacts chemically with silicon. Assuming the rate of reaction is proportional to the concentration of the species at the silicon surface, the flux F_2 is given by

$$F_2 = \kappa C_s,$$ (4)

where κ is the surface reaction rate constant for oxidation. At the steady state, $F_1 = F_2 = F$. Combining Eqs. 3 and 4 gives

$$F = \frac{DC_0}{x + (D/\kappa)}.$$ (5)

The reaction of the oxidizing species with silicon forms silicon dioxide. Let C_1 be the number of molecules of the oxidizing species in a unit volume of the oxide. There are 2.2×10^{22} silicon dioxide molecules/cm^3 in the oxide, and we add one oxygen molecule

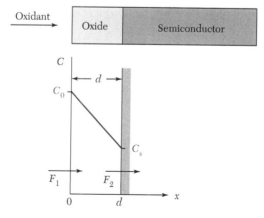

Fig. 5 Basic model for the thermal oxidation of silicon.[2]

(O_2) to each silicon dioxide molecule, whereas we add two water molecules (H_2O) to each silicon dioxide molecule. Therefore, C_1 for oxidation in dry oxygen is 2.2×10^{22} cm^{-3}, and for oxidation in water vapor it is twice this number (4.4×10^{22} cm^{-3}). Thus, the growth rate of the oxide layer thickness is given by

$$\frac{dx}{dt} = \frac{F}{C_1} = \frac{DC_0/C_1}{x + (D/\kappa)}. \tag{6}$$

We can solve this differential equation subject to the initial condition, $x(0) = d_0$, where d_0 is the initial oxide thickness; d_0 can also be regarded as the thickness of oxide layer grown in an earlier oxidation step. Solving Eq. 6 yields the general relationship for the oxidation of silicon,

$$x^2 + \frac{2D}{\kappa}x = \frac{2DC_0}{C_1}(t + \tau), \tag{7}$$

where $\tau \equiv (d_0^2 + 2Dd_0/\kappa)C_1/2DC_0$, which represents a time coordinate shift to account for the initial oxide layer d_0.

The oxide thickness after an oxidizing time t is given by

$$x = \frac{D}{k}\left[\sqrt{1 + \frac{2C_0\kappa^2(t + \tau)}{DC_1}} - 1\right]. \tag{8}$$

For small values of t, Eq. 8 reduces to

$$x \cong \frac{C_0\kappa}{C_1}(t + \tau), \tag{9}$$

and for larger values of t, it reduces to

$$x \cong \sqrt{\frac{2DC_0}{C_1}(t + \tau)}. \tag{10}$$

During the early stages of oxide growth, when surface reaction is the rate-limiting factor, the oxide thickness varies linearly with time. As the oxide layer becomes thicker, the oxidant must diffuse through the oxide layer to react at the silicon-silicon dioxide interface and the reaction becomes diffusion limited. The oxide growth then becomes proportional to the square root of the oxidizing time, which results in a parabolic growth rate.

Equation 7 is often written in a more compact form:

$$\boxed{x^2 + Ax = B(t + \tau).} \tag{11}$$

where $A = 2D/\kappa$, $B = 2DC_0/C_1$ and $B/A = \kappa C_0/C_1$. Using this form, Eqs. 9 and 10 can be written as

$$\boxed{x = \frac{B}{A}(t + \tau)} \tag{12}$$

for the linear region and as

$$x^2 = B(t + \tau).$$ (13)

for the parabolic region. For this reason, the term B/A is referred to as the linear rate constant and B is the parabolic rate constant. Experimentally measured results agree with the predictions of this model over a wide range of oxidation conditions. For wet oxidation, the initial oxide thickness d_0 is very small, or $\tau \cong 0$. However, for dry oxidation, the extrapolated value of d_0 at $t = 0$ is about 20 nm.

The temperature dependence of the linear rate constant B/A is shown in Fig. 6 for both dry and wet oxidation and for (111)- and (100)-oriented silicon wafers.[2] The linear rate constant varies as exp $(-E_a/kT)$, where the activation energy E_a is about 2 eV for both dry and wet oxidation. This closely agrees with the energy required to break silicon-silicon bonds, 1.83 eV/molecule. Under a given oxidation condition, the linear rate constant depends on crystal orientation. This is because the rate constant is related to the rate of incorporation of oxygen atoms into the silicon. The rate depends on the surface bond structure of silicon atoms, making it orientation dependent. Because the density of available bonds on the (111)-plane is higher than that on the (100)-plane, the linear rate constant for (111)-silicon is larger.

Figure 7 shows the temperature dependence of the parabolic rate constant B, which can also be described by exp $(-E_a/kT)$. The activation energy E_a is 1.24 eV for dry oxidation. The comparable activation energy for oxygen diffusion in fused silica is 1.18 eV. The corresponding value for wet oxidation, 0.71 eV, compares favorably with the value of 0.79 eV for the activation energy of diffusion of water in fused silica. The parabolic rate constant is independent of crystal orientation. This independence is expected because

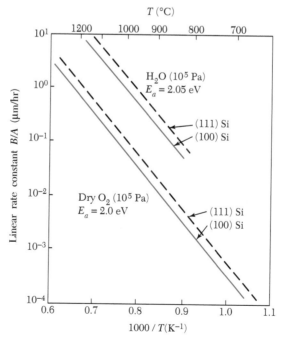

Fig. 6 Linear rate constant versus temperature.[2]

it is a measure of the diffusion process of the oxidizing species through a random network layer of amorphous silica.

Although oxides grown in dry oxygen have the best electrical properties, considerably more time is required to grow the same oxide thickness at a given temperature in dry oxygen than in water vapor. For relatively thin oxides such as the gate oxide in a MOSFET (typically ≤ 20 nm), dry oxidation is used. However, for thicker oxides such as field oxides (≥20 nm) in MOS integrated circuits, and for bipolar devices, oxidation in water vapor (or steam) is used to provide both adequate isolation and passivation.

Figure 8 shows the experimental results of silicon dioxide thickness as a function of reaction time and temperature for two substrate orientations.[3] Under a given oxidation condition, the oxide thickness grown on a (111)-substrate is larger than that grown on a (100)-substrate because of the larger linear rate constant of the (111)-orientation. Note that for a give temperature and time, the oxide film obtained using wet oxidation is about 5–10 times thicker than that using dry oxidation.

11.1.2 Thin Oxide Growth

Relative slow growth rates must be used to reproducibly grow thin oxide films of precise thickness. Various approaches to achieve such slower growth rates have been reported, including growth in dry O_2 at atmospheric pressure and lower temperatures (800°–900°C); growth at pressures lower than atmospheric pressure; growth in a reduced partial pressures of O_2 by using a diluent inert gas, such as N_2, Ar, or He, together with the gas containing the oxidizing species; and the use of composite oxide films with the gate-oxide films consisting of a layer of thermally grown SiO_2 and an overlayer of chemical-vapor

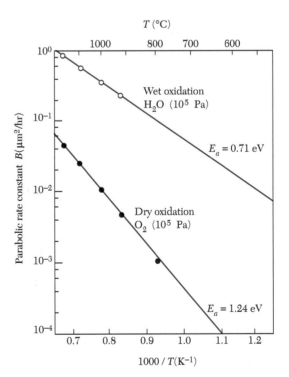

Fig. 7   Parabolic rate constant versus temperature.[2]

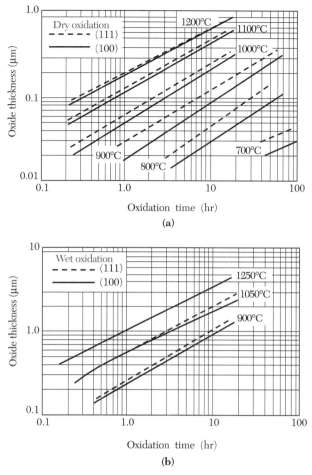

Fig. 8 Experimental results of silicon dioxide thickness as a function of reaction time and temperature for two substrate orientations. (*a*) Growth in dry oxygen. (*b*) Growth in steam.[3]

deposition (CVD) SiO_2. However, the mainstream approach for gate oxides 10–15 nm thick is to grow the oxide film at atmospheric pressure and lower temperatures (800°–900°C). With this approach, processing using modern *vertical* oxidation furnaces can grow reproducible, high-quality 10 nm oxides to within 0.1 nm across the wafer.

We noted earlier that for dry oxidation, there is an apparently rapid oxidation that gives rise to an initial oxide thickness d_0 of about 20 nm. Therefore, the simple model presented in Section 11.1.1 is not valid for dry oxidation with an oxide thickness ≤ 20 nm. For ultralarge-scale integration (ULSI), the ability to grow thin (5 ~ 20 nm), uniform, high-quality reproducible gate oxides has become increasingly important. We briefly consider the growth mechanisms of such thin oxides.

In the early stage of growth in dry oxidation, there is a large compressive stress in the oxide layer that reduces the oxygen diffusion coefficient in the oxide. As the oxide becomes thicker, the stress will be reduced due to the viscous flow of silica and the diffusion coefficient will approach its stress-free value. Therefore, for thin oxides, the value of D/κ may be sufficiently small that we can neglect the term Ax in Eq. 11 and obtain

$$x^2 - d_0{}^2 = Bt, \tag{14}$$

where d_0 is equal to $\sqrt{2DC_0\tau/C_1}$, which is the initial oxide thickness when time is extrapolated to zero, and B is the parabolic rate constant defined previously. We therefore expect the initial growth in dry oxidation to follow a parabolic form.

11.2 DIELECTRIC DEPOSITION

Deposited dielectric films are used mainly for insulation and passivation of discrete devices and integrated circuits. There are three commonly used deposition methods: atmospheric-pressure CVD, low-pressure CVD (LPCVD), and plasma-enhanced chemical vapor deposition (PECVD, or plasma deposition). PECVD is an energy-enhanced CVD method, in which plasma energy is added to the thermal energy of a conventional CVD system. Considerations in selecting a deposition process are the substrate temperature, the deposition rate and film uniformity, the morphology, the electrical and mechanical properties, and the chemical composition of the dielectric films.

The reactor for atmospheric-pressure CVD is similar to the one shown in Fig. 2, except that different gases are used at the gas inlet. In a hot-wall, reduced-pressure reactor as shown in Fig. 9a, the quartz tube is heated by a three-zone furnace, and gas is introduced at one end and pumped out at the opposite end. The semiconductor wafers are held vertically in a slotted quartz boat.[4] The quartz tube wall is hot because it is adjacent to the

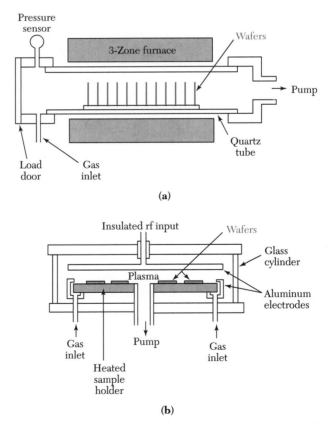

Fig. 9 Schematic diagrams of chemical-vapor deposition reactors. (*a*) Hot-wall, reduced-pressure reactor. (*b*) Parallel-plate plasma deposition reactor.[4] rf, radio frequency.

furnace, in contrast to a cold-wall reactor such as the horizontal epitaxial reactor that uses radio frequency (rf) heating.

The parallel-plate, radial-flow, PECVD reactor shown in Fig. 9*b* consists of a cylindrical glass or aluminum chamber sealed with aluminum endplates. Inside are two parallel aluminum electrodes. An rf voltage is applied to the upper electrode, whereas the lower electrode is grounded. The rf voltage causes a plasma discharge between the electrodes. Wafers are placed on the lower electrode, which is heated between 100° and 400°C by resistance heaters. The reaction gases flow through the discharge from outlets located along the circumference of the lower electrode. The main advantage of this reactor is its low deposition temperature. However, its capacity is limited, especially for large-diameter wafers, and the wafers may become contaminated if loosely adhering deposits fall on them.

11.2.1 Silicon Dioxide

CVD silicon dioxide does not replace thermally grown oxides because the best electrical properties are obtained with thermally grown films. CVD oxides are used instead to complement the thermal oxides. A layer of undoped silicon dioxide is used to insulate multilevel metallization, to mask ion implantation and diffusion, and to increase the thickness of thermally grown field oxides. Phosphorus-doped silicon dioxide is used both as an insulator between metal layers and as a final passivation layer over devices. Oxides doped with phosphorus, arsenic, or boron are used occasionally as diffusion sources.

Deposition Methods

Silicon dioxide films can be deposited by several methods. For low-temperature deposition (300°–500°C), the films are formed by reacting silane, dopant, and oxygen. The chemical reactions for phosphorus-doped oxides are

$$SiH_4 + O_2 \xrightarrow{\ 450°C\ } SiO_2 + 2H_2, \tag{15}$$

$$4PH_3 + 5O_2 \xrightarrow{\ 450°C\ } 2P_2O_5 + 6H_2. \tag{16}$$

The deposition process can be performed either at atmospheric pressure in a CVD reactor or at reduced pressure in an LPCVD reactor (Fig. 9*a*). The low deposition temperature of the silane-oxygen reaction makes it a suitable process when films must be deposited over a layer of aluminum.

For intermediate-temperature deposition (500°–800°C), silicon dioxide can be formed by decomposing tetraethylorthosilicate, $Si(OC_2H_5)_4$, in an LPCVD reactor. The compound, abbreviated TEOS, is vaporized from a liquid source. The TEOS compound decomposes as follows:

$$Si(OC_2H_5)_4 \xrightarrow{\ 700°C\ } SiO_2 + by\text{-}products, \tag{17}$$

forming both SiO_2 and a mixture of organic and organosilicon by-products. Although the higher temperature required for the reaction prevents its use over aluminum, it is suitable for polysilicon gates requiring a uniform insulating layer with good step coverage. The good step coverage is a result of enhanced surface mobility at higher temperatures. The oxides can be doped by adding small amounts of the dopant hydrides (phosphines, arsine, or diborane), similar to the process in epitaxial growth.

The deposition rate as a function of temperature varies as $e^{-E_a/kT}$, where E_a is the activation energy. The E_a of the silane-oxygen reaction is quite low: about 0.6 eV for undoped oxides and almost zero for phosphorus doped oxide. In contrast, E_a for the TEOS reaction is much higher: about 1.9 eV for undoped oxide and 1.4 eV when phosphorus doping compounds are present. The dependence of the deposition rate on TEOS partial pressure is proportional to $(1 - e^{-P/P_0})$, where P is the TEOS partial pressure and P_0 is about 30 Pa. At low TEOS partial pressures, the deposition rate is determined by the rate of the surface reaction. At high partial pressures, the surface becomes nearly saturated with adsorbed TEOS and the deposition rate becomes essentially independent of TEOS pressure.[4]

Recently, atmospheric-pressure and low-temperature CVD processes using TEOS and ozone (O_3) have been proposed,[5] as shown in Fig. 10. This CVD technology produces oxide films with high conformality and low viscosity under low deposition temperature. In addition, the shrinkage of oxide film during annealing is also a function of ozone concentration, as shown in Fig. 11. Because of their porosity, O_3-TEOS CVD oxides are often accompanied by plasma-assisted oxides to permit planarization in ULSI processing.

For high-temperature deposition (900°C), silicon dioxide is formed by reacting dichlorosilane, $SiCl_2H_2$, with nitrous oxide at reduced pressure:

$$SiCl_2H_2 + 2N_2O \xrightarrow{900°C} SiO_2 + 2N_2 + 2HCl. \tag{18}$$

This deposition gives excellent film uniformity and is sometimes used to deposit insulating layers over polysilicon.

Properties of Silicon Dioxide

Deposition methods and properties of silicon dioxide films are listed[4] in Table 1. In general, there is a direct correlation between deposition temperature and film quality. At higher temperatures, deposited oxide films are structurally similar to silicon dioxide that has been thermally grown.

The lower densities occur in films deposited below 500°C. Heating deposited silicon dioxide at temperatures between 600° and 1000°C causes densification, during which the oxide thickness decreases, whereas the density increases to 2.2 g/cm³. The refractive index of silicon dioxide is 1.46 at a wavelength of 0.6328 μm. Oxides with lower indices are porous, such as the oxide from the silane-oxygen deposition, which has a refractive index of 1.44. The porous nature of the oxide also is responsible for the lower dielectric strength, which is the applied electric field that will cause a high current to flow in the oxide film. The etch rates of oxides in a hydrofluoric acid solution depend on deposition

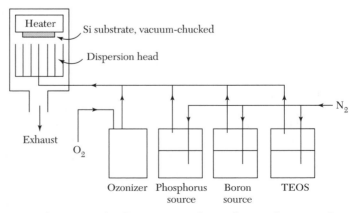

Fig. 10 Experimental apparatus for the O_3–TEOS chemical-vapor deposition (CVD) system.

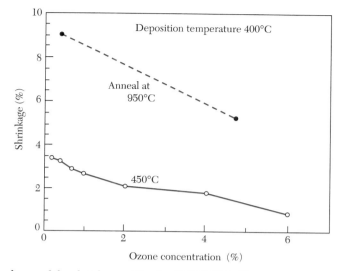

Fig. 11 Dependence of the shrinkage of the O$_3$–TEOS CVD film on ozone concentration using annealing. (Courtesy of SAMCO Company, Japan.)

temperature, annealing history, and dopant concentration. Usually higher-quality oxides are etched at lower rates.

Step Coverage

Step coverage relates the surface topography of a deposited film to the various steps on the semiconductor substrate. In the illustration of ideal, or conformal, step coverage shown in Fig. 12a, film thickness is uniform along all surfaces of the step. The uniformity of the film thickness, regardless of topography, is due to the rapid migration of reactants after adsorption on the step surfaces.[6]

Figure 12b shows an example of nonconformal step coverage, which results when the reactants adsorb and react without significant surface migration. In this instance, the deposition rate is proportional to the arrival angle of the gas molecules. Reactants arriving along the top horizontal surface come from many different angles and ϕ_1, the arrival

TABLE 1 Properties of SiO$_2$ Films

Property	Thermally grown at 1000°C	SiH$_4$ + O$_2$ at 450°C	TEOS at 700°C	SiCl$_2$H$_2$ + N$_2$O at 900°C
Composition	SiO$_2$	SiO$_2$ (H)	SiO$_2$	SiO$_2$(Cl)
Density (g/cm^3)	2.2	2.1	2.2	2.2
Refractive index	1.46	1.44	1.46	1.46
Dielectric strength (10^6 V/cm)	>10	8	10	10
Etch rate (Å /min) (100:1 H$_2$O:HF)	30	60	30	30
Etch rate (Å /min) (buffered HF)	440	1200	450	450
Step coverage	—	Nonconformal	Conformal	Conformal

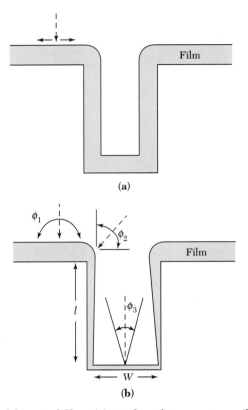

Fig. 12 Step coverage of deposited films. (*a*) Conformal step coverage. (*b*) Nonconformal step coverage.[4]

angle, varies in two dimensions, from 0 to 180°, whereas reactants arriving at the top of a vertical wall have an arrival angle ϕ_2 that varies from 0° to 90°. Thus, the film thickness on the top surface is double that of a wall surface. Further down the wall, ϕ_3 is related to the width of the opening, and the film thickness is proportional to

$$\phi_3 \cong \arctan \frac{W}{l}, \tag{19}$$

where l is the distance from the top surface and W is the width of the opening. This type of step coverage is thin along the vertical walls, with a possible crack at the bottom of step caused by self-shadowing.

Silicon dioxide formed by TEOS decomposition at reduced pressure gives a nearly conformal coverage due to rapid surface migration. Similarly, the high-temperature dichlorosilane-nitrous oxide reaction also results in conformal coverage. However, during silane-oxygen deposition, no surface migration takes place and the step coverage is determined by the arrival angle. Most evaporated or sputtered materials have a step coverage similar to that in Fig. 12*b*.

P-Glass Flow

A smooth topography is usually required for the deposited silicon dioxide used as an insulator between metal layers. If the oxide used to cover the lower metal layer is concave, circuit failure may result from an opening that may occur in the upper metal layer during deposition. Because phosphorus-doped silicon dioxide (P-glass) deposited at low temperatures becomes soft and flows upon heating, it provides a smooth surface and is often used to insulate adjacent metal layers. This process is called P-glass flow.

Figure 13 shows four cross sections of scanning electron microscope photographs of P-glass covering a polysilicon step.[6] All samples are heated in steam at 1100°C for 20 min. Figure 13a shows a sample of glass that contains a negligibly small amount of phosphorus and does not flow. Note the concavity of the film and that the corresponding angle θ is about 120°. Figures 13b, 13c, and 13d show samples of P-glass with progressively higher phosphorus contents up to 7.2 wt% (weight percent). In these samples the decreasing step angles of the P-glass layer indicate how flow increases with phosphorus concentration. P-glass flow depends on annealing time, temperature, phosphorus concentration, and the annealing ambient.[6]

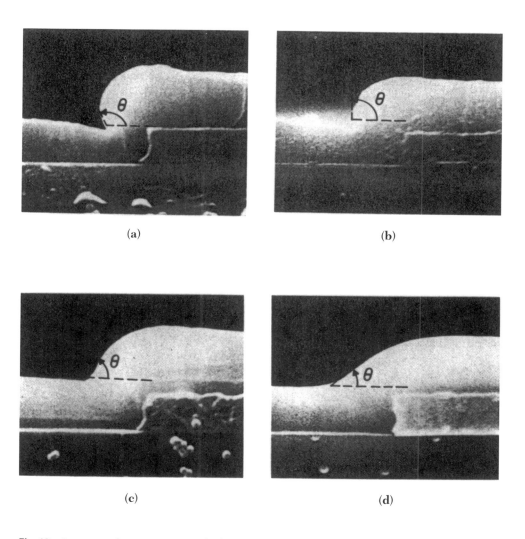

(a)

(b)

(c)

(d)

Fig. 13 Scanning-electron micrographs (10,000×) of samples annealed in steam at 1100°C for 20 minutes for the following weight percent of phosphorus.[6] (a) 0 wt%; (b) 2.2 wt%; (c) 4.6 wt%; and (d) 7.2 wt%.

The angle θ as a function of weight percent of phosphorus as shown in Fig. 13 can be approximated by

$$\theta \cong 120° \left(\frac{10 - \text{wt}\%}{10} \right). \tag{20}$$

If we want an angle smaller than 45° we require a phosphorus concentration larger than 6 wt%. However, at concentrations above 8 wt% , the metal film (e.g., aluminum) may be corroded by the acid products formed during the reaction between the phosphorus in the oxide and atmospheric moisture. Therefore, the P-glass flow process uses phosphorus concentrations of 6–8 wt%.

11.2.2 Silicon Nitride

It is difficult to grow silicon nitride by thermal nitridation (e.g., with ammonia, NH_3) because of its low growth rate and high growth temperature. However, silicon nitride films can be deposited by an intermediate-temperature (750°C) LPCVD process or a low-temperature (300°C) plasma-assisted CVD process.[7,8] The LPCVD films are of stoichiometric composition (Si_3N_4) with high density (2.9–3.1 g/cm^3). These films can be used to passivate devices because they serve as good barriers to the diffusion of water and sodium. The films also can be used as masks for the selective oxidation of silicon because silicon nitride oxidizes very slowly and prevents the underlying silicon from oxidizing. The films deposited by plasma-assisted CVD are not stoichiometric and have a lower density (2.4–2.8 g/cm^3). Because of the low deposition temperature, silicon nitride films can be deposited over fabricated devices and serve as their final passivation. The plasma-deposited nitride provides excellent scratch protection, serves as a moisture barrier, and prevents sodium diffusion.

In the LPCVD process, dichlorosilane and ammonia react at reduced pressure to deposit silicon nitride at temperatures between 700° and 800°C. The reaction is

$$3SiCl_2H_2 + 4NH_3 \xrightarrow{\sim 750°C} Si_3N_4 + 6HCl + 6H_2. \tag{21}$$

Good film uniformity and high wafer throughout (the number of wafers processed per hour) are advantages of the reduced-pressure process. As in the case of oxide deposition, silicon nitride deposition is controlled by temperature, pressure, and reactant concentration. The activation energy for deposition is about 1.8 eV. The deposition rate increases with increasing total pressure or dichlorosilane partial pressure and decreases with an increasing ammonia-to-dichlorosilane ratio.

Silicon nitride deposited by LPCVD is an amorphous dielectric containing up to 8 atomic percent (at %) hydrogen. The etch rate in buffered HF is less than 1 nm/min. The film has a very high tensile stress of approximately 10^{10} $dynes/cm^2$, which is nearly 10 times that of TEOS-deposited SiO_2. Films thicker than 200 nm may crack because of the very high stress. The resistivity of silicon nitride at room temperature is about 10^{16} Ω-cm. Its dielectric constant is 6 and its dielectric strength is 10^7 V/cm.

In the plasma-assisted CVD process, silicon nitride is formed either by reacting silane and ammonia in an argon plasma or by reacting silane in a nitrogen discharge. The reactions are as follows:

$$SiH_4 + NH_3 \xrightarrow{300°C} SiNH + 3H_2, \tag{22a}$$

$$2SiH_4 + N_2 \xrightarrow{300°C} 2SiNH + 3H_2. \tag{22b}$$

The products depend strongly on deposition conditions. The radial-flow, parallel-plate reactor (Fig. 9b) is used to deposit the films. The deposition rate generally increases with increasing temperature, power input, and reactant gas pressure.

Large concentrations of hydrogen are contained in plasma-deposited films. The plasma nitride (also referred to as SiN) used in semiconductor processing generally contains 20–25 at % hydrogen. Films with low tensile stress (~2×10^9 dynes/cm^2) can be prepared by plasma deposition. Film resistivities range from 10^5 to 10^{21} Ω-cm, depending on silicon-to-nitrogen ratio, whereas dielectric strengths are between 1×10^6 and 6×10^6 V/cm.

11.2.3 Low-Dielectric–Constant Materials

As devices continue to scale down to the deep submicron region, they require multilevel interconnection architecture to minimize the time delay due to parasitic resistance (R) and capacitance (C). The gain in device speed at the gate level will be offset by the propagation delay at the metal interconnects because of the increased RC time constant, as shown in Fig. 14. For example, in devices with gate length of 250 nm or less, up to 50% of the time delay is due to the RC delay of long interconnections.[9] Therefore, the device interconnection network becomes a limiting factor in determining chip performance such as device speed, cross talk, and power consumption of ULSI circuits.

In order to reduce the RC time constant of ULSI circuits, interconnection materials with low resistivity and interlayer films with low capacitance are required. On the low-capacitance topic ($C = \varepsilon_i A/d$, where ε_i is the dielecric permittivity, A the area, and d the

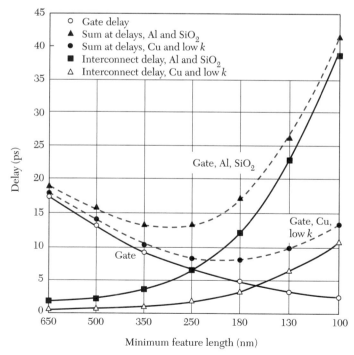

Fig. 14 Calculated gate and interconnect delay versus technology generation. The dielectric constant for the low-k material is 2.0. Both Al and Cu interconnects are 0.8 μm thick and 43 μm long.

thickness of the dielectric film), it is not easy to lower the parasitic capacitance by increasing thickness d of the interlayer dielectric (which makes gap filling more difficult), or decreasing wiring height and area A (which results in the increase of interconnect resistance). Therefore, materials with low dielectric constant (low-k) are required. The ε_i is equal to the multiplication of k and ε_0, where k and ε_0 are the dielectric constant and the vacuum permittivity, respectively.

Material Options

The properties of the interlayer dielectric film and how they are formed have to meet the following requirements: low dielectric constant, low residual stress, high planarization capability, high capability for gap filling, low deposition temperature, simplicity of process, and ease of integration.

A substantial number of low-k materials have been synthesized for the intermetal dielectric in ULSI circuits. Some of the promising low-k materials are shown in Table 2. These materials can be either inorganic or organic and can be deposited by either CVD or spin-on techniques.[9]

▶ **EXAMPLE 2**

Estimate the intrinsic RC value of two parallel Al wires 0.5 μm × 0.5 μm in cross section, 1 mm in length, and separated by a polyimide ($k \sim 2.7$) dielectric layer that is 0.5 μm thick. The resistivity of Al is 2.7 μΩ–cm.

SOLUTION

$$RC = \left(\rho \frac{\ell}{t_m^2} \right) \times \left(\varepsilon_i \frac{t_m \times \ell}{\text{spacing width}} \right) = \left(2.7 \times 10^{-6} \times \frac{1 \times 10^{-1}}{0.25 \times 10^{-8}} \right) \times$$

$$\left(8.85 \times 10^{-14} \times 2.7 \times \frac{0.5 \times 10^{-4} \times 10^{-1}}{0.5 \times 10^{-4}} \right) = 2.57 \text{ ps.} \quad ◀$$

TABLE 2   Low-k Materials

Determinant	Materials	Dielectric constant
Vapor-phase deposition polymers	Fluorosilicate glass (FSG)	3.5–4.0
	Parylene N	2.6
	Parylene F	2.4–2.5
	Black diamond (C-coped oxide)	2.7–3.0
	Fluorinated hydrocarbon	2.0–2.4
	Teflon-AF	1.93
Spin-on polymers	HSQ/MSQ	2.8–3.0
	Polyimide	2.7–2.9
	SiLK (aromatic hydrocarbon polymer)	2.7
	PAE [poly(arylene ethers)]	2.6
	Fluorinated amorphous carbon	2.1
	Xerogels (porous silica)	1.1–2.0

11.2.4 High-Dielectric–Constant Materials

High-k materials are required for ULSI circuits, especially for dynamic random access memory (DRAM). The storage capacitor in a DRAM has to maintain a certain value of capacitance for proper operation (e.g., 40 fF). For a given capacitance ($\varepsilon_i A/d$), usually a minimum d is selected to meet the conditions of the maximum allowed leakage current and the minimum required breakdown voltage. The area of the capacitor can be increased by using stacked or trench structures. These structures are considered in Chapter 14. However, for a planar structure, area A is reduced with increasing DRAM density. Therefore, the dielectric constant of the film must be increased.

Several high-k materials have been proposed, including titanates such as barium strontium titanate (BST) and lead zirconium titanate (PZT). They are shown in Table 3. In addition, there are titanates doped with one or more acceptors, such as alkaline earth metals, or doped with one or more donors, such as rare earth elements. The tantalum oxide (Ta_2O_5) has a dielectric constant in a range of 20–30. As a reference, the dielectric constant of Si_3N_4 is in a range of 6–7 and that for SiO_2 is 3.9. A Ta_2O_5 film can be deposited by a CVD process using gaseous $TaCl_5$ and O_5 as the starting materials.

► **EXAMPLE 3**

A DRAM capacitor has the following parameters: capacitance C = 40 fF, cell size A = 1.28 μm², and dielectric constant k = 3.9 for silicon dioxide. If we replace SiO_2 with Ta_2O_5 (k = 25) without changing thickness, what is the equivalent cell area of the capacitor?

SOLUTION

$$C = \frac{\varepsilon_i A}{d},$$

$$\frac{3.9 \times 1.28}{d} = \frac{25 \times A}{d},$$

∴ Equivalent cell size $A = \dfrac{3.9}{25} \times 1.28 = 0.2\ \mu m^2$. ◄

TABLE 3 High-k Materials

	Materials	Dielectric constant
Binary	Ta_2O_5	25
	TiO_2	40
	Y_2O_3	17
	Si_3N_4	7
Paraelectric perovskite	$SrTiO_3$ (STO)	140
	$(Ba_{1-x}Sr_x)\,TiO_3$(BST)	300–500
	$Ba(Ti_{1-x}Zr_x)O_3$(BZT)	300
	$(Pb_{1-x}La_x)(Zr_{1-y}Ti_y)O_3$(PLZT)	800–1000
	$Pb(Mg_{1/3}Nb_{2/3})O_3$(PMN)	1000–2000
Ferroelectric perovskite	$Pb(Zr_{0.47}Ti_{0.53})O_3$(PZT)	>1000

▶ 11.3 POLYSILICON DEPOSITION

Using polysilicon as the gate electrode in MOS devices is a significant development in MOS technology. One important reason is that polysilicon surpasses aluminum for electrode reliability. Figure 15 shows the maximum time to breakdown for capacitors with both polysilicon and aluminum electrodes.[10] The polysilicon is clearly superior, especially for thinner gate oxides. The inferior time to breakdown of aluminum electrode is due to the migration of aluminum atoms into the thin oxide under an electrical field. Polysilicon is also used as a diffusion source to create shallow junctions and to ensure ohmic contact to crystalline silicon. Additional uses include the manufacture of conductors and high-value resistors.

A low-pressure reactor (Fig. 9a) operated between 600° and 650°C is used to deposit polysilicon by pyrolyzing silane according to the following reaction.

$$SiH_4 \xrightarrow{600°C} Si + 2H_2. \tag{23}$$

Of the two most common low-pressure processes one operates at a pressure of 25–130 Pa using 100% silane, whereas the other process involves a diluted mixture of 20%–30% silane in nitrogen at the same total pressure. Both processes can deposit polysilicon on hundreds of wafers per run with good uniformity (i.e., thickness within 5%)

Figure 16 shows the deposition rate at four deposition temperatures. At low silane partial pressure, the deposition rate is proportional to the silane pressure.[4] At higher silane

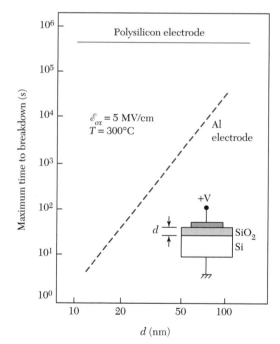

Fig. 15 Maximum time to breakdown versus oxide thickness for a polysilicon electrode and an aluminum electrode.[10]

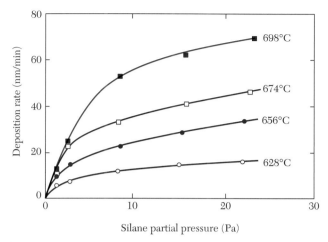

Fig. 16   Effect of silane concentration on the polysilicon deposition rate.[4]

concentrations, saturation of the deposition rate occurs. Deposition at reduced pressure is generally limited to temperatures between 600° and 650°C. In this temperature range, the deposition rate varies as exp $(\sim E_a/kT)$, where the activation energy E_a is 1.7 eV, which is essentially independent of the total pressure in the reactor. At higher temperatures, gas-phase reactions that result in a rough, loosely adhering deposit become significant and silane depletion will occur, causing poor uniformity. At temperatures much lower than 600°C, the deposition rate is too slow to be practical.

Process parameters that affect the polysilicon structure are deposition temperature, dopants, and the heat cycle applied following the deposition step. A columnar structure results when polysilicon is deposited at a temperature of 600°–650°C. This structure is comprised of polycrystalline grains ranging in size from 0.03 to 0.3 µm at a preferred orientation of (110). When phosphorus is diffused at 950°C, the structure changes to crystallite and grain size increases to a size between 0.5 and 1.0 µm. When temperature is increased to 1050°C during oxidation, the grains reach a final size of 1–3 µm. Although the initially deposited film appears amorphous when deposition occurs below 600°C, growth characteristics similar to the polycrystalline-grain columnar structure are observed after doping and heating.

Polysilicon can be doped by diffusion, ion implantation, or the addition of dopant gases during deposition, referred to as in situ doping. The implantation method is most commonly used because of its lower processing temperatures. Figure 17 shows the sheet resistance of single crystal silicon and of 500 nm polysilicon doped with phosphorus and antimony using ion implantation.[11] The ion implantation process is considered in Chapter 13. Implant dose, annealing temperature, and annealing time all influence the sheet resistance of implanted polysilicon. Carrier traps at the grain boundaries cause a very high resistance in the lightly implanted polysilicon. As Fig. 17 illustrates, resistance drops rapidly, approaching that of implanted single crystal silicon, as the carrier traps become saturated with dopants.

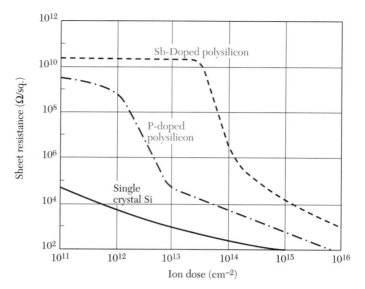

Fig. 17   Sheet resistance versus ion dose into 500 nm polysilicon at 30 keV.[11]

▶ 11.4 METALLIZATION

11.4.1   Physical-Vapor Deposition

The most common methods of physical-vapor deposition (PVD) of metals are evaporation, e-beam evaporation, plasma spray deposition, and sputtering. Metals and metal compounds such as Ti, Al, Cu, TiN, and TaN can be deposited by PVD. Evaporation occurs when a source material is heated above its melting point in an evacuated chamber. The evaporated atoms then travel at high velocity in straight-line trajectories. The source can be molten by resistance heating, by rf heating, or with a focused electron beam. Evaporation and e-beam evaporation were used extensively in earlier generations of integrated circuits, but they have been replaced by sputtering for ULSI circuits.

In ion-beam sputtering, a source of ions is accelerated toward the target and impinged on its surface. Figure 18*a* shows the standard sputtering system. The sputtered material deposits on a wafer that is placed facing the target. The ion current and energy can be independently adjusted. Since the target and wafer are placed in a chamber that has lower pressure, more target material and less contamination are transferred to the wafer.

One method to increase the ion density and, hence, the sputter-deposition rate is to use a third electrode that provides more electrons for ionization. Another method is to use a magnetic field, such as electron cyclotron resonance (ECR), to capture and spiral electrons, increasing their ionizing efficiency in the vicinity of the sputtering target. This technique, referred to as magnetron sputtering, has found widespread applications for the deposition of aluminum and its alloys at a rate that can approach 1 μm/min.

Long-through sputtering is another technique used to control the angular distribution. Figure 18*b* shows the long through-sputtering system. In standard sputtering configurations there are two primary reasons for a wide angular distribution of incident flux at the surface: the use of a small target to substrate separation, d_{ts}, and scattering of the flux by the working gas as the flux travels from the target to the substrate. These two factors are linked because a small d_{ts} is needed to achieve good throughput, uniformity, and film properties when there is substantial gas scattering. A solution to this problem is to sputter at very low pressures, a capability that has been developed using a variety of sys-

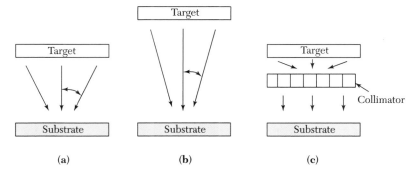

Fig. 18 (*a*) Standard sputtering, (*b*) long-through sputtering, and (*c*) sputtering with a collimator.

tems, which can sustain the magnetron plasma under more rarefied conditions. These systems allow for sputtering at working pressures of less than 0.1 Pa. At these pressures, gas scattering is less important, and the target-to-substrate distance can be greatly increased. From a simple geometrical argument, this allows the angular distribution to be greatly narrowed, which permits more deposition at the bottom of high-aspect features such as contact holes.

Contact holes with large aspect ratio are difficult to fill with material, mainly because scattering events cause the top opening of the hole to seal before appreciable material has deposited on its floor. This problem can be overcome by collimating the sputtered atoms by placing an array of collimating tubes just above the wafer to restrict the depositing flux to normal ±5°. Sputtering with a collimator is shown in Fig. 18*c*. Atoms whose trajectory is more than 5° from normal are deposited on the inner surface of the collimators.

11.4.2 Chemical-Vapor Deposition

CVD is attractive for metallization because it offers coatings that are conformal, has good step coverage, and can coat a large number of wafers at a time. The basic CVD setup is the same as that used for deposition of dielectrics and polysilicon (see Fig. 9*a*). Low-pressure CVD (LPCVD) is capable of producing conformal step coverage over a wide range of topographical profiles, often with lower electrical resistivity than that from PVD.

One of the major new applications of CVD metal deposition for integrated circuit production is in the area of refractory-metal deposition. For example, tungsten's low electrical resistivity (5.3 $\mu\Omega$-cm) and its refractory nature make it a desirable metal for use in integrated circuit fabrication.

CVD-W

Tungsten is used both as a contact plug and as a first-level metal. Tungsten can be deposited by using WF_6 as the W source gas, since it is a liquid that boils at room temperature. WF_6 can be reduced by silicon, hydrogen, or silane. The basic chemistry for CVD-W is as follows:

$$WF_6 + 3H_2 \rightarrow W + 6HF \text{ (hydrogen reduction)}, \tag{24}$$

$$2WF_6 + 3Si \rightarrow 2W + 3SiF_4 \text{ (silicon reduction)}, \tag{25}$$

$$2WF_6 + 3SiH_4 \rightarrow 2W + 3SiF_4 + 6H_2 \text{ (silane reduction)}. \tag{26}$$

On a Si contact, the selective process starts from a silicon reduction process. This process provides a nucleation layer of W grown on Si but not on SiO$_2$. The hydrogen reduction process can deposit W rapidly on the nucleation layer, forming the plug. The hydrogen reduction process provides excellent conformal coverage of the topography. This process, however, does not have perfect selectivity, and the HF gas by-product of the reaction is responsible for the encroachment of the oxide, as well as for the rough surface of deposited W films.

The silane reduction process gives a high deposition rate and much smaller W grain size than that obtained with the hydrogen reduction process. In addition, the problems of encroachment and rough W surface are eliminated because there is no HF by-product generation. Usually, a silane reduction process is used as the first step in blanket W deposition to serve as a nucleation layer and to reduce junction damage. After the silane reduction, hydrogen reduction is used to grow the blanket W layer.

CVD TiN

TiN is widely used as a diffusion barrier-metal layer in metallization and can be deposited by sputtering from a compound target or by CVD. The CVD TiN can provide better step coverage than PVD methods in deep submicron technology. CVD TiN can be deposited,[12–14] using TiCl$_4$ with NH$_3$, H$_2$/N$_2$, or NH$_3$/H$_2$:

$$6\mathrm{TiCl_4} + 8\mathrm{NH_3} \rightarrow 6\ \mathrm{TiN} + 24\mathrm{HCl} + \mathrm{N_2}, \tag{27}$$

$$2\mathrm{TiCl_4} + \mathrm{N_2} + 4\mathrm{H_2} \rightarrow 2\mathrm{TiN} + 8\mathrm{HCl}, \tag{28}$$

$$2\mathrm{TiCl_4} + 2\mathrm{NH_3} + \mathrm{H_2} \rightarrow 2\ \mathrm{TiN} + 8\mathrm{HCl}. \tag{29}$$

The deposition temperature is about 400°–700°C for NH$_3$ reduction and is higher than 700°C for the N$_2$/H$_2$ reaction. The higher the deposition temperature, the better the TiN film and the less Cl incorporated in TiN (~5%).

11.4.3 Aluminum Metallization

Aluminum and its alloys are used extensively for metallization in integrated circuits. The Al film can be deposited by a PVD or CVD method. Because aluminum and its alloys have low resistivities, 2.7 μΩ-cm for Al and up to 3.5 μΩ-cm for its alloys, these metals satisfy the low-resistance requirements. Aluminum also adheres well to silicon dioxide. However, the use of aluminum in integrated circuits with shallow junctions often creates problems, such as spiking and eletromigration. We consider the problems of aluminum metallization and their solutions in this section.

Junction Spiking

Figure 19 shows the phase diagram of the Al-Si system at 1 atm.[15] The phase diagram relates these two components as a function of temperature. The Al-Si system exhibits eutectic characteristics; that is, the addition of either component lowers the system's melting point below that of either metal. Here, the minimum melting temperature, called eutectic temperature, is 577°C, corresponding to a 11.3% Si and 88.7% Al composition. The melting points of pure aluminum and pure silicon are 660°C and 1412°C, respectively. Because of the eutectic characteristics, during aluminum deposition the temperature on the silicon substrate must be limited to less than 577°C.

The inset of Fig. 19 also shows the solid solubility of silicon in aluminum. For example, the solubility of silicon in aluminum is 0.25 wt% at 400°C, 0.5 wt% at 450°C, and 0.8 wt% at 500°C. Therefore, wherever aluminum contacts silicon, the silicon will dis-

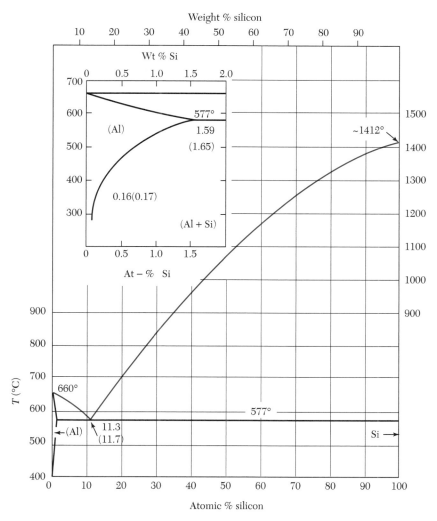

Fig. 19 Phase diagram of the aluminum–silicon system.[15]

solve into the aluminum during annealing. The amount of silicon dissolved will depend not only on the solubility at the annealing temperature but also on the volume of aluminum to be saturated with silicon. Consider a long aluminum metal line in contact with an area ZL of silicon as shown in Fig. 20. After an annealing time t, the silicon will diffuse a distance of approximately $\sqrt{Dt}$ along the aluminum line from the edge of the contact, where D is the diffusion coefficient given by $4 \times 10^{-2} \exp(-0.92/kT)$ for silicon diffusion in deposited aluminum films. Assuming that this length of aluminum is completely saturated with silicon, the volume of silicon consumed is then

$$\text{Vol} \cong 2\sqrt{Dt}\,(HZ)S\left(\frac{\rho_{\text{Al}}}{\rho_{\text{Si}}}\right), \tag{30}$$

where Al and ρ_{Si} are the densities of aluminum and silicon, respectively, and S is the solubility of silicon in aluminum at the annealing temperature.[16] If the consumption takes

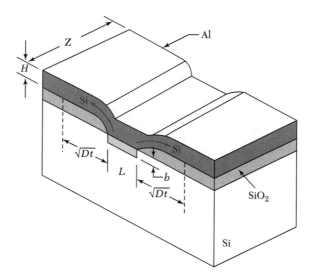

Fig. 20 Diffusion of silicon in aluminum metallization.[16]

place uniformly over the contact area A (where $A = ZL$ for uniform dissolution), the depth to which silicon would be consumed is

$$b \cong 2\sqrt{Dt}\left(\frac{HZ}{A}\right)S\left(\frac{\rho_{Al}}{\rho_{Si}}\right). \tag{31}$$

▷ **EXAMPLE 4**

For $T = 500°C$, $t = 30$ min, $ZL = 16$ μm², $Z = 5$ μm, and $H = 1$ μm. Find the depth b, assuming uniform dissolution.

SOLUTION   The diffusion coefficient of silicon in aluminum at 500°C is about 2×10^{-8} cm²/s; thus, $\sqrt{Dt}$ is 60 μm. The density ratio is 2.7/2.33 = 1.16.

At 500°C, S is 0.8 wt%. From Eq. 31 we have

$$b = 2 \times 60\left(\frac{1 \times 5}{16}\right)0.8\% \times 1.16 = 0.35 \text{ μm}.$$

Aluminum will fill a depth of $b = 0.35$ μm from which silicon is consumed. If at the contact point there is a shallow junction whose depth is less than b, the diffusion of silicon into aluminum can short-circuit the junction.   ◀

In a practical situation, the dissolution of silicon does not take place uniformly but rather at only a few points. The effective area in Eq. 31 is less than the actual contact area; hence b is much larger. Figure 21 illustrates the actual situation in the p–n junction area of aluminum penetrating the silicon at only the few points where spikes are formed. One way to minimize aluminum spiking is to add silicon to the aluminum by coevaporation until the amount of silicon contained by the alloy satisfies the solubility requirement. Another method is to introduce a barrier metal layer between the aluminum and the silicon substrate (Fig. 22). This barrier metal layer must meet the following requirements: it forms low contact resistance with silicon, it will not react with aluminum, and

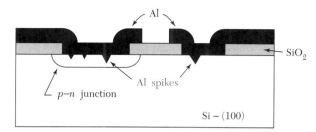

Fig. 21 Schematic view of aluminum films contacting silicon. Note the aluminum spiking in the silicon.

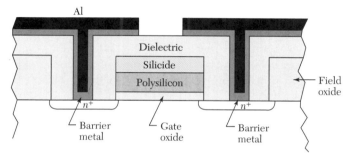

Fig. 22 Cross-sectional view of a MOSFET with a barrier metal between the aluminum and silicon and a composite gate electrode of silicide and polysilicon.

its deposition and formation are compatible with the overall process. Barrier metals such as titanium nitride (TiN) have been evaluated and found to be stable for contact annealing temperatures up to 550°C for 30 min.

Electromigration

In Chapter 6 we discussed scaled-down devices. As the device becomes smaller, the corresponding current density becomes larger. High current densities can cause device failure due to electromigration. The term electromigration refers to the transport of mass (i.e., atoms) in metals under the influence of current. It occurs by the transfer of momentum from the electrons to the positive metal ions. When a high current passes through thin metal conductors in integrated circuits, metal ions in some regions will pile up and voids will form in other regions. The pileup can short-circuit adjacent conductors, whereas the voids can result in an open circuit.

The mean time to failure (MTF) of a conductor due to electromigration can be related to the current density J and the activation energy E_a by

$$\text{MTF} \sim \frac{1}{J^2} \, \exp\!\left(\frac{E_a}{kT} \right). \tag{32}$$

Experimentally, a value of $E_a \cong 0.5$ eV is obtained for deposited aluminum. This indicates that low-temperature, grain-boundary diffusion is the primary vehicle of material transport, since $E_a \cong 1.4$ eV would characterize the self-diffusion of single-crystal aluminum. The electromigration resistance of aluminum conductors can be increased by using several techniques. These techniques include alloying with copper (e.g., Al with 0.5% Cu), encapsulating the conductor in a dielectric, or incorporating oxygen during film deposition.

11.4.3 Copper Metallization

It is well known that both high conductivity wiring and low–dielectric-constant insulators are required to lower the RC time delay of the interconnect network. Copper is the obvious choice for a new interconnection metallization because it has higher conductivity and higher electromigration resistance than aluminum. Copper can be deposited by PVD, CVD, and electrochemical methods. However, the use of Cu as an alternative material to Al in ULSI circuits has drawbacks, such as its tendency to corrode under standard chip manufacture conditions, its lack of a feasible dry-etching method or a stable self-passivating oxide similar to Al_2O_3 on Al, and its poor adhesion to dielectric materials, such as SiO_2 and low-k polymers. In this section, we discuss the copper metallization techniques.

Several different techniques for fabrication of multilevel Cu interconnects have been reported.[17,18] The first method is a conventional method to pattern the metal lines followed by dielectric deposition. The second method is to pattern the dielectric layer first and fill copper metal into trenches. This step is followed by chemical mechanical polishing, discussed in Section 11.4.5, to remove the excess metal on the top surface of dielectric and leave Cu material in the holes and trenches. This method is also known as a damascene process.

Damascene technology

The approach for fabricating a copper-polyimide interconnect structure is by the "damascene" or "dual damascene process" process. Figure 23 shows the dual damascene sequences for an advanced Cu interconnection structure. For a typical damascene structure, trenches for metal lines are defined and etched in the interlayer dielectric (ILD) and then followed by metal deposition of TaN/Cu . The TaN layer serves as a diffusion barrier layer and prevents copper from penetrating the low-k polyimide. The excess copper metal on the surface is removed to obtain a planar structure with metal inlays in the dielectric.

For the dual damascene process, the vias and trenches in the dielectric are defined using two lithography and reactive ion etching (RIE) steps before depositing the Cu metal (Fig. 23a–c). Then a Cu chemical-mechanical polishing process is used to remove the metal on the top surface, leaving the planarized wiring and via imbedded in the insulator.[19] One special benefit of dual damascene is that the via plug is now of the same material as the metal line and the risk of via electromigration failure is reduced.

▷ **EXAMPLE 5**

If we replace Al with Cu wire associated with some low-k dielectric ($k = 2.6$) instead of SiO_2 layer, what percentage of reduction of RC time constant will be achieved? (The resistivity of Al is 2.7 $\mu\Omega$-cm, and the resistivity of Cu is 1.7 $\mu\Omega$-cm.)

SOLUTION

$$\frac{1.7}{2.7} \times \frac{2.6}{3.9} \times 100\% = 42\%.$$ ◀

11.4.5 Chemical-Mechanical Polishing

In recent years, the development of chemical-mechanical polishing (CMP) has become increasingly important for multilevel interconnection because it is the only technology

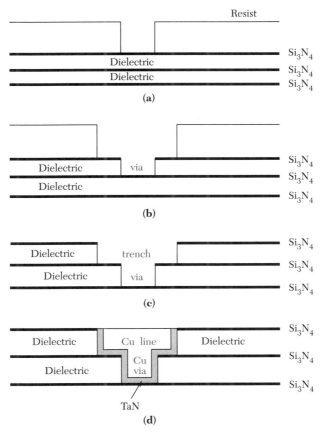

Fig. 23 Process sequence used to fabricate a Cu line-stud structure using dual damascene. (*a*) Resist stencil applied; (*b*) reactive ion etching dielectric and resist patterning; , (*c*) trench and via definition; and (*d*) Cu depositions followed by chemical-mechanical polishing (CMP).

that allows global planarization (i.e., to make a flat surface across the whole wafer). It offers many advantages over other types of technologies—better global planarization over large or small structures, reduced defect density, and the avoidance of plasma damage. Three CMP approaches are summarized in Table 4.

The CMP process consists of moving the sample surface against a pad that carries slurry between the sample surface and the pad. Abrasive particles in the slurry cause mechanical damage on the sample surface, loosening the material for enhanced chemical attack or fracturing off the pieces of surface into a slurry where they dissolve or are swept away. The process is tailored to provide an enhanced material removal rate from

TABLE 4   Three Methods of Chemical-Mechanical Polishing (CMP)

Method	Wafer facing	Platen movement	Slurry feeding
Rotary CMP	Down	Rotary against rotating wafer carrier	Dripping to pad surface
Orbital CMP	Down	Orbital against rotating wafer carrier	Through the pad surface
Linear CMP	Down	Linear against rotating wafer carrier	Dripping to pad surface

high points on surfaces, thus affecting the planarization because most chemical actions are isotropic. Mechanical grinding alone, theoretically, may achieve the desired planarization but is not desirable because of extensive associated damage to the material surfaces. There are three main parts of the process: the surface to be polished, the pad—the key media enabling the transfer of mechanical action to the surface being polished, and the slurry—which provides both chemical and mechanical effects. Figure 24 shows the CMP setup. [20]

▷ **EXAMPLE 6**

The oxide removal rate and the removal rate of a layer underneath the oxide (called a stop layer) are $1r$ and $0.1r$, respectively. To remove 1 μm oxide and 0.01 μm stop layer, the total removal time is 5.5 min. Find the oxide removal rate.

SOLUTION

$$\frac{1}{1r} + \frac{0.01}{0.1r} = 5.5$$

$$\frac{1.1}{1r} = 5.5, \qquad r = 0.2 \text{ μm / min.} \qquad \blacktriangleleft$$

11.4.6 Silicide

Silicon forms many stable metallic and semiconducting compounds with metals. Several metal silicides show low resisivity and high thermal stability, making them suitable for ULSI application. Silicides such as $TiSi_2$ and $CoSi_2$ have reasonably low resistivities and are generally compatible with integrated-circuit processing. Silicides become important metallization materials as devices become smaller. One important application of silicide is for the MOSFET gate electrode either alone or with doped polysilicon (polycide) above the gate oxide. Table 5 shows a comparison of titanium silicide and cobalt silicide.

Metal silicides have been used to reduce the contact resistance of the source and drain, the gate electrodes, and the interconnections. The self-aligned metal silicide technology (salicide) has been proven to be a highly attractive technique for improving the performance of submicron devices and circuits. The self-aligned process uses the silicide gate electrode as the mask to form the source and drain electrodes of a MOSFET (e.g., by ion implantation, considered in Chapter 13). This process can minimize the overlaps of these electrodes and thus reduce the parasitic capacitances.

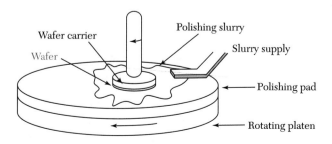

Fig. 24 A schematic of a CMP polisher.

TABLE 5   A Comparsion of TiSi$_2$ and CoSi$_2$ Films

Properties	TiSi$_2$	CoSi$_2$
Resistivity	13–16	22–28
Silicide/metal ratio	2.37	3.56
Silicide/Si ratio	1.04	0.97
Reactive to native oxide	Yes	No
Silicidation temperature (°C)	800–850	550–900
Film stress (dyne/cm^2)	1.5×10^{10}	1.2×10^{10}

Figure 25 shows the polycide and salicide process. A typical polycide formation sequence is shown in Fig. 25*a*. For sputter deposition, a high-temperature, high-purity compound target is used to ensure the quality of the silicide. The most commonly used silicides for the polycide process are WSi$_2$, TaSi$_2$, and MoSi$_2$. They are all refractory, thermally stable, and resistant to processing chemicals. A self-aligned silicide process is illustrated in Fig. 25*b*. In the process, the polysilicon gate is patterned without any silicide, and a sidewall spacer (silicon oxide or silicon nitride) is formed to prevent shorting the gate to the source and drain during the silicidation process. A metal layer, either Ti or Co, is blanket-sputtered on the entire structure, followed by silicide sintering. Silicide is formed, in principle, only where the metal is in contact with Si. A wet chemical wash then rinses off the unreacted metal, leaving only the silicide. This technique eliminates

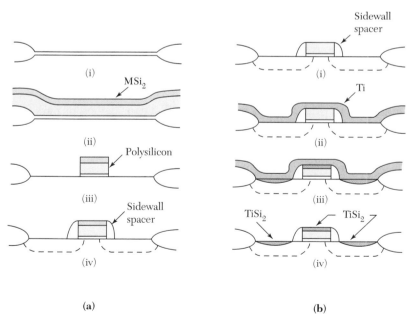

Fig. 25   Polycide and salicide processes. (*a*) Polycide structure: (i) gate oxide; (ii) polysilicon and silicide deposition, (iii) pattern polycide; and (iv) lightly doped drain (LDD) implant, sidewall formation, and *S/D* implant. (*b*) Salicide structure: (i) gate patterning (polysilicon only), LDD, sidewall, and *S/D* implant; (ii) metal (Ti,Co) deposition; (iii) anneal to form salicide; and (iv) selective (wet) etch to remove unreacted metal.

the need to pattern the composite polycide gate structure and adds silicide to the source/drain area to reduce the contact resistance.

The silicides are promising materials for ULSI circuits because of their low resistivity and excellent thermal stability. Cobalt silicide has been widely investigated recently because of its lowest resistivity and high-temperature thermal stability. However, cobalt is sensitive to native oxide as well as an oxygen-contained environment, and a large amount of silicon is consumed during silicidation.

▷ **EXAMPLE 7**

Calculate the thickness of the cobalt silicide. For a desired sheet resistance of 0.6 Ω/sq, the resistivity is 18 $\mu\Omega$-cm.

SOLUTION   The resistivity is equal to the product of the sheet resistance and the film thickness:

$$\rho = R_s \times t.$$

Then

$$t = \frac{\rho}{R_s} = \frac{18 \times 10^{-6}}{0.6} = 3 \times 10^{-5} \text{ cm} = 300 \text{ nm}. \qquad \blacktriangleleft$$

▷ SUMMARY

Modern semiconductor device fabrication requires the use of thin films. Currently, there are four important types of films—thermal oxides, dielectric layers, polycrystalline silicon, and metal films. The major issues related to film formation are low-temperature processing, step coverage, selective deposition, uniformity, film quality, planarization, throughput, and large-wafer capacity.

Thermal oxidation offers the best quality for the Si-SiO$_2$ interface and has the lowest interface trap density. Therefore, it is used to form the gate oxide and the field oxide. LPCVD of dielectrics and polysilicon offer conformal step coverage. In contrast, PVD and atmospheric-pressure CVD generally result in noncomformal step coverage. CMP offers global planarization and reduces defect density. Conformal step coverage and planarization are also required for precise pattern transfer at the deep-submicron lithography level. Pattern transfer technology is discussed in the next chapter.

To minimize the *RC* time delay due to parasitic resistance and capacitance, the silicide process for ohmic contacts, copper metallization for interconnects, and low–dielectric-constant materials for interlayer films are extensively used to meet the requirements of the multilevel interconnect structures of ULSI circuits. In addition, we have investigated high-dielectric-constant materials to improve the gate insulator performance and to increase the capacitance per unit area for DRAM.

▷ REFERENCES

1.   E. H. Nicollian and J. R. Brews, *MOS Physics and Technology*, Wiley, New York, 1982.

2.   B. E. Deal and A. S. Grove, "General Relationship for the Thermal Oxidation of Silicon," *J. Appl. Phys.*, **36**, 3770 (1965).

3.   J. D. Meindl, et al., "Silicon Epitaxy and Oxidation," in F. Van de wiele, W. L. Engl, and P. O. Jespers, Eds., *Process and Device Modeling for Integrated Circuit Design*, Noorhoff, Leyden, 1977.

4. For a discussion on film deposition, see, for example, A.C. Adams, "Dielectric and Polysilicon Film Deposition," in S. M. Sze, Ed., *VLSI Technology*, McGraw-Hill, New York, 1983.

5. K. Eujino, et al., "Doped Silicon Oxide Deposition by Atmospheric Pressure and Low Temperature Chemical Vapor Deposition Using Tetraethoxysilane and Ozone," *J. Electrochem. Soc.*, **138**, 3019 (1991).

6. A. C. Adams and C. D. Capio, "Planarization of Phosphorus-Doped Silicon Dioxide," *J. Electrochem. Soc.*, **127**, 2222 (1980).

7. T. Yamamoto et al., "An Advanced 2.5nm Oxidized Nitride Gate Dielectric for Highly Reliable 0.25 μm MOSFETs," *Symp. on VLSI Technol. Dig. of Tech. Pap*, 1997, p. 45.

8. K. Kumar, et al., "Optimization of Some 3 nm Gate Dielectrics Grown by Rapid Thermal Oxidation in a Nitric Oxide Ambient," *Appl. Phys. Lett.*, **70**, 384 (1997).

9. T. Homma, "Low Dielectric Constant Materials and Methods for Interlayer Dielectric Films in Ultralarge-Scale Integrated Circuit Multilevel Interconnects," *Mater. Sci. Eng.*, **23**, 243 (1998).

10. H. N. Yu, et al., "1 μm MOSFET VLSI Technology. Part I—An Overview," *IEEE Trans. Electron Devices*, **ED-26**, 318 (1979).

11. J. M. Andrews, "Electrical Conduction in Implanted Polycrystalline Sillicon," *J. Electron. Mater.*, **8**, 3, 227 (1979).

12. M. J. Buiting, A. F. Otterloo, and A. H. Montree, "Kinetical Aspects of the LPCVD of Titanium Nitride from Titanium Tetrachloride and Ammonia," *J. Electrochem. Soc.*, **138**, 500 (1991).

13. R. Tobe, et al., "Plasma-Enhanced CVD of TiN and Ti Using Low-Pressure and High-Density Helicon Plasma," *Thin Solid Film*, **281–282**, 155 (1996).

14. J. Hu, et al., "Electrical Properties of Ti/TiN Films Prepared by Chemical Vapor Deposition and Their Applications in Submicron Structures as Contact and Barrier Materials, " *Thin Solid Film*, **308**, 589 (1997).

15. M. Hansen and A. Anderko, *Constitution of Binary Alloys*, McGraw-Hill, New York, 1958.

16. D. Pramanik and A. N. Saxena, "VLSI Metallization Using Aluminum and Its Alloys," *Solid State Tech.*, **26**, No. 1, 127 (1983), **26**. No. 3, 131 (1983).

17. C. L. Hu, and J. M. E. Harper, "Copper Interconnections and Reliability," *Matter. Chem. Phys.*, **52**, 5 (1998)

18. P. C. Andricacos, et al., "Damascene Copper Electroplating for Chip Interconnects," *193rd Meet. Electrochem. Soc.*, 1998, p. 3

19. J. M. Steigerwald, et al., "Chemical Mechanical Planarization of Microelectronic Materials," Wiley, New York, 1997.

20. L. M. Cook, et al.,*Theoretical and Practical Aspects of Dielectric and Metal CMP*, Semicond. Int., p. 141 (1995).

▶ PROBLEMS (* DENOTES DIFFICULT PROBLEMS)

FOR SECTION 11.1   THERMAL OXIDATION

1. A *p*-type <100>-oriented, silicon wafer with a resistivity of 10 Ω-cm is placed in a wet oxidation system to grow a field oxide of 0.45 μm at 1050°C. Determine the time required to grow the oxide.

*2. After the first oxidation as given in Prob. 1, a window is opened in the oxide to grow a gate oxide at 1000°C for 20 minutes in dry oxidation. Find the thicknesses of the gate oxide and the total field oxide.

3. Show that Eq. 11 reduces to $x^2 = Bt$ for long times and to $x = B/A(t + \tau)$ for short times.

4. Determine the diffusion coefficient D for dry oxidation of <100>-oriented silicon samples at 980°C and 1 atm.

FOR SECTION 11.2   DIELECTRIC DEPOSITION

5. (a) In a plasma-deposited silicon nitride that contains 20 at% hydrogen and has a silicon-to-nitrogen ratio (Si/N) of 1.2, find x and y in the empirical formula of SiN_xH_y. (b) If the variation of film resistivity with Si/N ratio is given by $5 \times 10^{28} \exp(-33.3\gamma)$ for $2 > \gamma > 0.8$, where γ is the ratio, find the resistivity of the film in (a).

6. The dielectric constants of SiO_2, Si_3N_4, and Ta_2O_5 are about 3.9, 7.6, and 25, respectively. What is the capacitance ratio for the capacitors with the Ta_2O_5 and oxide/nitride/oxide dielectrics for the same dielectric thickness, provided the oxide/nitride/oxide has thickness ratio 1:1:1 for the oxide to the nitride?

7. In Prob. 6, if BST with a dielectric constant of 500 is chosen to replace Ta_2O_5, calculate the area reduction ratio to maintain the same capacitance if the two films have the same thickness.

8. In Prob. 6, calculate the equivalent thickness of the Ta_2O_5 in terms of SiO_2 thickness if both have the same capacitance. Assume the actual thickness of Ta_2O_5 is $3t$.

9. In a silane-oxygen reaction to deposit undoped SiO_2 film, the deposition rate is 15 nm/min at 425°C. What temperature is required to double the deposition rate?

10. The P-glass flow process requires temperatures above 1000°C. As device dimensions become smaller in ULSI, we must use lower temperatures. Suggest methods to obtain a smooth topography at < 900°C for deposited silicon dioxide that can be used as an insulator between metal layers.

FOR SECTION 11.3   POLYSILICON DEPOSITION

11. Why is silane more often used for polysilicon deposition than silicon chloride?

12. Explain why the deposition temperature for polysilicon films is moderately low, usually between 600°–650°C.

FOR SECTION 11.4   METALLIZATION

13. An e-beam evaporation system is used to deposit aluminum to form MOS capacitors. If the flatband voltage of the capacitance is shifted by 0.5 V because of e-beam radiation, find the number of fixed oxide charges (the silicon dioxide thickness is 50 nm). How can these charges be removed?

14. A metal line ($L = 20\ \mu m$, $W = 0.25\ \mu m$) has a sheet resistance 5 Ω/sq. Calculate the resistance of the metal line.

15. Calculate the thickness of the $TiSi_2$ and $CoSi_2$, where the initial Ti and Co film thickness are 30 nm.

16. Compare the advantages and disadvantages of $TiSi_2$ and $CoSi_2$ for salicide applications.

17. A dielectric material is placed between the two parallel metal lines. The length $L = 1$ cm, width $W = 0.28\ \mu m$, thickness $T = 0.3\ \mu m$, and spacing $S = 0.36\ \mu m$. (a) Calculate the RC time delay. The metal is Al with a resistivity of 2.67 μΩ-cm and the dielectric is oxide with dielectric constant 3.9. (b) Calculate the RC time delay. The metal is Cu with a resistivity of 1.7 μΩ-cm and the dielectric is organic polymer with dielectric constant 2.8. (c) Compare the results in (a) and (b). How much can we decrease the RC time delay?

18. Repeat Prob. 17 (a) and (b) if the fringing factor for the capacitors is 3. The fringing factor is due to the spreading of the electric-field lines beyond the length and the width of the metal lines.

*19. To avoid electromigration problems, the maximum allowed current density in an aluminum runner is about 5×10^5 A/cm². If the runner is 2 mm long, 1 μm wide, and nominally 1 μm thick, and if 20% of the runner length passes over steps and is only 0.5 μm thick there, find the total resistance of the runner if the resistivity is 3×10^{-6} Ω-cm. Find the maximum voltage that can be applied across the runner.

*20. To use Cu for wiring one must overcome several obstacles: the diffusion of Cu through SiO_2, adhesion of Cu to SiO_2, and corrosion of Cu. One way to overcome these obstacles is to use a cladding/adhesion layer (e.g. Ta or TiN) to protect the Cu wires. Consider a cladded Cu wire with a square cross section of 0.5 μm $\times$ 0.5 μm and compare it with a layered TiN/Al/TiN wire of the same size, with the top and bottom TiN layers 40 nm and 60 nm thick, respectively. What is the maximum thickness of the cladding layer if the resistance of the cladded Cu wire and the TiN/Al/TiN wire is the same?

12

Lithography and Etching

Lithography is the process of transferring patterns of geometric shapes on a mask to a thin layer of radiation-sensitive material (called resist) covering the surface of a semi-conductor wafer.[1] These patterns define the various regions in an integrated circuit such as the implantation regions, the contact windows, and the bonding-pad areas. The resist patterns defined by the lithographic process are not permanent elements of the final device but only replicas of circuit features. To produce circuit features, these resist patterns must be transferred once more into the underlying layers comprising the device. The pattern transfer is accomplished by an etching process that selectively removes unmasked portions of a layer.[2] A brief description of the pattern transfer has been given in Section 4.1.

Specifically, we cover the following topics:

- The importance of a clean room for lithography.

- The most widely used lithographic method—the optical lithography and its resolution-enhancement techniques.

- Advantages and limitations of other lithographic methods.

- Mechanisms for wet chemical etching of semiconductors, insulator, and metal films.

- Plasma-assisted etching (also called dry etching) for high-fidelity pattern transfer.

- Microelectromechanical systems (MEMS) formed by orientation-dependent etching, sacrificial etching, or LIGA (lithography, electroplating, and molding) process.

▶ 12.1 OPTICAL LITHOGRAPHY

The vast majority of lithographic equipment for integrated-circuit (IC) fabrication is optical equipment using ultraviolet light ($\lambda \cong 0.2$–0.4 μm). In this section we consider the exposure tools, the masks, the resists , and resolution-enhancement techniques used for optical lithography. We also consider the pattern transfer process, which serves as a basis for other lithographic systems. We first briefly consider the *clean room*, because all lithographic processes must be performed in an ultraclean environment.

12.1.1 The Clean Room

An IC fabrication facility requires a clean processing room, especially in the area used for lithography. The need for such a clean room arises because dust particles in the air can settle on semiconductor wafers and lithographic masks and can cause defects in the devices, which result in circuit failure. For example, a dust particle on a semiconductor surface can disrupt the single-crystal growth of an epitaxial film, causing the formation of dislocations. A dust particle incorporated into the gate oxide can result in enhanced conductivity and cause device failure due to low breakdown voltage. The situation is even more critical in the lithographic area. When dust particles adhere to the surface of a photomask, they behave as opaque patterns on the mask, and these patterns will be transferred to the underlying layer along with the circuit patterns on the mask. Figure 1 shows three dust particles on a photomask.[3] Particle 1 may result in the formation of a pinhole in the underlying layer. Particle 2 is located near a pattern edge and may cause a constriction of current flow in a metal runner. Particle 3 can lead to a short circuit between the two conducting regions and render the circuit useless.

In a clean room, the total number of dust particles per unit volume must be tightly controlled along with the temperature and humidity. Figure 2 shows the particle-size distribution curves for various *classes* of clean rooms. We have two systems to define the classes of clean room.[4] For the English system, the numerical designation of the class is taken from the maximum allowable number of particles 0.5 μm and larger, per *cubic foot*. For the metric system, the class is taken from the logarithm (base 10) of the maximum allowable number of particles 0.5 μm and larger, per *cubic meter*. For example, a class 100 clean room (English system) has a dust count of 100 particles/ft³ with particle diameters of 0.5 μm and larger, whereas a class M 3.5 clean room (metric system) has a dust count of $10^{3.5}$ or about 3500 particles/m³ with particle diameters of 0.5 μm or large. Since

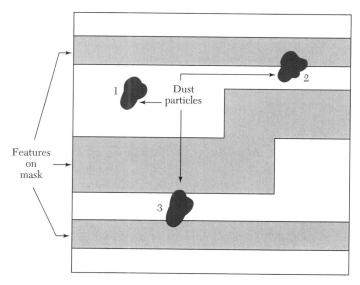

Fig. 1 Various ways in which dust particles can interfere with photomask patterns.[3]

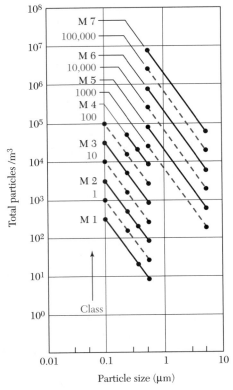

Fig. 2 Particle-size distribution curve for English (- - -) and metric (—) classes of clean rooms.[4]

100 particles/ft³ = 3500 particles/m³, a class 100 in English system corresponds to a class M 3.5 in the metric system.

Since the number of dust particles increases as particle size decreases, a more stringent control of the clean room environment is required when the minimum feature lengths of ICs are reduced to the deep-submicron range. For most IC fabrication areas, a class 100 clean room is required, that is, the dust count must be about four orders of magnitude lower than that of ordinary room air. However, for the lithography area, a class 10 clean room or one with a lower dust count is required.

▶ **EXAMPLE 1**

If we expose a 200-mm wafer for 1 minute to an air stream under a laminar-flow condition at 30 m/min, how many dust particles will land on the wafer in a class 10 clean room?

SOLUTION For a class 10 clean room, there are 350 particles (0.5 μm and larger) per cubic meter. The air volume that goes over the wafer in 1 minute is

$$(30 \text{ m / min}) \times \pi \left(\frac{0.2 \text{ m}}{2} \right)^2 \times 1 \text{ minute} = 0.942 \text{ m}^3.$$

The number of dust particles (0.5 μm and larger) contained in the air volume is

$350 \times 0.942 = 330$ particles.

Therefore, if there are 400 IC chips on the wafer, the particle count amounts to one particle on each of 82% of the chips. Fortunately, only a fraction of the particles that land adhere to the wafer surface, and of those only a fraction are at a circuit location critical enough to cause a failure. However, the calculation indicates the importance of the clean room. ◀

12.1.2 Exposure Tools

The pattern transfer process is accomplished by using a lithographic exposure tool. The performance of an exposure tool is determined by three parameters: resolution, registration, and throughput. *Resolution* is the minimum feature dimension that can be transferred with high fidelity to a resist film on a semiconductor wafer. *Registration* is a measure of how accurately patterns on successive masks can be aligned (or overlaid) with respect to previously defined patterns on the wafer. *Throughput* is the number of wafers that can be exposed per hour for a given mask level.

There are basically two optical exposure methods: shadow printing and projection printing.[5,6] Shadow printing may have the mask and wafer in direct contact with one another as in *contact printing*, or in close proximity as in *proximity printing*. Figure 3a shows a basic setup for contact printing where a resist-coated wafer is brought into physical contact with a mask, and the resist is exposed by a nearly collimated beam of ultraviolet light through the back of the mask for a fixed time. The intimate contact between resists and mask provides a resolution of ~1 μm. However, contact printing suffers a major drawback caused by dust particles. A dust particle or a speck of silicon dust on the wafer can be imbedded into the mask when the mask makes contact with the wafer. The imbedded particle causes permanent damage to the mask and results in defects in the wafer with each succeeding exposure.

To minimize mask damage, the proximity exposure method is used. Figure 3b shows the basic setup. It is similar to the contact printing method except that there is a small gap, 10–50 μm, between the wafer and the mask during exposure. The small gap, however, results in optical diffraction at feature edges on the photomask, that is, when light

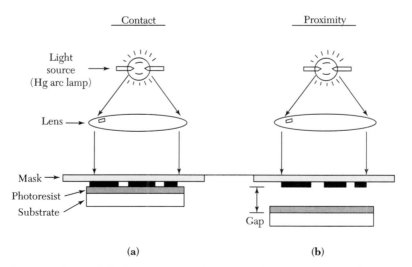

Fig. 3 Schematic of optical shadow printing techniques[1]: (*a*) contact printing, (*b*) proximity printing.

passes by the edges of an opaque mask feature, fringes are formed and some light penetrates into the shadow region. As a result, the resolution is degraded to the 2–5 μm range.

In shadow printing, the minimum linewidth [or critical dimension (CD)] that can be printed is roughly

$$CD \cong \sqrt{\lambda g}, \tag{1}$$

when λ is the wavelength of the exposure radiation and g is the gap between the mask and the wafer and includes the thickness of the resist. For $\lambda = 0.4$ μm and $g = 50$ μm, the CD is 4.5 μm. If we reduce λ to 0.25 μm (wavelength range of 0.2–0.3 μm is in the deep-UV spectral region) and g to 15 μm, CD becomes 2 μm. Thus, there is an advantage in reducing both λ and g. However, for a given distance g, any dust particle with a diameter larger than g potentially can cause mask damage.

To avoid the mask damage problem associated with shadow printing, projection-printing exposure tools have been developed to project an image of the mask patterns onto a resist-coated wafer many centimeters away from the mask. To increase resolution, only a small portion of the mask is exposed at a time. The small image area is scanned or stepped over the wafer to cover the entire wafer surface. Figure 4a shows a 1:1 wafer scan projection system.[6,7] A narrow, arc-shaped image field ~1 mm in width serially transfers the slit image of the mask onto the wafer. The image size on the wafer is the same as that on the mask.

The small image field can also be stepped over the surface of the wafer by two-dimensional translations of the wafer only, whereas the mark remains stationary. After the exposure of one chip site, the wafer is moved to the next chip site and the process is repeated.

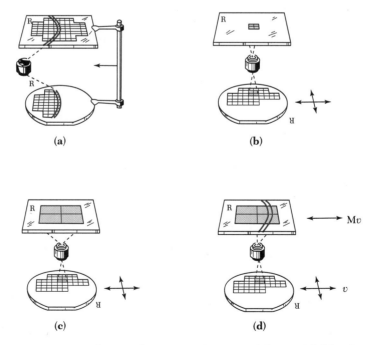

Fig. 4 Image partitioning techniques for projection printing: (*a*) annual-field wafer scan, (*b*) 1:1 step-and-repeat, (*c*) *M*:1 reduction step-and repeat, and (*d*) *M*:1 reduction step-and-scan.[6,7]

Figure 4*b* and 4*c* show the partitioning of the wafer image by *step-and-repeat projection* with a ratio of 1:1 or at a demagnification ratio *M*:1 (e.g., 10:1 for a 10 times reduction on the wafer), respectively. The demagnification ratio is an important factor in our ability to produce both the lens and the mask from which we wish to print. The 1:1 optical systems are easier to design and fabricate than a 10:1 or a 5:1 reduction systems, but it is much more difficult to produce defect-free masks at 1:1 than it is at a 10:1 or a 5:1 demagnification ratio.

Reduction projection lithography can also print larger wafers without redesigning the stepper lens as long as the field size (i.e., the exposure area onto the wafer per se) of the lens is large enough to contain one or more IC chips. When the chip size exceeds the field size of the lens, further partitioning of the image on the reticle is necessary. In Fig. 4*d* the image field on the reticle can be a narrow, arc-shaped for *M*:1 step-and-scan projection lithography. For the step-and-scan system, we have two-dimensional translations of the wafer with speed *v*, and one-dimensional translation of the mask with *M* times that of the wafer speed.

The resolution of a projection system is given by

$$l_m = k_1 \frac{\lambda}{\text{NA}}, \qquad (2)$$

where λ is again the exposure wavelength, k_1 is a process dependent factor, and NA is the numerical aperture, which is given by

$$\text{NA} = \bar{n}\sin\theta \qquad (3)$$

with $\bar{n}$ the index of refraction in the image medium (usually air, where $\bar{n} = 1$), and θ the half-angle of the cone of light converging to a point image at the wafer, as shown[5] in Fig. 5. Also shown in the figure is the depth of focus (DOF), which can be expressed as

$$\text{DOF} = \frac{\pm l_m/2}{\tan\theta} \approx \frac{\pm l_m/2}{\sin\theta} = k_2 \frac{\lambda}{(\text{NA})^2}, \qquad (4)$$

where k_2 is another process-dependent factor.

Equation 2 indicates that resolution can be improved (i.e., smaller l_m) by either reducing the wavelength or increasing NA or both. However, Eq. 4 indicates that the DOF

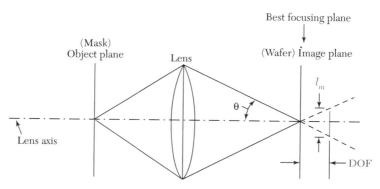

Fig. 5   Simple image system.[5]

degrades much more rapidly by increasing NA than by decreasing λ. This explains the trend toward shorter-wavelength sources in optical lithography.

The high-pressure mercury-arc lamp is widely used in exposure tools because of its high intensity and reliability. The mercury-arc spectrum is composed of several peaks, as shown in Fig. 6. The terms G-line, H-line, and I-line refer to the peaks at 436 nm, 405 nm, and 365 nm, respectively. I-line lithography with 5:1 step-and-repeat projection can offer a resolution of 0.3 μm with resolution enhancement techniques (see Section 12.1.6). Advanced exposure tools such as the 248 nm lithographic system using a KrF excimer laser, the 193 nm lithographic system using an ArF excimer laser, and the 157 nm lithographic system using a F_2 excimer laser have been developed for mass production with a resolution of 0.18 μm (180 nm), 0.10 μm (100 nm), and 0.07 μm (70 nm), respectively.

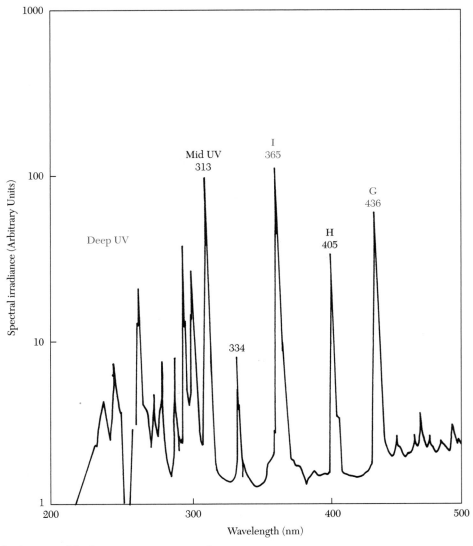

Fig. 6 Typical high-pressure mercury-arc lamp spectrum.

12.1.3 Masks

Masks used for IC manufacturing are usually reduction reticles. The first step in mask making is to use a computer-aided design (CAD) system in which designers can completely describe the circuit patterns electrically. The digital data produced by the CAD system then drives a pattern generator, which is an electron-beam lithographic system (see Section 12.2.1) that transfers the patterns directly to electron-sensitized mask. The mask consists of a fused silica substrate covered with a chrominum layer. The circuit pattern is first transferred to the electron-sensitized layer (electron resist), which is transferred once more into the underlying chrominum layer for the finished mask. The details of pattern transfer are considered in Section 12.1.5.

The patterns on a mask represent one level of an IC design. The composite layout is broken into mask levels that correspond to the IC process sequence such as the isolation region on one level, the gate region on another, and so on. Typically, 15–20 different mask levels are required for a complete IC process cycle.

The standard-size mask substrate is a fused silica plate 15×15 cm square, 0.6 cm thick. The size is needed to accommodate the lens field sizes for 4:1 or 5:1 optical exposure tools, whereas the thickness is required to minimize pattern placement errors due to substrate distortion. The fused silica plate is needed for its low coefficient of thermal expansion, its high transmission at shorter wavelengths, and its mechanical strength. Figure 7 shows a mask on which patterns of geometric shapes have been formed. A few secondary-chip sites, used for process evaluation, are also included in the mask.

One of the major concerns about masks is the defect density. Mask defects can be introduced during the manufacture of the mask or during subsequent lithographic processes. Even a small mask-defect density has a profound effect on the final IC yield. The *yield* is defined as the ratio of good chips per wafer to the total number of chips per wafer. As a first-order approximation, the yield Y for a given masking level can be expressed as

$$Y \cong e^{-DA}, \tag{5}$$

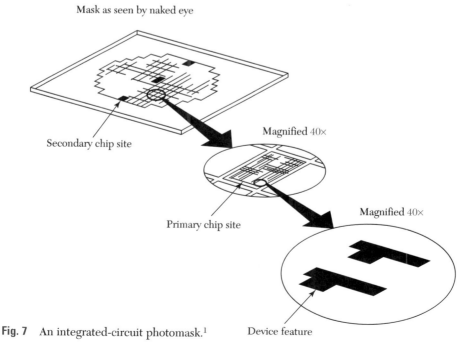

Fig. 7 An integrated-circuit photomask.[1]

where D is the average number of "fatal" defects per unit area and A is the area of an IC chip. If D remains the same for all mask levels (e.g., $N = 10$ levels), then the final yield becomes

$$Y \cong e^{-NDA}. \tag{6}$$

Figure 8 shows the mask-limit yield for a 10-level lithographic process as a function of chip size for various values of defect densities. For example, for $D = 0.25$ defect/cm², the yield is 10% for a chip size of 90 mm², and it drops to about 1% for a chip size of 180 mm². Therefore, inspection and cleaning of masks are important to achieve high yields on large chips. Of course, an ultraclean processing area is mandatory for lithographic processing.

12.1.4 Photoresist

The photoresist is a radiation-sensitive compound. Photoresists can be classified as positive and negative, depending on how they respond to radiation. For positive resists, the exposed regions become more soluble and thus more easily removed in the development process. The net result is that the patterns formed (also called images) in the positive resist are the same as those on the mask. For negative resists, the exposed regions become less soluble, and the patterns formed in the negative resist are the reverse of the mask patterns.

Positive photoresists consist of three components: a photosensitive compound, a base resin, and an organic solvent. Prior to exposure, the photosensitive compound is insoluble in the developer solution. After exposure, the photosensitive compound absorbs radiation in the exposed pattern areas, changes its chemical structure, and becomes soluble in the developer solution. After development, the exposed areas are removed.

Negative photoresists are polymers combined with a photosensitive compound. After exposure, the photosensitive compound absorbs the optical energy and converts it into

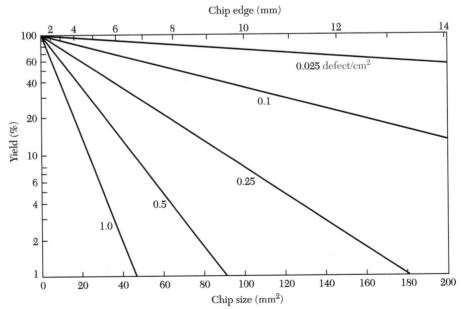

Fig. 8 Yield for a 10-mask lithographic process with various defect densities per level.

chemical energy to initiate a polymer linking reaction. This reaction causes cross linking of the polymer molecules. The cross-linked polymer has a higher molecular weight and becomes insoluble in the developer solution. After development, the unexposed areas are removed. One major drawback of a negative photoresist is that in the development process, the whole resist mass swells by absorbing developer solvent. This swelling action limits the resolution of negative photoresists.

Figure 9a shows a typical exposure response curve and image cross section for a positive resist.[1] The response curve describes the percentage of resist remaining after exposure and development versus the exposure energy. Note that the resist has a finite solubility in its developer, even without exposure to radiation. As the exposure energy increases, the solubility gradually increases until at a threshold energy E_T, the resist becomes completely soluble. The sensitivity of a positive resist is defined as the energy required to produce complete solubility in the exposed region. Thus, E_T corresponds to the sensitivity. In addition to E_T, a parameter γ, the contrast ratio, is defined to characterize the resist:

$$\gamma \equiv \left[\ln\left(\frac{E_T}{E_1} \right) \right]^{-1} ,$$

(7)

where E_1 is the energy obtained by drawing the tangent at E_T to reach 100% resist thickness as shown in Fig. 9a. A larger γ implies a higher solubility of the resist with an incremental increase of exposure energy and results in sharper images.

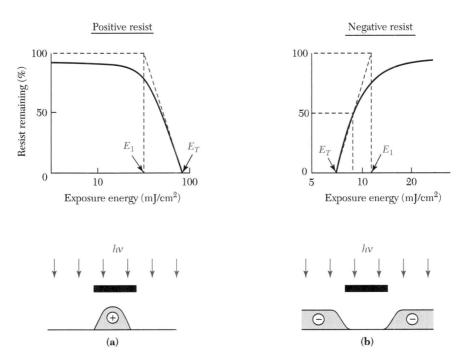

Fig. 9   Exposure-response curve and cross section of the resist image after development.[1] (a) Positive photoresist; (b) negative photoresist.

The image cross section in Fig. 9a illustrates the relationship between the edges of a photomask image and the corresponding edges of the resist images after development. The edges of the resist image are generally not at the vertically projected positions of the mask edges because of *diffraction*. The edge of the resist image corresponds to the position where the total absorbed optical energy equals the threshold energy E_T.

Figure 9b shows the exposure-response curve and image cross section for a negative resist. The negative resist remains completely soluble in the developer solution for exposure energies lower than the threshold energy E_T. Above E_T, more of the resist film remains after development. At exposure energies twice the threshold energy, the resist film becomes essentially insoluble in the developer. The sensitivity of a negative resist is defined as the energy required to retain 50% of the original resist film thickness in the exposed region. The parameter γ is defined similarly to γ in Eq. 7 except that E_1 and E_T are interchanged. The image cross section for the negative resist (Fig. 9b) is also influenced by the diffraction effect.

▶ **EXAMPLE 2**

Find the parameter γ for the photoresists shown in Fig. 9

SOLUTION   For the positive resist, $E_T = 90$ mJ/cm^2 and $E_1 = 45$ mJ/cm^2:

$$\gamma = \left[\ln\left(\frac{E_1}{E_T} \right) \right]^{-1} = \left[\ln\left(\frac{90}{45} \right) \right]^{-1} = 1.4.$$

For the negative resist, $E_T = 7$ mJ/cm^2 and $E_1 = 12$ mJ/cm^2:

$$\gamma = \left[\ln\left(\frac{E_T}{E_1} \right) \right]^{-1} = \left[\ln\left(\frac{12}{7} \right) \right]^{-1} = 1.9.$$ ◀

For deep UV lithography (e.g., 248 and 193 nm), we cannot use conventional photoresists because these resists require a high-dose exposure in deep UV, which will cause lens damage and lower throughput. The chemical-amplified resist (CAR) has been developed for the deep UV process. CAR consists of a photo-acid generator, a resin polymer, and a solvent. CAR is very sensitive to deep UV radiation and the exposed and unexposed regions differ greatly in their solubility in the developer solution.

12.1.5   Pattern Transfer

Figure 10 illustrates the steps to transfer IC patterns from a mask to a silicon wafer that has an insulating SiO$_2$ layer formed on its surface.[8] The wafer is placed in a clean room, which, typically, is illuminated with yellow light, since photoresists are not sensitive to wavelengths greater than 0.5 μm. To ensure satisfactory adhesion of the resist the surface must be changed from hydrophilic to hydrophobic. This change can be made by the application of an adhesion promoter, which can provide a chemically compatible surface for the resist. The most common adhesion promoter for silicon ICs is hexa-methylene-di-siloxane (HMDS). After the application of this adhesion layer, the wafer is held on a vacuum spindle, and 2–3 cc of liquidous resist is applied to the center of wafer. The wafer is then rapidly accelerated up to a constant rotational speed, which is maintained for about

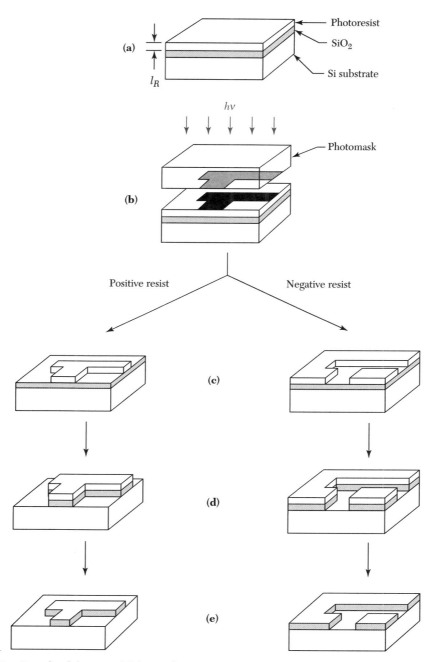

Fig. 10   Details of the optical lithographic pattern transfer process.[8]

30 seconds. Spin speed is generally in the range of 1000–10,000 rpm to coat a uniform film about 0.5 to 1 μm thick, as shown in Fig. 10a. The thickness of photoresist is correlated with its viscosity.

After the spinning step, the wafer is given a soft bake (typically at 90°–120°C for 60–120 seconds) to remove the solvent from the photoresist film and to increase resist adhesion to the wafer. The wafer is aligned with respect to the mask in an optical lithographic system, and the resist is exposed to UV light, as shown in Fig. 10*b*. If a positive photoresist is used, the exposed resist is dissolved in the developer, as shown in the left side of Fig. 10*c*. The photoresist development is usually done by flooding the wafer with the developer solution. The wafer is then rinsed and dried. After development, a postbaking at ~100°–180°C may be required to increase the adhesion of the resist to the substrate. The wafer is then put in an ambient that etches the exposed insulation layer but does not attack the resist, as shown in Fig. 10*d*. Finally, the resist is stripped (e.g., using solvent or plasma oxidation), leaving behind an insulator image (or pattern) that is the same as the opaque image on the mask (left side of Fig. 10*e*).

For the negative photoresist, the procedures described are also applicable, except that the unexposed areas are removed. The final insulator image (right side of Fig. 10*e*) is the reverse of the opaque image on the mask.

The insulator image can be used as a mask for subsequent processing. For example, an ion implantation can be done to dope the exposed semiconductor region, but not the area covered by the insulator. The dopant pattern is a duplicate of the design pattern on the photomask for a negative photoresist or is its complementary pattern for a positive photoresist. The complete circuit is fabricated by aligning the next mask in the sequence to the previous pattern and repeating the lithographic transfer process.

A related pattern transfer process is the liftoff technique, shown in Fig. 11. A positive resist is used to form the resist pattern on the substrate (Fig. 11*a* and 11*b*). The film (e.g., aluminum) is deposited over the resist and the substrate (Fig. 11*c*); the film thickness must be smaller than that of the resist. Those portions of the film on the resist are removed by selectively dissolving the resist layer in an appropriate liquid etchant so that the overlying film is lifted off and removed (Fig. 11*d*). The liftoff technique is capable of high resolution and is used extensively for discrete devices such as high-power

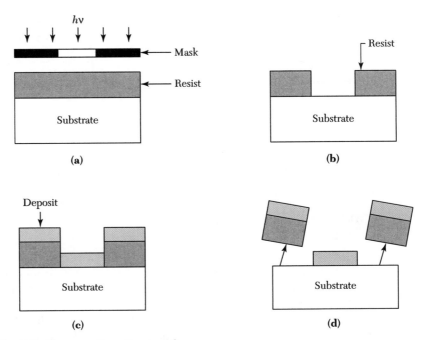

Fig. 11   Liftoff process for patter transfer.

MESFETs. However, it is not as widely applicable for ultralarge-scale integration, in which dry etching is the preferred technique.

12.1.6 Resolution Enhancement Techniques

Optical lithography has been continuously challenged to provide better resolution, greater depth of focus (DOF), and wider exposure latitude in IC processing. These challenges have been met by reducing the wavelength of the exposure tools and developing new resists. In addition, many resolution–enhancement techniques have been developed to extend the capability of optical lithography to even smaller feature lengths.

An important–resolution enhancement technique is the phase-shifting mask (PSM). The basic concept is shown[9] in Fig. 12. For a conventional mask, the electric field has the same phase at every aperture (clear area) in Fig. 12a. Diffraction and the limited resolution of the optical system spread the electric field at the wafer, as shown by the dotted lines. Interference between waves diffracted by the adjacent apertures enhances the field between them. Because the intensity I is proportional to the square of the electric field, it becomes difficult to separate the two images that are projected close to one another. The phase-shift layer that covers adjacent apertures reverses the sign of the electric field, as shown in Fig. 12b. Because the intensity at the mask is unchanged, the electric field of the images at the wafer can be cancelled. Therefore, images that are projected close to one another can be separated. A 180° phase change can be obtained by using a trans

parent layer of thickness $d = \lambda / 2(\bar{n} - 1)$, where $\bar{n}$ is the refractive index and λ is the wavelength, which covers one aperture, as shown in Fig. 12b.

Another resolution-enhancement technique is the optical proximity correction (OPC), which uses modified shapes of adjacent subresolution geometry to improve imaging capability. For example, a square contact hole with dimensions near the resolution limit

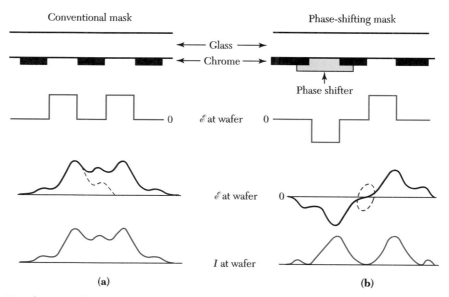

Fig. 12 The principle of phase-shift technology. (*a*) Conventional technology; (*b*) phase-shift technology.[9]

will print nearly as a circle. Modifying the contact-hole pattern with additional geometrics at the corners will help to print a more accurate square hole.

▶ 12.2 NEXT-GENERATION LITHOGRAPHIC METHODS

Why is optical lithography so widely used and what makes it such a promising method? The reasons are that it has high throughput, good resolution, low cost, and ease in operation.

However, due to deep-submicron IC process requirements, optical lithography has some limitations not yet solved. Although we can use PSM or OPC to extend its useful span, the complexity of mask production and mask inspection can not be easily resolved. In addition, the cost of the masks is very high. Therefore, we need to find postoptical lithography to process deep-submicron or even nanometer ICs.

Various types of next-generation lithographic methods for IC fabrication are discussed in this section. We consider electron-beam lithography, extreme UV lithography, X-ray lithography, and ion-beam lithography. We also consider the differences among these lithographic methods.

12.2.1 Electron-Beam Lithography

Electron-beam lithography is primarily used to produce photomasks. Relatively few tools are dedicated to direct exposure of the resist by a focused electron beam without a mask. Figure 13 shows a schematic of an electron-beam lithography system.[10] The electron gun is a device that can generate a beam of electrons with a suitable current density. A tung-

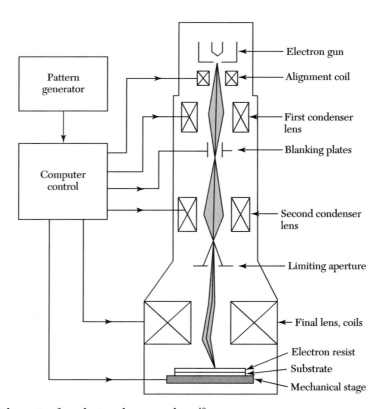

Fig. 13 Schematic of an electron-beam machine.[10]

sten thermionic-emission cathode or single-crystal lanthanum hexa-boride (LaB$_6$) is used for the electron gun. Condenser lenses are used to focus the electron beam to a spot size 10–25 nm in diameter. Beam-blanking plates that turn the electron beam on and off, and beam deflection coils are computer controlled and operated at MHz or higher rates to direct the focused electron beam to any location in the scan field on the substrate. Because the scan field (typically 1 cm) is much smaller than the substrate diameter, a precision mechanical stage is used to position the substrate to be patterned.

The advantages of electron-beam lithography include the generation of submicron resist geometries, highly automated and precisely controlled operation, greater depth of focus than that available from optical lithography, and direct patterning on a semiconductor wafer without using a mask. The disadvantage is that electron beam lithographic machines have low throughput—approximately 10 wafers per hour at less than 0.25 μm resolution. This throughput is adequate for the production of photomasks, for situations that require small numbers of custom circuits, and for design verification. However, for maskless direct writing, the machine must have the highest possible throughput and therefore the largest beam diameter possible consistent with the minimum device dimensions.

There are basically two ways to scan the focused electron beam: raster scan and vector scan.[11] In a raster scan system, resist patterns are written by a beam that moves through a regular mode, vertically oriented, as shown in Fig. 14a. The beam scans sequentially over every possible location on the mask and is blanked (turned off) where no exposure is required. All patterns on the area to be written must be subdivided into individual addresses, and a given pattern must have a minimum incremental interval that is evenly divisible by the beam address size.

In the vector scan system, as shown in Fig. 14b, the beam is directed only to the requested pattern features and jumps from feature to feature, rather than scanning the whole chip, as in raster scan. For many chips, the average exposed region is only 20% of the chip area, so we can save time by using a vector scan system.

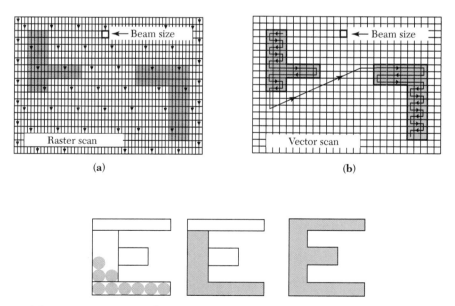

Fig. 14 (a) Raster scan writing scheme; (b) vector scan writing schemes; and (c) shapes of electron beam: round, variable, cell projection.[12]

Figure 14c shows several types of electron beam employed for e-beam lithography: Gaussian spot beam (round beam), variable-shaped beam, and cell projection. In variable-shaped beam system, the patterning beam has a rectangular cross section of variable size and aspect ratio. It offers the advantage of exposing several address units simultaneously. Therefore, the vector scan method using variable-shaped beam has higher throughout than the conventional Gaussian sport beam. It is also possible to pattern a complex geometric shape in one exposure with an electron beam system; this is called cell projection as shown in the far right of Fig. 14c. The cell projection technique[12] is particularly suitable for highly repetitive designs, as in an MOS memory cells, since several memory cell patterns can be exposed at once. Cell projection has not yet achieved the throughput of optical exposure tools.

SCALPEL System

A new approach for electron-beam technology is the electron-beam *projection* lithography. An example is the SCALPEL (*S*cattering with *a*ngular *l*imitation *p*rojection *e*lectron-beam *l*ithography) system.[13] SCALPEL combines high resolution with the wide process latitude inherent in electron-beam lithography (depth of focus about 20–30 μm compared with <1 μm for KrF lithography) and the high throughput of a projection system. SCALPEL's masks are expected to be considerably less expensive than optical masks, which require optical proximity correction, and phase-shift masks for the 100 nm generation, because SCALPEL adopts a 4× projection reduction.

The system has an initial design for a 3 mm field, which would be capable of delivering a throughput of 200 mm wafers at a rate of 30 per hour. As shown in Fig. 15, SCALPEL uses a scattering mask to conduct the reduction projection lithography through a scanner. The mask is uniformly illuminated by a parallel beam of 100 keV electrons

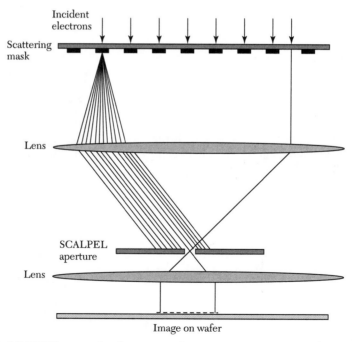

Fig. 15 Basic SCALPEL principle of operation showing contrast generation by differentiating more- or less-scattered electrons.[13]

over an area 1 mm × 1 mm. The key aspect of SCALPEL that differentiates it from previous attempts at projection electron-beam lithography is the mask design. The mask is comprised of a thin (100–150 nm), low-atomic-number membrane (SiN$_x$), and a thin (30–60 nm) high-atomic-number pattern layer (Cr/W).

Although the mask is almost completely electron transparent at the energies used (100 keV), contrast is generated by utilizing the difference in electron-scattering characterics between the membrane and patterned materials. The membrane of low-atomic-number material scatters electrons weakly with small angles, whereas the pattern layer of high-atomic-number material scatters most electrons strongly with high angles. A SCALPEL aperture in the back-focal plane of the projection optics blocks the strongly scattered electrons, forming a high-contrast aerial image on the wafer.

Electron Resist

Electron resists are polymers. The behavior of an electron-beam resist is similar to that of a photoresist, that is, a chemical or physical change is induced in the resist by irradiation. This change allows the resist to be patterned. For a positive electron resist, the polymer-electron interaction causes chemical bonds to be broken (chain scission) to form shorter molecular fragments, as shown[14] in Fig. 16a. As a result, the molecular weight is reduced in the irradiated area, which can be dissolved subsequently in a developer solution that attacks the low-molecular-weight material. Common positive electron resists include poly-methyl methacrylate (PMMA) and poly-butene-1 sulfone (PBS). Positive electron resists can achieve resolution of 0.1 μm or better.

For a negative electron resist, the irradiation causes radiation-induced polymer linking, as shown in Fig. 16b. The cross linking creates a complex three-dimensional structure with a molecular weight higher than that of the nonirradiated polymer. The nonirradiated resist can be dissolved in a developer solution that does not attack the high-molecular-weight material. Poly-glycidyl methacrylate-co-ethyl acrylate (COP) is a common

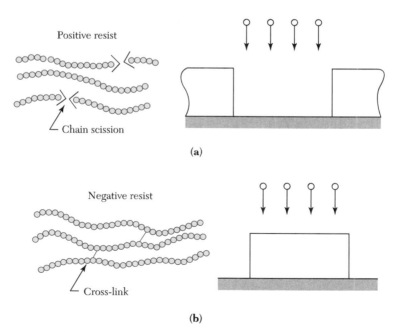

Fig. 16 Schematic of positive and negative resists used in electron-beam lithography. [14]

negative electron resist. COP, like most negative photoresists, also swells during development so the resolution is limited to about 1 μm.

The Proximity Effect
In optical lithography, the resolution is limited by diffraction of light. In electron-beam lithography, the resolution is not limited by diffraction (because the wavelengths associated with electrons of a few keV and higher energies are less than 0.1 nm) but by electron scattering. When electrons penetrate the resist film and underlying substrate, they undergo collisions. These collisions lead to energy losses and path changes. Thus, the incident electrons spread out as they travel through the material until either all of their energy is lost or they leave the material because of backscattering.

Figure 17a shows computed electron trajectories of 100 electrons with initial energy of 20 keV incident at the origin of a 0.4 μm PMMA film on a thick silicon substrate.[15] The electron beam is incident along the z-axis, and all trajectories have been projected onto the xz plane. This figure shows qualitatively that the electrons are distributed in an oblong pear-shaped volume with a diameter on the same order of magnitude as the electron penetration depth (~3.5 μm). Also, there are many electrons that undergo backscat-

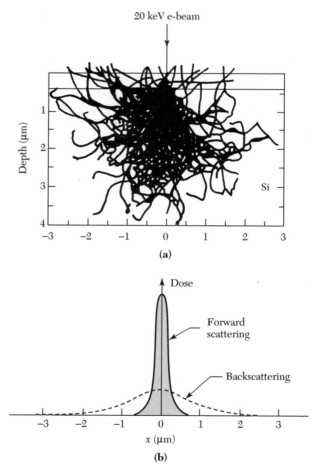

Fig. 17 (a) Simulated trajectories of 100 electrons in PMMA for a 20-keV electron beam.[15] (b) Dose distribution for forward scattering and backscattering at the resist-substrate interface.

tering collisions and travel backward from the silicon substrate into the PMMA resist film and leave the material.

Figure 17*b* shows the normalized distributions of the forward scattering and backscattering electrons at the resist-substrate interface. Because of the backscattering, electrons effectively can irradiate several micrometers away from the center of the exposure beam. Since the dose of a resist is given by the sum of the irradiations from all surrounding areas, the electron-beam irradiation at one location will affect the irradiation in neighboring locations. This phenomenon is called the *proximity effect*. The proximity effect places a limit on the minimum spacings between pattern features. To correct for the proximity effect, patterns are divided into smaller segments. The incident electron dose in each segment is adjusted so that the integrated dose from all its neighboring segments is the correct exposure dose. This approach further decreases the throughput of the electron-beam system, because of the additional computer time required to expose the subdivided resist patterns.

12.2.2 Extreme-Ultraviolet Lithography

Extreme-ultraviolet (EUV) lithography is a promising next-generation lithographic technology to extend the minimum linewidths to 30 nm without throughput loss.[16] Figure 18 shows a schematic diagram of an EUV lithographic system. A laser-produced plasma or a synchrotron radiation can serve as the EUV source of $\lambda = 10$–14 nm EUV light. The EUV radiation is reflected by a mask that is produced by patterning an absorber material deposited on a multilayer-coated flat silicon or glass-plate mask blank. EUV radiation is reflected from the nonpatterned regions (i.e., nonabsorbing regions) of the mask through a 4× reduction camera and imaged into a thin layer of resist on the wafer.

Since the EUV radiation beam is narrow, the mask must be scanned by the beam to illuminate the entire pattern field that describes the circuit mask layer. Also, for a 4× four-mirror (i.e., the one-paraboloid, two-ellipsoid, and one-plane mirrors) reduction camera, the wafer must be scanned at one-fourth the mask speed in a direction opposite to the mask movement to reproduce the image field on all chip sites on the wafer surface. A

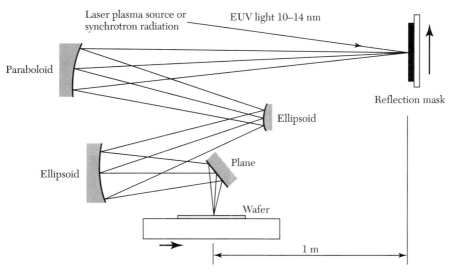

Fig. 18 Schematic representation of an extreme-ultraviolet (EUV) lithography system.[16]

precision system is required to perform the chip-site alignment and to control the wafer and mask stage movements and the exposure dose during the scanning process.

EUV lithography is capable of printing 50-nm features with PMMA resist using 13-nm radiation. However, the production of EUV exposure tools has a number of challenges. Since EUV is strongly absorbed in all materials, the lithography process must be performed in vacuum. The camera must use reflective-lens elements, and the mirrors must be coated with multilayer coatings that produce distributed quarter-wave Bragg reflectors. In addition, the mask blank must also be multilayer coated to maximize its reflectivity at λ = 10–14 nm.

12.2.3 X-Ray Lithography

X-ray lithography[17] (XRL) is a potential candidate to succeed optical lithography for the fabrication of integrated circuits at 100 nm. The synchrotron storage ring is the choice of X-ray source for high-volume manufacturing. It can provide a large amount of collimated flux and can easily accommodate 10–20 exposure tools.

XRL uses a shadow printing method similar to optical proximity printing. Figure 19 shows a schematic XRL system. The X-ray wavelength is about 1 nm, and the printing is through a 1× mask in close proximity (10–40 μm) to the wafer. Since X-ray absorption depends on the atomic number of the material and most materials have low transparency at $\lambda \cong 1$ nm, the mask substrate must be a thin membrane (1–2 μm thick) made of low-atomic-number material, such as silicon carbide or silicon. The pattern itself is defined in a thin (~ 0.5 μm) relatively high-atomic-number material, such as tantalum, tungsten, gold, or one of their alloys, which is supported by the thin membrane.

Masks are the most difficult and critical element of an XRL system, and the construction of an X-ray mask is much more complicated than that of a photomask. To avoid absorption of the X-rays between the source and mask, the exposure generally takes place in a helium environment. The X rays are produced in vacuum, which is separated from the helium by a thin vacuum window (usually of beryllium). The mask substrate will absorb 25%–35% of the incident flux and must therefore be cooled. An X-ray resist 1 μm thick

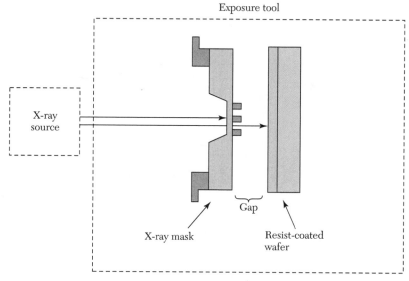

Fig. 19 Schematic representation of a proximity X-ray lithography system.[17]

will absorb about 10% of the incident flux, and there are no reflections from the substrate to create standing waves, so antireflection coatings are unnecessary.

We can use electron beam resists as X-ray resists because when an X-ray is absorbed by an atom, the atom goes to an excited state with the emission of an electron. The excited atom returns to its ground state by emitting an X-ray having a different wavelength than the incident X-ray. This X-ray is absorbed by another atom, and the process repeats. Since all the processes result in the emission of electrons, a resist film under X-ray irradiation is equivalent to one being irradiated by a large number of secondary electrons from any of the other processes. Once the resist film is irradiated, chain cross linking or chain scission will occur, depending on the type of resist.

12.2.4   Ion-Beam Lithography

Ion-beam lithography can achieve higher resolution than optical, X-ray, or electron-beam lithographic techniques because ions have a higher mass and therefore scatter less than electrons. The most important application is the repair of masks for optical lithography, a task for which commercial systems are available.

Figure 20 shows the computer-simulated trajectories of 50 H^+ ions implanted at 60 keV into PMMA and various substrates.[18] Note that the spread of the ion beam at a depth of 0.4 µm is only 0.1 µm in all cases (compare with Fig. 17a for electrons). The backscattering is completely absent for the silicon substrate, and there is only a small amount of backscattering for the gold substrate. However, ion-beam lithography may suffer from random (or stochastic) space-charge effect, causing broadening of the ion beam.

There are two types of ion-beam lithography systems: a scanning focused-beam system and a mask-beam system. The former system is similar to the electron-beam machine (Fig. 13), in which the ion source can be Ga^+ or H^+. The latter system is similar to an optical 5× reduction projection step-and-repeat system, which projects 100 keV light ions such as H^+_2 through a stensil mask.

12.2.5   Comparison of Various Lithographic Methods

The lithographic methods discussed above all have 100 nm or better resolution. A comparison of various lithographic technologies is shown in Table 1. However, each method

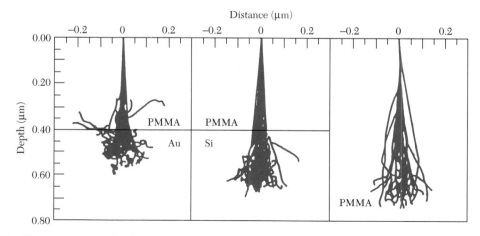

Fig. 20   Trajectories of 60 keV H^+ ions traveling through PMMA into Au, Si, and PMMA.[18]

TABLE 1   Comparison of Various Lithographic Technologies

		Optical 248/193 nm	SCALPEL[a]	EUV[a]	X-ray	Ion beam
Exposure Tool	Source	Laser	Filament	Laser plasma	Synchrotron	Multicusp
	Diffraction limited	Yes	No	Yes	Yes	No
	Optics	Refractive	Refractive	Refractive	No optics	Full-field refractive
	Step and scan	Yes	Yes	Yes	Yes	Stepper
	Throughput of 200 mm wafers/hr	40	30–35	20–30	30	30
Mask	Demagnification	4X	4X	4X	1X	4X
	Optical proximity correction	Yes	No	Yes	Yes	No
	Radiation path	Transmission	Transmission	Reflection	Transmission	Stencil
Resist	Single or Multilayer	Single	Single	Surface Imaging	Single	Single
	Chemical amplified resist	Yes	Yes	No	Yes	No

[a]SCALPEL, Scattering with angular limitation projection electron-beam lithoraph; EUV, extreme ultraviolet.

has its own limitations: the diffraction effect in optical lithography, the proximity effect in electron-beam lithography, mask fabrication complexities in X-ray lithography, difficulty in mask blank production for EUV lithography, and stochastic space charge in ion projection lithography.

For IC fabrication, many mask levels are involved. However, it is not necessary to use the same lithographic method for all levels. A mix-and-match approach can take advantage of the unique features of each lithographic process to improve resolution and to maximize throughput. For example, a 4:1 SCALPEL or EUV method can be used for the most critical mask levels, whereas 4:1 or 5:1 optical system can be used for the rest.

According to the Roadmap of Semiconductor Industry Association, IC manufacturing technology will reach the 50 nm generation around 2010. With each new technology generation, lithography has become an even more important key driver for the semiconductor industry because of the requirements of smaller feature size and tighter overlay tolerance. In addition, lithography tool costs have become higher relative to the total equipment costs for IC manufacturing facility. Currently, the technology development of next-generation lithography is conducted by multinational research projects or by industrial partners.

▶ 12.3 WET CHEMICAL ETCHING

Wet chemical etching is used extensively in semiconductor processing. Starting from the sawed semiconductor wafers, chemical etchants are used for lapping and polishing to give an optically flat, damage-free surface. Prior to thermal oxidation or epitaxial growth, the semiconductor wafers are chemically cleaned to remove contamination that results from handling and storing. Wet chemical etchings are especially suitable for blanket etches (i.e., over the whole wafer surface) of polysilicon, oxide, nitride, metals, and III-V compounds.

The mechanisms for wet chemical etching involve three essential steps, as illustrated in Fig. 21: the reactants are transported by diffusion to the reacting surface, chemical reactions occur at the surface, and the products from the surface are removed by diffusion. Both agitation and the temperature of the etchant solution will influence the etch rate, which is the amount of film removed by etching per unit time. In IC processing, most wet chemical etchings proceed by immersing the wafers in a chemical solution or by spraying the wafers with the etchant solution. For immersion etching, the wafer is immersed in the etch solution, and mechanical agitation is usually required to ensure etch uniformity and a consistent etch rate. Spray etching has gradually replaced immersion etching because it greatly increases the etch rate and uniformity by constantly supplying fresh etchant to the wafer surface.

For semiconductor production lines, highly uniform etch rates are important. Etch rates must be uniform across a wafer, from wafer to wafer, from run to run, and for any variations in feature sizes and pattern densities. Etch rate uniformity is given by:

$$\text{Etch rate uniformity }(\%) = \frac{(\text{maximum etch rate} - \text{minimum etch rate})}{\text{maximum etch rate} + \text{minimum etch rate}} \times 100\%. \qquad (8)$$

▷ **EXAMPLE 3**

Calculate the Al average etch rate and etch rate uniformity on a 200 mm diameter silicon wafer, assuming the etch rates at the center, left, right, top, and bottom of the wafer are 750, 812, 765, 743, and 798 nm/min, respectively.

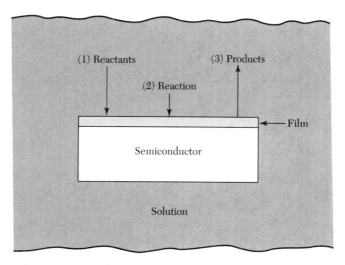

Fig. 21 Basic mechanisms in wet chemical etching.

SOLUTION

Al average etch rate = (750 + 812 + 765 + 743 + 798) ÷ 5 = 773.6 nm/min.

Etch rate uniformity = (812 – 743) ÷ (812 + 743) × 100% = 4.4 %.   ◀

12.3.1 Silicon Etching

For semiconductor materials, wet chemical etching usually proceeds by oxidation, followed by the dissolution of the oxide by a chemical reaction. For silicon, the most commonly used etchants are mixtures of nitric acid (HNO_3) and hydrofluoric acid (HF) in water or acetic acid (CH_3COOH). Nitric acid oxidizes silicon to form a SiO_2 layer.[19] The oxidation reaction is

$$Si + 4HNO_3 \rightarrow SiO_2 + 2H_2O + 4NO_2. \tag{9}$$

Hydrofluoric acid is used to dissolve the SiO_2 layer. The reaction is:

$$SiO_2 + 6HF \rightarrow H_2SiF_6 + 2H_2O. \tag{10}$$

Water can be used as a diluent for this etchant. However, acetic acid is preferred because it reduces the dissolution of the nitric acid.

Some etchants dissolve a given crystal plane of single-crystal silicon much faster than another plane; this results in orientation-dependent etching.[20] For a silicon lattice, the (111)-plane has more available bonds per unit area than the (110)- and (100)-planes; therefore, the etch rate is expected to be slower for the (111)-plane. A commonly used orientation-dependent etch for silicon consists of a mixture of KOH in water and isopropyl alcohol. For example, a solution with 19 wt % KOH in deionized (DI) water at about 80°C removes the (100)-plane at a much higher rate than the (110)- and (111)-planes. The ratio of the etch rates for the (100)-, (110)-, and (111)-planes is 100:16:1.

Orientation-dependent etching of <100>-oriented silicon through a patterned silicon dioxide mask creates precise V-shaped grooves,[10] the edges being (111)-planes at an angle of 54.7° from the (111)-surface, as shown at the left of Fig. 22a. If the window in the mask is sufficiently large or if the etching time is short, a U-shaped groove will be formed, as shown at the right of Fig. 22a. The width of the bottom surface is given by

$$W_b = W_0 - 2l\cot 54.7°$$

or

$$\boxed{W_b = W_0 - \sqrt{2}\, l,} \tag{11}$$

where W_0 is the width of the window on the wafer surface and l is the etched depth. If <$\bar{1}$10>-oriented silicon is used, essentially straight-walled grooves with sides of (111)-planes can be formed, as shown in Fig. 22b. We can use the large orientation dependence in the etch rates to fabricate device structures with submicron feature lengths.

12.3.2 Silicon Dioxide Etching

The wet etching of silicon dioxide is commonly achieved in a dilute solution of HF with or without the addition of ammonium fluoride (NH_4F). Adding NH_4F is referred to as a buffered HF solution (BHF), also called buffered-oxide-etch (BOE). The addition of NH_4F to HF controls the pH value and replenishes the depletion of the fluoride ions,

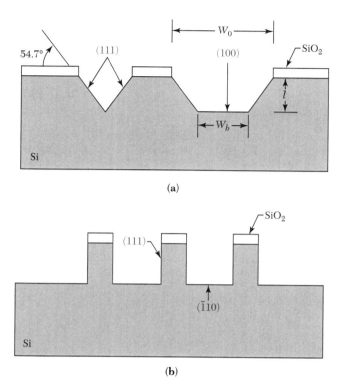

Fig. 22 Orientation-dependent etching. (*a*) Through window patterns on <100>-oriented silicon; (*b*) through window patterns on <$\overline{1}$10>-oriented silicon.[20]

thus maintaining stable etching performance. The overall reaction for SiO_2 etching is the same as that in Eq. 10. The etch rate of SiO_2 etching depends on etchant solution, etchant concentration, agitation, and temperature. In additional, density, porosity, microstructure, and the presence of impurities in the oxide also influence the etch rate. For example, a high concentration of phosphorus in the oxide results in a rapid increase in the etch rate, and a loosely structured chemical-vapor deposition (CVD) or sputtered oxide exhibits a faster etch rate than thermally grown oxide.

Silicon dioxide can also be etched in vapor-phase HF. Vapor-phase-HF oxide-etch technology has a potential for submicron feature etching because the process can be well controlled.

12.3.3 Silicon Nitride and Polysilicon Etching

Silicon nitride films are etchable at room temperature in concentrated HF or buffered HF and in a boiling H_3PO_4 solution. Selective etching of nitride to oxide is done with 85% H_3PO_4 at 180°C because this solution attacks silicon dioxide very slowly. The etch rate is typically 10 nm/min for silicon nitride, but less than 1 nm/min for silicon dioxide. However photoresist adhesion problems are encountered when etching nitride with boiling H_3PO_4 solution. Better patterning can be achieved by depositing a thin oxide layer on top of the nitride film before resist coating. The resist pattern is transferred to the oxide layer, which then acts as a mask for subsequent nitride etching.

Etching polysilicon is similar to etching single-crystal silicon. However, the etch rate is considerably faster because of grain boundaries. The etch solution is usually modified to ensure that it does not attack the underlying gate oxide. Dopant concentrations and temperature may affect the etch rate of polysilicon.

12.3.4 Aluminum Etching

Aluminum and aluminum alloy films are generally etched in heated solutions of phosphoric acid, nitric acid, acetic acid, and DI water. The typical etchant is a solution of 73% H_3PO_4, 4% HNO_3, 3.5% CH_3COOH, and 19.5% DI water at 30°–80°C. The wet etching of aluminum proceeds as follows: HNO_3 oxidizes aluminum, and H_3PO_4 then dissolves the oxidized aluminum. The etch rate depends on etchant concentration,

TABLE 2 Etchants for Insulators and Conductors

Material	Etchant composition	Etch rate (nm/min)
SiO_2	28 ml of HF 170 ml of HF ⎱Buffered HF 113 g of NH_4F	100
	15 ml of HF 10 ml of HNO_3 ⎱P – etch 300 ml of H_2O	12
Si_3N_4	Buffered HF	0.5
Al	H_3PO_4	10
	4 ml of HNO_3 3.5 ml of CH_3COOH 73 ml of H_3PO_4 19.5 ml of H_2O	30
Au	4 g KI	1000
Mo	1 g of I_2 40 ml of H_2O 5 ml of H_3PO_4 2 ml of HNO_3 4 ml of CH_3COOH 150 ml of H_2O	500
Pt	1 ml of HNO_3 7 ml of HCl 8 ml of H_2O	50
W	34 g of KH_2PO_4 13.4 g of KOH 33 g of $K_3Fe(CN)_6$ H_2O to make 1 liter	160

temperature, agitation of the wafers, and impurities or alloys in the aluminum film. For example, the etch rate is reduced when copper is added to the aluminum.

Wet etching of insulating and metal films is usually done with the similar chemicals that dissolve these materials in bulk form and involve their conversion into soluble salts or complexes. Generally, film materials will be etched more rapidly than their bulk counterparts. Also, the etch rates are higher for films that have a poor microstructure, built-in stress, or departure from stoichiometry, or that have been irradiated. Some useful etchants for insulating and metal films are listed in Table 2.

12.3.5   Gallium Arsenide Etching

A wide variety of etches has been investigated for gallium arsenide; however, few of them are truly isotropic.[21] This is because the surface activities of the (111)-Ga and (111)-As faces are very different. Most etches give a polished surface on the arsenic face, but the gallium face tends to show crystallographic defects and etches more slowly. The most commonly used etchants are the H_2SO_4-H_2O_2-H_2O and H_3PO_4-H_2O_2-H_2O systems. For an etchant with an 8:1:1 volume ratio of H_2SO_4:H_2O_2:H_2O, the etch rate is 0.8 μm/min for the <111>-Ga face and 1.5 μm/min for all other faces. For an etchant with 3:1:50 volume ratio of H_3PO_4:H_2O_2:H_2O, the etch rate is 0.4 μm/min for ⟨111⟩-Ga face and 0.8 μm for all other faces.

12.4 DRY ETCHING

In pattern-transfer operations, a resist pattern is defined by a lithographic process to serve as a mask for etching of its underlying layer (Fig. 23a).[22] Most of the layer materials (e.g., SiO_2, Si_3N_4, and deposited metals) are amorphous or polycrystalline thin films. If they are etched in a wet chemical etchant, the etch rate is generally isotropic (i.e., the lateral and vertical etch rates are the same), as illustrated in Fig. 23b. If h_f is the thickness of the layer material and l the lateral distance etched underneath the resist mask, we can define the degree of anisotropy A_f by

$$A_f \equiv 1 - \frac{1}{h_f} = 1 - \frac{R_l t}{R_v t} = 1 - \frac{R_l}{R_v}, \qquad (12)$$

where t is the time and R_l and R_v are the lateral and vertical etch rates, respectively. For isotropic etching, $R_l = R_v$ and $A_f = 0$.

The major disadvantage of wet chemical etching for pattern transfer is the undercutting of the layer underneath the mask, resulting in a loss of resolution in the etched pattern. In practice, for isotropic etching, the film thickness should be about one-third or less of the resolution required. If patterns are required with resolutions much smaller than the film thickness, anisotropic etching (i.e., $1 \geq A_f > 0$) must be used. In practice, the value of A_f is chosen to be close to unity. Figure 23c shows the limiting case where $A_f = 1$, corresponding to $l = 0$ (or $R_l = 0$).

To achieve a high-fidelity transfer of the resist patterns required for ultralarge-scale integration processing ($A_f = 1$), dry etching methods have been developed. Dry etching is synonymous with plasma-assisted etching, which denotes several techniques that use plasma in the form of low-pressure discharges. Dry-etch methods include plasma etching,

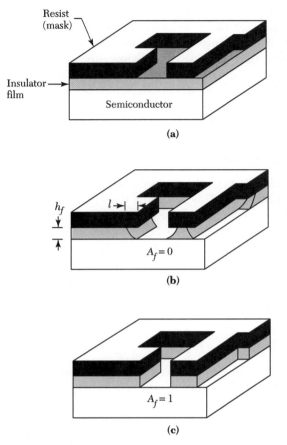

Fig. 23 Comparison of wet chemical etching and dry etching for pattern transfer. [22]

reactive ion etching (RIE), sputter etching, magnetically enhanced RIE (MERIE), reactive ion beam etching, and high-density plasma (HDP) etching.

12.4.1 Plasma Fundamentals

Plasma is a fully or partially ionized gas composed of equal numbers of positive and negative charges and a different number of unionized molecules. Plasma is produced when an electric field of sufficient magnitude is applied to a gas, causing the gas to break down and become ionized. The plasma is initiated by free electrons that are released by some means, such as field emission from a negatively biased electrode. The free electrons gain kinetic energy from the electric field. In the course of their travel through the gas, the electrons collide with gas molecules and lose their energy. The energy transferred in the collision causes the gas molecules to be ionized (i.e., to free electrons). The free electrons gain kinetic energy from the field, and the process continues. Therefore, when the applied voltage is larger than the breakdown potential, a sustained plasma is formed throughout the reaction chamber.

The electron concentrations in the plasma for dry etchings are relatively low, typically on the order of 10^9–10^{12} cm^{-3}. At a pressure of 1 Torr, the concentrations of gas molecules are 10^4–10^7 times higher than the electron concentrations. This results in an average gas temperature in the range of 50°–100°C. Therefore, the plasma-assisted dry etching is a low-temperature process.

▶ **EXAMPLE 4**

The electron densities in a RIE system and HDP system range from 10^9–10^{10} and 10^{11}–10^{12} cm^{-3}, respectively. Assuming the RIE chamber pressure is 200 mTorr and HDP chamber pressure is 5 mTorr, calculate the ionization efficiency in RIE reactors and HDP reactors at room temperature. The ionization efficiency is the ratio of the electron density to the density of molecules.

SOLUTION

$$PV = nRT$$

where P is the pressure in atm (1 atm = 760,000 mTorr), V the volume in liters, n the number of moles, R the gas constant (0.082 liter·atm/mol-K), and T the absolute temperature in K, respectively.

For a RIE system,

$$n/V = P/RT = (200/760,000)/(0.082 \times 300) = 1.06 \times 10^{-5} \text{ (mol/liter)},$$

$$= 1.06 \times 10^{-5} \times 6.02 \times 10^{23} \div 1000,$$

$$= 6.38 \times 10^{15} \text{ (cm}^{-3}).$$

$$\text{Ionization effeciency} = (10^9 \sim 10^{10})/(6.38 \times 10^{15}),$$

$$= 1.56 \times 10^{-7} \sim 1.56 \times 10^{-6}.$$

For a HDP system,

$$n/V = P/RT = (5/760,000)/(0.082 \times 300) = 2.66 \times 10^{-7} \text{ (mol/liter)},$$

$$= 2.66 \times 10^{-7} \times 6.02 \times 10^{23} \div 1000,$$

$$= 1.6 \times 10^{14} \text{ (cm}^{-3}).$$

$$\text{Ionization effeciency} = (10^{11} \sim 10^{12})/(1.6 \times 10^{14}),$$

$$= 6.25 \times 10^{-4} \sim 6.25 \times 10^{-3}.$$

Therefore, HDP has much higher ionization efficiency than RIE. ◀

12.4.2 Etch Mechanism, Plasma Diagnostics, and End-Point Control

Plasma etching is a process in which a solid film is removed by a chemical reaction with ground-state or excited-state neutral species. Plasma etching is often enhanced or induced by energetic ions generated in a gaseous discharge. The basic etch mechanism, plasma diagnostic, and end-point control are introduced briefly in this section.

Etch Mechanism

Plasma etching proceeds in five steps, as illustrated in Fig. 24: the etchant species is generated in the plasma; the reactant is then transported by diffusion through a stagnant gas layer to the surface. The reactant is adsorbed on the surface; a chemical reaction (along with physical effects such as ion bombardment) follows to form volatile compounds; the compounds are desorbed from the surface, diffused into the bulk gas, and pumped out by the vacuum system.[23]

Plasma etching is based on the generation of plasma in a gas at low pressure. Two basic methods are used—physical methods and chemical methods. The former includes sputter etching and the latter includes pure chemical etching. In physical etching, positive

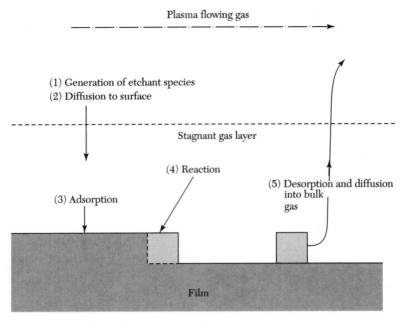

Fig. 24　Basic steps in a dry-etching processing.[23]

ions bombard the surface at high speed; small amounts of negative ions formed in the plasma cannot reach the wafer surface and therefore play no direct role in plasma etching. In chemical etching, neutral reactive species generated by the plasma interact with the material surface to form volatile products. Chemical- and physical-etch mechanisms have different characteristics. Chemical etching exhibits a high etch rate, and good selectivity (i.e., the ratio of etch rates for different materials) produces low ion-bombardment–induced damage but yields isotropic profiles. Physical etching can yield anisotropic profiles, but it is associated with low etch selectivity and high bombardment-induced damage. Combinations of chemical and physical etching give anisotropic etch profiles, reasonably good selectivity, and moderate bombardment-induced damage. An example is the RIE, which uses a physical method to assist chemical etching or creates reactive ions to participate in chemical etching

Plasma Diagnostics
Most processing plasmas emit radiation in that range from infrared to ultraviolet. A simple analytical technique is to measure the intensity of these emissions versus wavelength with the aid of optical emission spectroscopy (OES). Using observed spectral peaks, it is usually possible to determine the presence of neutral and ionic species by correlating these emissions with previously determined spectral series. Relative concentrations of the species can be obtained by correlating changes in intensity with the plasma parameter. The emission signal derived from the primary etchant or by-product begins to rise or fall at the end of the etch cycle.

End-Point Control
Dry etching differs from the wet chemical etching, in that dry etching does not have enough etch selectivity to the underlying layer. Therefore, the plasma reactor must be equipped with a monitor that indicates when the etching process is to be terminated

(i.e., an end point detection system). Laser interferometry of the wafer surface is used to continuously monitor etch rates and to determine the end point. During etching, the intensity of laser light reflected off a thin film surface oscillates. This oscillation occurs because of the phase interference between the light reflected from the outer and inner interfaces of the etching layer. This layer must therefore be optically transparent or semi-transparent to observe the oscillation. Figure 25 shows a typical signal from a silicide/polycrystalline Si gate etch. The period of the oscillation is related to the change in film thickness by

$$\Delta d = \lambda / 2\overline{n} \qquad (13)$$

where Δd is the change in film thickness for one period of reflected light, λ is the wavelength of the laser light, and $\overline{n}$ is the refractive index of the etching layer. For example, Δd for polysilicon is 80 nm, measured by using a helium-neon laser source for which $\lambda = 632.8$ nm.

12.4.3 Reactive Plasma-Etching Techniques and Equipment

Plasma reactor technology in the IC industry has changed dramatically since the first application of plasma processing to photoresist stripping. A reactor for plasma etching contains a vacuum chamber, pump system, power supply generators, pressure sensors, gas flow control units, and end-point detector. Table 3 shows the similarities and differences in the types of etching equipment that are commercially available. A comparison of pressure operating ranges and ion energies for different types of reactors is shown in Fig. 26. Each etch tool is designed empirically and uses a particular combination of pressure, electrode configuration and type, and source frequency to control the two primary etch mechanisms – chemical and physical. Higher etch rates and tool automation are required for most etchers used in manufacturing.

Reactive Ion Etcher (RIE)
RIE has been extensively used in the microelectronic industry. In a parallel-plate diode system, a radio frequency (rf) capacitive-coupled bottom electrode holds the wafer. This allows the grounded electrode to have a significantly larger area because it is, in fact, the chamber itself. The larger grounded area combined with the lower operating pressure

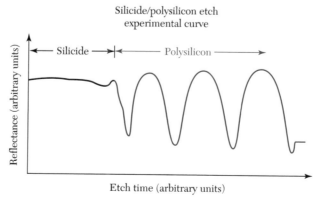

Fig. 25 The relative reflectance of the etching surface of a composite silicide/poly-Si layer. The end point of the etch is indicated by the cessation of the reflectance oscillation.

TABLE 3   Etch Mechanisms, Pressure Ranges of Plasma Reactors

Etch tool configuration	Etch mechanism	Pressure range (Torr)
Barrel etching	Chemical	0.1–10
Downstream plasma etching	Chemical	0.1–10
Reactive ion etching (RIE)	Chemical and physical	0.01–1
Magnetic enhanced RIE	Chemical and physical	0.01–1
Magnetic confinement triode RIE	Chemical and physical	0.001–0.1
Electron cyclotron resonance plasma etch	Chemical and physical	0.001–0.1
Inductively coupled plasma or transformer-coupled plasma	Chemical and physical	0.001–0.1
Surface wave coupled plasma or helicon plasma etching	Chemical and physical	0.001–0.1

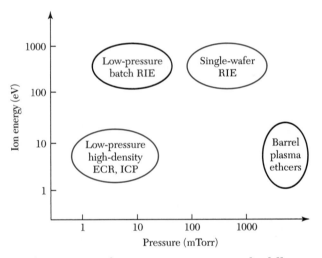

Fig. 26 Comparison of ion energy and operating pressure ranges for different types of plasma reactors.

(< 500 mTorr) causes the wafers to be subjected to a heavy bombardment of energetic ions from the plasma as a result of the large, negative self-bias at the wafer surface.

The etch selectivity of this system is relatively low compared with traditional barrel-etch systems because of strong physical sputtering. However, selectivity can be improved by choosing the proper etch chemistry, for example, by polymerizing the silicon surface with fluorocarbon polymers to obtain selectivity of SiO_2 over silicon. Alternatively, a triode-configuration RIE etch, as shown in Fig. 27, can separate plasma generation from ion transport. Ion energy is controlled through a separate bias on the wafer electrode, thereby minimizing the loss of selectivity and the ion-bombardment–induced damage observed in most traditional RIE systems.

Electron Cyclotron Resonance Plasma Etcher
Most parallel-plate plasma etchers, except triode RIE, do not provide the ability to control plasma parameters, such as electron energy, plasma density, and reactant density inde-

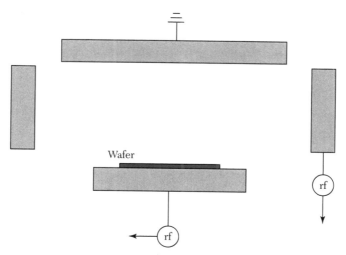

Fig. 27 Schematic of a triode reactive ion etch reactor. The ion energy is separately controlled by a bias voltage on the bottom electrode. rf, radio frequency.

pendently. As a result, ion-bombardment–induced damage becomes a serious problem. The electron cyclotron resonance (ECR) reactor combines microwave power with a static magnetic field to force electrons to circulate around the magnetic field lines at an angular frequency. When this frequency equals to the applied microwave frequency, a resonance coupling occurs between the electron energy and the applied electric field that results in a high degree of dissociation and ionization (10^{-2} for ECR compared with 10^{-6} for RIE). Figure 28 shows an ECR reaction chamber configuration. Microwave power is coupled through a microwave window into the ECR source region. The magnetic field is supplied from the magnetic coils. The ECR plasma systems can also be used in thin film deposition. The high efficiency in exciting the reactants in ECR plasmas allows the deposition of films at room temperature without the need for thermal activation.

Other High-Density Plasma Etchers

As feature sizes for ultralarge-scale integration (ULSI) continue to decrease, the limits of the conventional RIE system are being approached. In addition to the ECR system, other types of high-density plasma (HDP) sources, such as the inductively coupled plasma (ICP) source, the transformer-coupled plasma (TCP) source, and the surface wave–coupled plasma (SWP) sources, have been developed. These etchers have high plasma density (10^{11}–10^{12} cm^{-3}), and low processing pressure (< 20 mTorr). In addition, they allow the wafer platen to be powered independently of the source, providing significant decoupling between the ion energy (wafer bias) and the ion flux (plasma density primarily driven by source power). The primary processing advantages of HDP sources are better critical dimension (CD) control, higher etching rates, and better selectivity.

In addition, HDP sources provide low substrate damage (because of independent biasing of the substrate and the side-electrode potentials) and high anisotropy (because of low pressure yet high active species density). However, because of their complexity and higher cost, the systems may not be used for less critical applications, such as spacer etching or planarization.[24] Figure 29 shows a TCP plasma reactor. A high-density, low-pressure plasma is generated by a flat spiral coil that is separated from the plasma by a dielectric plate on the top of the reactor. The wafer is located away from the coil, so it is not affected by the electromagnetic field generated by the coil. There is little plasma

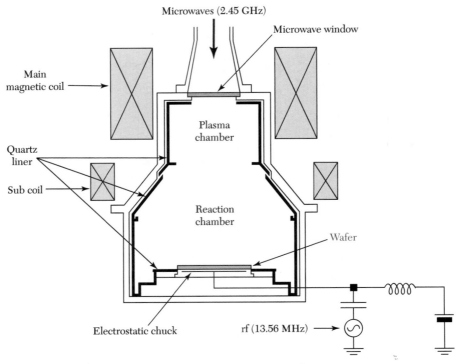

Fig. 28 Schematic of an electron cyclotron resonance reactor.[24]

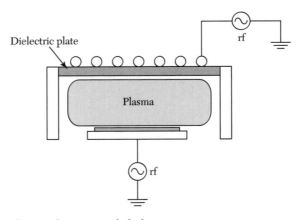

Fig. 29 Schematic of a transformer-coupled plasma reactor.

density loss because plasma is generated only a few mean free paths away from the wafer surface. Therefore, a high-density plasma and high etch rates are achieved.

Clustered Plasma Processing
Semiconductor wafers are processed in clean rooms to minimize exposure to ambient particulate contamination. As device dimensions shrink, particulate contamination becomes a more serious problem. To minimize particulate contamination, clustered plasma tools uses a wafer handler to pass wafers from one process chamber to another in a vacuum

environment. The clustered plasma processing tools can also increase throughput. Figure 30 shows the multilayer metal interconnect (TiW/AlCu/TiW) etching process with clustered tools in an AlCu etch chamber, a TiW etch chamber, and a strip passivation chamber. The clustered tools provide an economic advantage by having a high chip yield because the wafer is exposed to less ambient contamination and is handled less.

12.4.4 Reactive Plasma–Etching Applications

Plasma etching has rapidly evolved from a simple and batch resist stripping to a large and single-wafer processing. The etching system continues to be improved from the conventional RIE tool to the high-density plasma tool for pattern transfer of deep-submicron devices. Besides the etching tool, etch chemistry also plays a critical role in the performance of etch process. Table 4 lists some etch chemistries for different etch processes. Developing an etch process usually means optimizing etch rate, selectivity, profile control, critical dimension, damage, etc., by adjusting a large number of process parameters.

Silicon Trench Etching
As device feature size decreases, a corresponding decrease is needed in the wafer surface area occupied by the isolation between circuit elements and the storage capacitor of a DRAM cell. This surface area can be reduced by etching trenches into the silicon substrate and filling them with suitable dielectric or conductive materials. Deep trenches, usually with a depth larger than 5 μm, are used mainly for forming storage capacitors. Shallow trenches, usually with a depth less than 1 μm, are often used for isolation.

Chlorine-based and bromine-based chemistries have a high silicon etch rate and high etch selectivity to the silicon dioxide mask. The combination of $HBr + NF_3 + SF_6 + O_2$ gas mixtures is used to form a trench capacitor with a depth of ~7 μm. It is also used for shallow trench isolation etching. Aspect ratio dependent etching (i.e., variation in etch rate with aspect ratio) is often observed in submicron-deep silicon trench etching caused

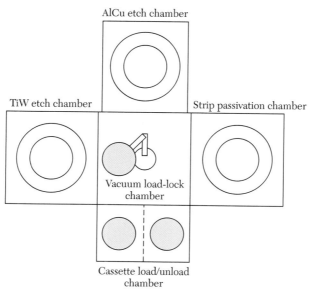

Fig. 30 Cluster reactive ion etch tool for multilayer metal (TiW/AlCu/TiW) interconnect etching.[2]

TABLE 4   Etch Chemistries of Different Etch Processes

Material being etched	Etching chemistry
Deep Si trench	$HBr/NF_3/O_2/SF_6$
Shallow Si trench	$HBr/Cl_2/O_2$
Poly Si	$HBr/Cl_2/O_2$, HBr/O_2, BCl_3/Cl_2, SF_6
Al	BCl_3/Cl_2, $SiCl_4/Cl_2$, HBr/Cl_2
AlSiCu	$BCl_3/Cl_2/N_2$
W	SF_6 only, NF_3/Cl_2
TiW	SF_6 only
WSi_2, $TiSi_2$, $CoSi_2$	CCl_2F_2/NF_3, CF_4/Cl_2, $Cl_2/N_2/C_2F_6$
SiO_2	$CF_4/CHF_3/Ar$, C_2F_6, C_3F_8, C_4F_8/CO, C_5F_8, CH_2F_2
Si_3N_4	CHF_3/O_2, CH_2F_2, CH_2CHF_2

by limited ion and neutral transport within the trench. Figure 31 shows the dependence of average silicon trench etch rate on aspect ratio. Trenches with large aspect ratios are etched more slowly than trenches with small aspect ratios.

Polysilicon and Polycide Gate Etching

Polysilicon or polycide (i.e., low-resistance metal silicides over polysilicon) is usually used as a gate material for MOS devices. Anisotropic etching and high etch selectivity to the gate oxide are the most important requirements for gate etching. For example, the selectivity required in 1G DRAM is more than 150 (i.e., the ratio of etch rates for polycide and gate oxide is 150:1). Achieving high selectivity and etch anisotropy at the same time is difficult for most ion-enhanced etching processes. Therefore, multistep processing is used in which different etch steps in the process are optimized for etch anisotropy and selectivity. On the other hand, the trend in plasma technology for anisotropic etching and high selectivity is to utilize a low-pressure, high-density plasma using a relatively low power. Most chlorine-based and bromine-based chemistries can be used for gate etching to achieve the required etch anisotropy and selectivity.

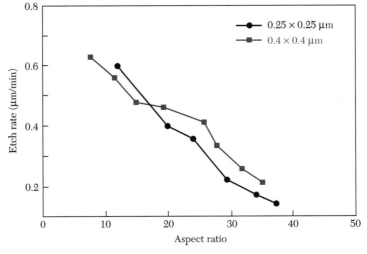

Fig. 31   Dependence of average silicon trench etch rate on aspect ratio.[2]

Dielectric Etching

The patterning of dielectrics, especially silicon dioxide and silicon nitride, is a key process in the manufacture of modern semiconductor devices. Because of their higher bonding energies, dielectric etching requires aggressive ion-enhanced, fluorine-based plasma chemistry. Vertical profiles are achieved by sidewall passivation, typically by introducing a carbon-containing fluorine species to the plasma (e.g., CF_4, CHF_3, C_4F_8). High ion-bombardment energies are required to remove this polymer layer from the oxide, as well as to mix the reactive species into the oxide surface to form SiF_X products.

A low-pressure, high-density plasma is advantageous for aspect-ratio–dependent etching. However, the HDP generates high-temperature electrons and subsequently generates a high degree of dissociation of ions and radicals. It generates far more active radicals and ions than RIE or MERIE plasmas . In particular, a high F concentration worsens the selectivity to silicon. Various methods were tried to enhance the selectivities in the high-density plasma. First, a parent gas with a high C/F ratio, such as C_2F_6, C_4F_8, or C_5F_8, has been tried. Also other methods to scavenge F radicals have been developed.[25]

Interconnect Metal Etching

Etching of a metallization layer is a very important step in IC fabrication. Aluminum, copper, and tungsten are the most popular materials used for interconnection. These materials usually require anisotropic etching. The reaction of aluminum with fluorine results in nonvolatile AlF_3, which has a vapor pressure of only 1 Torr at 1240°C. Chlorine-based (e.g., Cl_2/BCl_3 mixture) chemistry has been widely used for aluminum etching. Chlorine has a very high chemical etch rate with aluminum and tends to produce an undercut during etching. Carbon-containing gas (e.g., CHF_3) or N_2 is added to form sidewall passivation during aluminum etching to obtain anisotropic etching.

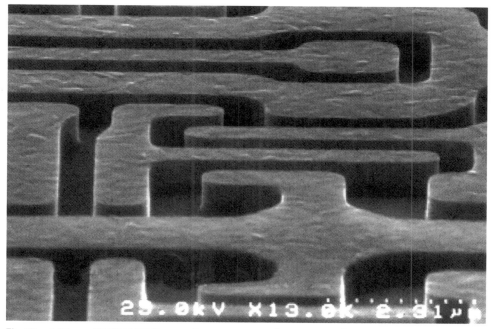

Fig. 32 0.35 μm TiN/Al/TiN lines and spaces on a wafer maintained at ambient for 72 hours after a microwave strip are not corroded.

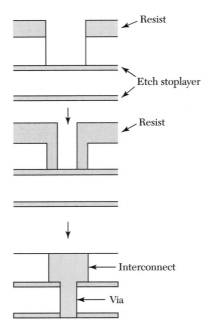

Fig. 33 Various process sequences for dual damascene process.

Exposure to the ambient is another problem in aluminum etching. Residue chlorine on the Al sidewall and the photoresist tends to react with atmosphere water to form HCl, which corrodes aluminum. An in-situ exposure of the wafer to a CF_4 discharge to exchange Cl with F and then to an oxygen discharge to remove the resist followed by immediate immersion in deionized water can eliminate Al corrosion. Figure 32 shows 0.35 µm TiN/Al/Ti lines and spaces on a wafer maintained at ambient for 72 hours. No corrosion is present even after prolonged exposure to the ambient.

Copper has drawn much attention as a new metallization material in ULSI circuits because of its low resistivity (~1.7 µohm-cm) and superior resistance to electromigration compared with Al or Al alloys. However, because of the low volatility of copper halides, plasma etching at room temperature is difficult. Process temperatures higher than 200°C are required to etch coppers films. Therefore, the damascene process is used to form Cu interconnection without dry etching. Damascene processing involves the creation of interconnect lines by first etching a trench or canal in a planar dielectric layer and then filling that trench with metal, such as aluminum or copper. In dual damascene processing (Fig. 33), a second level is involved where a series of holes (i.e., contacts or vias) are etched and filled in addition to the trench. After filling, the metal and dielectric are planarized by chemical-mechanical processing (CMP). The advantage of damascene processing is that it eliminates the need for metal etch. This is an important concern as the industry moves from aluminum to copper interconnections.

Low-pressure CVD (LPCVD) tungsten (W) has been widely used for filling contact holes and first-level metallization because of its excellent deposition conformability. Both fluorine- and chlorine-based chemistries etch W and form volatile etch products. An important tungsten etch processes is the blanket W etchback to form a W plug. The blanket LPCVD W is deposited on top of a TiN barrier layer, as shown in Fig. 34. A two-step process is usually used. First, 90% of the W is etched at a high etch rate, and then the

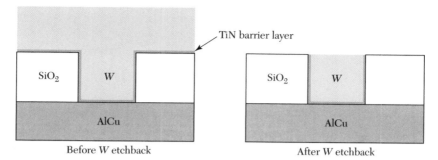

Before W etchback   After W etchback

Fig. 34 Formation of tungsten plug in a contact hole by depositing blanket low-pressure chemical-vapor deposition W and then using reaction ion etching etchback.

etch rate is reduced to remove the remaining W with an etchant that has a high W-to-TiN selectivity.

12.5 MICROELECTROMECHANICAL SYSTEMS

There has been rapidly growing interest in microelectromechanical systems (MEMS), since the late 1980s when a spinning micromotor made from polysilicon was fabricated on a silicon chip.[26,27] The manufacture of silicon MEMS adopted many of the highly developed technologies of silicon integrated circuits. This approach has enabled MEMS products to be produced at low cost by using batch fabrication, just as for ICs. In addition to the IC fabrication process, some new specialized techniques have been developed for MEMS. We consider three specialized etching techniques: bulk micromachining, surface micromachining, and the LIGA process.

12.5.1 Bulk Micromachining

In bulk micromachining, the device (e.g., a sensor or an actuator) is shaped by etching a large single-crystal substrate. The films are patterned on the bulk to define the isolation and transducer functions. Orientation-dependent wet chemical etching technique provide a high-resolution etch and tight dimensional control. Often, a bulk-micromachined device uses two-sided processing, creating a self-isolated structure with one side exposed to the measured variables, such as mechanical or chemical signals while the other side is enclosed in a clean package. Two-sided structures are very robust for operation in environments hostile to microelectronic devices. Simple mechanical devices such as diaphragm pressure sensors, membranes, and cantilever-beam piezoresistive acceleration sensors are fabricated commercially by this technique. Figure 35 illustrates a fabrication process of a simple silicone rubber membrane.[28]

12.5.2 Surface Micromachining

Surface-micromachined devices are constructed entirely from thin films. There are several differences and trade-offs between structures made from bulk and thin film materials. Typical dimensions for bulk-micromachined sensors are in the millimeter range, but surface-micromachined devices are of micrometer dimensions. Surface micromachining

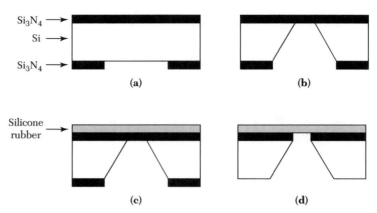

Fig. 35 Fabrication process of simple silicone rubber membrane. (*a*) Nitride deposition and patterning; (*b*) KOH etching; (*c*) silicone rubber spin coating; snf (*d*) nitride removal on back side. [28]

permits the fabrication of structurally complex devices by stacking and patterning layers or "building blocks" of thin films whereas multilayered bulk devices are difficult to construct. Free-standing and movable parts can be fabricated using sacrificial layers. Figure 36 illustrates how sacrificial etching techniques can be used to create an electrostatic micromotor with well-defined, submicron tolerance between the rotor and the center hub.[27]

12.5.3 LIGA Process

LIGA is a German acronym for lithographic, galvanoformung, abformung.[29] It consists of three basic processing steps: lithography, electroplating, and molding. The LIGA process is based upon X-ray radiation from a synchrotron. The process can produce microstructure with lateral dimensions in the micrometer range and structural heights of several hundred micrometers from a variety of materials. Its potential applications cover microelectronics, sensors, microoptics, micromechanics, and biotechnology.

An example of LIGA process is shown in Fig. 37. A thick X-ray resist ranging from 300 μm to more than 500 μm in thickness is deposited on a substrate with an electrically conductive surface. Lithographic patterning is done with extended exposure from highly collimated X-ray radiation through an X-ray mask, as shown Fig. 37*a*. The flower-shaped trench structure is formed in the thick resist after developer treatment (Fig. 37*b*). Metal is then electroplated on the exposed bottom conductive surface, filling the trench space and covering the top surface of the resist (Fig. 37*c*). The metal structure is formed after removing the resist (Fig. 37*d*). This structure can be used repeatedly as a mold insert for injection molding to form multiple plastic replicas of the original plating base (Fig. 37*e*). The plating base replicas, in turn, can be used to electroplate many metal structures as the final products, as shown in Fig. 37*f* and *g*.

The distinct advantage of the LIGA process is the ability to create three-dimensional structures as thick as bulk-micromachined devices while retaining the same degree of design flexibility as surface micromachining. However, the initial synchnotron radiation process is a very costly step, and the mold-separation steps may result in degradation of the original mold insert.

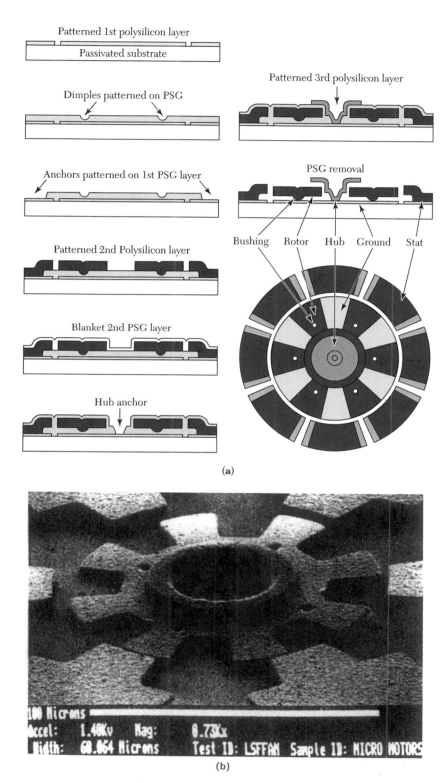

Patterned 1st polysilicon layer

Passivated substrate

Dimples patterned on PSG

Anchors patterned on 1st PSG layer

Patterned 2nd Polysilicon layer

Blanket 2nd PSG layer

Hub anchor

Patterned 3rd polysilicon layer

PSG removal

Bushing   Rotor   Hub   Ground   Stat

(a)

100 Microns
Accel: 1.40Kv   Mag: 0.73Kx
Width: 60.064 Microns   Test ID: LSFFAM   Sample ID: MICRO MOTORS

(b)

Fig. 36 (a) Sacrificial process flow for an electrostatic micromorotor. (b) Photograph of a micromotor.[27] PSG, phosphosilicate glass.

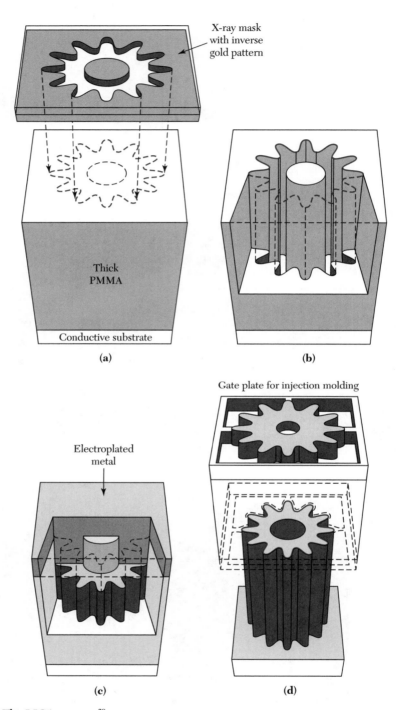

Fig. 37 The LIGA process.[29]

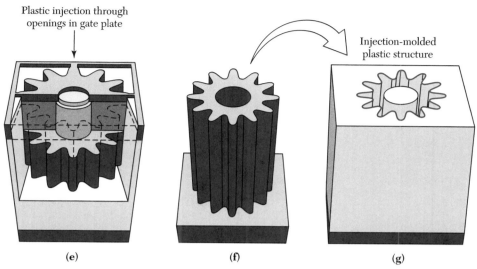

Plastic injection through openings in gate plate

Injection-molded plastic structure

(e)   (f)   (g)

Fig. 37 *(continued)*

► SUMMARY

The continued growth of the semiconductor industry is a direct result of the capability of transferring smaller and smaller circuit patterns onto semiconductor wafers. The two major processes to transfer patterns are lithography and etching.

Currently, the vast majority of lithographic equipment are optical systems. We have considered various exposure tools, masks, photoresists, and the clean room for optical lithography. The primary factor limiting resolution in optical lithography is diffraction. However, because of advancements in excimer lasers, photoresist chemistry, and resolution enhancement techniques such as the PSM and OPC, optical lithography will remain the mainstream technology, at least to the 100 nm generation.

Electron-beam lithography is the technology of choice for mask making and nanofabrication, in which new device concepts are explored. Other lithographic processing technologies are EUV, X-ray, and ion-beam lithography. Although all these have 100 nm or better resolution, each process has its own limitation: proximity effect in electron-beam lithography, mask blank production difficulties in EUV lithography, mask-fabrication complexity in X-ray lithography, and stochastic space charge in ion-beam lithography.

At the present time, no obvious successor to optical lithography can be identified unambiguously. However, a mix-and-match approach can take advantage of the unique features of each lithography process to improve resolution and to maximize throughput.

Wet chemical etching is used extensively in semiconductor processing. It is particularly suitable for blanket etching. We have discussed wet chemical etching processes for silicon and gallium arsenide, insulators, and metal interconnections. Wet chemical etching was used for pattern transfer. However the undercutting of the layer underneath the mask has resulted in loss of resolution in the etched pattern.

Dry etching methods are used to achieve high-fidelity pattern transfer. Dry etching is synonymous with plasma-assisted etching. We have considered plasma fundamentals and various dry-etching systems, which have grown from relatively simple, parallel-plate configurations to complex chambers with multiple-frequency generators and a variety of process-control sensors.

The challenges for future etching technology are high etch selectivity, better critical-dimensioned control, low aspect-ratio–dependent etching, and low plasma-induced damage. Low-pressure, high-density plasma reactors are necessary to meet these requirements. As processing evolves from 200 mm to 300 mm and even larger wafers, continued improvements are required for etch uniformity within the wafer. New gas chemistries must be developed to provide the improved selectivity necessary for advanced integration schemes.

Microelectromechanical system (MEMS) is an emerging field. MEMS has adopted the lithographic and etching technologies from IC fabrication. Specialized etching techniques have also been developed for MEMS: bulk micromachining using an orientation-dependent etching process, surface micromachining using sacrifical layers, and the LIGA process using X-ray lithography with highly collimated radiation.

▶ REFERENCES

1. For a more detailed discussion on lithography, see (a) K. Nakamura, "Lithography," in C. Y. Chang and S. M. Sze, Eds., *ULSI Technology*, McGraw-Hill, New York, 1996. (b) P. Rai-Choudhurg, *Handbook of Microlithography, Micromachining, and Microfabrication*, Vol. 1, SPIE, Washington, DC, 1997. (c) D. A. McGillis, "Lithography," in S. M. Sze. Ed., *VLSI Technology*, McGraw-Hill, New York, 1983.

2. For a more detailed discussion on etching, see Y. J. T. Liu, " Etching," in C. Y. Chang and S. M. Sze, Eds., *ULSI Technology*, McGraw-Hill, New York, 1996.

3. J.M. Duffalo and J. R. Monkowski, " Particulate Contamination and Device Performance," *Solid State Technol.* **27**, 3, 109 (1984).

4. H. P. Tseng and R. Jansen, "Cleanroom Technology," in C. Y. Chang and S. M. Sze, Eds., *ULSI Technology*, McGraw-Hill, New York, 1996.

5. M. C. King, " Principles of Optical Lithography," in N. G. Einspruch, Ed., *VLSI Electronics*, Vol. 1, Academic, New York, 1981.

6. J. H. Bruning, "A Tutorial on Optical Lithography," in D. A. Doane, et al., Eds., *Semiconductor Technology*, Electrochemical Soc., Penningston, 1982.

7. R. K. Watts and J. H. Bruning, "A Review of Fine-Line Lithographic Techniques: Present and Future," *Solid State Technol.*, **24**, 5,99 (1981).

8. W. C. Till and J. T. Luxon, *Integrated Circuits, Materials, Devices, and Fabrication*, Princeton-Hall, Englewood Cliffs, NJ, 1982.

9. M. D. Levenson, N. S. Viswanathan, and R. A. Simpson, "Improving Resolution in Photolithography with a Phase-Shift Mask," *IEEE Trans. Electron Devices*, **ED-29,** 18–28 (1982).

10. D. P. Kern, et al., "Practical Aspects of Microfabrication in the 100-nm Region," *Solid State Technol.*, **27**, 2, 127 (1984).

11. J. A. Reynolds, "An Overview of e-Beam Mask-Making, "*Solid State Technol.*, **22**, 8, 87 (1979).

12. Y. Someda, et al. "Electron-Beam Cell Projection Lithography: Its Accuracy and Its Throughput," *J. Vac. Sci. Technol.*, **B12** (6), 3399 (1994).

13. J. A. Liddle, et al., "The Scattering with Angular Limitation in Projection Electron-Beam Lithography (SCALPEL) System," *Jpn. J. Appl. Phys.*, **34**, 6663 (1995).

14. W. L. Brown, T. Venkatesan, and A. Wagner, "Ion Beam Lithography," *Solid State Technol.*, **24**, 8, 60 (1981).

15. D. S. Kyser and N. W. Viswanathan, *J. Vac. Sci. Technol.*, **12**, 1305 (1975).

16. Charles Gwyn, et al., *Extreme Ultraviolet Lithography-White Paper*, Sematech, Next Generation Lithography Workshop, Colorado Springs, Dec. 7–10, 1998.

17. J. P. Silverman, *Proximity X-Ray Lithography-White Paper*, Sematech, Next Generation Lithography Workshop, Colorado Spring, Dec. 7-10. 1998.

18. L. Karapiperis, et al., "Ion Beam Exposure Profiles in PMMA-Computer Simulation," *J. Vac. Sci. Technol.*, **19**, 1259 (1981).

19. H. Robbins and B. Schwartz, "Chemical Etching of Silicon II, the System HF, HNO_3, H_2O and $HC_2H_3O_2$, " *J. Electrochem. Soc.*, **107**, 108 (1960).

20. K. E. Bean, "Anisotropic Etching in Silicon," *IEEE Trans. Electron Devices*, **ED-25**, 1185 (1978).

21. S. Iida and K. Ito, "Selective Etching of Gallium Arsenide Crystal in H_2SO_4-H_2O_2-H_2O System," *J. Electrochem. Soc.*, **118**, 768 (1971).

22. E. C. Douglas, "Advanced Process Technology for VLSI Circuits," *Solid State Technol.*, **24**, 5, 65 (1981).

23. J. A. Mucha and D. W. Hess, " Plasma Etching," in L. F. Thompson and C. G. Willson, Eds., *Microcircuit Processing: Lithography and Dry Etching*, American Chemical Society, Washington, D. C., 1984.

24. M. Armacost et. al., "Plasma-Etching Processes for ULSI Semiconductor Circuits," *IBM J. Res. Dev.*, **43**, 39 (1999).

25. C. O. Jung, et al., "Advanced Plasma Technology in Microelectronics," *Thin Solid Films*, **341**, 112, (1999).

26. C. H. Mastrangelo and W. C. Tang, " Semiconductor Sensor Technology," in S. M. Sze, Ed., *Semiconductor Sensors*, Wiley, New York, 1994.

27. L. S. Fan, Y. C. Tai, and R. S. Muller, "IC-Processed Electrostatic Micromotors," in *IEEE Int. Electron Devices Meeting* (IEDM), p. 666, 1988.

28. X. Yang, et al., "A MEMS Thermopenumatic Silicone Rubber Membrane Valve, " *Sens. Actuators*, **A64**, 101, (1998).

29. W. Ehrfeld, et al. "Fabrication of Microstructures Using the LIGA Process," *Proc. IEEE Micro Robots and Teleoperators Workshop*, Hyannis, MA, Nov. 1987.

▶ PROBLEMS (* DENOTES DIFFICULT PROBLEMS)

FOR SECTION 12.1   OPTICAL LITHOGRAPHY

1. For a class 100 clean room, find the number of dust particles per cubic meter with particle sizes (a) between 0.5 and 1 μm, (b) between 1 and 2 μm, and (c) above 2 μm.

2. Find the final yield for a nine-mask–level process in which the average fatal defect density per cm^2 is 0.1 for four levels, 0.25 for four levels, and 1.0 for one level. The chip area is 50 mm^2.

3. An optical lithographic system has an exposure power of 0.3 mW/cm^2. The required exposure energy for a positive photoresist is 140 mJ/cm^2 and for a negative photoresist is 9 mJ/cm^2. Assuming negligible times for loading and unloading wafers, compare the wafer throughput for positive photoresist and negative photoresist.

4. (a) For an ArF-excimer laser 193 nm optical lithographic system with NA = 0.65, k_1 = 0.60, and k_2 = 0.50, what are the theoretical resolution and depth-of-focus for this tool? (b) What can we do in practice to adjust NA, k_1, and k_2 parameters to improve resolution? (c) What parameter does the phase-shift mask (PSM) technique change to improve resolution?

5. The plots in Fig. 9 are called *response curves* in microlithography. (a) What are the advantages and disadvantages of using resists with high γ values? (b) Conventional resists cannot be used for 248 nm or 193 nm lithography. Why not?

FOR SECTION 12.2   NEXT-GENERATION LITHOGRAPHIC METHODS

6. (a) Explain why a shaped beam promises higher throughput than a Gaussian beam in e-beam lithography. (b) How can alignment be performed for e-beam lithography? Why is alignment in X-ray lithography so difficult? (c) What are the potential advantages of X-ray lithography over e-beam lithography?

7. Why has the operating mode of optical lithographic systems evolved from proximity printing to 1:1 projection printing and finally to 5:1 projection step-and-repeat? (b) Is it possible to build a step-and-scan X-ray lithographic system? Why or why not?

FOR SECTION 12.3 WET CHEMICAL ETCHING

8. If the mask and the substrate can not be etched by a particular etchant, sketch the edge profile of an isotropically etched feature in a film of thickness h_f for (a) etching just to completion, (b) 100% overetch, and (c) 200% overetch.

9. A <100>-oriented silicon crystal is etched in a KOH solution through a 1.5 μm × 1.5 μm window defined in silicon dioxide. The etch rate normal to (100)-planes is 0.6 μm/min. The etch rate ratios are 100:16:1 for the (100):(110):(111)-planes. Show the etched profile after 20 seconds, 40 seconds, and 60 seconds.

10. Repeat the previous problem, a $<\bar{1}10>$-oriented silicon is etched with a thin SiO_2 mask in KOH solution. Show the etched pattern profiles on $<\bar{1}10>$-Si.

11. A <100>-oriented silicon wafer 150 mm in diameter is 625 μm thick. The wafer has 1000 μm × 1000 μm ICs on it. The IC chips are to be separated by orientation-dependent etching. Describe two methods for doing this and calculate the fraction of the surface area that is lost in these process.

FOR SECTION 12.4 DRY ETCHING

*12. The average distance traveled by particles between collisions is called the mean free path (λ), $\lambda \cong 5 \times 10^{-3}/P$(cm), where P is pressure in Torr. In typical plasmas of interest, the chamber pressure ranges from 1 Pa to 150 Pa. What are the corresponding density of gas molecules (cm^{-3}) and the mean free path?

13. Fluorine (F) atoms etch Si at a rate given by

$$\text{Etch Rate (nm/min)} = 2.86 \times 10^{-13} n_F \times T^{1/2} e^{-E_a/RT}$$

where n_F is the concentration of F atoms (cm^{-3}), T the temperature (K), and E_a and R the activation energy (2.48 kcal/mol) and gas constant (1.987 cal-K), respectively. If n_F is 3×10^{15}, calculate the etch rate of Si at room temperature.

14. Repeat the previous problem, SiO_2 etched by F atoms could also be expressed by

$$\text{Etch rate (nm/min)} = 0.614 \times 10^{-13} n_F \times T^{1/2} e^{-E_a/RT}$$

where n_F is $3 \times 10^{15}(cm^{-3})$ and E_a is 3.76 kcal/mol. Calculate the etch rate of SiO_2 and etch selectivity of SiO_2 over Si at room temperature.

15. A multiple-step etch process is required for etching a polysilicon gate with thin gate oxide. How do you design an etch process that has no micromasking, has an anisotropic etch profile, and is selective to thin gate oxide?

16. Find the etch selectivity required to etch a 400-nm polysilicon layer without removing more than 1 nm of its underlying gate oxide, assuming that the polysilicon is etched with a process having a 10% etch-rate uniformity.

17. A 1 μm Al film is deposited over a flat field oxide region and patterned with photoresist. The metal is then etched with a mixture of BCl_3/Cl_2 gases at a temperature of 70°C in a Helicon etcher. The selectivity of Al over photoresist is maintained at 3. Assuming a 30% overetch, what is the minimum photoresist thickness required to ensure that the top metal surface is not attacked?

18. In an ECR plasma, a static magnetic field B forces electrons to circulate around the magnetic field lines at an angular frequency, ω_e, which is given by

$$\omega_e = qB/m_e,$$

where q is the electronic charge, and m_e the electron mass. If the microwave frequency is 2.45 GHz, what is the required magnetic field.

19. What are the major distinctions between the traditional reactive ion etching and high density plasma etching (ECR, ICP, etc.)?

20. Describe how to eliminate the corrosion issues in Al lines after etching with chlorine-based plasma.

Impurity Doping

Impurity doping is the introduction of controlled amounts of impurity dopants into semiconductors. The practical use of impurity doping has been mainly to change the electrical properties of the semiconductors. *Diffusion* and *ion implantation* are the two key methods of impurity doping. Until the early 1970s, impurity doping was done mainly by diffusion at elevated temperatures, as shown in Fig. l*a*. In this method the dopant atoms are placed on or near the surface of the wafer by deposition from the gas phase of the dopant or by using doped-oxide sources. The doping concentration decreases monotonically from the surface, and the profile of the dopant distribution is determined mainly by the temperature and diffusion time.

Since the early 1970s, many doping operations have been performed by ion implantation, as shown in Fig. 1*b*. In this process the dopant ions are implanted into the semiconductor by means of an ion beam. The doping concentration has a peak distribution inside the semiconductor and the profile of the dopant distribution is determined mainly by the ion mass and the implanted-ion energy. Both diffusion and ion implantation are used for fabricating discrete devices and integrated circuits because these processes generally complement each other. [1,2] For example, diffusion is used to form a deep junction (e.g., a twin well in CMOS) whereas ion implantation is used to form a shallow junction (e.g., a source/drain junction of a MOSFET).

Specifically, we cover the following topics:

- The movement of impurity atoms in the crystal lattice under high temperature and high concentration-gradient conditions.

- Impurity profiles for constant diffusivity and concentration-dependent diffusivity.

- The impact of lateral diffusion and impurity redistribution on device characteristics.

- The process and advantages of ion implantation.

- Ion distributions in the crystal lattice and how to remove lattice damages caused by ion implantation.

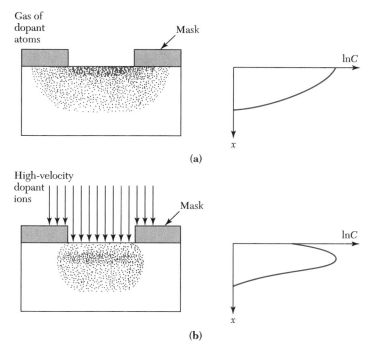

Fig. 1 Comparison of (*a*) diffusion and (*b*) ion-implantation techniques for the selective introduction of dopants into the semiconductor substrate.

- Implantation-related processes such as masking, high-energy implantation, and high-current implantation.

13.1 BASIC DIFFUSION PROCESS

Diffusion of impurities is typically done by placing semiconductor wafers in a carefully controlled high-temperature quartz-tube furnace and passing a gas mixture that contains the desired dopant through it. The temperature usually ranges between 800° and 1200°C for silicon and 600° and 1000°C for gallium arsenide. The number of dopant atoms that diffuse into the semiconductor is related to the partial pressure of the dopant impurity in the gas mixture.

For diffusion in silicon, boron is the most popular dopant for introducing a *p*-type impurity, whereas arsenic and phosphorus are used extensively as *n*-type dopants. These three elements are highly soluble in silicon, as they have solubilities above $5 \times 10^{20} \, \text{cm}^{-3}$ in the diffusion temperature range. These dopants can be introduced in several ways, including solid sources (e.g., BN for boron, As_2O_3 for arsenic, and P_2O_5 for phosphorus), liquid sources (BBr_3, $AsCl_3$, and $POCl_3$), and gaseous sources (B_2H_6, AsH_3, and PH_3). However, liquid sources are most commonly used. A schematic diagram of the furnace and gas flow arrangement for a liquid source is shown in Fig. 2. This arrangement is similar to that used for thermal oxidation. An example of the chemical reaction for phosphorus diffusion using a liquid source is

$$4POCl_3 + 3O_2 \rightarrow 2P_2O_5 + 6Cl_2 \uparrow. \tag{1}$$

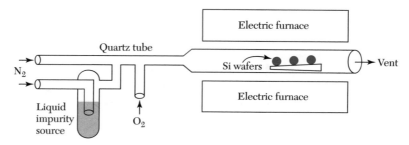

Fig. 2 The schematic diagram of a typical open-tube diffusion system.

The P_2O_5 forms a glass on silicon wafer and is then reduced to phosphorus by silicon,

$$2P_2O_5 + 5Si \rightarrow 4P + 5SiO_2, \tag{2}$$

and the phosphorus is released and diffuses into the silicon and Cl_2 is vented.

For diffusion in gallium arsenide, the high vapor pressure of arsenic requires special methods to prevent the loss of arsenic by decomposition or evaporation.[2] These methods include diffusion in sealed ampules with an overpressure of arsenic and diffusion in an open-tube furnace with a doped-oxide capping layer (e.g., silicon nitride). Most of the studies on p-type diffusion have been confined to the use of zinc in the forms of Zn-Ga-As alloys and $ZnAs_2$ for the sealed-ampule approach or $ZnO-SiO_2$ for the open-tube approach. The n-type dopants in gallium arsenide include selenium and tellurium.

13.1.1 Diffusion Equation

Diffusion in a semiconductor can be visualized as atomic movement of the diffusant (dopant atoms) in the crystal lattice by vacancies or interstitials. Figure 3 shows the two basic atomic diffusion models in a solid.[1,3] The open circles represent the host atoms occupying the equilibrium lattice positions. The solid dots represent impurity atoms. At elevated temperatures, the lattice atoms vibrate around the equilibrium lattice sites. There is a finite probability that a host atom acquires sufficient energy to leave the lattice site and to become an interstitial atom, thereby creating a vacancy. When a neighboring impurity atom migrates

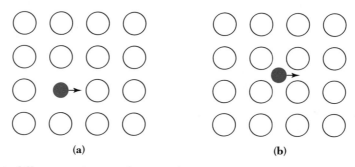

(a)   **(b)**

Fig. 3 Atomic diffusion mechanisms for a two-dimensional lattice.[1,3] (*a*) Vacancy mechanism; (*b*) interstitial mechanism.

to the vacancy site, as illustrated in Fig. 3a, the mechanism is called *vacancy diffusion*. If an interstitial atom moves from one place to another without occupying a lattice site (Fig. 3b), the mechanism is *interstitial diffusion*. An atom smaller than the host atom often moves interstitially.

The basic diffusion process of impurity atoms is similar to that of charge carriers (electrons and holes), discussed in Chapter 3. Accordingly, we define a flux F as the number of dopant atoms passing through a unit area in a unit time and C as the dopant concentration per unit volume. From Eq. 27 in Chapter 3, we have

$$F = -D\frac{\partial C}{\partial x}, \tag{3}$$

where we have substituted C for the carrier concentration and the proportionality constant D is the diffusion coefficient or diffusivity. Note that the basic driving force of the diffusion process is the concentration gradient dC/dx. The flux is proportional to the concentration gradient, and the dopant atoms will move (diffuse) away from a high-concentration region toward a lower-concentration region.

If we substitute Eq. 3 into the one-dimensional continuity equation, Eq. 56, given in Chapter 3 under the condition that no materials are formed or consumed in the host semiconductor (i.e., $G_n = R_n = 0$), we obtain

$$\frac{\partial C}{\partial t} = -\frac{\partial F}{\partial x} = \frac{\partial}{\partial x}\left(D\frac{\partial C}{\partial x}\right). \tag{4}$$

When the concentration of the dopant atoms is low, the diffusion coefficient can be considered to be independent of doping concentration, and Eq. 4 becomes

$$\frac{\partial C}{\partial t} = D\frac{\partial^2 C}{\partial x^2}. \tag{5}$$

Equation 5 is often referred to as *Fick's diffusion equation*.

Figure 4 shows the measured diffusion coefficients for low concentrations of various dopant impurities in silicon and gallium arsenide.[4,5] The logarithm of the diffusion coefficient plotted against the reciprocal of the absolute temperature gives a straight line in most of the cases. This implies that over the temperature range, the diffusion coefficients can be expressed as

$$D = D_0 \exp\left(\frac{-E_a}{kT}\right), \tag{6}$$

where D_0 is the diffusion coefficient in cm²/s extrapolated to infinite temperature and E_a is the activation energy in eV.

For the interstitial diffusion model, E_a is related to the energies required to move dopant atoms from one interstitial site to another. The values of E_a are found to be between 0.5 and 2 eV in both silicon and gallium arsenide. For the vacancy diffusion model, E_a is related to both the energies of motion and the energies of formation of vacancies. Thus,

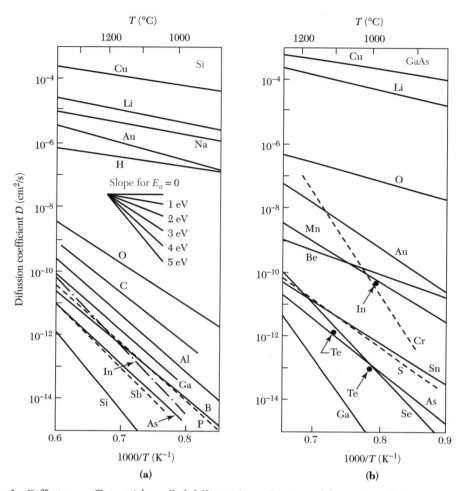

Fig. 4 Diffusion coefficient (also called diffusivity) as a function of the reciprocal of temperature for (a) silicon and (b) gallium arsenide.[4,5]

E_a for vacancy diffusion is larger than that for interstitial diffusion, usually between 3 and 5 eV.

For fast diffusants, such as Cu in Si and GaAs, shown in the upper portion of Fig. 4a and 4b, the measured activation energies are less than 2 eV, and interstitial atomic movement is the dominant diffusion mechanism. For slow diffusants, such as As in Si and GaAs, shown in the lower portion of Fig. 4a and 4b, E_a is larger than 3 eV, and vacancy diffusion is the dominant mechanism.

13.1.2 Diffusion Profiles

The diffusion profile of the dopant atoms is dependent on the initial and boundary conditions. In this subsection we consider two important cases, namely, constant–surface-concentration diffusion and constant–total-dopant diffusion. In the first case, impurity atoms are transported from a vapor source onto the semiconductor surface and diffused into the semiconductor wafers. The vapor source maintains a constant level of surface concentration during the entire diffusion period. In the second case, a fixed amount of dopant is deposited onto the semiconductor surface and is subsequently diffused into the wafers.

Constant–Surface-Concentration Diffusion
The initial condition at $t = 0$ is

$$C(x, 0), \tag{7}$$

which states that the dopant concentration in the host semiconductor is initially zero. The boundary conditions are

$$C(0, t) = C_s \tag{8a}$$

and

$$C(\infty, t) = 0 \tag{8b}$$

where C_s is the surface concentration (at $x = 0$), which is independent of time. The second boundary condition states that at large distances from the surface there are no impurity atoms.

The solution of the diffusion equation (Eq. 5) that satisfies the initial and boundary conditions is given by [6]

$$C(x,t) = C_s \operatorname{erfc}\left(\frac{x}{2\sqrt{Dt}}\right), \tag{9}$$

where erfc is the complementary error function and $\sqrt{Dt}$ is the diffusion length. The definition of erfc and some properties of the function are summarized in Table 1. The diffusion profile for the constant–surface-concentration condition is shown in Fig. 5a, where we plot, on both linear (upper) and logarithmic (lower) scales, the normalized concentration as a function of depth for three values of the diffusion length $\sqrt{Dt}$ corresponding to three consecutive diffusion times and a fixed D for a given diffusion temperature. Note that as the time progresses, the dopant penetrates deeper into the semiconductor.

The total number of dopant atoms per unit area of the semiconductor is given by

$$Q(t) = \int_0^\infty C(x,t)dx. \tag{10}$$

Substituting Eq. 9 into Eq. 10 yields

$$Q(t) = \frac{2}{\sqrt{\pi}} C_s \sqrt{Dt} \cong 1.13\, C_s \sqrt{Dt}. \tag{11}$$

This expression can be interpreted as follows. The quantity $Q(t)$ represents the area under one of the diffusion profiles of the linear plot in Fig. 5a. These profiles can be approximated by triangles with height C_s and base $2\sqrt{Dt}$. This leads to $Q(t) \cong C_s\sqrt{Dt}$, which is close to the exact result obtained from Eq. 11.

A related quantity is the gradient of the diffusion profile dC/dx. The gradient can be obtained by differentiating Eq. 9:

$$\left.\frac{dC}{dx}\right|_{x,t} = -\frac{C_s}{\sqrt{\pi Dt}} e^{-x^2/4Dt}. \tag{12}$$

TABLE 1   Error Function Algebra

$$\text{erf}(x) \equiv \frac{2}{\sqrt{\pi}} \int_0^x e^{-y^2} dy$$

$$\text{erfc}(x) \equiv 1 - \text{erf}(x)$$

$$\text{erf}(0) = 0$$

$$\text{erf}(\infty) = 1$$

$$\text{erf}(x) \cong \frac{2}{\sqrt{\pi}} x \quad \text{for } x \ll 1$$

$$\text{erfc}(x) \cong \frac{1}{\sqrt{\pi}} \frac{e^{-x^2}}{x} \quad \text{for } x \gg 1.$$

$$\frac{d}{dx}\text{erf}(x) = \frac{2}{\sqrt{\pi}} e^{-x^2}$$

$$\frac{d^2}{dx^2}\text{erf}(x) = -\frac{4}{\sqrt{\pi}} x e^{-x^2}$$

$$\int_0^x \text{erfc}(y')dy' = x\,\text{erfc}(x) + \frac{1}{\sqrt{\pi}}(1 - e^{-x^2})$$

$$\int_0^\infty \text{erfc}(x)dx = \frac{1}{\sqrt{\pi}}$$

▶ **EXAMPLE 1**

For a boron diffusion in silicon at 1000°C, the surface concentration is maintained at $10^{19}\,\text{cm}^{-3}$ and the diffusion time is 1 hour. Find $Q(t)$ and the gradient at $x = 0$ and at a location where the dopant concentration reaches $10^{15}\,\text{cm}^{-3}$.

SOLUTION   The diffusion coefficient of boron at 1000°C, as obtained from Fig. 4, is about $2 \times 10^{14}\,\text{cm}^2/\text{s}$, so that the diffusion length is

$$\sqrt{Dt} = \sqrt{2 \times 10^{-14} \times 3600} = 8.48 \times 10^{-6}\,\text{cm},$$

$$Q(t) = 1.13 C_s \sqrt{Dt} = 1.13 \times 10^{19} \times 8.48 \times 10^{-6} = 9.5 \times 10^{13}\,\text{atoms}/\text{cm}^2.$$

$$\left.\frac{dC}{dx}\right|_{x=0} = -\frac{C_s}{\sqrt{\pi Dt}} = \frac{-10^{19}}{\sqrt{\pi} \times 8.48 \times 10^{-6}} = -6.7 \times 10^{23}\,\text{cm}^{-4}.$$

When $C = 10^{15}\,\text{cm}^{-3}$, the corresponding distance x_j is given by Eq. 9, or

$$x_j = 2\sqrt{Dt}\ \text{erfc}^{-1}\left(\frac{10^{15}}{10^{19}}\right) = 2\sqrt{Dt}\ (2.75) = 4.66 \times 10^{-5}\,\text{cm} = 0.466\ \mu\text{m},$$

$$\left.\frac{dC}{dx}\right|_{x=0.466\ \mu m} = -\frac{C_s}{\sqrt{\pi Dt}} e^{-x_j^2/4Dt} = -3.5 \times 10^{20}\,\text{cm}^{-4}. \qquad ◀$$

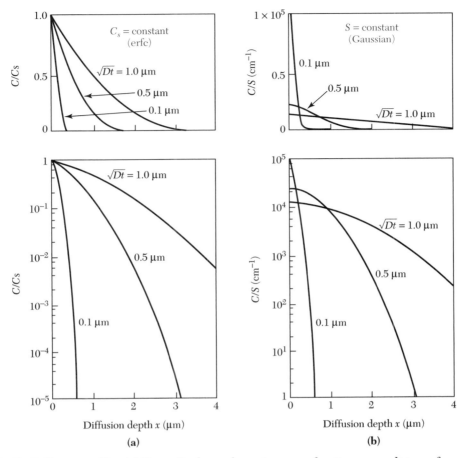

Fig. 5 Diffusion profiles. (*a*) Normalized complementary error function versus distance for successive diffusion times. (*b*) Normalized Gaussian function versus distance.

Constant–Total-Dopant Diffusion

For this case, a fixed (or constant) amount of dopant is deposited onto the semiconductor surface in a thin layer, and the dopant subsequently diffuses into the semiconductor. The initial condition is the same as in Eq. 7. The boundary conditions are

$$\int_0^\infty C(x,t)dx = S \tag{13a}$$

and

$$C(\infty, t) = 0 \tag{13b}$$

where S is the total amount of dopant per unit area.

The solution of the diffusion equation, Eq. 5, that satisfies the above conditions is

$$C(x,t) = \frac{S}{\sqrt{\pi Dt}}\exp\left(-\frac{x^2}{4Dt}\right). \tag{14}$$

This expression is the Gaussian distribution. Since the dopant will move into the semi-conductor as time increases, to keep the total dopant S constant, the surface concentration must decrease. This is indeed the case, since the surface concentration is given by Eq. 14 with $x = 0$:

$$C_s(t) = \frac{S}{\sqrt{\pi Dt}}. \tag{15}$$

Figure 5b shows the dopant profile for a Gaussian distribution where we plot the normalized concentration (C/S) as a function of the distance for three increasing diffusion lengths. Note the reduction of the surface concentration as the diffusion time increases. The gradient of the diffusion profile is obtained by differentiating Eq. 14 and is

$$\left.\frac{dC}{dx}\right|_{x,t} = -\frac{xS}{2\sqrt{\pi}(Dt)^{3/2}} e^{-x^2/4Dt} = -\frac{x}{2Dt}C(x,t). \tag{16}$$

The gradient (or slope) is zero at $x = 0$ and at $x = \infty$ and the maximum gradient occurs at $x = \sqrt{2Dt}$.

In integrated-circuit processing, a two-step diffusion process is commonly used, in which a *predeposition* diffused layer is first formed under a constant–surface-concentration condition. This step is followed by a *drive-in* diffusion (also called *redistribution* diffusion) under a constant–total-dopant condition. For most practical cases, the diffusion length $\sqrt{Dt}$ for the predeposition diffusion is much smaller than the diffusion length for the drive-in diffusion. Therefore, we can consider the predeposition profile as a delta function at the surface, and we can regard the extent of the penetration of the predeposition profile to be negligibly small compared with that of the final profile that results from the drive-in step.

▶ **EXAMPLE 2**

Arsenic was predeposited by arsine gas and the resulting total amount of dopant per unit area is 1×10^{14} atoms/cm². How long would it take to drive the arsenic in to a junction depth of 1 μm? Assume a background doping of 1×10^{15} atoms/cm³, and a drive-in temperature of 1200°C. For As diffusion, $D_0 = 24$ cm²/s, and $E_a = 4.08$ eV.

SOLUTION

$$D = D_0 \exp\left(\frac{-E_a}{kT}\right) = 24 \ \exp\left(\frac{-4.08}{8.614 \times 10^{-5} \times 1473}\right) = 2.602 \times 10^{-13} \ \text{cm}^2/\text{s},$$

$$x_j^2 = 10^{-8} = 4Dt \ \ln\left(\frac{S}{C_B\sqrt{\pi Dt}}\right) = 1.04 \times 10^{-12} \ t \ \ln\left(\frac{1.106 \times 10^5}{\sqrt{t}}\right),$$

$t \cdot \log t - 10.09t + 8350 = 0.$

The solution to the above equation can be determined by the cross point of equation $y = t \cdot \log t$ and $y = 10.09t - 8350$.

Therefore, $t = 1190$ seconds $\cong 20$ minutes. ◀

13.1.3 Evaluation of Diffused Layers

The results of a diffusion process can be evaluated by three measurements—the junction depth, the sheet resistance, and the dopant profile of the diffused layer. The junction depth can be delineated by cutting a groove into the semiconductor and etching the surface with a solution (e.g., 100 cm^3 HF and a few drops of HNO_3 for silicon) that stains the p-type region darker than the n-type region, as illustrated in Fig. 6a. If R_0 is the radius of the tool used to form the groove, then the junction depth x_j is given by

$$x_j = \sqrt{R_0^2 - b^2} - \sqrt{R_0^2 - a^2}, \tag{17}$$

where a and b are indicated in the figure. In addition, if R_0 is much larger than a and b, then

$$x_j \cong \frac{a^2 - b^2}{2R_0}. \tag{18}$$

The junction depth x_j as illustrated in Fig. 6b is the position where the dopant concentration equals the substrate concentration C_B, or

$$C(x_j) = C_B. \tag{19}$$

Thus, if the junction depth and C_B are known, the surface concentration C_s and the impurity distribution can be calculated, provided the diffusion profile follows one or the other simple equation derived in Section 13.1.2.

The resistance of a diffused layer can be measured by the four-point probe technique described in Chapter 3. The *sheet resistance R* is related to the junction depth x_j, the carrier mobility μ (which is a function of the total impurity concentration), and the impurity distribution $C(x)$ by the following expression:[7]

$$R = \frac{1}{q \int_0^{x_j} \mu C(x)\,dx}. \tag{20}$$

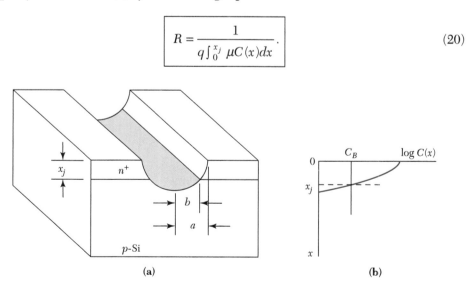

(a)   (b)

Fig. 6 Junction-depth measurement. (*a*) Grooving and staining; (*b*) position in which dopant and substrate concentration are equal.

For a given diffusion profile, the average resistivity $\overline{\rho} = Rx_j$ is uniquely related to the surface concentration C_s and the substrate-doping concentration for an assumed diffusion profile. Design curves relation C_s and $\overline{\rho}$ have been calculated for simple diffusion profiles, such as the erfc or Gaussian distribution.[8] To use these curves correctly we must be sure that the diffusion profiles agree with the assumed profiles. For low concentration and deep diffusions, the diffusion profiles generally can be represented by the aforementioned simple functions. However, as we discuss in the next section, for high concentration and shallow diffusions, the diffusion profiles cannot be represented by these simple functions.

The diffusion profile can be measured using the capacitance-voltage technique described in Chapter 7. The majority carrier profile, which is equal to the impurity profile if impurities are fully ionized, can be determined by measuring the reverse-bias capacitance of a p-n junction or a Schottky barrier diode as a function of the applied voltage. A more elaborate method is the secondary-ion–mass spectroscope (SIMS) technique, which measures the total impurity profile. In the SIMS technique, an ion beam sputters material off the surface of a semiconductor, and the ion component is detected and mass analyzed. This technique has high sensitivity to many elements, such as boron and arsenic, and it is an ideal tool for providing the precision needed for profile measurements in high-concentration or shallow-junction diffusions.[9]

13.2 EXTRINSIC DIFFUSION

The diffusion profiles described in Section 13.1 are for constant diffusivities. These profiles occur when the doping concentration is lower than the intrinsic-carrier concentration $n_i(T)$ at the diffusion temperature. For example, at $T = 1000°C$, n_i equals 5×10^{18} cm^{-3} for silicon and 5×10^{17} cm^{-3} for gallium arsenide. The diffusivity at low concentrations is often referred to as the intrinsic diffusivity. Doping profiles that have concentrations less than n_i (T) are in the *intrinsic* diffusion region as indicated in the left side of Fig. 7. In this region, the resulting dopant profiles of sequential or simultaneous diffusions of n- and p-type impurities can be determined by superposition, that is, the diffusions can be treated independently. However, when the impurity concentration, including both the substrate and the dopant, is greater than $n_i(T)$, the semiconductor becomes extrinsic and the diffusivity is considered to be extrinsic. In the extrinsic diffusion region the diffusivity becomes concentration dependent.[10] In the extrinsic diffusion region the diffusion profiles are more complicated, and there are interactions and cooperative effects among the sequential or simultaneous diffusions.

13.2.1 Concentration-Dependent Diffusivity

As mentioned previously, when a host atom acquires sufficient energy from the lattice vibration to leave its lattice site, a vacancy is created. Depending on the charges associated with a vacancy, we can have a neutral vacancy V^0, an acceptor vacancy V^-, a double-charged acceptor vacancy V^{2-}, a donor vacancy V^+, and so forth. We expect that the vacancy density of a given charge state (i.e., the number of vacancies per unit volume, C_V) has a temperature dependence similar to that of the carrier density (see to Eq. 28 in Chapter 2), that is ,

$$C_V = C_i \exp\left(\frac{E_F - E_i}{kT}\right), \tag{21}$$

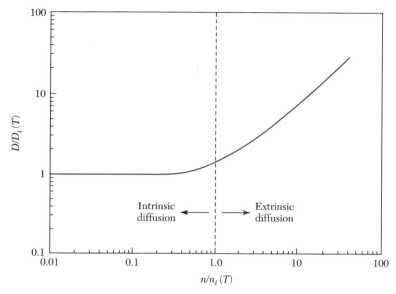

Fig. 7 Donor-impurity diffusivity versus electron concentration showing regions of intrinsic and extrinsic diffusion.[10]

where C_i is the intrinsic vacancy density, E_F is the Fermi level, and E_i is the intrinsic Fermi level.

If the dopant diffusion is dominated by the vacancy mechanism, the diffusion coefficient is expected to be proportional to the vacancy density. At low doping concentrations ($n < n_i$), the Fermi level coincides with the intrinsic Fermi level ($E_F = E_i$). The vacancy density is equal to C_i and is independent of doping concentration. The diffusion coefficient, which is proportional to C_i, also is independent of doping concentration. At high concentrations ($n > n_i$), the Fermi level will move toward the conduction band edge (for donor-type vacancies), and the term [exp ($E_F - E_i$)/kT] becomes larger than unity. This causes C_V to increase, which in turn causes the diffusion coefficient to increase, as shown in the right side of Fig. 7.

When the diffusion coefficient varies with dopant concentration, Eq. 4 should be used as the diffusion equation instead of Eq. 5, in which D is independent of C. We consider the case where the diffusion coefficient can be written as

$$D = D_s \left(\frac{C}{C_s} \right)^\gamma,$$ (22)

where C_s is the surface concentration, D_s is the diffusion coefficient at the surface, and γ is a parameter to describe the concentration dependence. For such a case, we can write the diffusion equation, Eq. 4, as an ordinary differential equation and solve it numerically.

Figure 8 shows the solutions[11] for a constant–surface-concentration diffusion with different values of γ. For $\gamma = 0$, we have the case of constant diffusivity and the profile is the same as that shown in Fig. 5a. For $\gamma > 0$ the diffusivity decreases as the concentration decreases, and increasingly steep and box-like concentration profiles result for increasing γ. Therefore, highly abrupt junctions are formed when diffusions are made into a background of an opposite impurity type. The abruptness of the doping profile

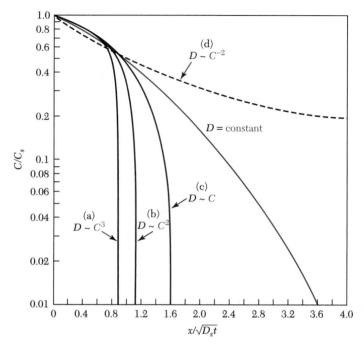

Fig. 8 Normalized diffusion profiles for extrinsic diffusion where the diffusion coefficient becomes concentration dependent.[10,11]

results in a junction depth virtually independent of the background concentration. Note that the junction depth (see Fig. 8) is given by

$$x_j = 1.6\sqrt{D_s t} \quad \text{for } D \sim C\,(\gamma = 1),$$

$$x_j = 1.1\sqrt{D_s t} \quad \text{for } D \sim C^2\,(\gamma = 2), \qquad (23)$$

$$x_j = 0.87\sqrt{D_s t} \quad \text{for } D \sim C^3\,(\gamma = 3).$$

In the case of $\gamma = -2$, the diffusivity increases with decreasing concentration, which leads to a concave profile, as opposed to to the convex profiles for other cases.

13.2.2 Diffusion Profiles

Diffusion in Silicon

The measured diffusion coefficients of boron and arsenic in silicon have a concentration dependence with $\gamma \cong 1$. Their concentration profiles are abrupt, as depicted in curve c of Fig. 8. For gold and platinum diffusion in silicon, γ is close to -2, and their concentration profiles have the concave shape shown in curve d of Fig. 8.

The diffusion of phosphorus in silicon is associated with the doubly charged acceptor vacancy V^{2-}, and the diffusion coefficient at high concentration varies as C^2. We would expect that the diffusion profile of phosphorus resembles that shown in curve b of Fig. 8. However, because of a *dissociation effect*, the diffusion profile exhibits anomalous behavior.

Figure 9 shows phosphorus diffusion profiles for various surface concentrations after diffusion into silicon for 1 hour at 1000°C.[12] When the surface concentration is low, cor-

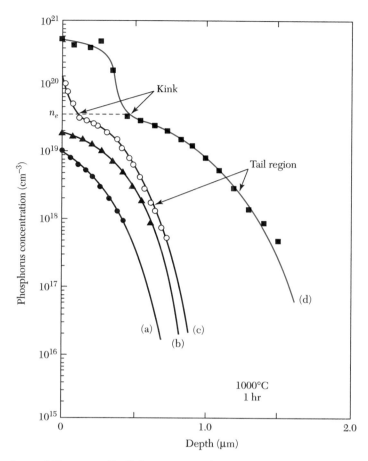

Fig. 9 Phosphorus diffusion profiles[12] for various surface concentrations after diffusion into silicon for 1 hour at 1000°C.

responding to the intrinsic diffusion region, the diffusion profile is given by an erfc (curve *a*). As the concentration increases, the profile begins to deviate from the simple expression (curves *b* and *c*). At very high concentration (curve *d*), the profile near the surface is indeed similar to that shown in curve *b* of Fig. 8. However, at concentration n_e, a kink occurs and is followed by a rapid diffusion in the tail region. The concentration n_e corresponds to a Fermi level 0.11 eV below the conduction band. At this energy level, the coupled impurity-vacancy pair (P^+V^{2-}) dissociates to P^+, V^-, and an electron. Thus, the dissociation generates a large number of singly charged acceptor vacancies V^-, which in turn enhances the diffusion in the tail region of the profile. The diffusivity in the tail region is over 10^{-12} cm²/s, which is about two orders of magnitude larger than the intrinsic diffusivity at 1000°C. Because of its high diffusivity, phosphorus is commonly used to form deep junctions, such as the *n*-tubs in CMOS.

Zinc Diffusion in Gallium Arsenide
We expect diffusion in gallium arsenide to be more complicated than that in silicon because the diffusion of impurities may involve atomic movements on both the gallium and arsenic sublattices. Vacancies play a dominant role in diffusion processes in gallium arsenide

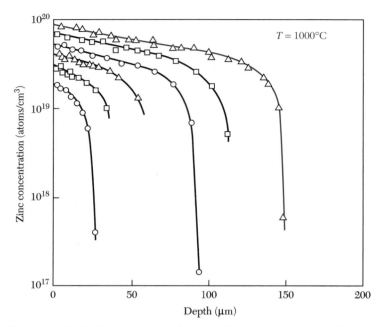

Fig. 10 Diffusion profiles[13] of zinc in GaAs after annealing at 1000°C for 2.7 hours. The different surface concentrations are obtained by maintaining the Zn source at temperatures in the range 600°–800°C.

because both *p*- and *n*-type impurities must ultimately reside in lattice sites. However, the charge states of the vacancies have not been established.

Zinc is the most extensively studied diffusant in gallium arsenide. Its diffusion coefficient is found to vary as C^2. Therefore, the diffusion profiles are steep, as shown [13] in Fig. 10, and resemble curve *b* of Fig. 8. Note that even for the case of the lowest surface concentration, the diffusion is in the extrinsic-diffusion region, because n_i for GaAs at 1000°C is less than 10^{18} cm^{-3}. As can be seen in Fig. 10, the surface concentration has a profound effect on the junction depth. The diffusivity varies linearly with the partial pressure of the zinc vapor, and the surface concentration is proportional to the square root of the partial pressure. Therefore, from Eq. 23, the junction depth is linearly proportional to the surface concentration.

▶ 13.3 DIFFUSION-RELATED PROCESSES

In this section we consider two processes in which diffusion plays an important role and the impact of these processes on device performances.

13.3.1 Lateral Diffusion

The one-dimensional diffusion equation discussed previously can describe satisfactorily the diffusion process, except at the edge of the mask window. Here the impurities will diffuse downward and sideways (i.e., laterally). In this case, we must consider a two-dimensional diffusion equation and use a numerical technique to obtain the diffusion profiles under different initial and boundary conditions.

Figure 11 shows the contours of constant doping concentration for a constant–surface-concentration diffusion condition assuming that the diffusivity is independent of concentration.[14] At the far right of the figure, the variation of the dopant concentration from $0.5\,C_s$ to $10^{-4}\,C_s$ (where C_s is the surface concentration) corresponds to the erfc distribution given by Eq. 9. The contours are in effect a map of the location of the junctions created by diffusing into various background concentrations. For example, at $C/C_s = 10^{-4}$ (i.e., the background doping is 10^4 times lower than the surface concentration), we see from this constant-concentration curve that the vertical penetration is about 2.8 µm, whereas the lateral penetration is about 2.3 µm (i.e., the penetration along the diffusion mask-semiconductor interface). Therefore, the lateral penetration is about 80% of the penetration in the vertical direction for concentrations three or more orders of magnitude below the surface concentration. Similar results are obtained for a constant–total-dopant diffusion condition. The ratio of lateral to vertical penetration is about 75%. For concentration-dependent diffusivities, the ratio is found to be reduced slightly, to about 65%–70%.

Because of the lateral-diffusion effect, the junction consists of a central plane (or flat) region with approximately cylindrical edges with a radius of curvature r_j, as shown in Fig. 11. In addition, if the diffusion mask contains sharp corners, the shape of the junction near the corner will be roughly spherical because of lateral diffusion. Since the electric-field intensities are higher for cylindrical and spherical junction regions, the avalanche breakdown voltages of such regions can be substantially lower than that of a plane junction having the same background doping. This junction " curvature effect" was discussed in Chapter 4.

13.3.2 Impurity Redistribution During Oxidation

Dopant impurities near the silicon surface will be redistributed during thermal oxidation. The redistribution depends on several factors. When two solid phases are brought together, an impurity in one solid will redistribute between the two solids until it reaches equilibrium. This is similar to our previous discussion on impurity redistribution in crystal growth

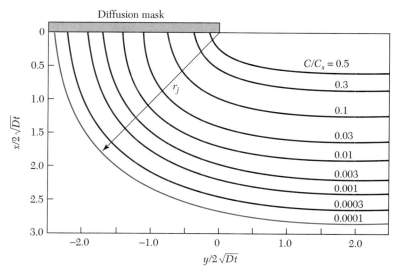

Fig. 11   Diffusion contours at the edge of an oxide window, where r_j is the radius of curvature.[14]

from the melt. The ratio of the equilibrium concentration of the impurity in the silicon to that in the silicon dioxide is called the *segregation coefficient* and is defined as

$$k = \frac{\text{equilibrium concentration of impurity in silicon}}{\text{equilibrium concentration of impurity in SiO}_2}. \tag{24}$$

A second factor that influences impurity distribution is that the impurity may diffuse rapidly through the silicon dioxide and escape to the gaseous ambient. If the diffusivity of the impurity in silicon dioxide is large, this factor will be important. A third factor in the redistribution process is that the oxide is growing, and thus, the boundary between the silicon and the oxide is advancing into the silicon as a function of time. The relative rate of this advance compared with the diffusion rate of the impurity through the oxide is important in determining the extent of the redistribution. Note that even if the segregation coefficient of an impurity k equals unity, some redistribution of the impurity in the silicon will still take place. As indicated in Fig. 3 of Chapter 11, the oxide layer will be about twice as thick as the silicon layer it replaced. Therefore, the same amount of impurity will now be distributed in a larger volume, resulting in a depletion of the impurity from the silicon.

Four possible redistribution processes are illustrated[6] in Fig. 12. These processes can be classified into two groups. In one group the oxide takes up the impurity (Fig. 12*a* and

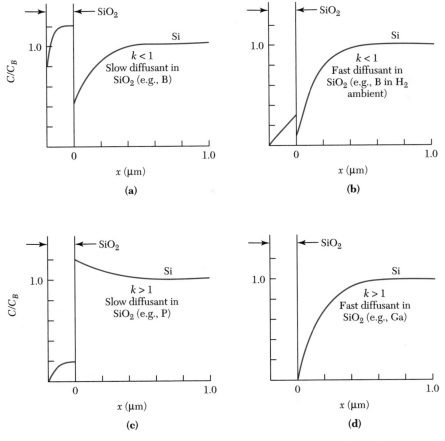

Fig. 12 Four different cases of impurity redistribution in silicon due to thermal oxidation.[6]

b for *k* < 1), and in the other the oxide rejects the impurity (Fig. 12*c* and *d* for *k* > 1). In each case, what happens depends on how rapidly the impurity can diffuse through the oxide. In group 1, the silicon surface is depleted of impurities; an example is boron with *k* approximately equal to 0.3. Rapid diffusion of the impurity through the silicon dioxide increases the amount of depletion; an example is boron-doped silicon heated in a hydrogen ambient because hydrogen in silicon dioxide enhances the diffusivity of boron. In group 2, *k* is greater than unity, so that the oxide rejects the impurity. If diffusion of the impurity through the silicon dioxide is relatively slow, the impurity piles up near the silicon surface; an example is phosphorus, with *k* approximately equal to 10. When diffusion through the silicon dioxide is rapid, so much impurity may escape from the solid to the gaseous ambient that the overall effect will be a depletion of the impurity; an example is gallium, with *k* approximately equal to 20.

The redistributed dopant impurities in silicon dioxide are seldom electrically active. However, redistribution in silicon has an important effect on processing and device performance. For example, nonuniform dopant distribution will modify the interpretation of the measurements of interface-trap properties (see Chapter 6), and the change of the surface concentration will modify the threshold voltage and device contact resistance (see Chapter 7).

▶ 13.4 RANGE OF IMPLANTED IONS

Ion implantation is the introduction of energetic, charged particles into a substrate such as silicon. Implantation energies are between 1 keV and 1 MeV, resulting in ion distributions with average depths ranging from 10 nm to 10 μm. Ion doses vary from 10^{12} ions/cm^2 for threshold voltage adjustment to 10^{18} ions/cm^2 for the formation of buried insulating layer. Note that the dose is expressed as the number of ions implanted into 1 cm^2 of the semiconductor surface area. The main advantages of ion implantation are its more precise control and reproducibility of impurity dopings and its lower processing temperature compared with those of the diffusion process.

Figure 13 shows schematically a medium-energy ion implantor.[15] The ion source has a heated filament to break up source gas such as BF_3 or AsH_3 into charged ions (B^+ or As^+). An extraction voltage, around 40 kV, causes the charged ions to move out of the ion-source chamber into a mass analyzer. The magnetic field of the analyzer is chosen such that only ions with the desired mass-to-charge ratio can travel through it without being filtered. The selected ions then enter the acceleration tube, where they are accelerated to the implantation energy as they move from high voltage to ground. Apertures

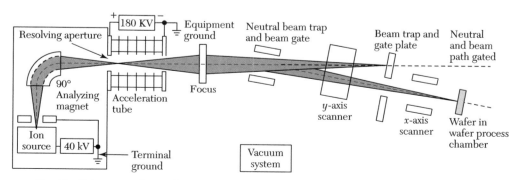

Fig. 13   Schematic of a medium-current ion implantor.

ensure that the ion beam is well collimated. The pressure in the implantor is kept below 10^{-4} Pa to minimize ion scattering by gas molecules. The ion beam is then scanned over the wafer surface using electrostatic deflection plates and is implanted into the semiconductor substrate.

The energetic ions lose their energies through collision with electrons and nuclei in the substrate and finally come to rest at some depth within the lattice. The average depth can be controlled by adjusting the acceleration energy. The dopant dose can be controlled by monitoring the ion current during implantation. The principle side effect is the disruption or damage of the semiconductor lattice due to ion collisions. Therefore, a subsequent annealing treatment is needed to remove these damages.

13.4.1 Ion Distribution

The total distance that an ion travels in coming to rest is called its *range R* and is illustrated[16] in Fig. 14a. The projection of this distance along the axis of incidence is called the *projected range R_p*. Because the number of collisions per unit distance and the energy lost per collision are random variables, there will be a spatial distribution of ions having

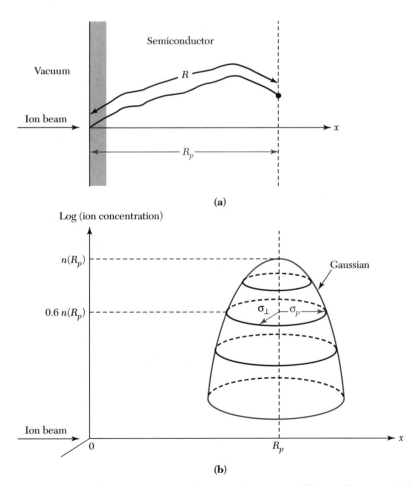

Fig. 14 (a) Schematic of the ion range R and projected range σ_p. (b) Two-dimensional distribution of the implanted ions.[16]

the same mass and the same initial energy. The statistical fluctuations in the projected range are called the *projected straggle* σ_p. There is also a statistical fluctuation along an axis perpendicular to the axis of incidence, called the *lateral straggle* $\sigma_\perp$.

Figure 14*b* shows the ion distribution. Along the axis of incidence, the implanted impurity profile can be approximated by a Gaussian distribution function:

$$n(x) = \frac{S}{\sqrt{2\pi}\sigma_p} \exp\left[-\frac{(x-R_p)^2}{2\sigma_p^2}\right],$$

(25)

where S is the ion dose per unit area. This equation is similar to Eq. 14 for constant–total-dopant diffusion, except that the quantity $4Dt$ is replaced by $2\sigma_p^2$ and the distribution is shifted along the x-axis by R_p. Thus, for diffusion, the maximum concentration is at $x = 0$, whereas for ion implantation the maximum concentration is at the projected range R_p. The ion concentration is reduced by 40% from its peak value at $(x - R_p) = \pm\sigma_p$, by one decade at $\pm2\sigma_p$, by two decades at $\pm3\sigma_p$, and by five decades at $\pm4.8\sigma_p$.

Along the axis perpendicular to the axis of incidence, the distribution is also a Gaussian function of the form $\exp(-y^2/2\sigma_\perp^2)$. Because of this distribution, there will be some lateral implantation.[17] However, the lateral penetration from the mask edge (on the order of $\sigma_\perp$) is considerably smaller than that from the thermal diffusion process discussed in Section 13.3.

13.4.2   Ion Stopping

There are two stopping mechanisms by which an energetic ion, on entering a semiconductor substrate (also called the target), can be brought to rest. The first is by transferring its energy to the target nuclei. This causes deflection of the incident ion and also dislodges many target nuclei from their original lattice sites. If E is the energy of the ion at any point x along its path, we can define a nuclear stopping power $S_n(E) \equiv (dE/dx)_n$ to characterize this process. The second stopping mechanism is by the interaction of the incident ion with the cloud of electrons surrounding the target's atoms. The ion loses energy in collisions with electrons through Coulombic interaction. The electrons can be excited to higher energy levels (excitation), or they can be ejected from the atom (ionization). We can define an electronic stopping power $S_e(E) \equiv (dE/dx)_e$ to characterize this process.

The average rate of energy loss with distance is given by a superposition of the above two stopping mechanisms:

$$\frac{dE}{dx} = S_n(E) + S_e(E).$$

(26)

If the total distance traveled by the ion before coming to rest is R, then

$$R = \int_0^R dx = \int_0^{E_0} \frac{dE}{S_n(E) + S_e(E)},$$

(27)

where E_0 is the initial ion energy. The quantity R has been defined previously as the range.

We can visualize the nuclear stopping process by considering the elastic collision between an incoming hard sphere (energy E_0 and mass M_1) and a target hard sphere (initial energy zero and mass M_2), as illustrated in Fig. 15. When the spheres collide, momentum is transferred along the centers of the spheres. The deflection angle θ and the velocities v_1 and v_2 can be obtained from the requirements for conservation of momentum and

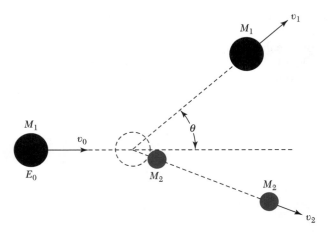

Fig. 15 Collision of hard spheres.

energy. The maximum energy loss is in a head-on collision. For this case, the energy loss by the incident particle M_1 or the energy transferred to M_2 is

$$\frac{1}{2}M_2v_2^2 = \left[\frac{4M_1M_2}{(M_1+M_2)^2}\right]E_0. \tag{28}$$

Since M_2 is usually of the same order of magnitude as M_1, a large amount of energy can be transferred in nuclear stopping process.

Detailed calculations show that the nuclear stopping power increases linearly with energy at low energies (similar to Eq. 28), and $S_n(E)$ reaches a maximum at some intermediate energy. At high energies, $S_n(E)$ becomes smaller because fast particles may not have sufficient interaction time with the target atoms to achieve effective energy transfer. The calculated values of $S_n(E)$ for arsenic, phosphorus, and boron in silicon at various energies are shown in Fig. 16 (solid lines, where the superscript indicates the atomic weight).[18] Note that heavier atoms, such as arsenic, have larger nuclear stopping power, that is, larger energy loss per unit distance.

The electronic stopping power is found to be proportional to the velocity of the incident ion, or

$$S_e(E) = k_e\sqrt{E} \tag{29}$$

where the coefficient k_e is a relatively weak function of atomic mass and atomic number. The value of k_e is approximately 10^7 (eV)$^{1/2}$/cm for silicon and 3×10^7 (eV)$^{1/2}$/cm for gallium arsenide. The electronic stopping power in silicon is plotted in Fig. 16 (dotted line). Also shown in the figure are the crossover energies, at which $S_e(E)$ equals $S_n(E)$. For boron, which has a relatively low ion mass compared with the target silicon atom, the crossover energy is only 10 keV. This means that over most of the implantation energy range of 1 keV to 1 MeV, the main energy loss mechanism is due to electronic stopping. On the other hand, for arsenic with relatively high ion mass, the crossover energy is 700 keV. Thus, nuclear stopping dominates over most of the energy range. For phosphorus, the crossover energy is 130 keV. For an E_0 less than 130 keV, nuclear stopping will dominate; for higher energies, electronic stopping will take over.

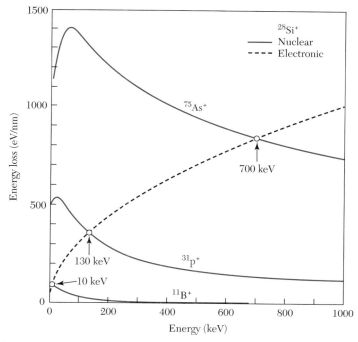

Fig. 16 Nuclear stopping power $S_n(E)$, and electronic stopping power $S_e(E)$ for As, P, and B in Si. The points of intersection of the curves correspond to the energy at which nuclear and electronic stopping are equal.[18]

Once $S_n(E)$ and $S_e(E)$ are known, we can calculate the range from Eq. 27. This in turn can give us the projected range and projected straggle with the help of the following approximate equations[15]:

$$R_p \cong \frac{R}{1 + (M_2 / 3M_1)}, \tag{30}$$

$$\sigma_p \cong \frac{2}{3} \left[\frac{\sqrt{M_1 M_2}}{M_1 + M_2} \right] R_p. \tag{31}$$

Figure 17a shows the projected range (R_p), the projected straggle (σ_p), and the lateral straggle ($\sigma_\perp$) for arsenic, boron, and phosphorus in silicon.[19] As expected, the larger the energy loss, the smaller the range. Also, the projected range and straggles increase with ion energy. For a given element at a specific incident energy, σ_p and $\sigma_\perp$ are comparable and usually within ±20%. Figure 17b shows the corresponding values for hydrogen, zinc, and tellurium in gallium arsenide.[17] If we compare Fig. 17a with Fig. 17b, we see that most of the popular dopants (except hydrogen) have larger projected ranges in silicon than they have in gallium arsenide.

▷ **EXAMPLE 3**

Assume 100 keV boron implants on a 200 mm silicon wafer at a dose of 5×10^{14} ions/cm². Calculate the peak concentration and the required ion beam current for 1 minute of implantation.

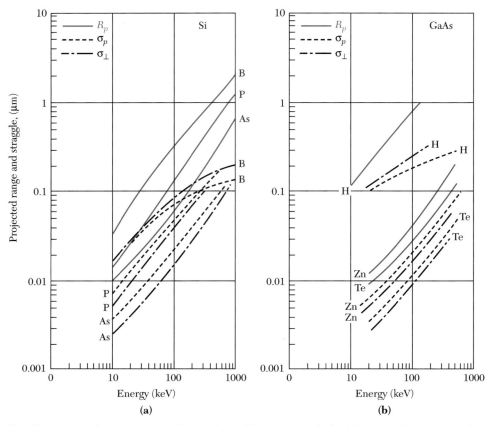

Fig. 17 Projected range, projected straggle, and lateral straggle for (a) B, P, and As in Si, and (b) H, Zn, and Te in GaAs.[17,19]

SOLUTION From Fig. 17a, we obtain 0.31 and 0.07 μm for the projected range and project straggle, respectively.

From Eq.25, $n(x) = \dfrac{S}{\sqrt{2\pi}\sigma_p} \exp\left[\dfrac{-(x-R_p)^2}{2\sigma_p^2}\right]$,

$$\frac{dn}{dx} = -\frac{S}{\sqrt{2\pi}\sigma_p} \frac{2(x-R_p)}{2\sigma_p^2} \exp\left[\frac{-(x-R_p)^2}{2\sigma_p^2}\right] = 0.$$

The peak concentration is at $x = R_p$, and $n(x) = 2.85 \times 10^{19}$ ions/cm³.

The total number of implanted ions $= Q = 5 \times 10^{14} \times \pi \times \left(\dfrac{20}{2}\right)^2 = 1.57 \times 10^{17}$ ions.

The required ion current $= I = \dfrac{qQ}{t} = \dfrac{1.6 \times 10^{-19} \times 1.57 \times 10^{17}}{60} = 4.19 \times 10^{-4}$ A.

$= 0.42$ mA.

13.4.3   Ion Channeling

The projected range and straggle of the Gaussian distribution discussed previously give a good description of the implanted ions in amorphous or fine-grain polycrystalline substrates. Both silicon and gallium arsenide behave as if they were amorphous semiconductors, provided the ion beam is misoriented from the low-index crystallographic direction (e.g.,<111>). In this situation, the doping profile described by Eq. 25 is followed closely near the peak and extended to one or two decades below the peak value. This is illustrated [16] in Fig. 18. However, even for a misorientation of 7° from the <111>-axis, there still is a tail that varies exponentially with distance as exp $(-x/\lambda)$, where λ is typically on the order of 0.1 μm.

The exponential tail is related to the ion-channeling effect. Channeling occurs when incident ions align with a major crystallographic direction and are guided between rows of atoms in a crystal. Figure 19 illustrates a diamond lattice viewed along a <110>-direction.[20] Ions implanted in the <110>-direction will follow trajectories that will not bring them close enough to a target atom to lose significant amounts of energy in nuclear collisions. Thus, for channeled ions, the only energy loss mechanism is electronic stopping, and the range of channeled ions can be significantly larger than it would be in an amorphous target. Ion channeling is particularly critical for low-energy implant and heavy ions.

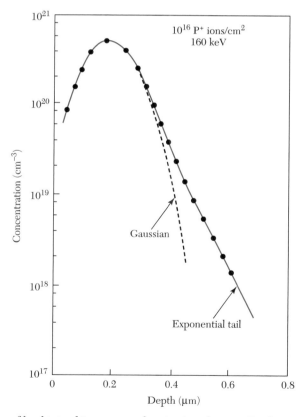

Fig. 18   Impurity profile obtained in a purposely misoriental target. Ion beam is incident 7° from the <111> -axis.[16]

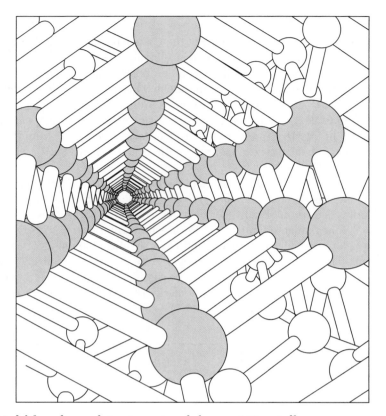

Fig. 19 Model for a diamond structure, viewed along a <110>-axis.[20]

Channeling can be minimized by several techniques: a blocking amorphous surface layer, misorientation of the wafer, and creating a damage layer in the wafer surface. The usual blocking amorphous layer is simply a thin layer of grown silicon dioxide (Fig. 20a). The layer randomizes the direction of the ion beam so that the ions enter the wafer at different angles and not directly down the crystal channels. Misorientation of the wafers 5°–10 ° off the major plane also has the effect of preventing the ions from entering the channels (Fig. 20b). With this method, most implantation machines tilt the wafer by 7° and then apply a 22° twist from the flat to prevent channeling. Predamaging the wafer surface with a heavy silicon or germanium implant creates a randomizing layer in the wafer surface (Fig. 20c). This method, however, increases the use of the expensive ion implantor.

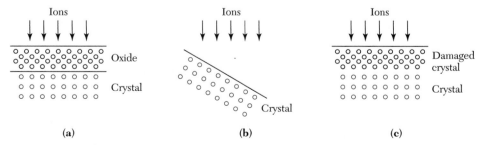

Fig. 20 (a) Implant through an amorphous oxide layer, (b) misorient the beam direction to all crystal axes, and (c) predamage on the crystal surface.

13.5 IMPLANT DAMAGE AND ANNEALING

13.5.1 Implant Damage

When energetic ions enter a semiconductor substrate, they lose their energy in a series of nuclear and electronic collisions and finally come to rest. The electronic-energy loss can be accounted for in terms of electronic excitations to higher energy levels or in the generation of electron-hole pairs. However, electronic collisions do not displace semiconductor atoms from their lattice positions. Only nuclear collisions can transfer sufficient energy to the lattice so that host atoms are displaced resulting in implant damage (also called lattice disorder).[21] These displaced atoms may possess large fractions of the incident energy, and they in turn cause cascades of secondary displacements of nearby atoms to form a *tree of disorder* along the ion path. When the displaced atoms per unit volume approach the atomic density of the semiconductor, the material becomes amorphous.

The tree of disorder for light ions is quite different from that for heavy ions. Much of the energy loss for light ions (e.g., $^{11}B^+$ in silicon) is due to electronic collisions (see Fig. 16), which do not cause lattice damage. The ions lose their energies as they penetrate deeper into the substrate. Eventually, the ion energy is reduced below the crossover energy (10 keV for boron) where nuclear stopping becomes dominant. Therefore, most of the lattice disorder occurs near the final ion position. This is illustrated in Fig. 21a.

We can estimate the damage by considering a 100 keV boron ion. Its projected range is 0.31 μm, (Fig.17a), and its initial nuclear energy loss is only 3 eV/Å (Fig. 16). Since the spacing between lattice planes in silicon is about 2.5 Å, this means that the boron ion will lose 7.5 eV for each lattice plane because of nuclear stopping. The energy required to displace a silicon atom from its lattice position is about 15 eV. Therefore, the incident boron ion does not release enough energy from nuclear stopping to displace a silicon atom when it first enters the silicon substrate. When the ion energy is reduced to about

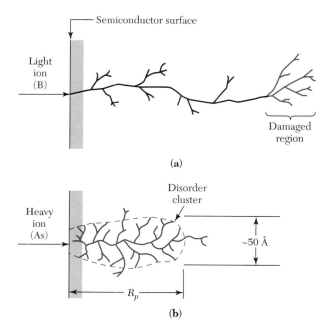

Fig. 21 Implantation disorder caused by (*a*) light ions and (*b*) heavy ions.[2,15]

50 keV (at a depth of 1500 Å), the energy loss due to nuclear stopping increases to 15 eV for each lattice plane (i.e., 6 eV/ Å), sufficient to create a lattice disorder. Assuming that one atom is displaced per lattice plane for the remaining ion range, we have 600 lattice atoms displaced (i.e., 1500 Å/2.5 Å). If each displaced atom moves roughly 25 Å from its original position, the damage volume is given by $V_D \cong \pi(25\text{Å})^2(1500\text{Å}) = 3 \times 10^{-18}\text{cm}^3$. The damage density is $600/V_D \cong 2 \times 10^{20}$ cm^{-3}, which is only 0.4% of the atoms. Thus, very high doses of light ions are needed to create an amorphous layer.

For heavy ions, the energy loss is primarily due to nuclear collisions; therefore, we expect substantial damage. Consider a 100 keV arsenic ion with a projected range of 0.06 μm or 60 nm. The average nuclear energy loss over the entire energy range is about 1320 eV/nm (Fig. 16). This means that the arsenic ion loses about 330 eV for each lattice plane on the average. Most of the energy is given to one primary silicon atom. Each primary atom will subsequently cause 22 displaced target atoms (i.e., 330 eV/15eV). The total number of displaced atom is 5280. Assuming a range of 2.5 nm for the displaced atoms, the damage volume is $V_D \cong \pi(2.5 \text{ nm})^2 (60 \text{ nm}) = 10^{-18}$ cm^3. The damage density is then $5280/V_D \cong 5 \times 10^{21}$ cm^{-3}, or about 10% of the total number of atoms in V_D. As a result of the heavy-ion implantation, the material has become essentially amorphous. Figure 21*b* illustrates the situation where the damage forms a disordered cluster over the entire projected range.

To estimate the dose required to convert a crystalline material to an amorphous form, we can use the criterion that the energy density is of the same order of magnitude as that needed for melting the material (i.e., 10^{21} keV/cm^3). For 100 keV arsenic ions, the dose required to make amorphous silicon is then

$$S = \frac{(10^{21} \text{ keV / cm}^3)R_p}{E_0} = 6 \times 10^{13} \text{ ions / cm}^2. \tag{32}$$

For 100 keV boron ions, the dose required is 3×10^{14} ions/cm^2 because R_p for boron is five times larger than for arsenic. However, in practice, higher doses (>10^{16} ions/cm^2) are required for boron implant into a target at room temperature because of the nonuniform distribution of the damage along the ion path.

13.5.2 Annealing

Because of the damaged region and the disorder cluster that result from ion implantation, semiconductor parameters such as mobility and lifetime are severely degraded. In addition, most of the ions as implanted are not located in substitutional sites. To activate the implanted ion and to restore mobility and other material parameters, we must anneal the semiconductor at an appropriate combination of time and temperature.

In conventional annealing, we use an open-tube batch-furnace system similar to that for thermal oxidation. This process requires long time and high temperature to remove the implant damages. However, conventional annealing may cause substantial dopant diffusion and cannot meet the requirement for shallow junctions and narrow doping profiles. Rapid thermal annealing (RTA) is an annealing process that employs a variety of energy sources with a wide range of times, from 100 seconds down to nanoseconds—all short compared with the conventional annealing. RTA can activate the dopant fully with minimal redistribution.

Conventional Annealing of B and P

Annealing characteristics depend on the dopant type and the dose involved. Figure 22 shows the annealing behaviors of boron and phosphorus implantation into silicon substrates.[19] The substrate is held at room temperature (T_s) during implantation. At a given

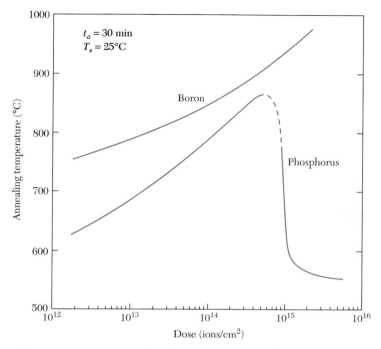

Fig. 22 Annealing temperature versus dose for 90% activation of boron and phosphorus.[1,19]

ion dose, the annealing temperature is defined as the temperature at which 90% of the implanted ions are activated by a 30 minute annealing in a conventional annealing furnace. For boron implantation, higher annealing temperatures are needed for higher doses. For phosphorus at lower doses, the annealing behavior is similar to that for boron. However, when the dose is greater than 10^{15} cm^{-2}, the annealing temperature drops to about 600°C. This phenomenon is related to the *solid-phase epitaxy* process. At phosphorus doses greater than 6×10^{14} cm^{-2}, the silicon surface layer becomes amorphous. The single-crystal semiconductor underneath the amorphous layer serves as a seeding area for recrystallization of the amorphous layer. The epitaxial-growth rate along the <100> direction is 10 nm/min at 550°C and 50 nm/min at 600°C, with an activation energy at 2.4 eV. Therefore, a 100–500 nm amorphous layer can be recrystallized in a few minutes. During the solid-phase epitaxial process, the impurity dopant atoms are incorporated into the lattice sites along with the host atoms; thus, full activation can be obtained at relatively low temperatures.

Rapid Thermal Annealing

The machine for RTA with a transient lamp heating is shown in Fig. 23. The temperature measured [22] from the heated wafer is usually from 600°–1100°C. A wafer is heated quickly under atmospheric conditions or at low pressure under isothermal conditions. Typical lamps in a RTA system are tungsten filaments or arc lamps. The processing chamber is made of either quartz, silicon carbide, stainless steel, or aluminum and has quartz windows through which the optical radiation passes to illuminate the wafer. The wafer holder is often made of quartz and contacts the wafer in a minimum number of places. A measurement system is placed in a control loop to set wafer temperature. The RTA system interfaces with a gas-handling system and a computer that controls system operation. Typically, wafer temperature in a RTA system is measured with a noncontact optical pyrometer that determines temperature from radiated infrared energy.

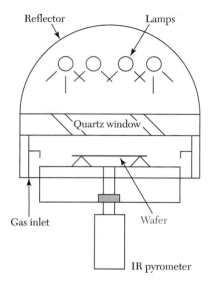

Fig. 23 Rapid thermal annealing system that is optically heated.

A comparison between conventional furnace and RTA technologies is shown in Table 2. To achieve short processing times using RTA, trade-offs must be made in temperature and process uniformity, temperature measurement and control, and wafer stress and throughput. In addition, there are concerns over the introduction of electrically active wafer defects during the very fast (100°–300°C/s) thermal transients. Rapid heating with temperature gradients in the wafers can cause wafer damage in the form of slip dislocations induced by thermal stress. On the other hand, conventional furnace processing brings with it significant problems, such as particle generation from the hot walls, limited ambient control in an open system, and a large thermal mass that restricts controlled heating times to tens of minutes. In fact, requirements on contamination, process control, and cost of manufacturing floor space have resulted in the paradigm shift to the RTA process.

TABLE 2 Technology Comparison

Determinant	Conventional furnace	Rapid thermal annealing
Process	Batch	Single-wafer
Furnace	Hot-wall	Cold-wall
Heating rate	Low	High
Cycle time	High	Low
Temperature monitor	Furnace	Wafer
Thermal budget	High	Low
Particle problem	Yes	Minimal
Uniformity and repeatability	High	Low
Throughput	High	Low

13.6 IMPLANTATION-RELATED PROCESSES

In this section we consider a few implantation-related processes, such as multiple implantation, masking, tilt-angle implantation, high-energy implantation, and high-current implantation.

13.6.1 Multiple Implantation and Masking

In many applications doping profiles other than the simple Gaussian distribution are required. One such case is the preimplantation of silicon with an inert ion to make the silicon surface region amorphous. This technique allows close control of the doping profile and permits nearly 100% dopant activation at low temperatures, as discussed previously. In such a case, a deep amorphous region may be required. To obtain this type of region, we must make a series of implants at varying ion energies and doses.

Multiple implantation can also be used to form a flat doping profile, as shown in Fig. 24. Here, four boron implants into silicon are used to provide a composite doping profile.[23] The measured carrier concentration and that predicted using range theory are shown in the figure. Other doping profiles, unavailable from diffusion techniques, can be obtained by using various combinations of impurity dose and implantation energy. Multiple implants have been used to preserve stoichiometry during the implantation and annealing of GaAs. This approach, whereby equal amount of gallium and *n*-type dopant (or arsenic and *p*-type dopant) are implanted prior to annealing, has resulted in higher carrier activation.

To form *p–n* junctions in selected areas of the semiconductor substrate, an appropriate mask should be used for the implantation. Because implantation is a low-temperature process, a large variety of masking materials can be used. The minimum thickness of masking material required to stop a given percentage of incident ions can be estimated from the range parameters for ions. The inset of Fig. 25 shows a profile of an implant in

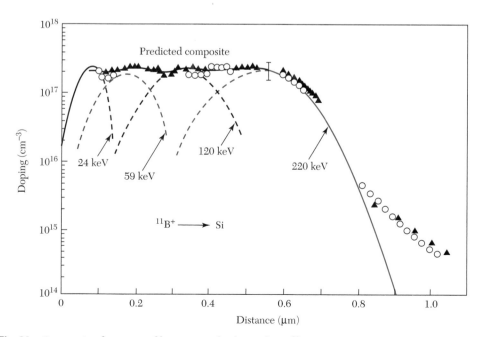

Fig. 24 Composite doping profile using multiple implants.[23]

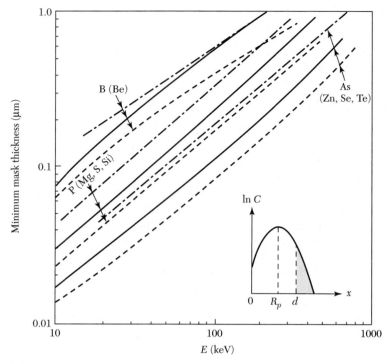

Fig. 25 Minimum thickness [24] of SiO_2 (—), Si_3N_4 (- - - -), and photoresist (– • – • –) to produce a masking effectiveness of 99.99%.

a masking material. The dose implanted in the region beyond a depth d (shown shaded) is given by integration of Eq. 25 as

$$S_d = \frac{S}{\sqrt{2\pi}\,\sigma_p} \int_d^\infty \exp\left[-\left(\frac{x - R_p}{\sqrt{2}\sigma_p}\right)^2\right] dx. \tag{33}$$

From Table 1 we can derive the expression

$$\int_x^\infty e^{-y^2}\,dy = \frac{\sqrt{\pi}}{2}\,\text{erfc}\,(x). \tag{34}$$

Therefore, the fraction of the dose that has "transmitted" beyond a depth d is given by the transmission coefficient T:

$$T \equiv \frac{S_d}{S} = \frac{1}{2}\,\text{erfc}\left(\frac{d - R_p}{\sqrt{2}\,\sigma_p}\right). \tag{35}$$

Once T is given, we can obtain the mask thickness d from Eq. 35 for any given R_p and σ_p.

The values of d to stop 99.99% of the incident ions ($T = 10^{-4}$) are shown in Fig. 25 for SiO_2, Si_3N_4, and a photoresist as masking materials.[19,24] Mask thicknesses given in this figure are for boron, phosphorus, and arsenic implanted into silicon. These mask thicknesses also can be used as guidelines for impurity masking in gallium arsenide. The dopants are shown in the parentheses. Since both R_p and σ_p vary approximately linearly with energy, the minimum thickness of the masking material also increases linearly with energy. In certain applications, instead of totally stopping the beam, the masks can be used as attenuators, which can provide an amorphous surface layer to the incident ion beam to minimize the channeling effect.

▷ **EXAMPLE 4**

When boron ions are implanted at 200 keV, what thickness of SiO_2 will be required to mask 99.996% of the implanted ions (R_p = 0.53 μm, σ_p = 0.093 μm)?

SOLUTION   The complementary error function in Eq. 35 can be approximated if the argument is large (see Table 1):

$$T \cong \frac{1}{2\sqrt{\pi}} \frac{e^{-u^2}}{u},$$

where the parameter u is given by $(d - R_p)/\sqrt{2}\sigma$. For $T = 10^{-4}$, we can solve the above equation to give $u = 2.8$. Thus,

$$d = R_p + 3.96\sigma_p = 0.53 + 3.96 \times 0.093 = 0.898 \text{ μm} \qquad \blacktriangleleft$$

13.6.2 Tilt-Angle Ion Implantation

In scaling devices to submicron dimensions, it is important also to scale dopant profiles vertically. We need to produce junction depths less than 100 nm, including diffusion during dopant activation and subsequent processing steps. Modern device structures, such as the lightly doped drain (LDD), require precise control of dopant distributions vertically and laterally.

It is the ion velocity perpendicular to the surface that determines the projected range of an implanted ion distribution. If the wafer is tilted at a large angle to the ion beam, then the effective ion energy is greatly reduced. Figure 26 illustrates this for 60 keV arsenic ions as a function of the tilt angle, showing that it is possible to achieve extremely shallow distributions using a high tilt angle (86°). In tilt-angle ion implantation, we should consider the shadow effect (inset in Fig. 26) for the patterned wafer. A lower tilt angle leads to a small shadow area. For example, if the height of patterned mask is 0.5 μm, with vertical sidewall, a 7° incident ion beam will induce a 61 nm shadow. This shadow effect may introduce an unexpected series resistance in the device.

13.6.3 High-Energy and High-Current Implantation

High-energy implantors, capable of energies as high as 1.5–5 MeV, are available and have been used for a number of novel applications. The majority of these depend on the ability to dope the semiconductor to many micrometers in depth, without the need for long diffusion times at high temperatures. High-energy implantors can also be used to produce low-resistivity buried layers. For example, a buried layer 1.5–3 μm below the surface for a CMOS device can be achieved by high-energy implantation.

High-current implantors (10–20 mA), operating in the 25–30 keV range, are routinely used for the predeposition step in diffusion technology because the amount of total dopant can be controlled precisely. After the predeposition, the dopant impurities can be driven in by a high-temperature diffusion step at the same time that implant damage at the surface region is annealed out. Another application is the threshold voltage adjustment in MOS devices. A precisely controlled amount of dopant (e.g., boron) is implanted through the gate oxide to the channel region[25] (Fig. 27a). Because the projected range of boron in silicon and silicon oxide are comparable, if we choose a suitable incident energy, the ions will penetrate just the thin gate oxide, not the thicker-field oxide. The threshold voltage will vary approximately linearly with the implanted dose. After boron implantation, polysilicon can be deposited and patterned to form the gate electrode of the MOSFET. The thin oxide surrounding the gate electrode is removed, and the source and drain regions are formed as shown in Fig. 27b by another high-dose arsenic implantation.

High-current implantors with energies in the 150–200 keV range are now available. A major use for these machines is to form high-quality silicon films, which are insulated

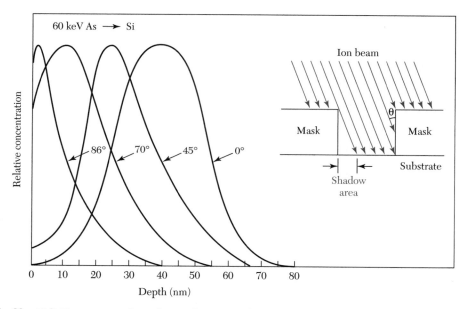

Fig. 26 60 keV arsenic implanted into silicon, as a function of beam tilt angle. Inset shows the shadow area for tilt-angle ion implantation.

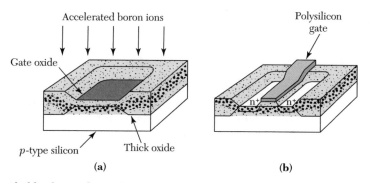

Fig. 27 Threshold voltage adjustment using boron ion implantation.[25]

from the substrate by implanting oxygen, creating an intervening layer of silicon dioxide. This separation by implantation of oxygen (SIMOX) is a key SOI (silicon-on-insulator) technology.

The SIMOX process uses a high-energy O^+ beam, typically in the 150 to 200 keV range, so that the oxygen ions have projected ranges of 100–200 nm. Additionally, a heavy dose, 1–2×10^{18} ions/cm², is used to produce an insulating layer of SiO_2 that is 100–500 nm thick. The use of SIMOX material leads to a significant reduction of source/drain capacitances in MOS devices. Moreover, it reduces coupling between devices and, thus, allows tighter packing without the problem of latchup. As a result, it is widely proposed as the material of choice for advanced, high-speed CMOS circuits.

▶ SUMMARY

Diffusion and ion implantation are the two key methods for impurity doping. We first considered the basic diffusion equation for constant diffusivity. We obtain the complementary error function (erfc) and the Gaussian function for the constant–surface-concentration case and constant–total-dopant case, respectively. The results of a diffusion process can be evaluated by measurements of the junction depth, the sheet resistance, and the dopant profile.

When the doping concentration is higher than the intrinsic carrier concentration n_i at the diffusion temperature, the diffusivity becomes concentration dependent. This dependence has profound effect on the resulting doping profile. For example, arsenic and boron diffusivity in silicon vary linearly with the impurity concentration. Their doping profiles are much more abrupt than the erfc profile. Phosphorus diffusivity in silicon varies as the square of concentration. This dependence and a dissociation effect give rise to a phosphorus diffusivity that is 100 times larger than its intrinsic diffusivity.

Lateral diffusion at the edge of a mask and impurity redistribution during oxidation are two processes in which diffusion can have an important impact on device performance. The former can substantially reduce the breakdown voltage, and the latter will influence the threshold voltage as well as the contact resistance.

The key parameters for ion implantation are the projected range R_p and its standard deviation σ_p also called the projected straggle. The implantation profile can be approximated by a Gaussian distribution with its peak located at R_p from the surface of the semiconductor substrate. The advantages of ion implantation process are more precise control of the amount of dopant, a more reproducible doping profile, and lower processing temperature compared with the diffusion process.

We considered R_p and σ_p for various elements in silicon and gallium arsenide and discussed the channeling effect and ways to minimize this effect. However, implantation may cause severe damage to the crystal lattice. To remove the implant damage and to restore mobility and other device parameters, we must anneal the semiconductor at an appropriate combination of time and temperature. Currently, the rapid thermal annealing (RTA) is preferred to conventional furnace annealing because RTA can remove implant damage without thermal broadening of the doping profile.

Ion implantation has wide applications for advanced semiconductor devices. These include (a) multiple implantation to form novel distributions, (b) selection of masking materials and thickness to stop a given percentage of incident ions from reaching the substrate, (c) tilt-angle implantation to form ultrashallow junctions, (d) high-energy implantation to form buried layers, and (e) high-current implantation for predeposition and threshold voltage adjustment and to form an insulating layer for SOI applications.

▶ REFERENCES

1. S. M. Sze, Ed., *VLSI Technology*, 2nd Ed., McGraw-Hill, New York, 1988, Ch. 7, 8.

2. S. K. Ghandhi, *VLSI Fabrication Principles*, 2nd Ed., Wiley, New York, 1994, Ch. 4, 6.

3. W. R. Runyan and K. E. Bean, *Semiconductor Integrated Circuit Processing Technology*, Addison-Wesley, Massachusetts, 1990, Ch. 8.

4. H. C. Casey, and G. L. Pearson, "Diffusion in Semiconductors," in J. H. Crawford, and L. M. Slifkin, Eds., *Point Defects in Solids*, Vol. 2, Plenum, New York, 1975.

5. J. P. Joly, "Metallic Contamination of Silicon Wafers," *Microelectron. Eng.*, **40**, 285 (1998).

6. A. S. Grove, *Physics and Technology of Semiconductor Devices*, Wiley, New York, 1967.

7. ASTM Method F374-88, "Test Method for Sheet Resistance of Silicon Epitaxial, Diffused, and Ion-implanted Layers Using a Collinear Four-Probe Array," **V10**, 249 (1993).

8. J. C. Irvin, "Evaluation of Diffused Layers in Silicon," *Bell Syst. Tech. J.*, **41**, 2 (1962).

9. ASTM Method E1438-91, "Standard Guide for Measuring Width of Interfaces in Sputter Depth Profiling Using SIMS," **V10**, 578 (1993).

10. R. B. Fair, "Concentration Profiles of Diffused Dopants," in F. F. Y. Wang, Ed., *Impurity Doping Processes in Silicon*, North-Holland, Amsterdam, 1981.

11. L. R. Weisberg and J. Blanc, "Diffusion with Interstitial-Substitutional Equilibrium, Zinc in GaAs," *Phys. Rev.*, **131**, 1548 (1963).

12. A. F. W. Willoughby, "Double-Diffusion Processes in Silicon," in F. F. Y. Wang, Ed., *Impurity Doping Processes in Silicon*, North-Holland, Amsterdam, 1981.

13. F. A. Cunnell and C. H. Gooch, *J. Phys. Chem. Solid*, **15**, 127 (1960).

14. D. P. Kennedy and R. R. O'Brien, "Analysis of the Impurity Atom Distribution Near the Diffusion Mask for a Planar *p-n* Junction," *IBM J. Res. Dev.*, **9**, 179 (1965).

15. I. Brodie and J. J. Muray, *The Physics of Microfabrication*, Plenum, New York, 1982.

16. J. F. Gibbons, "Ion Implantation," in S. P. Keller, Ed., *Handbook on Semiconductors*, Vol. 3, North-Holland, Amsterdam, 1980.

17. S. Furukawa, H. Matsumura, and H. Ishiwara, "Theoretical Consideration on Lateral Spread of Implanted Ions," *Jpn. J. Appl. Phys.*, **11**, 134 (1972).

18. B. Smith, *Ion Implantation Range Data for Silicon and Germanium Device Technologies*, Research Studies, Forest Grove, OR., 1977.

19. K. A. Pickar, "Ion Implantation in Silicon," in R. Wolfe, Ed., *Applied Solid State Science*, Vol. 5, Academic, New York, 1975.

20. L. Pauling and R. Hayward, *The Architecture of Molecules*, Freeman, San Francisco, 1964.

21. D. K. Brice, "Recoil Contribution to Ion Implantation Energy Deposition Distribution," *J. Appl. Phys.*, **46**, 3385 (1975).

22. C. Y. Chang and S. M. Sze, Eds., *ULSI Technology*, McGraw-Hill, New York, 1996, Ch. 4.

23. D. H. Lee and J. W. Mayer, "Ion-Implanted Semiconductor Devices," *Proc. IEEE*, **62**, 1241 (1974).

24. G. Dearnaley, et al., *Ion Implantation*, North-Holland, Amsterdam, 1973.

25. W. G. Oldham, "The Fabrication of Microelectronic Circuit," in *Microelectronics*, Freeman, San Francisco, 1977.

▶ PROBLEMS (* DENOTES DIFFICULT PROBLEMS)

FOR SECTION 13.1   BASIC DIFFUSION PROCESS

1. Calculate the junction depth and the total amount of dopant introduced after boron pre-deposition performed at 950°C for 30 minutes in a neutral ambient. Assume the substrate is *n*-type silicon with $N_D = 1.8 \times 10^{16}$ cm^{-3} and the boron surface concentration is $C_s = 1.8 \times 10^{20}$ cm^{-3}.

2. If the sample in Prob. 1 is subjected to a neutral drive-in at 1050°C for 60 minutes, calculate the diffusion profile and the junction depth.

3. Assume the measured phosphorus profile can be represented by a Gaussian function with a diffusivity $D = 2.3 \times 10^{-13}$ cm^2/s. The measured surface concentration is 1×10^{18} atoms/cm^3, and the measured junction depth is 1 μm at a substrate concentration of 1×10^{15}. Calculate the diffusion time and the total dopant in the diffused layer.

*4. To avoid wafer warp due to a sudden reduction in temperature, the temperature in a diffusion furnace is decreased linearly from 1000°C to 500°C in 20 minutes. What is the effective diffusion time at the initial diffusion temperature for a phosphorus diffusion in silicon?

*5. For a low-concentration phosphorus drive-in diffusion in silicon at 1000°C, find the percentage change of surface concentration for 1% variation in diffusion time and temperature.

6. If arsenic is diffused into a thick slice of silicon doped with 10^{15} boron atoms/cm^3 at a temperature of 1100°C for 3 hours, what is the final distribution of arsenic if the surface concentration is held fixed at 4×10^{18} atoms/cm^3? What are the diffusion length and junction depth?

FOR SECTION 13.2   EXTRINSIC DIFFUSION

7. If arsenic is diffused into a thick slice of silicon doped with 10^{15} boron atoms/cm^3 at a temperature of 900°C for 3 hours, what is the final distribution of arsenic if the surface concentration is held fixed at 4×10^{18} atoms/cm^3? What is the junction depth? Assume

$$D = D_0 e^{\frac{-E_a}{kT}} \times \frac{n}{n_i}, \ D_0 = 45.8 \text{ cm}^2/\text{s}, \ E_a = 4.05 \text{ eV}, \ x_j = 1.6\sqrt{Dt}.$$

8. Explain the meaning of intrinsic and extrinsic diffusion.

FOR SECTION 13.3   DIFFUSION RELATED PROCESSES

9. Define the segregation coefficient.

10. Assume that the Cu concentration in SiO$_2$ layer is 5×10^{13} atoms/cm^3 after vapor phase decomposition and is measured with atomic absorption spectometry. The Cu concentration in the Si layer is 3×10^{11} atoms/cm^3 afer HF/H$_2$O$_2$ dissolution. Calculate the segregation coefficient of Cu in SiO$_2$/Si layers.

FOR SECTION 13.4   RANGE OF IMPLANTED IONS

11. In a 200 mm wafer boron-ion–implantation system, assume the beam current is 10 μA. For the p-channel transistor, calculate the implant time required to reduce threshold voltage from –1.1 V to – 0.5 V. Assume that the implanted acceptors form a sheet of negative charge just below the Si surface, and the oxide thickness is 10 nm.

12. Assume that a 100 mm diameter GaAs wafer is uniformly implanted with 100 keV zinc ions for 5 minutes with a constant ion beam current of 10 μA. What are the ion dose per unit area and the peak ion concentration?

13. A silicon p-n junction is formed by implanting boron ions at 80 keV through a window in an oxide. If the boron dose is 2×10^{15} cm^{-2} and the n-type substrate concentration is 10^{15} cm^{-3}, find the location of the metallurgical junction.

14. A threshold-voltage adjustment implantation is made through a 25 nm gate oxide. The substrate is a <100>-oriented p-type silicon with a resistivity of 10 Ω-cm. If the incremental threshold voltage due to a 40 keV boron implantation is 1 V, what is the total implanted dose per unit area? Estimate the location of the peak boron concentration.

*15. For the substrate in Prob. 14, what percentage of the total dose is in the silicon?

FOR SECTION 13.5   IMPLANT DAMAGE AND ANNEALING

16. If a 50 keV boron ion is implanted into the silicon substrate, calculate the damage density. Assume silicon atom density is 5.02×10^{22} atoms/cm³, the silicon displacement energy is 15 eV, the range is 2.5 nm, and the spacing between silicon lattice plane is 0.25 nm.

17. Explain why high-temperature RTA is preferable to low-temperature RTA for defect-free shallow-junction formation.

18. Estimate the implant dose required to reduce a p-channel threshold voltage by 1 V if the gate oxide is 4 nm thick. Assume that the implant voltage is adjusted so that the peak of the distribution occurs at the oxide-silicon interface. Thus, half of the implant goes into the silicon. Further, assume that 90% of the implanted ions in the silicon are electrically activated by the annealing process. These assumptions allow 45% of the implanted ions to be used for threshold adjusting. Also assume that all of the charge in the silicon is effectively at the silicon-oxide interface.

FOR SECTION 13.6   IMPLANTATION-RELATED PROCESSES

19. We would like to form 0.1 μm deep, heavily doped junctions for the source and drain regions of a submicron MOSFET. Compare the options that are available to introduce and activate dopant for this application. Which option would you recommend and why?

20. When an arsenic implant at 100 keV is used and the photoresist thickness is 400 nm, find the effectiveness of the resist mask in preventing the transmission of ions (R_p = 0.6 μm, σ_P = 0.2 μm). If the resist thickness is changed to 1 μm, calculate the masking efficiency.

21. With reference to Ex. 4, what thickness of SiO_2 is required to mask 99.999% of the implanted ions?

Integrated Devices

Microwave, photonic, and power applications generally employ discrete devices. For example, an IMPATT diode is used as a microwave generator, an injection laser as an optical source, and a thyristor as a high-power switch. However, most electronic systems are built on the integrated circuit (IC), which is an ensemble of both active (e.g., transistor) and passive devices (e.g., resistor, capacitor, and inductor) formed on and within a single-crystal semiconductor substrate and interconnected by a metallization pattern.[1] ICs have enormous advantages over discrete devices connected by wire bondings. The advantages includes (a) reduction of the interconnection parasitics, because an IC with multilevel metallization can substantially reduce the overall wiring length, (b) full utilization of semiconductor wafer's "real estate," because devices can be closely packed within an IC chip, and (c) drastic reduction in processing cost, because wire bonding is a time-consuming and error-prone operation.

In this chapter we combine the basic processes described in previous chapters to fabricate active and passive components in an IC. Because the key element of an IC is the transistor, specific processing sequences are developed to optimize its performance. We consider three major IC technologies associated with the three transistor families: the bipolar transistor, the MOSFET, and the MESFET.

Specifically, we cover the following topics:

- The design and fabrication of IC resistor, capacitor, and inductor.
- The processing sequence for standard bipolar transistor and advanced bipolar devices.
- The processing sequence for MOSFET with special emphasis on CMOS and memory devices.
- The processing sequence for high-performance MESFET and monolithic microwave IC.
- The major challenges for future microelectronics, including ultrashallow junction, ultrathin oxide, new interconnection materials, low power dissipation, and isolation.

Figure 1 illustrates the interrelationship between the major process steps used for IC fabrication. Polished wafers with a specific resistivity and orientation are used as the starting material. The film formation steps include thermally grown oxide films, deposited polysilicon, dielectric, and metal films (Chapter 11). Film formation is often followed by lithography (Chapter 12) or impurity doping (Chapter 13). Lithography is generally followed by etching, which in turn is often followed by another impurity doping or film formation. The final IC is made by sequentially transferring the patterns from each mask, level by level, onto the surface of the semiconductor wafer.

After processing, each wafer contains hundreds of identical rectangular chips (or dice), typically between 1 and 20 mm on each side, as shown in Fig. 2a. The chips are separated by sawing or laser cutting; Figure 2b shows a separated chip. Schematic top views of a single MOSFET and a single bipolar transistor are shown in Fig. 2c to give some perspective of the relative size of a component in an IC chip. Prior to chip separation, each chip is electrically tested. Defective chips are usually marked with a dab of black ink. Good chips are selected and packaged to provide an appropriate thermal, electrical, and interconnection environment for electronic applications.[2]

IC chips may contain from a few components (transistors, diodes, resistors, capacitors, etc.) to as many as a billion or more. Since the invention of the monolithic IC in 1959, the number of components on a state-of-the-art IC chip has grown exponentially. We usually refer to the complexity of an IC as small-scale integration (SSI) for up to 100 components per chip, medium-scale integration (MSI) for up to 1000 components per chip, large-scale integration (LSI) for up to 100,000 components per chip, very-large-scale integrated (VLSI) for up to 10^7 components per chip, and ultra large-scale integration (ULSI) for larger numbers of components per chip. In Section 14.3, we show two ULSI chips, a 32-bit microprocessor chip, which contains over 42 million components, and a 1 Gbit dynamic random access memory (DRAM) chip, which contains over 2 billion components.

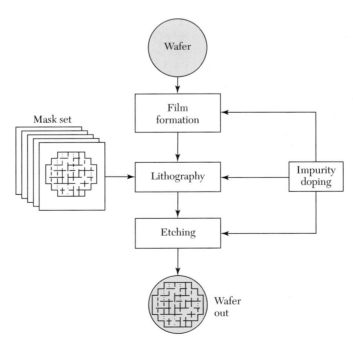

Fig. 1 Schematic flow diagram of integrated-circuit fabrication.

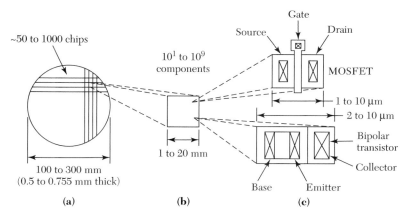

Fig. 2 Size comparison of a wafer to individual components. (*a*) Semiconductor wafer. (*b*) Chip. (*c*) MOSFET and bipolar transistor.

► 14.1 PASSIVE COMPONENTS

14.1.1 The Integrated-Circuit Resistor

To form an IC resistor, we can deposit a resistive layer on a silicon substrate, then pattern the layer by lithography and etching. We can also define a window in a silicon dioxide layer grown thermally on a silicon substrate and then implant (or diffuse) impurities of the opposite conductivity type into the wafer. Figure 3 shows the top and cross-sectional views of two resistors formed by the latter approach: one has a meander shape and the other has a bar shape.

Consider the bar-shaped resistor first. The differential conductance dG of a thin layer of the p-type material that is of thickness dx parallel to the surface and at a depth x (as shown by the B-B cross section) is

$$dG = q\mu_p p(x)\frac{W}{L}\,dx, \tag{1}$$

where W is the width of the bar, L is the length of the bar (we neglect the end contact areas for the time being), μ_p is mobility of hole, and $p(x)$ is the doping concentration. The total conductance of the entire implanted region of the bar is given by

$$G = \int_0^{x_j} dG = q\frac{W}{L}\int_0^{x_j}\mu_p p(x)dx, \tag{2}$$

where x_j is the junction depth. If the value of μ_p, which is a function of the hole concentration, and the distribution of $p(x)$ are known, the total conductance can be evaluated from Eq. 2. We can write

$$G \equiv g\frac{W}{L}, \tag{3}$$

where $g \equiv q\int_0^{x_j}\mu_p p(x)dx$ is the conductance of a square resistor pattern, that is, $G = g$ when $L = W$.

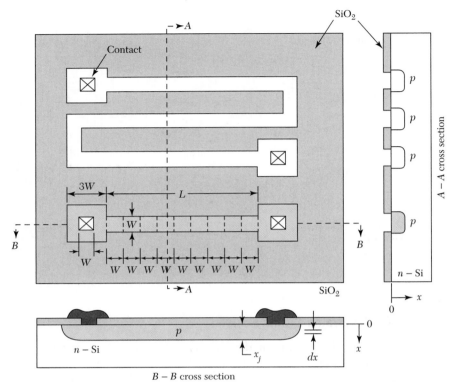

Fig. 3 Integrated-circuit resistors. All narrow lines in the large square area have the same width W, and all contacts are the same size.

The resistance is therefore given by

$$R \equiv \frac{1}{G} = \frac{L}{W}\left(\frac{1}{g}\right),$$

(4)

where $1/g$ usually is defined by the symbol $R_\square$ and is called the sheet resistance. The sheet resistance has units of ohms but is conventionally specified in units of ohms per square ($\Omega/\square$).

Many resistors in an integrated circuit are fabricated simultaneously by defining different geometric patterns in the mask such as those shown in Fig. 3. Since the same processing cycle is used for all these resistors, it is convenient to separate the resistance into two parts: the sheet resistance $R_\square$, determined by the implantation (or diffusion) process; and the ratio L/W, determined by the pattern dimensions. Once the value of $R_\square$ is known, the resistance is given by the ratio L/W, or the number of squares (each square has an area of $W \times W$) in the resistor pattern. The end contact areas will introduce additional resistance to the IC resistors. For the type shown in Fig. 3, each end contact corresponds to approximately 0.65 square. For the meander-shape resistor, the electric-field lines at the bends are not spaced uniformly across the width of the resistor but are crowded toward the inside corner. A square at the bend does not contribute exactly 1 square, but rather 0.65 square.

► **EXAMPLE 1**

Find the value of a resistor 90 μm long and 10 μm wide, such as the bar-shaped resistor in Fig. 3. The sheet resistance is 1 kΩ /□.

SOLUTION   The resistor contains 9 squares. The two end contacts correspond to 1.3 □. The value of the resistor is $(9 + 1.3) \times 1 \text{ k}\Omega/\square = 10.3 \text{ k}\Omega$.   ◄

14.1.2   The Integrated-Circuit Capacitor

There are basically two types of capacitors used in integrated circuits: MOS capacitors and p–n junctions. The MOS (metal-oxide-semiconductor) capacitor can be fabricated by using a heavily doped region (such as an emitter region) as one plate, the top metal electrode as the other plate, and the intervening oxide layer as the dielectric. The top and cross-sectional views of a MOS capacitor are shown in Fig. 4a. To form a MOS capacitor, a thick oxide layer is thermally grown on a silicon substrate. Next, a window is lithographically defined and then etched in the oxide. Diffusion or ion implantation is used to form a p^+-region in the window area, whereas the surrounding thick oxide serves as a mask. A thin oxide layer is then thermally grown in the window area, followed by a metallization step. The capacitance per unit area is given by

$$C = \frac{\varepsilon_{ox}}{d} \text{ F/cm}^2,$$ (5)

where ε_{ox} is the dielectric permittivity of silicon dioxide (the dielectric constant $\varepsilon_{ox}/\varepsilon_0$ is 3.9) and d is the thin-oxide thickness. To increase the capacitance further, insulators with higher dielectric constants are being studied, such as Si_3N_4, and Ta_2O_5, with dielectric constants of 7 and 25, respectively. The MOS capacitance is essentially independent of the applied voltage, because the lower plate of the capacitor is made of heavily doped material. This also reduces the series resistance associated with it.

A p–n junction is sometimes used as a capacitor in an integrated circuit. The top and cross sectional views of an n^+-p junction capacitor are shown in Fig. 4b. The detailed

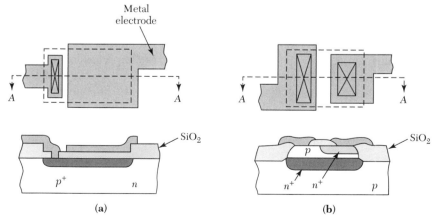

Fig. 4   (a) Integrated MOS capacitor. (b) Integrated p–n junction capacitor.

fabrication process is considered in Section 14.2, because this structure forms part of a bipolar transistor. As a capacitor, the device is usually reverse biased, that is, the p-region is reverse-biased with respect to the n^+-region. The capacitance is not a constant but varies as $(V_R + V_{bi})^{-1/2}$, where V_R is the applied reverse voltage and V_{bi} is the built-in potential. The series resistance is considerably higher than that of a MOS capacitor because the p-region has higher resistivity than does the p^+-region.

▶ **EXAMPLE 2**

What is the stored charge and the number of electrons on an MOS capacitor with an area of 4 μm^2, for (a) a dielectric of 10 nm thick SiO_2 and (b) a 5 nm thick Ta_2O_5. The applied voltage is 5 V for both cases.

SOLUTION

(a) $Q = \varepsilon_{ox} \times A \times \dfrac{V_s}{d} = 3.9 \times 8.85 \times 10^{-14} \ F/cm \times 4 \times 10^{-8} \ cm^2 \times \dfrac{5V}{1 \times 10^{-6} \ cm} = 6.9 \times 10^{-14} C$

or

$Q_s = 6.9 \times 10^{-14} \ C/q = 4.3 \times 10^5$ electrons.

(b) Changing the dielectric constant from 3.9 to 25 and the thickness from 10 nm to 5 nm, we obtain $Q_s = 8.85 \times 10^{-13} C$, and $Q_s = 8.85 \times 10^{-13} \ C/q = 5.53 \times 10^6$ electrons. ◀

14.1.3 The Integrated-Circuit Inductor

IC inductors have been widely used in III-V based monolithic microwave integrated circuits (MMIC)[3]. With the increased speed of silicon devices and advancement in multilevel interconnection technology, IC inductors have started to receive more and more attentions in silicon-based radio frequency (rf) and high-frequency applications. Many kinds of inductors can be fabricated using IC processes. The most popular method is the thin-film spiral inductor. Figure 5a and b shows the top-view and the cross section of a silicon-based, two-level–metal spiral inductor. To form a spiral inductor, a thick oxide is thermally grown or deposited on a silicon substrate. The first metal is then deposited and defined as one end of the inductor. Next, another dielectric is deposited onto the metal 1. A via hole is defined lithographically and etched in the oxide. Metal 2 is deposited and the via hole is filled. The spiral patterned can be defined and etched on the metal 2 as the second end of the inductor.

To evaluate the inductor, an important figure of merit is the quality factor, Q. The Q is defined as $Q = L\omega/R$, where L, R, and ω are the inductance, resistance, and frequency, respectively. The higher the Q values, the lower the loss from resistance, hence the better the performance of the circuits. Figure 5c shows the equivalent circuit model. R_1 is the inherent resistivity of the metal, C_{p1} and C_{p2} are the coupling capacitances between the metal lines and the substrate, and R_{sub1} and R_{sub2} are the resistances of the silicon substrate associated with the metal lines, respectively. The Q increases linearly with frequency initially and then drops at higher frequencies because of parasitic resistances and capacitances.

There are some approaches to improve the Q value. The first is to use low-dielectric-constant materials (<3.9) to reduce the C_p. The other is to use a thick film metal or low-resistivity metals (e.g., Cu, Au to replace Al) to reduce the R_1. The third approach

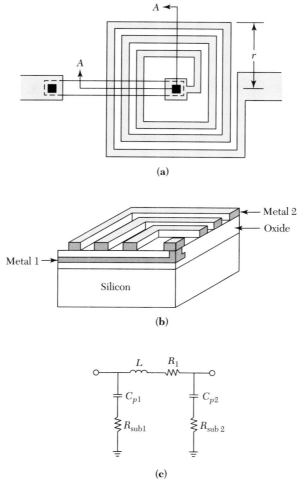

Fig. 5 (*a*) Schematic view of a spiral inductor on a silicon substrate. (*b*) Perspective view along A-A'. (*c*) An equivalent circuit model for an integrated inductor.

uses an insulating substrate (e.g., silicon-on-sapphire, silicon-on-glass, or quartz) to reduce R_{sub}.

To obtain the exact value of a thin-film inductor, complicated simulation tool, such as computer aided design, must be employed for both circuit simulation and inductor optimization. The model for thin-film inductor must take into account the resistance of the metal, the capacitance of the oxide, line-to-line capacitance, the resistance of the substrate, the capacitance to the substrate, and the inductance and mutual inductance of the metal lines. Hence, it is more difficult to calculate the integrated inductance compared with the integrated capacitors or resistors. However, a simple equation to estimate the square planar spiral inductor is given as[3]

$$L \approx \mu_0 n^2 r \approx 1.2 \times 10^{-6} n^2 r, \qquad (6)$$

where μ_0 is the permeability in vacuum ($4\pi \times 10^{-7}$ H/m), L is in henries, n is the number of turns, and r is the radius of the spiral in meters.

► **EXAMPLE 3**

For an integrated inductor with an inductance of 10 nH, what is the required radius if the number of turns is 20?

SOLUTION According to the Eq. 6,

$$r = \frac{10 \times 10^{-9}}{1.2 \times 10^{-6} \times 20^2} = 2.08 \times 10^{-5} \,(\text{m}) = 20.8 \,\mu\text{m}. \qquad \blacktriangleleft$$

► ## 14.2 BIPOLAR TECHNOLOGY

For IC applications, especially for VLSI and ULSI, the size of bipolar transistors must be reduced to meet the high-density requirement. Figure 6 illustrates the reduction in the size of the bipolar transistor in recent years.[4] The main differences in a bipolar transistor in an IC compared with a discrete transistor are that all electrode contacts are located on the *top surface* of the IC wafer, and each transistor must be electrically *isolated* to prevent interactions between devices. Prior to 1970, both the lateral and vertical isolations were provided by p–n junctions (Fig. 6a) and the lateral p-isolation region was always reverse biased with respect to the n-type collector. In 1971, thermal oxide was used for lateral isolation, resulting in a substantial reduction in device size (Fig. 6b), because the base and collector contacts abut the isolation region. In the mid-1970s, the emitter extended to the walls of the oxide, resulting in an additional reduction in area (Fig. 6c). At the present time, all the lateral and vertical dimensions have been scaled down and emitter stripe widths have dimensions in the submicron region (Fig. 6d).

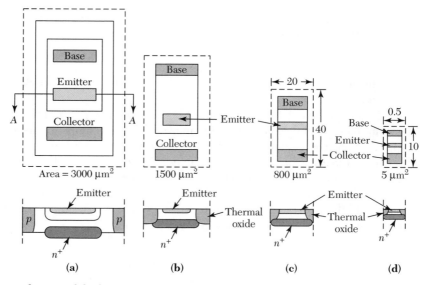

Fig. 6 Reduction of the horizontal and vertical dimensions of a bipolar transistor. (a) Junction isolation. (b) Oxide isolation. (c and d) Scaled oxide isolation.[4]

14.2.1 The Basic Fabrication Process

The majority of bipolar transistor used in ICs are of the *n-p-n* type because the higher mobility of minority carriers (electrons) in the base region results in higher-speed performance than can be obtained with *p-n-p* types. Figure 7 shows a perspective view of an *n-p-n* bipolar transistor, in which lateral isolation is provided by oxide walls and vertical isolation is provided by the n^+-*p* junction. The lateral oxide isolation approach reduces not only the device size but also the parasitic capacitance because of the smaller dielectric constant of silicon dioxide (3.9, compared with 11.9 for silicon). We consider the major process steps that are used to fabricate the device shown in Fig. 7.

For an *n-p-n* bipolar transistor, the starting material is a *p*-type lightly doped ($\sim10^{15}$ cm^{-3}), ⟨111⟩- or ⟨100⟩-oriented, polished silicon wafer. Because the junctions are formed inside the semiconductor, the choice of crystal orientation is not as critical as for MOS devices. The first step is to form a buried layer. The main purpose of this layer is to minimize the series resistance of the collector. A thick oxide (0.5–1 μm) is thermally grown on the wafer, and a window is then opened in the oxide. A precisely controlled amount of low-energy arsenic ions (~30 keV, $\sim10^{15}$ cm^{-2}) is implanted into the window region to serve as a predeposit (Fig. 8a). Next, a high temperature ($\sim1100°$C) drive-in step forms the n^+-buried layer, which has a typical sheet resistance of 20 Ω/□.

The second step is to deposit an *n*-type epitaxial layer. The oxide is removed and the wafer is placed in an epitaxial reactor for epitaxial growth. The thickness and the doping concentration of the epitaxial layer are determined by the ultimate use of the device. Analog circuits (with their higher voltages for amplification) require thicker layer (~10 μm) and lower dopings ($\sim5\times10^{15}$ cm^{-3}), whereas digital circuits (with their lower voltages for switching) require thinner layers (~3 μm) and higher dopings ($\sim2\times10^{16}$ cm^{-3}). Figure 8b shows a cross-sectional view of the device after the epitaxial process. Note that there is some outdiffusion from the buried layer into the epitaxial layer. To minimize the outdiffusion, a low-temperature epitaxial process should be employed, and low-diffusivity impurities should be used in the buried layer (e.g., As).

The third step is to form the lateral oxide isolation region. A thin-oxide pad (~50 nm) is thermally grown on the epitaxial layer, followed by a silicon-nitride deposition (~100 nm). If nitride is deposited directly onto the silicon without the thin-oxide pad, the nitride may cause damages to the silicon surface during the subsequent high-temperature steps. Next, the nitride-oxide layers and about half of the epitaxial layer are etched using a photoresist

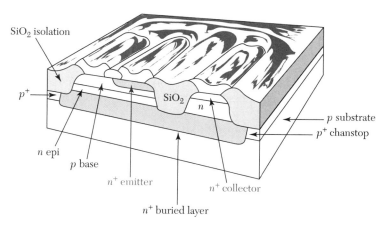

Fig. 7 Perspective view of an oxide-isolated bipolar transistor.

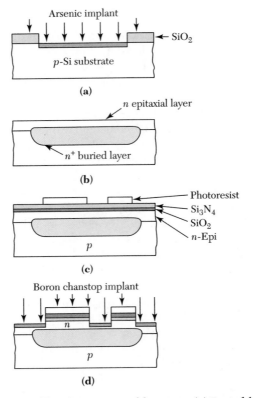

Fig. 8 Cross-sectional views of bipolar transistor fabrication. (*a*) Buried-layer implantation. (*b*) Epitaxial layer. (*c*) Photoresist mask. (*d*) Chanstop implant.

as mask (Fig. 8*c* and 8*d*). Boron ions are then implanted into the exposed silicon areas (Fig. 8*d*).

The photoresist is removed and the wafer is placed in an oxidation furnace. Since the nitride layer has a very low oxidation rate, thick oxides will be grown only in the areas not protected by the nitride layer. The isolation oxide is usually grown to a thickness such that the top of the oxide becomes coplanar with the original silicon surface to minimize the surface topography. This oxide isolation process is called local oxidation of silicon (LOCOS). Figure 9*a* shows the cross section of the isolation oxide after the removal of the nitride layer. Becaue of segregation effects, most of the implanted boron ions are pushed underneath the isolation oxide to form a p^+-layer. This is called p^+ channel stop (or chanstop), because the high concentration of p-type semiconductor will prevent surface inversion and eliminate possible high-conductivity paths (or channels) among neighboring buried layers.

The fourth step is to form the base region. A photoresist is used as a mask to protect the right half of the device; then, boron ions ($\sim 10^{12}$ cm^{-2}) are implanted to form the base regions, as shown in Fig. 9*b*. Another lithographic process removes all the thin-pad oxide except a small area near the center of the base region (Fig. 9*c*).

The fifth step is to form the emitter region. As shown in Fig. 9*d*, the base contact area is protected by a photoresist mask; then, a low-energy, high-arsenic–dose ($\sim 10^{16}$ cm^{-2}) implantation forms the n^+-emitter and the n^+-collector contact regions. The photoresist

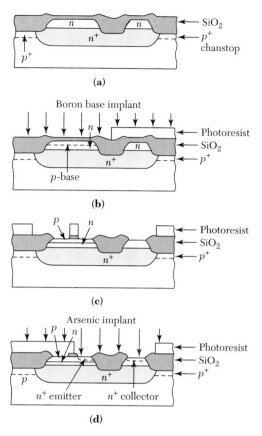

Fig. 9 Cross-section views of bipolar transistor fabrication. (*a*) Oxide isolation. (*b*) Base implant. (*c*) Removal of thin oxide. (*d*) Emitter and collector implant.

is removed; and a final metallization step forms the contacts to the base, emitter, and collector as shown in Fig. 7.

In this basic bipolar process, there are six film formation operations, six lithographic operations, four ion implantations, and four etching operations. Each operation must be precisely controlled and monitored. Failure of any one of the operations generally will render the wafer useless.

The doping profiles of the completed transistor along a coordinate perpendicular to the surface and passing through the emitter, base, and collector are shown in Fig. 10. The emitter profile is abrupt because of the concentration-dependent diffusivity of arsenic. The base doping profile beneath the emitter can be approximated by a Gaussian distribution for a limited-source diffusion. The collector doping is given by the epitaxial doping level ($\sim 2 \times 10^{16}$ cm^{-3}) for a representative switching transistor; however, at larger depths, the collector doping concentration increases because of outdiffusion from the buried layer.

14.2.2 Dielectric Isolation

In the isolation scheme described previously for the bipolar transistor, the device is isolated from other devices by the oxide layer around its periphery and is isolated from its common substrate by a n^+–p junction (buried layer). In high-voltage applications, a different

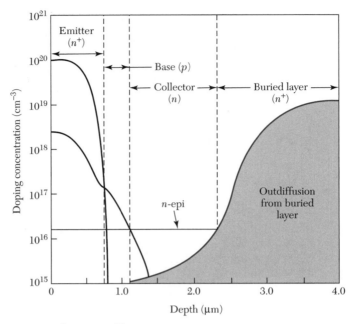

Fig. 10 *n-p-n* transistor doping profiles.

approach, called dielectric isolation, is used to form insulating tubs to isolate a number of pockets of single-crystal semiconductors. In this approach the device is isolated from both its common substrate and its surrounding neighbors by a dielectric layer.

A process sequence for the dielectric isolation is shown in Fig. 11. An oxide layer is formed inside a <100>-oriented *n*-type silicon substrate using high-energy oxygen ion implantation (Fig. 11*a*). Next, the wafer undergoes a high-temperature annealing process so that the implanted oxygen will react with silicon to form the oxide layer. The damage resulting from implantation is also annealed out in this process (Fig. 11*b*). After this, we can obtain an *n*-silicon layer that is fully isolated on an oxide [namely, silicon-on-insulator, (SOI)]. This process is called SIMOX (*se*paration by *im*planted *ox*ygen). Since the top silicon is so thin, the isolation region is easily formed by the LOCOS process illustrated in Fig. 8*c* or by etching a trench (Fig. 11*c*) and refilling it with oxide (Fig. 11*d*). The other processes are almost the same as those from Fig. 8*c* through Fig. 9 to form the *p*-type base, *n*+-emitter, and collector.

The main advantage of this technique is its high breakdown voltage between the emitter and the collector, which can be in excess of several hundreds volts. This technique is also compatible with modern CMOS integration. This CMOS-compatible process is very useful for mixed high-voltage and high-density IC.

14.2.3 Self-Aligned Double-Polysilicon Bipolar Structure

The process shown in Fig. 9*c* needs another lithographic process to define an oxide region to separate the base and emitter contact regions. This gives rise to a large inactive device area within the isolated boundary, which increases not only the parasitic capacitances but also the resistance that degrades the transistor performance. The most effective way to reduce these effects is by using the self-aligned structure.

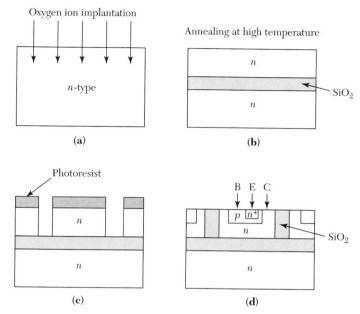

Fig. 11 Process sequence for dielectric isolation bipolar device using silicon-on-insulator for high-voltage application. (*a*) Oxygen ion implantation. (*b*) Annealing at high temperature to form the isolation dielectric. (*c*) Trench isolation formed by a dry-etching process. (*d*) Base, emitter, and collector formation.

The most widely used self-aligned structure is the double-polysilicon structure with the advanced isolation provided by a trench refilled with polysilicon,[5] shown in Fig. 12. Figure 13 shows the detail sequence of the steps for the self-aligned double-polysilicon (*n*-*p*-*n*) bipolar structure.[6] The transistor is built on an *n*-type epitaxial layer. A trench of 5.0 μm in depth is etched by reactive ion etching through the n^+-subcollector region into the p^-substrate region. A thin layer of thermal oxide is then grown and serves as the screen oxide for the channel stop implant of boron at the bottom of the trench. The trench is then filled with undoped polysilicon and capped by a thick planar field oxide.

The first polysilicon layer is deposited and heavily doped with boron. The p^+-polysilicon (called poly 1) will be used as a solid-phase diffusion source to form the extrinsic base region and the base electrode. This layer is covered with a chemical-vapor deposition

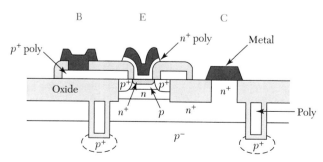

Fig. 12 Cross-section of a self-aligned, double-polysilicon bipolar transistor with advanced trench isolation.[5]

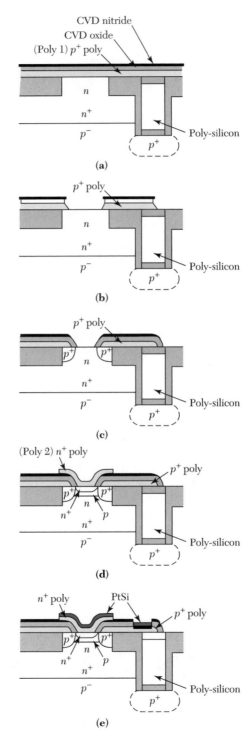

Fig. 13 Process sequence for fabricating double-polysilicon self-aligned *n-p-n* transistors.[6]

(CVD) oxide and nitride (Fig. 13a). The emitter mask is used to pattern the emitter-area regions, and a dry-etch process is used to produce an opening in the CVD oxide and poly 1 (Fig. 13b). A thermal oxide is then grown over the etched structure, and a relatively thick oxide (approximately 0.1–0.4 μm) is grown on the vertical sidewalls of the heavily doped poly. The thickness of this oxide determines the spacing between the edges of the base and emitter contacts. The extrinsic p^+ base regions are also formed during the thermal-oxide growth step as a result of the outdiffusion of boron from the poly 1 into the substrate (Fig. 13c). Because boron diffuses laterally as well as vertically, the extrinsic base region will be able to make contact with the intrinsic base region that is formed next, under the emitter contact.

Following the oxide-grown step, the intrinsic base region is formed using ion implantation of boron (Fig. 13d). This serves to self-align the intrinsic and extrinsic base regions. After the contact is cleaned to remove any oxide layer, the second polysilicon layer is deposited and implanted with As or P. The n^+–polysilicon (called poly 2) is used as a solid-phase diffusion source to form the emitter region and the emitter electrode. A shallow emitter region is then formed through dopant outdiffusion from poly 2. A rapid thermal anneal for the base and emitter outdiffusion steps facilitates the formation of shallow emitter-base and collector-base junctions. Finally, Pt film is deposited and sintered to form PtSi over the n^+-polysilicon emitter and the p^+-polysilicon base contact (Fig. 13e).

This self-aligned structure allows the fabrication of emitter regions smaller than the minimum lithographic dimension. When the sidewall-spacer oxide is grown, it fills the contact hole to some degree because the thermal oxide occupies a larger volume than the original volume of polysilicon. Thus, an opening 0.8 μm wide will shrink to about 0.4 μm if sidewall oxide a 0.2 μm thick is grown on each side.

14.3 MOSFET TECHNOLOGY

At present, the MOSFET is the dominant device used in ULSI circuits because it can be scaled to smaller dimensions than other types of devices. The dominant technology for MOSFET is the CMOS (complementary MOSFET) technology, in which both n-channel and p-channel MOSFETs (called NMOS and PMOS, respectively) are provided on the same chip. CMOS technology is particular attractive for ULSI circuits because it has the lowest power consumption of all IC technology.

Figure 14 shows the reduction in the size of the MOSFET in recent years. In the early 1970s, the gate length was 7.5 μm and the corresponding device area was about 6000 μm². As the device is scaled down, there is a drastic reduction in the device area. For a MOSFET with a gate length of 0.5 μm, the device area shrinks to less than 1% of the early MOSFET. We expect that device miniaturization will continue. The gate length will be less than 0.10 μm in the early twenty-first century. We consider the future trends of the devices in Section 14.5.

14.3.1 The Basic Fabrication Process

Figure 15 shows a perspective view of an n-channel MOSFET prior to its final metallization.[7] The top layer is a phosphorus-doped silicon dioxide (P-glass) that is used as an insulator between the polysilicon gate and the gate metallization and also as a gettering layer for mobile ions. Compare Fig. 15 with Fig. 7 for the bipolar transistor and note that a MOSFET is considerably simpler in its basic structure. Although both devices use

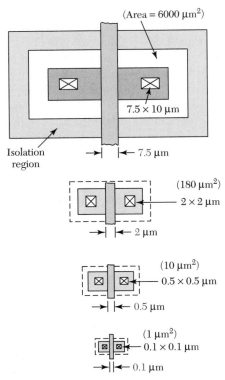

Fig. 14 Reduction in the area of the MOSFET as the gate length (minimum feature length) is reduced.

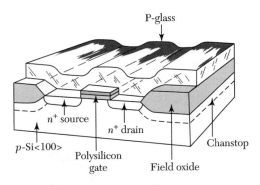

Fig. 15 Perspective view of an n-channel MOSFET.[7]

lateral oxide isolation, there is no need for vertical isolation in the MOSFET, whereas a buried-layer n^+-p junction is required in the bipolar transistor. The doping profile in a MOSFET is not as complicated as that in a bipolar transistor and the control of the dopant distribution is also less critical. We consider the major process steps that are used to fabricate the device shown in Fig. 15.

To process an n-channel MOSFET (NMOS), the starting material is a p-type, lightly doped ($\sim 10^{15}$ cm^{-3}), $\langle 100 \rangle$-oriented, polished silicon wafer. The $\langle 100 \rangle$-orientation is preferred over $\langle 111 \rangle$ because it has an interface-trap density that is about one-tenth that of $\langle 111 \rangle$. The first step is to form the oxide isolation region using LOCOS technology. The process sequence for this step is similar to that for the bipolar transistor. A thin-pad oxide (~35 nm) is thermally grown, followed by a silicon nitride (~150 nm) deposition (Fig. 16a).[7] The active device area is defined by a photoresist mask and a boron chanstop layer is then implanted through the composite nitride-oxide layer (Fig. 16b). The nitride layer not covered by the photoresist mask is subsequently removed by etching. After stripping the photoresist, the wafer is placed in an oxidation furnace to grow an oxide (called the field oxide), where the nitride layer is removed, and to drive in the boron implant. The thickness of the field oxide is typically 0.5–1 μm.

The second step is to grow the gate oxide and to adjust the threshold voltage (see Section 6.2.3). The composite nitride-oxide layer over the active device area is removed, and a thin-gate oxide layer (less than 10 nm) is grown. For an enhancement-mode n-channel device, boron ions are implanted in the channel region, as shown in Fig. 16c, to increase the threshold voltage to a predetermined value (e.g., + 0.5V). For a depletion-mode n-channel device, arsenic ions are implanted in the channel region to decrease the threshold voltage (e.g., –0.5V).

The third step is to form the gate. A polysilicon is deposited and is heavily doped by diffusion or implantation of phosphorus to a typical sheet resistance of 20–30 Ω/□. This resistance is adequate for MOSFETs with gate lengths larger than 3 μm. For smaller

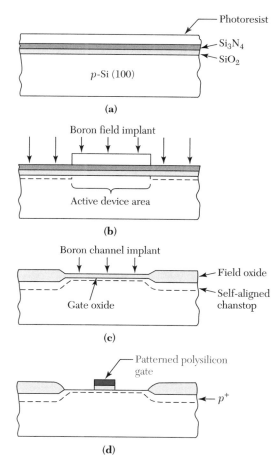

Fig. 16 Cross-sectional view of NMOS fabrication sequence.[7] (a) Formation of SiO_2, Si_3N_4, and photoresist layer. (b) Boron implant. (c) Field oxide. (d) Gate.

devices, polycide, a composite layer of metal silicide and polysilicon such as W-polycide, can be used as the gate materials to reduce the sheet resistance to about 1 $\Omega/\square$.

The fourth step is to form the source and drain. After the gate is patterned (Fig. 16d), it serves as a mask for the arsenic implantation (~30 keV, ~5 × 10^{15} cm^{-2}) to form the source and drain (Fig. 17a), which are self-aligned with respect to the gate[7]. At this stage, the only overlapping of the gate is due to lateral straggling of the implanted ions (for 30 keV As, $\sigma_\perp$ is only 5 nm). If low-temperature processes are used for subsequent steps to minimize lateral diffusion, the parasitic gate-drain and gate-source coupling capacitances can be much smaller than the gate-channel capacitance.

The last step is the metallization. A phosphorus-doped oxide (P-glass) is deposited over the entire wafer and is flowed by heating the wafer to give a smooth surface topography (Fig. 17b). Contact windows are defined and etched in the P-glass. A metal layer, such as aluminum, is then deposited and patterned. A cross-section view of the completed MOSFET is shown in Fig. 17c, and the corresponding top view is shown in Fig. 17d. The gate contact is usually made outside the active device area to avoid possible damage to the thin-gate oxide.

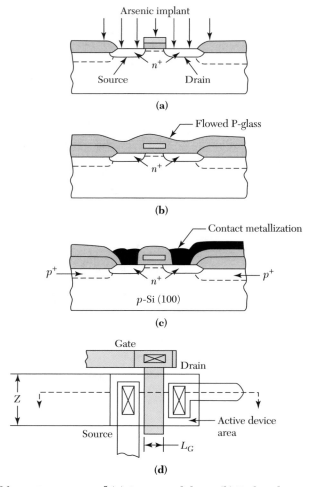

Fig. 17 NMOS fabrication sequence.[7] (a) Source and drain. (b) P-glass deposition. (c) Cross section of the MOSFET. (d) Top view of the MOSFET.

▷ **EXAMPLE 4**

What is the maximum gate-to-source voltage that a MOSFET with a 5 nm gate oxide can withstand. Assume that the oxide breaks down at 8 MV/ cm and the substrate voltage is zero.

SOLUTION

$V = \mathscr{E} \times d = 8 \times 10^6 \times 5 \times 10^{-7} = 4$ V. ◀

14.3.2 Memory Devices

Memories are devices that can store digital information (or data) in terms of *bits* (*bi*nary digi*ts*). Various memory chips have been designed and fabricated using NMOS technology. For most large memories, the random access memory (RAM) organization is preferred. In a RAM, memory cells are organized in a matrix structure and data can be accessed (i.e., stored, retrieved, or erased) in random order, independent of their physical locations. A static random access memory (SRAM) can retain stored data indefinitely as long as the power supply is on. The SRAM is basically a flip-flop circuit that can store one bit of information. A SRAM cell has four enhancement-mode MOSFETs and two depletion-mode MOSFETs. The depletion-mode MOSFETs can be replaced by resistors formed in undoped polysilicon to minimize power consumption.[8]

To reduce the cell area and power consumption, the dynamic random access memory (DRAM) has been developed. Figure 18*a* shows the circuit diagram of the one-transistor DRAM cell in which the transistor serves as a switch and one bit of information can be stored in the storage capacitor. The voltage level on the capacitor determines the state of the cell. For example, +1.5 V may be defined as logic 1 and 0 V defined as logic 0. The stored charge will be removed typically in a few milliseconds mainly because of the leakage current of the capacitors; thus, dynamic memories require periodic "refreshing" of the stored charge.

Figure 18*b* shows the layout of a DRAM cell, and Fig. 18*c* shows the corresponding cross section through AA′. The storage capacitor uses the channel region as one plate, the polysilicon gate as the other plate, and the gate oxide as the dielectric. The row line is a metal track to minimize the delay due to parasitic resistance (R) and parasitic capacitance (C), the RC delay. The column line is formed by n+-diffusion. The internal drain region of the MOSFET serves as a conductive link between the inversion layers under the storage gate and the transfer gate. The drain region can be eliminated by using the double-level polysilicon approach shown in Fig. 18*d*. The second polysilicon electrode is separated from the first polysilicon capacitor plate by an oxide layer that is thermally grown on the first-level polysilicon before the second electrode has been defined. The charge from the column line can therefore be transmitted directly to the area under the storage gate by the continuity of inversion layers under the transfer and storage gates.

To meet the requirements of high-density DRAM, the DRAM structure has been extended to the third dimension with stacked or trench capacitors. Figure 19*a* shows a simple trench cell structure.[9] The advantage of the trench type is that the capacitance of the cell could be increased by increasing the depth of the trench without increasing the surface area of silicon occupied by the cell. The main difficulties of making trench-type cells are the etching of the deep trench, which needs a rounded bottom corner and the growth of a uniform thin dielectric film on trench walls. Figure 19*b* shows a stacked cell structure. The storage capacitance increases as a result of stacking the storage capacitor on top of the access transistor. The dielectric is formed using the thermal oxidation

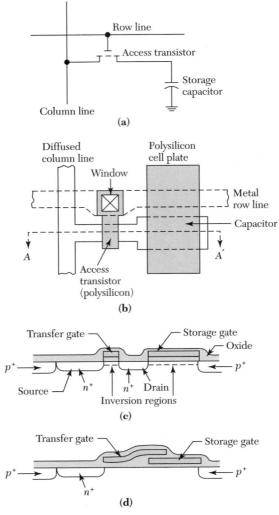

Fig. 18 Single-transistor dynamic random access memory (DRAM) cell with a storage capacitor.[8] (*a*) Circuit diagram. (*b*) Cell layout. (*c*) Cross section through *A-A'*. (*d*) Double-level polysilicon.

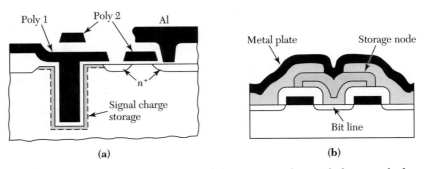

Fig. 19 (*a*) DRAM with a trench cell structure.[9] (*b*) DRAM with a single-layer stacked-capacitor cell.

or CVD nitride methods between the two-polysilicon plates. Hence, the stacked cell process is easier than the trench type process.

Figure 20 shows a 1 Gb DRAM chip. This memory chip uses 0.18 μm design rules. Trench capacitors and its peripheral circuits are in CMOS, which are considered in Section 14.3.3. The memory chip has an area of 390 mm² (14.3 mm × 27.3 mm) that contains over 2 billion components and operates at 2.5 V. This 1 Gb DRAM is mounted in an 88-pin ceramic package, which can provide adequate heat dissipation.

Both SRAM and DRAM are volatile memories, that is, they lose their stored data when power is switched off. Nonvolatile memories, on the other hand, can retain their data. Figure 21a shows a floating-gate nonvolatile memory, which is basically a conventional MOSFET that has a modified gate electrode. The composite gate has a regular (control gate) and a floating gate which is surrounded by insulators. When a large positive voltage is applied to the control gate, charge will be injected from the channel region through the gate oxide into the floating gate. When the applied voltage is removed, the injected charge can be stored in the floating gate for a long time. To remove this charge, a large negative voltage must be applied to the control gate, so that the charge will be injected back into the channel region.

Another version of the nonvolatile memory is the metal-nitride-oxide-semiconductor (MNOS) type shown inn Fig. 21b. When a positive gate voltage is applied, electrons can tunnel through the thin oxide layer (~2 nm) and be captured by the traps at the oxide-nitride interface, and thus become stored charges there. The equivalent circuit for both types of nonvolatile memories can be represented by two capacitors in series for the gate structure, as illustrated in Fig. 21c. The charge stored in the capacitor C_1 causes a shift in the threshold voltage, and the device remains at the higher threshold voltage-state (logic 1). For a well-designed memory device, the charge retention time can be over 100 years.

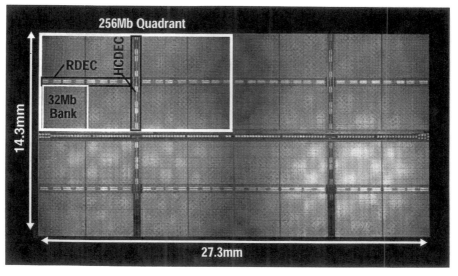

Fig. 20 A 1 Gb DRAM that contains over 2 billion components. (Photography courtesy of IBM/Siemens, 1999 IEEE Int. Solid State Circuit Conference.)

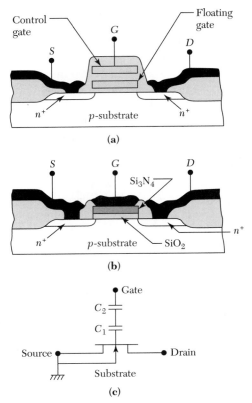

Fig. 21 Nonvolatile memory devices. (*a*) Floating-gate, nonvolatile memory. (*b*) MNOS nonvolatile memory. (*c*) Equivalent circuit of either type of nonvolatile memory.

To erase the memory (e.g., the store charge) and return the device to a lower threshold voltage state (logic 0), a gate voltage or other means (such as ultraviolet light) can be used.

The nonvolatile semiconductor memory (NVSM) has been extensively used in portable electronics systems, such as cellular phones and the digital cameras. Another interesting application is the chip card, also called IC card.

The top photo in Fig. 22 shows an IC card. The diagram at the bottom of Fig. 22 illustrates the nonvolatile memory device that stores the data that can be read and written through the bus to a central processing unit (CPU). In contrast to the limited volume (1 kbytes) inside a conventional magnetic tape card, the size of the nonvolatile memory can be increased to 16 kbytes, 64 kbytes, or even larger depending on the applications (e.g., you can store personal photos or finger prints). Through the IC card read/write machines, the data can be used in numerous applications, such as telecommunications (card telephone, mobile radio), payment transactions (electronic purse, credit card), pay television, transport (electronic ticket, public transport), health care (patient-data card), and access control. The IC card will play a central role in the global information and service society of the future.[10]

14.3.3 CMOS Technology

Figure 23*a* shows a CMOS inverter. The gate of the upper PMOS device is connected to the gate of the lower NMOS device. Both devices are enhancement-mode MOSFETs

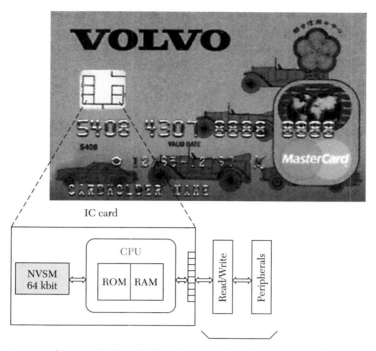

Fig. 22 An integrated-circuit (IC) card. The data stored in the NVSM can be accessed through the bus of the central processing unit (CPU). There are several metal pads connecting to the read/write machine. (Photography courtesy of Retone Information System Co., LTD.)

with the threshold voltage V_{Tp} less than zero for the PMOS device and V_{Tn} greater than zero for the NMOS device (typically the threshold voltage is about $1/4\ V_{DD}$). When the input voltage V_i is at ground or at small positive values, the PMOS device is turned on (the gate-to-ground potential of PMOS is $-V_{DD}$, which is more negative than V_{Tp}), and the NMOS device is off. Hence, the output voltage V_o is very close to V_{DD} (logic 1). When the input is at V_{DD}, the PMOS (with $V_{GS} = 0$) is turned off, and the NMOS is turned on ($V_i = V_{DD} > V_{Tn}$). Therefore, the output voltage V_o equals zero (logic 0). The CMOS inverter has a unique feature: in either logic state, one device in the series path from V_{DD} to ground is nonconductive. The current that flows in either steady state is a small leakage current, and only when both devices are on during switching does a significant current flow through the CMOS inverter. Thus, the average power dissipation is small, on the order of nanowatts. As the number of components per chip increases, the power dissipation becomes a major limiting factor. The low power consumption is the most attractive feature of the CMOS circuit.

Figure 23b shows a layout of the CMOS inverter, and Fig. 23c shows the device cross section along the A-A' line. In the processing, a p-tub (also called a p-well) is first implanted and subsequently driven into the n-substrate. The p-type dopant concentration must be high enough to overcompensate the background doping of the n-substrate. The subsequent processes for the n-channel MOSFET in the p-tub are identical to those described previously. For the p-channel MOSFET, $^{11}B^+$ or $^{49}(BF_2)^+$ ions are implanted into the n-substrate to form the source and drain regions. A channel implant of $^{75}As^+$ ions may be used to adjust the threshold voltage and a n^+-chanstop is formed underneath the field

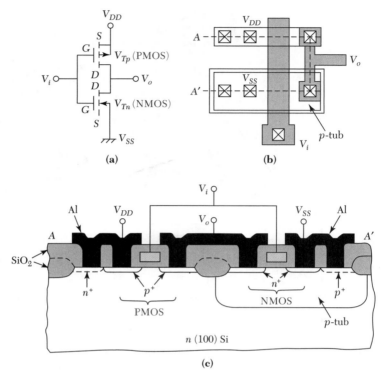

Fig. 23 Complementary MOS (CMOS) inverter. (*a*) Circuit diagram. (*b*) Circuit layout. (*c*) Cross section along dotted *A-A'* line of (*b*).

oxide around the *p*-channel device. Because of the *p*-tub and the additional steps needed to make the *p*-channel MOSFET, the number of steps to make a CMOS circuit is essentially double that to make an NMOS circuit. Thus, we have a trade-off between the complexity of processing and a reduction in power consumption.

Instead of the *p*-tub described above, an alternate approach is to use an *n*-tub formed in *p*-type substrate, as shown in Fig. 24*a*. In this case, the *n*-type dopant concentration must be high enough to overcompensate for the background doping of the p-substrate (i.e., $N_D > N_A$). In both the *p*-tub and the *n*-tub approach, the channel mobility will be degraded because mobility is determined by the total dopant concentration ($N_A + N_D$). A recent approach using two separated tubs implanted into a lightly doped substrate is shown in Fig. 24*b*. This structure is called a *twin* tub.[1] Because no overcompensation is needed in either of the twin tubs, higher channel mobility can be obtained.

All CMOS circuits have the potential for a troublesome problem called latchup that is associated with parasitic bipolar transistors (to see how this problem can occur, see Chapter 5). An effective processing technique to eliminate latchup problem is to use the deep-trench isolation, as shown[11] in Fig. 24*c*. In this technique, a trench with a depth deeper than the well is formed in the silicon by anisotropic reactive sputter etching. An oxide layer is thermally grown on the bottom and walls of the trench, which is then refilled by deposited polysilicon or silicon dioxide. This technique can eliminate latchup because the *n*-channel and *p*-channel devices are physically isolated by the refilled trench. The detailed steps for trench isolation and some related CMOS processes are now considered.

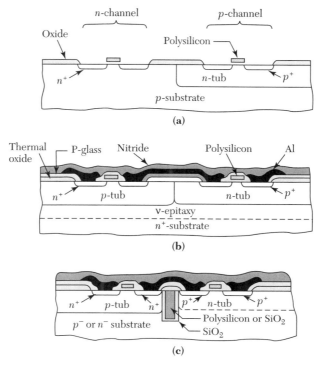

Fig. 24   Various CMOS structures. (*a*) n-tub. (*b*) Twin tub[1]. (*c*) Refilled trench.[11]

Well-Formation Technology

The well of a CMOS can be a single well, a twin well, or a retrograde well. The twin-well process exhibits some disadvantages, e.g., it needs high temperature processing (above 1050°C) and a long diffusion time (longer than 8 hours) to achieve the required depth of 2–3 μm. In this process, the doping concentration is highest at the surface and decreases monotonically with depth. To reduce the process temperature and time, high-energy implantation is used, i.e., implanting the ion to the desired depth instead of diffusion from the surface. Since the depth is determined by the implantion energy, we can design the well depth with different implantation energy. The profile of the well in this case can have a peak at a certain depth in the silicon substrate. This is called a retrograde well. Figure 25 shows a comparison of the impurity profiles in the retrograde well and the conventional thermal diffused well.[12] The energy for the n- and p-type retrograde wells is around 700 keV and 400 keV, respectively. As mentioned above, the advantage of the high-energy implantation is that it can form the well under low-temperature and short-time conditions, hence, it can reduce the lateral diffusion and increase the device density. The retrograde well can offer some additional advantages over the conventional well: (a) because of high doping near the bottom, the well resistivity is lower than that of the conventional well and the latchup problem can be minimized, (b) the chanstop can be formed at the same time as the retrograde well implantation, reducing processing steps and time, (c) higher well doping in the bottom can reduce the chance of punchthrough from the drain to the source.

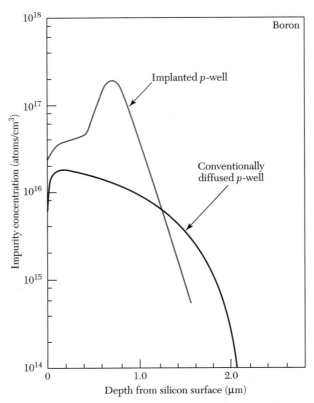

Fig. 25 Retrograded p-well implanted impurity concentration profile. Also shown is a conventionally diffused well.[12]

Advanced Isolation Technology

The conventional isolation process (Section 14.3.1) has some disadvantages that make it unsuitable for deep-submicron (0.25 μm and smaller) fabrications. The high-temperature oxidation of silicon and long oxidation time result in the encroachment of the chanstop implantation (usually boron for n-MOSFET) to the active region and cause V_T shift. The area of the active region is reduced because of the lateral oxidation. In addition, the field-oxide thickness in submicron-isolation spacings is significantly less than the thickness of field oxide grown in wider spacings. The trench isolation technology can avoid these problems and has become the mainstream technology for isolation. Figure 26 shows the process sequence for forming a deep (larger than 3 μm) but narrow (less than 2 μm) trench-isolation structure. There are four steps: patterning the area, trench etching and oxide growth, refilling with dielectric materials such as oxide or undoped polysilicon, and planarization. This deep trench isolation can be used in both advanced CMOS and bipolar devices and for the trench-type DRAM. Since the isolation material is deposited by CVD, it does not need a long-time or a high-temperature process, and it eliminates the lateral oxidation and boron encroachment problems.

Another example is the shallow trench (depth is less than 1 μm) isolation for CMOS, shown in Fig. 27. After patterning (Fig. 27a), the trench area is etched (Fig. 27b) and then re-filled with oxide (Fig. 27c). Before refilling, a chanstop implantation can be performed. Since the oxide has over filled the trench, the oxide on the nitride should be removed. Chemical-mechanical polishing (CMP) is used to remove the oxide on the nitride

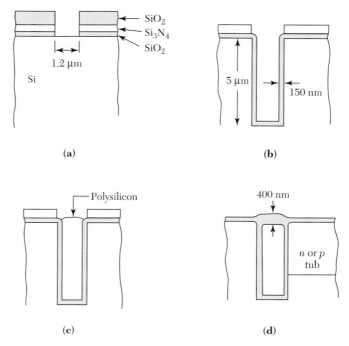

Fig. 26 Process sequence for forming a deep, narrow-trench, isolation structure. (*a*) Trench mask patterning. (*b*) Trench etching and oxide growth. (*c*) Polysilicon deposition to fill the trench. (*d*) planarization.

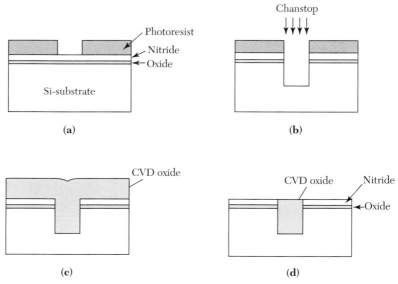

Fig. 27 A shallow trench isolation for CMOS. (*a*) Patterning with photoresist on nitride/oxide films. (*b*) Dry etching and chanstop implantation. (*c*) Chemical-vapor deposition (CVD) oxide to refill. (*d*) Planar surface after chemical-mechanical polishing (CMP).

and to get a flat surface (Fig. 27*d*). Due to its high resistance to polishing, the nitride acts as a stop-layer for the CMP process. After the polishing, the nitride layer and the oxide layer can be removed by H_3PO_4 and HF, respectively. This initial planarization step at the beginning is helpful for the subsequent polysilicon patterning and planarizations of the multilevel interconnection processes.

Gate-Engineering Technology

If we use n^+-polysilicon for both PMOS and NMOS gates, the threshold voltage for PMOS ($V_{TP} \cong -0.5$ to -1.0 V) has to be adjusted by boron implantation. This makes the channel of the PMOS a buried type, shown in Fig. 28*a*. The buried-type PMOS suffers serious short-channel effects as the device size shrinks to 0.25 μm and less. The most noticeable phenomena for short-channel effects are the V_T roll-off, drain-induced barrier lowering (DIBL), and the large leakage current at the off state so that even with the gate voltage at zero, leakage current flows through source and drain. To alleviate this problem, one can change n^+-polysilicon to p^+-polysilicon for PMOS. Due to the work function difference (there is a 1.0 eV difference from n^+- to p^+-polysilicon), one can obtain a surface p-type channel device without the boron V_T adjustment implantation. Hence, as the technology shrinks to 0.25 μm and less, dual-gate structures are required, i.e., p^+-polysilicon gate for PMOS, and n^+-polysilicon for NMOS (Fig. 28*b*). A comparison of V_T for the surface channel and the buried channel is shown in Fig. 29. We note that the V_T of surface channel rolls off slowly in the deep-submicron regime compared with the buried-channel device. This makes the surface-channel device with the p^+-polysilicon suitable for deep-submicron device operation.

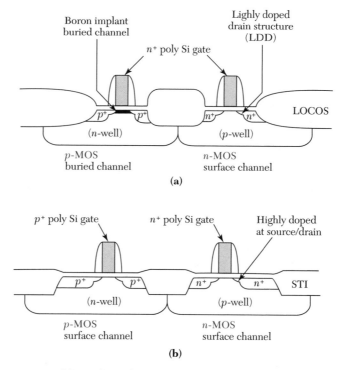

Fig. 28 (*a*) A conventional long-channel CMOS structure with a single-polysilicon gate (n^+). (*b*) Advanced CMOS structures with dual-polysilicon gates.

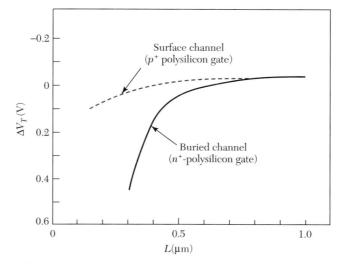

Fig. 29   The V_T roll-off for a buried type channel and for a surface type channel. The V_T drops very quickly as the channel length becomes less than 0.5 μm.

To form the p^+-polysilicon gate, ion implantation of BF_2^+ is commonly used. However, boron penetrates easily from the polysilicon through the oxide into the silicon substrate at high temperatures, resulting in a V_T shift. This penetration is enhanced in the presence of a F-atom. There are methods to reduce this effect: use of rapid thermal annealing to reduce the time at high temperatures and, consequently, the diffusion of boron; use of nitrided oxide to suppress the boron penetration, since boron can easily combine with nitrogen and becomes less mobile; and the making of a multilayer of polysilicon to trap the boron atoms at the interface of the two layers.

Figure 30 shows a microprocessor chip (Pentium 4) that has an area of about 200 mm² and contains 42 million components. This ULSI chip is fabricated using 0.18 μm CMOS technology with a six-level aluminum metallization.

14.3.4   BiCMOS Technology

BiCMOS is a technology that combines both CMOS and bipolar device structures in a single IC. The reason to combine these two different technologies is to create an IC chip that has the advantages of both CMOS and bipolar devices. As we know that CMOS exhibits advantages in power dissipation, noise margin, and packing density, whereas bipolar shows advantages in switching speed, current drive capability, and analog capability. As a result, for a given design rule, BiCMOS can have a higher speed than CMOS, better performance in analog circuits than CMOS, a lower power-dissipation than bipolar, and a higher component density than bipolar. Figure 31 shows the comparison of a BiCMOS and a CMOS logic gates. For a CMOS inverter, the current to drive (or to charging) the next loading, C_L, is the drain current I_{DS}. For a BiCMOS inverter, the current is $h_{fe}I_{DS}$, where h_{fe} is the current-gain of the bipolar transistor and I_{DS} is the base current of the bipolar transistor and is equal to the drain current of M_2 in the CMOS. Since h_{fe} is much larger than 1, the speed can be substantially enhanced.

BiCMOS has been widely used in many applications. In the early days, it was used in SRAM. At the present time, BiCMOS technology has been successfully developed for transceiver, amplifier, and oscillator applications in wireless-communication equipment.

Fig. 30 Micrograph of a 32-bit microprocessor chip, Pentium 4. (Photography courtesy of Intel Corporation.)

Most of the BiCMOS processes are based on the CMOS process, with some modifications, such as adding masks for bipolar transistor fabrication. The following example is for a high-performance BiCMOS process based on the twin-well CMOS process, shown[13] in Fig. 32.

The initial material is a p-type silicon substrate, and then an n^+-buried layer is formed to reduce the collector's resistance. The buried p-layer is formed through ion implantation to increase the doping level to prevent punchthrough. A lightly doped n-epi layer is grown on the wafer and a twin-well process for the CMOS is performed. To achieve high performance of the bipolar transistor, four additional masks are needed. They are the

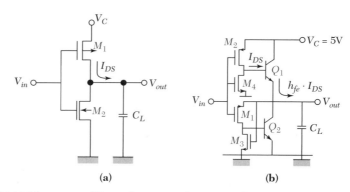

Fig. 31 (*a*) CMOS logic gate. (*b*) Bipolar CMOS (BiCMOS) logic gate.

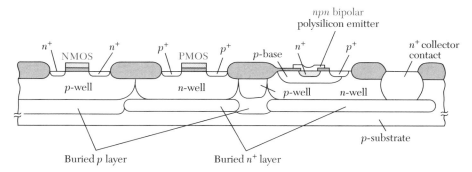

Fig. 32 Optimized BiCMOS device structure. Key features include self-aligned p and n^+ buried layers for improved packing density, separately optimized n- and p-well (twin-well CMOS) formed in an epitaxial layer with intrinsic background doping, and a polysilicon emitter for improved bipolar performance.[13]

buried n^+-mask, the collector deep-n^+-mask, the base p-mask, and the poly-emitter mask. In other processing steps, the p^+-region for base contact can be formed with the p^+-implant in the source/drain implantation of the PMOS, and the n^+-emitter can be formed with the source/drain implantation of the NMOS. The additional masks and longer processing time compared with a standard CMOS are the main drawbacks of BiCMOS. The additional cost should be justified by the enhanced performances of BiCMOS.

14.4 MESFET TECHNOLOGY

Recent advances in gallium arsenide processing techniques in conjunction with new fabrication and circuit approaches have made possible the development of "silicon-like" gallium arsenide IC technology. There are three inherent advantages of gallium arsenide compared with silicon: higher electron mobility, which results in lower series resistance for a given device geometry; higher drift velocity at a given electric field, which improves device speed; and the ability to be made semiinsulating, which can provide a lattice-matched dielectric-insulated substrate. However, gallium arsenide also has three disadvantages: a very short minority-carrier lifetime; lack of a stable, passivating native oxide; and crystal defects that are many orders of magnitude higher than in silicon. The short minority-carrier lifetime and the lack of high-quality insulating films have prevented the development of bipolar devices and delayed MOS technology using gallium arsenide. Thus, the emphasis of gallium arsenide IC technology is in the MESFET area, in which our main concerns are the majority carriers transport and the metal-semiconductor contact.

A typical fabrication sequence[14] for a high-performance MESFET is shown in Fig. 33. A layer of GaAs is epitaxially grown on a semiinsulating GaAs substrate, followed by an n^+-contact layer (Fig. 33a). A mesa etch step is performed for isolation (Fig. 33b), and a metal layer is evaporated for the source and drain ohmic contacts (Fig. 33c). A channel recess etch is followed by a gate recess etch and gate evaporation (Fig. 33d and e). After a liftoff process that removes the photoresist, shown in Fig. 33e, the MESFET is completed (Fig. 33f).

The n^+-contact layer reduces the source and drain ohmic contact resistances. Note that the gate is offset toward the source to minimize the source resistance. The epitaxial layer is thick enough to minimize the effect of surface depletion on the source and drain resistance. The gate electrode has maximal cross-sectional area with a minimal foot

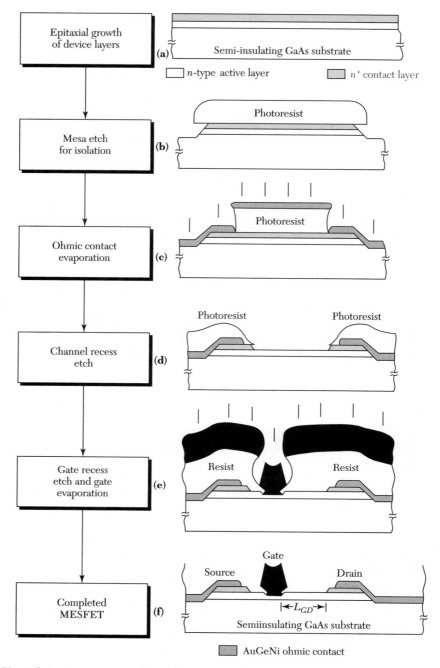

Epitaxial growth of device layers	(a)
Mesa etch for isolation	(b)
Ohmic contact evaporation	(c)
Channel recess etch	(d)
Gate recess etch and gate evaporation	(e)
Completed MESFET	(f)

Fig. 33 Fabrication sequence of a GaAs MESFET.[14]

print, which provides low gate resistance and minimal gate length. In addition, the length L_{GD} is designed to be greater than the depletion width at gate-drain breakdown.

A representative fabrication sequence for a MESFET integrated circuit is shown [15] in Fig. 34. In this process, n^+-source and drain regions are self-aligned to the gate of each MESFET. A relatively light channel implant is used for the enhancement-mode switching

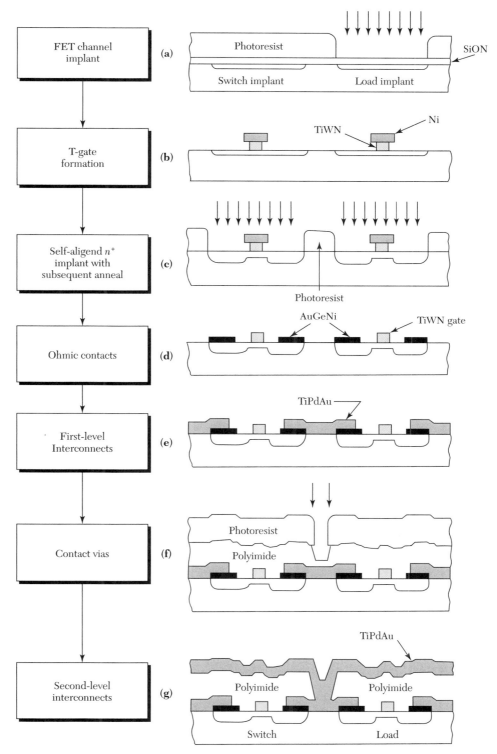

Fig. 34 Fabrication process for MESFET direct-coupled FET logic (DCFL) with active loads. Note that the n^+-source and drain regions are self-aligned to the gate.[15]

device and a heavier implant is used for the depletion-mode load device. A gate recess is usually not used for such digital IC fabrication because the uniformity of each depth has been difficult to control, leading to an unacceptable variation of the threshold voltage. This process sequence can also be used for a monolithic microwave integrated circuit (MMIC). Note that the gallium arsenide MESFET processing technology is similar to the silicon-based MOSFET processing technology.

Gallium arsenide ICs with complexities up to the large-scale integration level (~10,000 components per chip) have been fabricated. Because of the higher drift velocity (~20% higher than silicon), gallium arsenide ICs will have a 20% higher speed than silicon ICs that use the same design rules. However, substantial improvements in crystal quality and processing technology are needed before gallium arsenide can seriously challenge the preeminent position of silicon in ULSI applications.

14.5 CHALLENGES FOR MICROELECTRONICS

Since the beginning of the integrated-circuit era in 1959, the minimum device dimension, also called the minimum feature length, has been reduced at an annual rate of about 13% (i.e., a reduction of 30% every 3 years). According to the prediction by the International Technology Roadmap for Semiconductors[16], the minimum feature length will shrink from 130 nm (0.13 µm) in the year 2002 to 35 nm (0.035 µm) around 2014, as shown in Table 1. Also shown in Table 1 is the DRAM size. The DRAM has increased its memory cell capacity four times every 3 years and 64 Gbit DRAM is expected to be available in year 2011 using 50 nm design rules. The table also shows that the wafer size will increase to 450 mm (18 in. diameter) in 2014. In addition to the feature size reduction, challenges come from the device level, material level, and system level, discussed in the following subsections.

14.5.1 Challenges for Integration

Figure 35 shows the trends of power supply voltage V_{DD}, threshold voltage V_T, and gate oxide thickness d versus channel length for CMOS logic technology.[17] From the figure, one can find that the gate oxide thickness will soon approach the tunneling-current limit of 2 nm. V_{DD} scaling will slow down because of nonscalable V_T (i.e., to a minimum V_T of about 0.3 V due to subthreshold leakage and circuit noise immunity). Some challenges of the 180 nm technology and beyond are shown[18] in Fig. 36. The most stringent requirements are as follows.

TABLE 1   The Technology Generation[16] from 1997 to 2014

Year of the first product shipment	1997	1999	2002	2005	2008	2011	2014
Feature size (nm)	250	180	130	100	70	50	35
DRAM[a] size (bit)	256M	1G	—	8G	—	64G	—
Wafer size (mm)	200	300	300	300	300	300	450
Gate oxide (nm)	3–4	1.9–2.5	1.3–1.7	0.9–1.1	<1.0	—	—
Junction depth (nm)	50–100	42–70	25–43	20–33	15--30	—	—

[a]DRAM, dynamic random access memory.

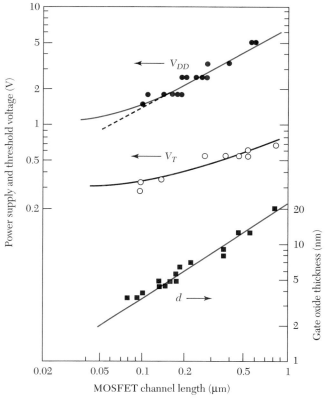

Fig. 35 Trends of power supply voltage V_{DD}, threshold voltage V_T, and gate oxide thickness d versus channel length for CMOS logic technologies. Points are collected from data published over recent years.[17]

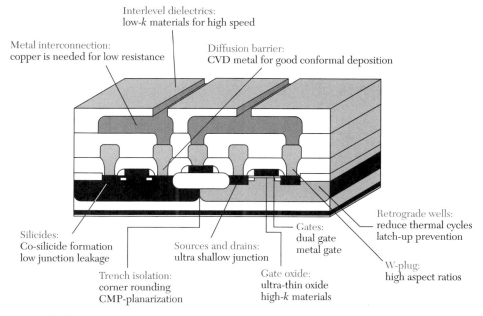

Fig. 36 Challenges for 180 nm and smaller MOSFET.[18]

Ultrashallow Junction Formation

As mentioned in Chapter 6, the short-channel effect happens as the channel length is reduced. This problem becomes critical as the device dimension is scaled down to 100 nm. To achieve an ultrashallow junction with low sheet resistance, low-energy (less than 1 keV) implantation technology with high dosage must be employed to reduce the short-channel effect. Table 1 shows the required junction depth versus the technology generation. The requirements of the junction for 100 nm are depths around 20–33 nm with a doping concentration of 1×10^{20} /cm^3.

Ultrathin Oxide

As the gate length shrinks below 130 nm, the oxide equivalent thickness of gate dielectric must be reduced around 2 nm to maintain the performance. However, if only SiO$_2$ (with a dielectric constant of 3.9) is used, the leakage through the gate becomes very high because of direct tunneling. For this reason, thicker high-k dielectric materials that have lower leakage current are suggested to replace oxide. Candidates for the short term are silicon nitride (with a dielectric constant of 7), Ta$_2$O$_5$ (25), and TiO$_2$ (60–100).

Silicide Formation

Silicide-related technology has become an integral part of submicron devices for reducing the parasitic resistance to improve device and circuit performance. The conventional Ti-silicide process has been widely use in 350–250 nm technology. However, the sheet resistance of a TiSi$_2$ line increases with decreasing line width, which limits the use of TiSi$_2$ in 100 nm CMOS applications and beyond. CoSi$_2$ or NiSi processes will replace TiSi$_2$ in the technology beyond 100 nm.

New Materials for Interconnection

To achieve high-speed operation, the RC time delay of the interconnection must be reduced. In Fig. 14 of Chapter 11 we have shown the delay as a function of feature size.[19] It is obvious that the gate delay decreases as the channel length decreases, meanwhile the delay resulting from interconnect increases significantly as the size decreases. This causes the total delay time to increase as the dimension of the device size scales down to 250 nm. Consequently, both high-conductivity metals, such as Cu, and low-dielectric-constant (low-k) insulators, such as organic (polyimide) or inorganic (F-doped oxide) materials offer major performance gains. Cu exhibits superior performance because of its high conductivity (1.7 μΩ-cm compared with 2.7 μΩ-cm of Al) and is 10–100 times more resistant to electromigration. The delay using the Cu and low-k material shows a significant decrease compared with that of the conventional Al and oxide. Hence, Cu with the low-k material is essential in multilevel interconnection for future deep-submicron technology.

Power Limitations

The power required merely to charge and discharge circuit nodes in an IC is proportional to the number of gates and the frequency at which they are switched (clock frequency). The power can be expressed as $P \cong 1/2 CV^2 \, nf$, when C is the capacitance per device, V is the applied voltage, n is the number of devices per chip, and f is the clock frequency. The temperature rise caused by this power dissipation in an IC package is limited by the thermal conductivity of the package material, unless auxiliary liquid or gas cooling is used. The maximum allowable temperature rise is limited by the bandgap of the semiconductor (~100°C for Si with a bandgap of 1.1 eV). For such a temperature rise, the maximum power dissipation of a typical high-performance package is about 10

W. As a result, we must limit either the maximum clock rate or the number of gates on a chip. As an example, in an IC containing 100 nm MOS devices with $C = 5 \times 10^{-2}$ fF, running at a 20 GHz clock rate, the maximum number of gates we can have is about 10^7 if we assume a 10% duty cycle. This is a design constraint fixed by basic material parameters.

SOI Integration

Mentioned in Section 14.2.2 was the isolation of the SOI wafer. Recently SOI technology has received more attention. The advantages of the SOI integration become significant as the minimum feature length approaches 100 nm. From the process point of view, SOI does not need the complex well structure and isolation processes. In addition, shallow junctions are directly obtained through the SOI film thickness. There is no risk of nonuniform interdiffusion of silicon and Al in the contact regions because of oxide isolation at the bottom of the junction. Hence, the contact barrier is not necessary. From the device point of view, the modern bulk silicon device needs high doping at the drain and substrate to eliminate short-channel effects and punch-through. This high doping results in high capacitance when the junction is reversed bias. On the other hand, in SOI, the maximum capacitance between the junction and substrate is the capacitance of the buried insulator whose dielectric constant is three times smaller than that of silicon (3.9 versus 11.9). Based on the ring oscillator performance, the 130 nm SOI CMOS technology can achieve 25% faster speed or require 50% less power compared to a similar bulk technology.[20] SRAM, DRAM, CPU, and rf CMOS have all been successfully fabricated using SOI technology. Therefore SOI is a key candidate for the future system-on-a-chip technology, considered in the following section.

► **EXAMPLE 5**

For an equivalent oxide thickness of 1.5 nm, what will be the physical thickness when high-k materials nitride ($\varepsilon_i/\varepsilon_0 = 7$), Ta_2O_5 (25), or TiO_2 (80) are used?

SOLUTION   For nitride,

$$\left(\frac{\varepsilon_{ox}}{1.5}\right) = \left(\frac{\varepsilon_{nitride}}{d_{nitride}}\right),$$

$$d_{nitride} = 1.5\left(\frac{7}{3.9}\right) = 2.69 \text{ nm.}$$

Using the same calculation, we obtain 9.62 nm for Ta_2O_5 and 10.77 nm for TiO_2.   ◄

14.5.2   System-On-A-Chip

The increased component density and improved fabrication technology have helped the realization of the system-on-a-chip (SOC), that is, an IC chip that contain a complete electronic system. The designers can build all the circuitry needed for a complete electronic system, such as a camera, radio, television, or personal computer (PC), on a single chip. Figure 37 shows the SOC application in the PC's mother-board. Components (11 chips in this case) once found on boards are becoming virtual components on the chip at the right.[21]

There are two obstacles in the realization of the SOC. The first is the huge complexity of the design. Since the component board is presently designed by different companies and different design tools, it is difficult to integrate them into one chip. The other

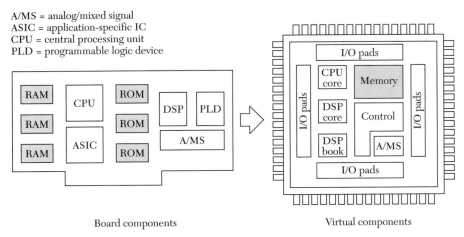

A/MS = analog/mixed signal
ASIC = application-specific IC
CPU = central processing unit
PLD = programmable logic device

Board components

Virtual components

Fig. 37 System-on-a-chip of a conventional personal computer mother-board.[21]

is the difficulty of fabrication. In general, the fabricating processes of the DRAM are significantly different from those of logic IC (e.g., CPU). Speed is the first priority for the logic, whereas leakage of the stored charge is the priority for memory. Therefore, multilevel interconnection using five to six levels of metals is essential for logic IC to improve the speed. However, DRAM needs only two to three levels. In addition, to increase the speed, a silicide process must be used to reduce the series resistance, and ultrathin gate oxide is needed to increase the drive current. These requirements are not critical for the memory.

To achieve the SOC goal, an embedded DRAM technology is introduced, i.e., to merge logic and DRAM into a single chip with compatible processes. Figure 38 shows the schematic cross section of the embedded DRAM, including the DRAM cells and the logic CMOS devices.[22] Some processing steps are modified as a compromise. The trench-type capaci-

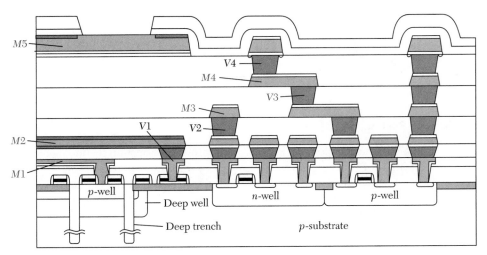

Fig. 38 Schematic cross section of the embedded DRAM including DRAM cells and logic MOSFETs. There is no height difference in the trench capacitor cell because of the DRAM cell structure. $M1$ to $M5$ are metal interconnections, and $V1$ to $V4$ are via holes.[22]

tor, instead of the stacked type, is used so that there is no height difference in the DRAM cell structure. In addition, multiple gate oxide thicknesses exist on the same wafer to accommodate multiple supply voltages and/or combine memory and logic circuits on one chip.

▶ SUMMARY

In this chapter we considered processing technologies for passive components, active devices, and IC. Three major IC technologies based on the bipolar transistor, the MOS-FET, and the MESFET were discussed in detail. It appears that the MOSFET will be the dominant technology at least until 2014 because of its superior performance compared with the bipolar transistor. For 100 nm CMOS technology, a good candidate is the combination of an SOI-substrate with interconnections using Cu and low-k materials.

Because the rapid reduction in feature length, the technology will soon reach its practical limit as the channel length is reduced to about 20 nm. What will be the device beyond the CMOS is the question being asked by research scientists. Major candidates include many innovative devices based on quantum mechanical effects. This is because when the lateral dimension is reduced to below 100 nm, depending on the materials and the temperature of operation, electronic structures will exhibit nonclassical behaviors. The operation of such devices will be on the scale of single-electron transport. This approach has been demonstrated by the single-electron memory cell. The realization of such systems with trillions of components will be a major challenge beyond CMOS.[23]

▶ REFERENCES

1. For a detailed discussion on IC process integration, see C. Y. Liu and W. Y. Lee, "Process Integration," in C. Y. Chang and S. M. Sze, Eds., *ULSI Technology*, McGraw-Hill, New York, 1996.

2. T. Tachikawa, "Assembly and Packaging," in C. Y. Chang and S. M. Sze, Eds., *ULSI Technology*, McGraw-Hill, New York, 1996.

3. T. H. Lee, *The Design of CMOS Radio-Frequency Integrated Circuits*, Cambridge Univ. Press, Cambridge, U.K., 1998, Ch. 2.

4. D. Rise, "Isoplanar-S Scales Down for New Heights in Performance," *Electronics*, **53**, 137 (1979).

5. T. C. Chen, et. al., "A submicrometer High-Performance Bipolar Technology," *IEEE Electron. Device Lett.*, **10**(8), 364, (1989).

6. G. P. Li et. al., "An Advanced High-Performance Trench-Isolated Self-Aligned Bipolar Technology," *IEEE Trans. Electron Devices*, **34**(10), 2246 (1987).

7. W. E. Beasle, J. C. C. Tsai, and R. D. Plummer, Eds., *Quick Reference Manual for Semiconductor Engineering*, Wiley, New York, 1985.

8. R. W. Hunt, "Memory Design and Technology," in M. J. Howes and D. V. Morgan, Eds., *Large Scale Integration*, Wiley, New York, 1981.

9. A. K. Sharma, *Semiconductor Memories—Technology, Testing, and Reliability*, IEEE, New York, 1997.

10. U. Hamann, "Chip Cards—The Application Revolution," *IEEE Tech. Dig. Int. Electron Devices Meet.*, p. 15, 1997.

11. R. D. Rung, H. Momose, and Y. Nagakubo, "Deep Trench Isolation CMOS Devices," *IEEE Tech. Dig. Int. Electron. Devices Meet.*, p. 237, 1982.

12. D. M. Bron, M. Ghezzo, and J. M. Primbley, "Trends in Advanced CMOS Process Technology," *Proc. IEEE*, p. 1646, (1986).

13. H. Higuchi, et al., "Performance and Structure of Scaled-Down Bipolar Devices Merge with CMOSFETs," *IEEE Tech.Dig. Int. Electron. Devices Meet.*, 694, 1984.

14. M. A. Hollis and R. A. Murphy, "Homogeneous Field-Effect Transistors," in S. M. Sze, Ed., *High-Speed Semiconductor Devices*, Wiley, New York, 1990.

15. H. P. Singh, et al., "GaAs Low Power Integrated Circuits for a High Speed Digital Signal Processor," *IEEE Trans. Electron Devices*, **36**, 240 (1989).

16. *International Technology Roadmap for Semiconductor (ITRS)*, Semiconductor Ind. Assoc., San Jose, 1999.

17. Y. Taur and E. J. Nowak, "CMOS Devices below 0.1 μm: How High Will Performance Go?" *IEEE Tech. Dig. Int. Electron Devices Meet.*, 215, 1997.

18. L. Peters, "Is the 0.18 μm Node Just a Roadside Attraction," *Semicond. Int.*, **22**, 46 (1999).

19. M. T. Bohr, "Interconnect Scaling—The Real Limiter to High Performance ULSI," *IEEE Tech. Dig. Int. Electron Devices Meet.*, p. 241, 1995.

20. E. Leobandung, et al.,"Scalability of SOI Technology into 0.13 μm 1.2 V CMOS Generation," *IEEE Int. Electron Devices Meet.*, p. 403, 1998.

21. B. Martin, "Electronic Design Automation," *IEEE Spectr.*, **36**, 61 (1999).

22. H. Ishiuchi, et al., " Embedded DRAM Technologies," *IEEE Tech. Dig. Int. Electron Devices Meet.*, p. 33, 1997.

23. S. Luryi, J. Xu, and A. Zaslavsky, Eds, *Future Trends in Microelectronics*, Wiley, New York, 1999.

▶ PROBLEMS (* DENOTES DIFFICULT PROBLEMS)

FOR SECTION 14.1   PASSIVE COMPONENTS

1. For a sheet resistance of 1 kΩ/□, find the maximum resistance that can be fabricated on a 2.5 × 2.5-mm chip using 2 μm lines with a 4 μm pitch (i.e., distance between the centers of the parallel lines).

2. Design a mask set for a 5 pF MOS capacitor. The oxide thickness is 30 nm. Assume that the minimum window size is 2 × 10 μm and the maximum registration errors are 2 μm.

3. Draw a complete step-by-step set of masks for the spiral inductor with three turns on a substrate.

4. Design a 10 nH square spiral inductor in which the total length of the interconnect is 350 μm; the spacing between turns is 2 μm.

FOR SECTION 14.2   BIPOLAR TECHNOLOGY

5. Draw the circuit diagram and device cross section of a clamped transistor.

6. Identify the purpose of the following steps in self-aligned double-polysilicon bipolar structure: (a) undoped polysilicon in trench in Fig. 13a, (b) the poly 1 in Fig. 13b, and (c) the poly 2 in Fig. 13d.

FOR SECTION 14.3   MOSFET TECHNOLOGY

*7. In NMOS processing, the starting material is a *p*-type 10 Ω-cm <100>-oriented silicon wafer. The source and drain are formed by arsenic implantation of 10^{16} ions/cm² at 30 keV through a gate oxide of 25 nm. (a) Estimate the threshold voltage change of the device. (b) Draw the doping profile along a coordinate perpendicular to the surface and passing through the channel region or the source region.

8. (a) Why is <100>-orientation preferred in NMOS fabrication? (b) What are the disadvantages if too thin a field oxide is used in NMOS devices? (c) What problems occur if a polysilicon gate is used for gate lengths less than 3 μm? Can another material be substituted for polysilicon? (d) How is a self-aligned gate obtained and what are its advantages? (e) What purpose does P-glass serve?

*9. For a floating-gate nonvolatile memory, the lower insulator has a dielectric constant of 4 and is 10 nm thick. The insulator above the floating gate has a dielectric constant of 10 and is 100 nm thick. If the current density J in the lower insulators is given by $J = \sigma \mathscr{E}$, where $\sigma = 10^{-7}$ S/cm, and the current in the other insulator is negligibly small, find the threshold voltage shift of the device caused by a voltage of 10 V applied to the control gate for (a) 0.25 μs, and (b) a sufficiently long time that J in the lower insulator becomes negligibly small.

10. Draw a complete step-by step set of masks for CMOS inverter shown in Fig. 23. Pay particular attention to the cross section shown in Fig. 23c for your scale.

*11. A 0.5 μm digital CMOS technology has 5 μm wide transistors. The minimum wire width is 1 μm and the metallization layer consists of 1 μm thick aluminum. Assume that μ_n is 400 cm^2/V-s, d is 10-nm, V_{DD} is 3.3 V, and the threshold voltage is 0.6 V. Finally, assume that the maximum voltage drop that can be tolerated is 0.1 V when a 1 μm^2 cross section aluminum wire is carrying the maximum current that can be supplied by the NMOS transistor. How long a wire can be allowed? Use a simple square-law, long-channel model to predict the MOS current drive (resistivity of aluminum is 2.7×10^{-8} Ω-cm).

12. Plot the cross-sectional views of a twin-tub CMOS structure of the following stages of processing: (a) n-tub implant, (b) p-tub implant, (c) twin-tub drive-in, (d) nonselective p^+-source/drain implant, (e) selective n^+-source/drain implant using photoresist as mask, and (f) P-glass deposition.

13. Why do we use a p^+-polysilicon gate for PMOS?

14. What is the boron penetration problem in p^+-polysilicon PMOS? How would you eliminate it?

15. To obtain a good interfacial property, a buffered layer is usually deposited between the high-k material and substrate. Calculate the effective oxide thickness if the stacked gate dielectric structure is (a) a buffered nitride of 0.5 nm and (b) a Ta$_2$O$_5$ of 10 nm.

16. Describe the disadvantages of LOCOS technology and the advantages of shallow-trench isolation technology.

FOR SECTION 14.4   MESFET TECHNOLOGY

17. What is the purpose for the polyimide used in Fig. 34f?

18. What is the reason that it is difficult to make bipolar transistor and MOSFET in GaAs?

FOR SECTION 14.5   CHALLENGES FOR MICROELECTRONICS

19. (a) Calculate the RC time constant of a aluminum runner 0.5 μm thick formed on a thermally grown SiO$_2$ 0.5 μm thick. The length and width of the runner are 1 cm and 1 μm, respectively. The resistivity of the runner is 10^{-5} Ω-cm. (b) What will be the RC time constant for a polysilicon runner ($R_\square = 30$ Ω/□) of identical dimension?

20. Why do we need multiple oxide thicknesses for a system-on-a-chip (SOC)?

21. Normally we need a buffered layer placed between a high-k Ta$_2$O$_5$ and the silicon substrate. Calculate the effective oxide thickness (EOT) when the stacked gate dielectric is Ta$_2$O$_5$ ($k = 25$) with a thickness of 75Å on a buffered nitride layer ($k = 7$ and a thickness of 10 Å). Also calculate EOT for a buffered oxide layer ($k = 3.9$, and a thickness of 5 Å).

Appendix A

List of Symbols

Symbol	Description	Unit
a	Lattice constant	Å
$\mathscr{B}$	Magnetic induction	Wb/m^2
c	Speed of light in vacuum	cm/s
C	Capacitance	F
$\mathscr{D}$	Electric displacement	C/cm^2
D	Diffusion coefficient	cm^2/s
E	Energy	eV
E_C	Bottom of conduction band	eV
E_F	Fermi energy level	eV
E_g	Energy bandgap	eV
E_V	Top of valence band	eV
$\mathscr{E}$	Electric field	V/cm
$\mathscr{E}_c$	Critical field	V/cm
$\mathscr{E}_m$	Maximum field	V/cm
f	Frequency	Hz(cps)
$F(E)$	Fermi–Dirac distribution function	
h	Planck constant	J·s
$h\nu$	Photon energy	eV
I	Current	A
I_C	Collector current	A
J	Current density	A/cm^2
J_{th}	Threshold current density	A/cm^2
k	Boltzmann constant	J/K
kT	Thermal energy	eV
L	Length	cm or μm
m_0	Electron rest mass	kg
m_n	Electron effective mass	kg
m_p	Hole effective mass	kg
$\bar{n}$	Refractive index	
n	Density of free electrons	cm^{-3}
n_i	Intrinsic carrier concentration	cm^{-3}

(continued)

Symbol	Description	Unit
N	Doping concentration	cm^{-3}
N_A	Acceptor-impurity density	cm^{-3}
N_C	Effective density of states in conduction band	cm^{-3}
N_D	Donor-impurity density	cm^{-3}
N_V	Effective density of states in valence band	cm^{-3}
p	Density of free holes	cm^{-3}
P	Pressure	Pa
q	Magnitude of electronic charge	C
Q_{it}	Interface-trapped charge	charges/cm^2
R	Resistance	Ω
t	Time	s
T	Absolute Temperature	K
v	Carrier velocity	cm/s
v_s	Saturation velocity	cm/s
v_{th}	Thermal velocity	cm/s
V	Voltage	V
V_{bi}	Built-in potential	V
V_{EB}	Emitter-base voltage	V
V_B	Breakdown voltage	V
W	Thickness	cm or μm
W_B	Base thickness	cm or μm
ε_0	Permittivity in vacuum	F/cm
ε_s	Semiconductor permittivity	F/cm
ε_{ox}	Insulator permittivity	F/cm
$\varepsilon_s/\varepsilon_0$ or $\varepsilon_{ox}/\varepsilon_0$	Dielectric constant	
τ	Lifetime or decay time	s
θ	Angle	rad
λ	Wavelength	μm or nm
ν	Frequency of light	Hz
μ_0	Permeability in vacuum	H/cm
μ_n	Electron mobility	cm^2/V·s
μ_p	Hole mobility	cm^2/V·s
ρ	Resistivity	Ω-cm
ϕ_{Bn}	Schottky barrier height on n-type semiconductor	V
ϕ_{Bp}	Schottky barrier height on p-type semiconductor	V
$q\phi_m$	Metal work function	eV
ω	Angular frequency ($2\pi f$ or $2\pi v$)	Hz
Ω	Ohm	Ω

Appendix B

International System of Units (SI Units)

Quantity	Unit	Symbol	Dimensions
Length§	meter	m	
Mass	kilogram	kg	
Time	second	s	
Temperature	kelvin	K	
Current	ampere	A	
Light intensity	candela	Cd	
Angle	radian	rad	
Frequency	hertz	Hz	1/s
Force	newton	N	kg-m/s^2
Pressure	pascal	Pa	N/m^2
Energy §	joule	J	N-m
Power	watt	W	J/s
Electric charge	coulomb	C	A·s
Potential	volt	V	J/C
Conductance	siemens	S	A/V
Resistance	ohm	Ω	V/A
Capacitance	farad	F	C/V
Magnetic flux	weber	Wb	V·s
Magnetic induction	tesla	T	Wb/m^2
Inductance	henry	H	Wb/A
Light flux	lumen	Lm	Cd-rad

§ It is more common in the semiconductor field to use cm for length and eV for energy (1 cm = 10^{-2} m, 1 eV = 1.6×10^{-19} J).

Appendix C

Unit Prefixes*

Multiple	Prefix	Symbol
10^{18}	exa	E
10^{15}	peta	P
10^{12}	tera	T
10^{9}	giga	G
10^{6}	mega	M
10^{3}	kilo	k
10^{2}	hecto	h
10	deka	da
10^{-1}	deci	d
10^{-2}	centi	c
10^{-3}	milli	m
10^{-6}	micro	μ
10^{-9}	nano	n
10^{-12}	pico	p
10^{-15}	femto	f
10^{-18}	atto	a

* Adopted by International Committee on Weights and Measures. (Compound prefixes should not be used, e.g., not $\mu\mu$ but p.)

Appendix D

Greek Alphabet

Letter	Lowercase	Uppercase
Alpha	α	A
Beta	β	B
Gamma	γ	Γ
Delta	δ	Δ
Epsilon	ε	E
Zeta	ζ	Z
Eta	η	H
Theta	θ	Θ
Iota	ι	I
Kappa	κ	K
Lambda	λ	Λ
Mu	μ	M
Nu	ν	N
Xi	ξ	Ξ
Omicron	o	O
Pi	π	Π
Rho	ρ	P
Sigma	σ	Σ
Tau	τ	T
Upsilon	υ	Υ
Phi	ϕ	Φ
Chi	χ	X
Psi	ψ	Ψ
Omega	ω	Ω

Appendix E

Physical Constants

Quantity	Symbol	Value
Angstrom unit	Å	$10 \text{ Å} = 1 \text{ nm} = 10^{-3} \text{ μm} = 10^{-7} \text{ cm} = 10^{-9} \text{ m}$
Avogadro constant	N_{av}	6.02214×10^{23}
Bohr radius	a_B	0.52917 Å
Boltzmann constant	k	$1.38066 \times 10^{-23} \text{ J/K } (R/N_{av})$
Elementary charge	q	$1.60218 \times 10^{-19} \text{ C}$
Electron rest mass	m_0	$0.91094 \times 10^{-30} \text{ kg}$
Electron volt	eV	$1 \text{ eV} = 1.60218 \times 10^{-19} \text{ J}$ $= 23.053 \text{ kcal/mol}$
Gas constant	R	$1.98719 \text{ cal/mol-K}$
Permeability in vacuum	μ_0	$1.25664 \times 10^{-8} \text{ H/cm } (4\pi \times 10^{-9})$
Permittivity in vacuum	ε_0	$8.85418 \times 10^{-14} \text{ F/cm } (1/\mu_0 c^2)$
Planck constant	h	$6.62607 \times 10^{-34} \text{ J·s}$
Reduced Planck constant	$\hbar$	$1.05457 \times 10^{-34} \text{ J·s } (h/2\pi)$
Proton rest mass	M_p	$1.67262 \times 10^{-27} \text{ kg}$
Speed of light in vacuum	c	$2.99792 \times 10^{10} \text{ cm/s}$
Standard atmophere		$1.01325 \times 10^5 \text{ Pa}$
Thermal voltage at 300 K	kT/q	0.025852 V
Wavelength of 1 eV quantum	λ	1.23984 μm

Appendix F

Properties of Important Element and Binary Compound Semiconductors at 300 K

Semiconductor		Lattice constant ($\mathring{A}$)	Bandgap (eV)	Band[a]	Mobility[b] (cm^2/V-s) μ_n	μ_p	Dielectric constant
Element	Ge	5.65	0.66	I	3900	1800	16.2
	Si	5.43	1.12	I	1450	505	11.9
IV-IV	SiC	3.08	2.86	I	300	40	9.66
III-V	AlSb	6.13	1.61	I	200	400	12.0
	GaAs	5.65	1.42	D	9200	320	12.4
	GaP	5.45	2.27	I	160	135	11.1
	GaSb	6.09	0.75	D	3750	680	15.7
	InAs	6.05	0.35	D	33000	450	15.1
	InP	5.86	1.34	D	5900	150	12.6
	InSb	6.47	0.17	D	77000	850	16.8
II-VI	CdS	5.83	2.42	D	340	50	5.4
	CdTe	6.48	1.56	D	1050	100	10.2
	ZnO	4.58	3.35	D	200	180	9.0
	ZnS	5.42	3.68	D	180	10	8.9
IV-VI	PbS	5.93	0.41	I	800	1000	17.0
	PbTe	6.46	0.31	I	6000	4000	30.0

[a]I, indirect, D, direct.

[b]The values are for drift mobilities obtained in the purest and most perfect materials available to date.

Appendix G

Properties of Si and GaAs at 300 K

Properties	Si	GaAs
Atoms/cm^3	5.02×10^{22}	4.42×10^{22}
Atomic weight	28.09	144.63
Breakdown field (V/cm)	~3×10^5	~ 4×10^5
Crystal structure	Diamond	Zincblende
Density (g/cm^3)	2.329	5.317
Dielectric constant	11.9	12.4
Effective density of states in conduction band, N_C(cm^{-3})	2.86×10^{19}	4.7×10^{17}
Effective density of states in valence band, N_V(cm^{-3})	2.66×10^{19}	7.0×10^{18}
Effective mass (conductivity)		
Electrons (m_n/m_0)	0.26	0.063
Holes (m_p/m_0)	0.69	0.57
Electron affinity, χ(V)	4.05	4.07
Energy gap (eV)	1.12	1.42
Index of refraction	3.42	3.3
Intrinsic carrier concentration(cm^{-3})	9.65×10^9	2.25×10^6
Intrinsic resistivity (Ω-cm)	3.3×10^5	2.9×10^8
Lattice constant (Å)	5.43102	5.65325
Linear coefficient of thermal expansion, $\Delta L/L \times T$ (°C^{-1})	2.59×10^{-6}	5.75×10^{-6}
Melting point (°C)	1412	1240
Minority-carrier lifetime (s)	3×10^{-2}	~10^{-8}
Mobility (cm^2/V·s)		
μ_n (electrons)	1450	9200
μ_p (holes)	505	320
Specific heat (J/g -°C)	0.7	0.35
Thermal conductivity(W/cm-K)	1.31	0.46
Vapor pressure (Pa)	1 at 1650°C	100 at 1050°C
	10^{-6} at 900°C	1 at 900°C

Appendix H

Derivation of the Density of States in a Semiconductor

To calculate the electron and hole concentrations in the conduction and valence bands, respectively, we need to know the density of states, that is, the number of allowed energy states per unit energy per unit volume (i.,e., in the units of number of states /eV/cm^3).

When electrons move back and forth along the x-direction in a semiconductor material, the movements can be described by standing-wave oscillations. The wavelength λ of a standing wave is related to the length of the semiconductor L by

$$\frac{L}{\lambda} = n_x, \tag{1}$$

where n_x is an integer. The wavelength can be expressed by de Broglie hypothesis:

$$\lambda = \frac{h}{p_x}, \tag{2}$$

where h is the Planck constant and p_x is the momentum in the x-direction. Substituting Eq. 2 into Eq. 1 gives

$$Lp_x = hn_x. \tag{3}$$

The incremental momentum dp_x required for a unity increase in n_x is

$$L\,dp_x = h. \tag{4}$$

For a three-dimensional cube of side L, we have

$$L^3\,dp_x\,dp_y\,dp_z = h^3. \tag{5}$$

The volume $dp_x\,dp_y\,dp_z$ in the momentum space for a unit cube ($L = 1$) is thus equal to h^3. Each incremental change in n corresponds to a unique set of integers (n_x, n_y, n_z), which in turn corresponds to an allowed energy state. Thus, the volume in momentum space for an energy state is h^3. The figure shows the momentum space in spherical coordinates. The volume between two concentric spheres (from p to $p + dp$) is $4\pi p^2 dp$. The number of energy states contained in the volume is then $2(4\pi p^2 dp)/h^3$, where the factor 2 accounts for the electron spins.

The energy E of the electron (here we only consider the kinetic energy) is given by

$$E = \frac{p^2}{2m_n} \tag{6}$$

or
$$p = \sqrt{2m_n E}, \tag{7}$$

where p is the total momentum (with components p_x, p_y, and p_z in Cartesian Coordinates) and m_n is the effective mass. From Eq. 7, we can substitute E for p and obtain

$$N(E)dE = \frac{8\pi p^2 dp}{h^3} = 4\pi \left(\frac{2m_n}{h^2} \right)^{\frac{3}{2}} E^{\frac{1}{2}} dE \tag{8}$$

and
$$N(E) = 4\pi \left(\frac{2m_n}{h^2} \right)^{\frac{3}{2}} E^{\frac{1}{2}}, \tag{9}$$

where $N(E)$ is called the density of states.

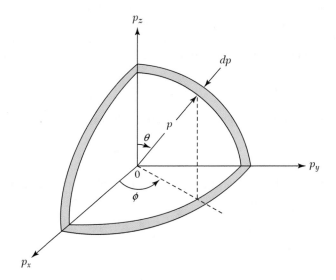

Appendix I

Derivation of Recombination Rate for Indirect Recombination

A schematic diagram showing the various transitions that occur in the recombination through recombination centers is shown in Fig. 12 of Chapter 3. If the concentration of centers in the semiconductor is N_t, the concentration of unoccupied centers is given by $N_t(1-F)$, where F is the Fermi distribution function for the probability that a center is occupied by an electron. In equilibrium,

$$F = \frac{1}{1 + e^{(E_t - E_F)/kT}} \tag{1}$$

where E_t is the energy level of the center and E_F is the Fermi level.

Therefore, the capture rate of an electron by a recombination center, Fig. 12a of Chapter 3, is given by

$$R_a \approx nN_t(1-F). \tag{2}$$

We designate the proportionality constant by the product $v_{th}\sigma_n$, so that

$$R_a = v_{th}\sigma_n nN_t(1-F). \tag{3}$$

The product $v_{th}\sigma_n$ may be visualized as the volume swept out per unit time by an electron with cross section σ_n. If the center lies within this volume, the electron will be captured by it.

The rate of emission of electrons from the center, Fig. 12b, is the inverse of the electron capture process. The rate is proportional to the concentration of centers occupied by electrons, that is, $N_t F$. We have

$$R_b = e_n N_t F. \tag{4}$$

The proportionality constant e_n is called the emission probability. At thermal equilibrium the rates of capture and emission of electrons must be equal ($R_a = R_b$). Thus, the emission probability can be expressed in terms of the quantities already defined in Eq. 3:

$$e_n = \frac{v_{th}\sigma_n n(1-F)}{F}. \tag{5}$$

Since the electron concentration in thermal equilibrium is given by

$$n = n_i e^{(E_F - E_i)/kT}, \tag{6}$$

we obtain

$$e_n = v_{th}\sigma_n n_i e^{(E_t - E_i)/kT}. \tag{7}$$

The transitions between the recombination center and valence band are analogous to those described above. The capture rate of a hole by an occupied recombination center, Fig. 12c, is given by

$$R_c = v_{th}\sigma_p p N_t F. \tag{8}$$

By arguments similar to those for electron emission, the rate of hole emission, Fig. 12d, is

$$R_d = e_p N_t (1 - F). \tag{9}$$

The emission probability e_p of a hole may be expressed in terms of v_{th} and σ_p by considering the thermal equilibrium condition for which $R_c = R_d$:

$$e_p = v_{th}\sigma_p n_i e^{(E_i - E_t)/kT}. \tag{10}$$

Let us now consider the nonequilibrium case in which an n-type semiconductor is illuminated uniformly to give a generation rate G_L. Thus in addition to the process shown in Fig. 12, electron-hole pairs are generated as a result of light. In steady state the electrons entering and leaving the conduction band must be equal. This is called the *principle of detailed balance*, and it yields

$$\frac{dn_n}{dt} = G_L - (R_a - R_b) = 0. \tag{11}$$

Similarly, in steady state the detailed balance of holes in valence band leads to

$$\frac{dp_n}{dt} = G_L - (R_c - R_d) = 0. \tag{12}$$

Under equilibrium conditions, that is, $G_L = 0$, $R_a = R_b$ and $R_c = R_d$. However, under state-state nonequilibrium conditions, $R_a \neq R_b$ and $R_c \neq R_d$. From Eqs. 11 and 12 we obtain

$$G_L = R_a - R_b = R_c - R_d \equiv U. \tag{13}$$

We can get the net recombination rate U from Eqs. 3, 4, 8, and 9:

$$U \equiv R_a - R_b = \frac{v_{th}\sigma_n\sigma_p N_t\left(p_n n_n - n_i^2\right)}{\sigma_p[p_n + n_i e^{(E_i - E_t)/kT}] + \sigma_n[n_n + n_i e^{(E_t - E_i)/kT}]}. \tag{14}$$

Appendix J

Calculation of the Transmission Coefficient for a Symmetric Resonant-Tunneling Diode

To calculate the transmission coefficient, we consider Fig. 13a of Chapter 8 where the five regions (I, II, III, IV, V) are specified by the coordinates (x_1, x_2, x_3, x_4). The Schrödinger equation for the electron in any region can be written as

$$-\frac{\hbar^2}{2m_i^*}\left(\frac{d^2\psi_i}{dx^2}\right) + V_i\psi_i = E\psi_i \qquad i = 1, 2, 3, 4, 5, \tag{1}$$

where $\hbar$ is the reduced Planck constant, m_i^* the effective mass in the ith region, E the incident energy, and V_i and ψ_i the potential energy and the wave function in the ith region, respectively. The wavefunction ψ_i can be expressed as

$$\psi_i(x) = A_i \exp(jk_ix) + B_i \exp(-jk_ix), \tag{2}$$

where A_i and B_i are constants to be determined from the boundary conditions, and $k_i = \sqrt{2m_i^*(E - V_i)}/\hbar$ Since the wavefunctions and their first derivatives (i.e., $\psi_i/m_i^* = \psi_{i+1}/m_{i+1}^*$) at each potential discontinuity must be continuous, we obtain the transmission coefficient (for identical effective mass across the five regions),

$$T_t = \frac{1}{1 + E_0^2(\sinh^2 \beta L_B)H^2/[4E^2(E_0 - E)^2]}, \tag{3}$$

where

$$H \equiv 2\left[E(E_0 - E)\right]^{1/2}\cosh \beta L_B \cos kL_W - (2E - E_0)\sinh \beta L_B \sin kL_W$$

and

$$\beta \equiv \frac{\sqrt{2m^*(E_0 - E)}}{\hbar}, \qquad k = \frac{\sqrt{2m^*E}}{\hbar}.$$

The resonant condition occurs when $H = 0$, and thus $T_t = 1$. The resonant-tunneling energy levels E_n can be calculated by solving the transcendental equation:

$$\frac{2\left[E(E_0 - E)\right]^{1/2}}{(2E - E_0)} = \tan kL_W \tanh \beta L_B \tag{4}$$

As a first-order estimate of the energy levels, one can use the results of a quantum well with infinite barrier height:

$$E_n \approx \left(\frac{\pi^2 \hbar^2}{2m^* L_W^2}\right) n^2. \tag{5}$$

For a double-barrier structure with finite barrier height and width, the energy level (for a given n) will be lower; however, it will have a similar dependence on the effective mass and well width, that is, E_n increases with decreasing m^* or L_W.

Appendix K

Basic Kinetic Theory of Gases

The ideal gas law states that

$$PV = RT = N_{av}kT, \tag{1}$$

where P is the pressure, V is the volume of one mole of gas, R is the gas constant (1.98 cal/mol-K, or 82 atm-cm^3/mol-K), T is the absolute temperature in K, N_{av} is the Avogadro constant (6.02×10^{23} molecules/mole), and k is the Boltzmann constant (1.38×10^{-23} J/K, or 1.37×10^{-22} atm-cm^2/K). Since real gases behave more and more like the ideal gas as the pressure is lowered, Eq.1 is valid for most vacuum processes. We can use Eq.1 to calculate the molecular concentration n (the number of molecules per unit volume):

$$n = \frac{N_{av}}{V} = \frac{P}{kT} \tag{2}$$

$$= 7.25 \times 10^{16} \frac{P}{T} \text{ molecules/cm}^3, \tag{2a}$$

where P is in Pa. The density ρ_d of a gas is given by the product of its molecular weight and its concentration:

$$\rho_d = \text{molecular weight} \times \left(\frac{P}{kT} \right). \tag{3}$$

The gas molecules are in constant motion and their velocities are temperature dependent. The distribution of velocities is described by the Maxwell-Boltzmann distribution law, which states that for a given speed v,

$$\frac{1}{n} \frac{dn}{dv} \equiv f_v = \frac{4}{\sqrt{\pi}} \left(\frac{m}{2kT} \right)^{3/2} v^2 \exp\left(-\frac{mv^2}{2kT} \right), \tag{4}$$

where m is the mass of a molecule. This equation states that if there are n molecules in the volume, there will be dn molecules having a speed between v and $v + dv$. The average speed is obtainable from Eq. 4:

$$v_{av} = \frac{\int_0^\infty v f_v dv}{\int_0^\infty f_v dv} = \frac{2}{\sqrt{\pi}} \sqrt{\frac{2kT}{m}}. \tag{5}$$

An important parameter for vacuum technology is the molecular *impingement rate*, that is, how many molecules impinge on a unit area per unit time. To obtain this parameter, first consider the distribution function f_{vx} for the velocities of molecules in the x-direction. This function can be expressed by an equation similar to Eq. 4:

$$\frac{1}{n}\frac{dn_x}{dv_x} \equiv f_{v_x} = \left(\frac{m}{2\pi kT}\right)^{1/2} v_x^2 \exp\left(\frac{-mv_x^2}{2kT}\right). \tag{6}$$

The molecular impingement rate ϕ is given by

$$\phi = \int_0^\infty v_x\,dn_x. \tag{7}$$

Substituting dn_x from Eq. 6 and integrating gives

$$\phi = n\sqrt{\frac{kT}{2\pi m}}. \tag{8}$$

The relationship between the impingement rate and the gas pressure is obtained by using Eq. 2:

$$\phi = P(2\pi mkT)^{-1/2} \tag{9}$$

$$= 2.64 \times 10^{20}\left(\frac{P}{\sqrt{MT}}\right), \tag{9a}$$

where P is the pressure in Pa and M is the molecular weight.

Appendix L

Answers to Selected Problems

Answers are provided for odd-numbered problems, which have numerical solutions.

CHAPTER 2

1. (a) 2.35 Å; (b) 6.78×10^{14} (100), 9.6×10^{15} (110), 7.83×10^{14} (111) atoms/cm^2.

3. 52% (simple cubic), 74% (fcc), 34% (diamond).

5. (643) plane

11. 0.583 eV (77 K), 0.569 eV (300 K), 0.557 eV (373 K).

13. 72.7 Å, 1154 Å.

15. 2.26×10^{16} cm^{-3}.

17. (a) 9.3×10^4 cm^{-3}, 0.26 eV; (b) 10^{15} cm^{-3}, 0.26 eV

19. $n = 10^{15}$ cm^{-3} ($N_D = 10^{15}$ cm^{-3}), 0.93×10^{17} (10^{17}), 0.27×10^{19}(10^{19}).

CHAPTER 3

1. 3.31×10^5 Ω-cm (Si), 2.92×10^8 Ω-cm (GaAs).

3. 167 cm^2/V·s.

5. 3.5×10^7 cm^{-3}, 400 cm^2/V·s.

7. 0.226 Ω-cm

9. $N_A = 50 N_D$.

11. (b) 259 V/cm.

13. (a) $\Delta n = 10^{11}$ cm^{-3}, $n = 10^{15}$ cm^{-3}, $p = 10^{11}$ cm^{-3}

19. $p(x) = 10^{14}(1 - 0.9e^{-x/L_p})$, $L_p = 31.6$ μm.

23. 0.403, 7.8×10^{-9}

25. 1.35×10^5 cm/s $< 9.5 \times 10^6$ cm/s (100 V/cm) 1.35×10^7 cm/s $\approx 9.5 \times 10^6$ cm/s (10^4 V/cm).

CHAPTER 4

1. 1.867 μm, $V_{bi} = 0.52$ V, $\mathscr{E}_m = 4.86 \times 10^3$ V/cm.

3. At 300 K, $V_{bi} = 0.714$ V, W = 0.97 μm, $\mathscr{E}_m = 1.47 \times 10^4$ V/cm.

5. For $N_D = 10^{15}$ cm^{-3}, $1 / C_j^2 = 1.187 \times 10^{16} (0.834 - V)$.

7. $N_D = 3.43 \times 10^{15}$ cm^{-3}.

9. 2.5×10^{17} cm^{-3}.

11. $N_A = 2.2 \times 10^{15}$ cm^{-3}, $N_D = 5.4 \times 10^{15}$ cm^{-3}.

13. 0.79 V.

15. 8.78×10^{-3} C/cm^2.

17. cross-sectional area is 8.6×10^{-5} cm^2.

19. (a) 587 V, (b) 42.8 V.

21. For V= +0.5 V, $V_{bi} = 1.1$ and 3.4×10^{-4} V; depletion widths = 3.82×10^{-5} and 1.27×10^{-8} cm.

CHAPTER 5

1. (a) 0.995, 199, (b) 2×10^{-6} A.

3. (a) 0.904 μm, (b) 2.54×10^{11} cm^{-3}.

5. (a)$I_E = 1.606 \times 10^{-5}$A, $I_C = 1.596 \times 10^{-5}$A, $I_B = 1.041 \times 10^{-7}$A; (b) $\beta_0 = 160$.

13. $\mu_E = 87.6$ cm^2/V-s, $\mu_B = 1186$, $\mu_c = 453$.

15. $\beta_0 = 50,000$.

17. 131.6.

21. $I_E = 1.715 \times 10^{-4}$A, $I_C = 1.715 \times 10^{-4}$A.

23. $f_T = 1.27$ GHz, $f_\alpha = 1.275$ GHz, $f_\beta = 2.55$ MHz.

25. 0.29.

27. 3.96×10^{-3} cm, 44.4 cm^2.

CHAPTER 6

5. 0.15 μm.

7. 0.59 V, 1.11×10^5 V/cm.

9. 2.32×10^{-2} V.

11. 7.74×10^{-2} V.

15. 3.42 V, 2.55×10^{-2} A.

17. 3.45×10^{-4} S.

19. 8×10^{11} cm^{-2}.

21. 1.7×10^{12} cm^{-2}.

23. 0.457 μm.

25. 0.83 V.

29. 49 nm.

31. 0.134 to 0.226 V.

33. 3.1×10^{-11} A.

35. 4.34 V.

CHAPTER 7

1. 0.54 eV, V_{bi} = 0.352 V.

3. ϕ_{Bn} = 0.64 V, V_{bi} = 0.463 V, W = 0.142 μm, $\mathscr{E}_m$ = 6.54×10^4 V/cm.

5. V_{bi} = 0.605 V, ϕ_m = 4.81 V.

7. 0.108 μm.

9. (b) −2.06V.

11. 0.152 μm, 0.0496 μm.

13. 5.8 nm.

15. 44.5 nm, −0.93 V.

CHAPTER 8

1. 11.25 nH.

3. 18.9 nm, 6.13×10^{-7} F/cm^2.

5. (a) 137 Ω; (b) 318.5 V.

7. (a) 74.8 V; (b) 2.2×10^5 V/cm; (c) 19 GHz.

9. (a) 10^{16} cm^{-3}; (b) 10 ps; (c) 2.02 W.

11. 3 meV, 11 meV.

CHAPTER 9

1. 1.57 mW.

3. (a) 40.58 cm^{-1}; (b) 36.6%.

7. 3.83 nA.

9. 0.0091 (°C)$^{-1}$.

11. 9.

13. 138 V, 90 ps.

15. V_m= 0.64 V, P_m = 52 mW.

17. 35.6 mW (R_s =0), 9 mW (R_s = 5Ω).

CHAPTER 10

1. At x = 0, C_s = 3×10^{16} cm^{-3}; at x = 0.9, C_s = 1.5×10^{17} cm^{-3}.

3. 0.75 g.

5. 6.56 m.

9. 24 cm.

11. ±30%, ±1%.

15. 2.14×10^{14} cm^{-3} at 900°C.

17. 4.68×10^4 cm/s.

19. 5.27×10^{14} atoms/cm^2.

CHAPTER 11

1. 44 min.

5. (a) x = 0.83, y = 0.46; (b) 2×10^{11} Ω-cm.

7. 0.0093.

9. 757°C.

13. 2.1×10^{11} cm^{-2}.

15. 71.1 nm for TiSi$_2$.

17. (a) 0.93 ns; (b) 0.42 ns; (c) 0.45.

19. 72 Ω, 0.18 V.

CHAPTER 12

1. (a) 2765; (b) 578; (c) 157.

3. 7 wafers/hr for positive resist, 120 for negative resist.

9. (a) W_b = 1.22 μm; (b) 0.93 μm; (c) 0.65 μm.

11. Etch from the top, wafer area lost = 127 cm^2; etch from the bottom, wafer area lost is small.

13. 224.7 nm/min.

17. 433.3 nm.

CHAPTER 13

1. 0.15 μm, 5.54×10^{14} atoms/cm^2.

3. 25 min, 3.4×10^{13} atoms/cm^2.

5. 16.9%.

7. x_j = 32.3 nm.

11. 6.7 s.

13. 0.53 μm.

15. 99.6%.

21. 0.927 μm.

CHAPTER 14

1. 781 MΩ.

3. 13 turns.

7. (a) 0.91 V; (b) peak concentration = 2.2×10^{21} cm^{-3}.

9. (a) 0.565 V; (b) 9.98 V.

11. 740 μm.

15. 1.84 nm.

19. 1.38 ns, 207 ns.

21. 17.3 Å, 16.7 Å.

Index

MOSi$_2$, 399
MOST, 186
Motor drives, 165
Multilevel metallization, 370, 379
Multiple implantation , 481, 485
Multiple-quantum-well (MQW), 310, 311, 329

n-channel MOSFET, 186, 194, 197, 199, 204–206, 213, 215, 216
Near-infrared region, 314
Negative differential mobility, 265, 267
Negative photoresist, 411–413, 416, 422, 449
Negative resistance, 5, 260, 261, 263, 279
Neutron, 342, 343
Neutral vacancy, 462
Neutron irradiation, 209, 342, 343, 366,
NH$_3$, 384, 392
NiSi, 524
Nitric acid, 428, 430
Nitrided oxide, 517
Nitrogen, 290, 291, 329
Nitrous oxide, 380, 382
Noise factor, 317, 318
Nonconformal step coverage, 381
Nonequilibrium situation, 60
Nondegenerate semiconductor, 39
Nonvolatile semiconductor memory (NVSM), 6, 216, 218, 510
Notebook computer, 6
n-tub, 512, 513, 529
Nuclear collisions, 477, 478
Nuclear fusion reaction, 318
Nuclear spectroscopy, 255
Nuclear stopping, 471–473, 477, 478
Nuclear stopping power, 471–473
Numerical aperture, 409
n-well, 197, 206

Ohmic contact, 2, 4, 87, 224, 225, 234, 236, 237, 251, 292, 311, 324, 400
On condition, 149

Open tube, 454, 478
Optical absorption, 282, 283, 285–287, 315, 330
Optical concentration, 326, 328
Optical diffraction, 407
Optical emission spectroscopy, 434
Optical fiber, 288, 295, 296, 298, 303, 307, 308, 311, 328, 329
Optical fiber communication systems, 6, 288, 296, 298, 303, 308, 309, 328
Optical gain, 301, 302
Optical lithographic system, 417–419, 422, 424–426, 447, 448, 450
Optical proximity correction(OPC), 417
Optical reading, 298, 328
Optical resonant cavity, 285
Opto-isolator, 288, 295, 311, 328
Orangic LED, 294
Organic semiconductors, 294
Organic solvent, 412
Orientation dependent, 375
  etching, 404, 428, 429, 448, 450
  wet chemical etching, 443
OR-NAND, 277, 278
Oscillator, 151, 517, 525
Oxidation, 85–87, 127
Oxide layer, 225
Oxide masking, 8
Oxide thickness, 187, 197, 201, 222, 223
Oxide trapped charge, 180–184, 222
Oxygen, 335, 351–354, 359
Oxygen ion implantation, 500, 501
Oxygen molecule, 373
Ozone (O$_3$), 380

P$^+$-polysilicon, 179, 180, 195, 196
Pb, 17, 20, 32
P$_2$O$_5$, 453
P-glass flow, 382–384, 402
Periodic table, 18, 19
Phase diagram, 343, 344
  of the Al-Si system, 392, 393
Phase-shifting mask (PSM), 417
Phosphine (PH$_3$), 356, 453

Phosphoric acid, 430
Phosphorus (P), 290, 335, 342, 353, 366, 379, 380, 383, 384, 389, 401, 453, 454, 464, 465, 469, 472, 473, 478, 479, 483, 485, 487
  doped silicon dioxide, 379, 383, 401
  doped oxide (P-glass), 506
Photo-acid generator, 414
Photoconductivity method, 351
Photoconductor, 311–313, 328, 331
Photocurrent gain, 313
Photodetector, 311, 328
Photodiode, 312–315, 317, 318, 328, 329, 331
Photomask, 405, 407, 411, 414–416, 419, 424
Photon energy, 282, 283, 285, 288, 312, 314
Photonic application, 352, 363
Photonic devices, 3, 10
Photoresist, 7, 8, 14, 15, 87, 414–416, 421, 422, 429, 435, 442, 447, 449, 482, 488
Photoresist adhesion, 429
Photosensitive compound, 430
Physical constants, 536
Physical etching, 433, 434
Physical vapor deposition, 87, 390
p-i-n photodiode, 315, 316, 331
Pinch-off voltage, 241–243, 253
Pinch-off point, 187, 188
Planarization, 390
Planar process, 8, 160, 225
Planar transmission lines, 256–257
Planes, 301–303, 325
Plasma damage, 397
Plasma diagnostic, 433
Plasma etching, 431, 433, 435–436, 439, 442, 449, 451
Plasma enhanced chemical vapor deposition (PECVD), 378
Plasma nitride, 385
Plasma reaction, 385
Plasma reactor, 211
Plasma spray deposition, 390
Plasma-assisted etching, 404, 431, 447